Developments in Petrophysics

Geological Society Special Publications
Series Editor A. J. FLEET

GEOLOGICAL SOCIETY SPECIAL PUBLICATION NO. 122

Developments in Petrophysics

EDITED BY

M. A. LOVELL & P. K. HARVEY
University of Leicester, UK

1997

Published by

The Geological Society

London

THE GEOLOGICAL SOCIETY

The Society was founded in 1807 as The Geological Society of London and is the oldest geological society in the world. It received its Royal Charter in 1825 for the purpose of 'investigating the mineral structure of the Earth'. The Society is Britain's national society for geology with a membership of around 8000. It has countrywide coverage and approximately 1000 members reside overseas. The Society is responsible for all aspects of the geological sciences including professional matters. The Society has its own publishing house, which produces the Society's international journals, books and maps, and which acts as the European distributor for publications of the American Association of Petroleum Geologists, SEPM and the Geological Society of America.

Fellowship is open to those holding a recognized honours degree in geology or cognate subject and who have at least two years' relevant postgraduate experience, or who have not less than six years' relevant experience in geology or a cognate subject. A Fellow who has not less than five years' relevant postgraduate experience in the practice of geology may apply for validation and, subject to approval, may be able to use the designatory letters C Geol (Chartered Geologist).

Further information about the Society is available from the Membership Manager, The Geological Society, Burlington House, Piccadilly, London W1V 0JU, UK. The Society is a Registered Charity, No. 210161.

Published by The Geological Society from:
The Geological Society Publishing House
Unit 7, Brassmill Enterprise Centre
Brassmill Lane
Bath BA1 3JN
UK
(Orders: Tel. 01225 445046
         Fax 01225 442836)

First published 1997

British Library Cataloguing in Publication Data
A catalogue record for this book is available from the British Library.

ISBN 1-897799-81-0

Typeset by Aarontype Ltd, Unit 47, Easton Business Centre, Felix Road, Bristol BS5 0HE, UK

Printed by The Alden Press, Osney Mead, Oxford, UK.

Distributors
USA
  AAPG Bookstore
  PO Box 979
  Tulsa
  OK 74101-0979
  USA
  (*Orders*: Tel. (918) 584-2555
            Fax (918) 560-2652)

Australia
  Australian Mineral Foundation
  63 Conyngham Street
  Glenside
  South Australia 5065
  Australia
  (*Orders*: Tel. (08) 379-0444
            Fax (08) 379-4634)

India
  Affiliated East-West Press PVT Ltd
  G-1/16 Ansari Road
  New Delhi 110 002
  India
  (*Orders*: Tel. (11) 327-9113
            Fax (11) 326-0538)

Japan
  Kanda Book Trading Co.
  Tanikawa Building
  3-2 Kanda Surugadai
  Chiyoda-Ku
  Tokyo 101
  Japan
  (*Orders*: Tel. (03) 3255-3497
            Fax (03) 3255-3495)

Contents

Preface

Petrophysics, or the physics of rocks impinges on many aspects of geology in its broadest sense from hydrocarbon reservoir evaluation (where its roots remain) through to evolution of the Earth's crust. It provides the focal point which enables geologists to exploit geophysical data through calibration, integration and, at times, understanding! Yet it remains largely undervalued and understated.

This Special Publication contains a selection of papers presented at a two-day meeting in September 1995 on Developments in Petrophysics at Burlington House, London. The meeting, held by the Borehole Research Group of the Geological Society of London in association with the London Petrophysical Society (SPWLA), attempted to draw together active petrophysicists from industry and academia to discuss areas of mutual interest. All 29 presentations at the meeting were in response to an open call for papers and so, within the constraints imposed by work loads and other factors, may be considered to represent a fair if perhaps not exhaustive snapshot of recent developments.

The presentations which have materialized as papers herewith cover a wide range of topics including technical developments and pore-scale effects, seismics, petrophysical prediction, electrical properties, and interpretation.

Many thanks go to Gail Williamson and Janette Thompson both for organization of the meeting and for persistence in coaxing authors, reviewers and editors; also to Angharad Hills at the Geological Society Publishing House for continuous support in the production of this volume. We additionally wish to thank the following independent referees who undertook refereeing of individual papers as part of their commitment towards the publication of this volume: S. Austin, A. Batchelor, J. D. Bennell, N. R. Brereton, H. Cambray, S. Cannon, J. Collier, B. Crawford, J. Doveton, P. Elkington, A. Foulds, P. K. Harvey, G. J. Henderson, D. Herrick, R. Hertzog, R. Hunter, A. Hurst, P. D. Jackson, F. I. W. Jones, A. Jupe, M. King, S. Luthi, D. McCann, C. McCann, B. Moss, J. Orr, S. Peacock, F. Rummel, C. Sayers, P. Scultheiss, I. Schweitzer, C. Snee, E. Stevens, P. Whattler, K. Whitworth, D. Woodhouse, P. Worthington.

M. A. Lovell & P. K. Harvey
Leicester University, December 1996

A non-contacting resistivity imaging method for characterizing whole round core while in its liner

P. D. JACKSON[1], D. G. GUNN[1], R. C. FLINT[1], D. BEAMISH[1],
P. I. MELDRUM[1], M. A. LOVELL[2], P. K. HARVEY[2] & A. PEYTON[3]

[1] British Geological Survey, Keyworth Nottingham, NG12 5GG, UK
[2] University of Leicester, University Road, Leicester LE1 7RH, UK
[3] UMIST, Manchester M60 1QD, UK

Abstract: Recent laboratory experimentation has shown that non-contacting whole-core resistivity imaging, with azimuthal discrimination, is feasible. It has shown the need for very sensitive coil pairs in order to provide resistivity measurements at the desired resolution. Independent high-resolution 'galvanic' resistivity estimations show the 'non-contacting' measurements to be directly proportional to the resistivity of core samples. The response of the technique to a variety of synthetic 'structures' is presented. A whole-core image of a dipping layer is used to demonstrate the three dimensional response of the technique and to show that the resolution of the measurements is of the order of 10 mm.

Experiments are described which show that the technique is capable of investigating to different depths within the whole round core. The results agree with theoretical predictions and indicate that the technique has the potential to assess invasion near the surface of the core.

The technique is intrinsically safe and has the potential to be packaged in a form that would be suitable for whole-core imaging at the well site, or laboratory, without taking core from their liners. Thus it is possible to acquire information crucial for core selection, in addition to acquiring resistivity data at a resolution not too far removed from that of the downhole imaging tools.

A review of research and development requirements in core analysis, sponsored by the Department of Trade and Industry Oil and Gas and Supplies Office (HMSO 1991), identified the potential of contacting resistivity imaging (Jackson et al. 1995; Lovell et al. 1995) but concluded that its widespread, routine, application was likely to be held back by the need to make direct electrical contact with the core. Non-contacting resistivity imaging was seen in the review to offer the possibility of portable, non-radiation-dependent scanning technology for use offshore. The primary purpose of such technology being to provide calibration data for downhole electrical images, identify core for special analyses and to lead to an understanding of heterogeneities of direct relevance to other logging measurements at the scale of sedimentary features.

The objective of the research reported here was to demonstrate the feasibility of a non-destructive, non-contacting resistivity method, for continuously logging oil industry core.

Contacting resistivity measurements are routinely made by the oil industry on cylindrical sub-samples taken from whole-round core. Only a small fraction of the core is sampled in this way, additional samples at different orientations are required to assess anisotropy, 'average

core-plug' values are obtained which are not representative of small-scale sedimentary features and the process is destructive to the core.

Routine, continuous high-resolution, non-destructive, non-contacting resistivity logging of oil-field core would, for the first time, provide core-resistivity datasets comparable to the corresponding downhole ones, that could be integrated with both downhole logs and borehole wall images to aid scaling up borehole datasets for use in reservoir studies.

Background

Contacting resistivity imaging of core (Jackson et al. 1995) has provided high-resolution images of the electrical resistivity of the central portion of cores using electrodes adjacent to the slab-surface surface of core which had been bisected by a single saw cut parallel to its long axis. These techniques have resolutions of 5 mm and have been used to investigate the control of fabric, heterogeneity and planar sedimentary features on the flow of fluids through sandstone samples which are an analogue for the Rotleigendes Formation of the southern North Sea (Lovell et al. 1995; Harvey et al. 1995).

From Lovell, M. A. & Harvey, P. K. (eds), 1997, *Developments in Petrophysics*, Geological Society Special Publication No. 122, pp. 1–10.

While these techniques are suitable for specialist laboratory investigations on core plugs and slabbed core, they are not suitable for use on core as it arrives at the well-site or laboratory, due to the presence of mud and irregular surfaces, which can typify whole-round core. Non-contacting methods, while having lower resolution, could be used on the entire suite of core that is collected, before it is removed from its liners, providing a spatially complete dataset for comparison to that obtained downhole. The present practice of subsampling core-plugs for resistivity measurements, provides an extremely sparse and incomplete data set.

Non-contacting electromagnetic methods are common in geophysical prospecting, but often use perturbations of the primary magnetic field to assess geological structure. These measurements are not directly related to the Earth's electrical resistivity. However, the currents induced in the Earth during such measurements are controlled by the resistivity of the material in which they flow and give rise to a secondary magnetic field which is directly proportional to the induced current. Therefore the secondary magnetic field, typically extremely small, is directly related to the electrical resistivity of the formation.

Although non-contacting resistivity induction logging has been used by the oil industry for almost 50 years (Doll 1949; Moran & Kuntz 1962), it is not used extensively for measurements on core material.

Gerland *et al.* (1993) used a modified instrument designed originally for assessing the electrical conductivity of small-diameter, cylindrical, ore-bearing hard-rock to assess the resistivity of marine sediments. They passed cylindrical core through a hole in the centre of a single coil. The arrangement produced resistivity values that appear to be averaged over 100 mm length of whole-core, sharp transitions in resistivity being smeared out over 100 mm.

Huang *et al.* (1993, 1995) describe the operation and use a system of coils which appear to have been scaled down from those first used for induction logging. They passed a coil array along the axis and through the centre of a special cylindrical sample holder, that contained the loose material under investigation.

The two approaches described above are suitable for imaging sedimentary structures within core because they respond to the whole of the core or sample over a depth interval of perhaps 100 mm, and have no azimuthal discrimination.

Non-contacting resistivity imaging of core

In order to overcome these problems, and to investigate azimuthally variable sedimentalogical structure within whole-round core, we investigated an approach based on scaling down conventional, electromagnetic prospecting which has been used successfully in engineering and environmental studies.

Electromagnetic geophysical surveying has traditionally used two coil systems, of one transmitter and one receiver, (e.g. Keller & Frischknecht 1966; Telford *et al.* 1976). The objective of such surveys is to identify the magnetic field at the receiver coil, which itself is a combination of the magnetic field produced by the transmitter and the magnetic field that arose due to currents induced in the Earth by the primary magnetic field, the primary field being created by current flowing in the transmitter coil.

At low frequencies, resistive terms dominate the governing equations and the secondary currents and their magnetic field are about 90° out of phase with the primary field. At very high frequencies skin effects dominate and the secondary fields can be 180° out of phase with the primary field. At low frequencies the induced currents can flow deeply within the Earth, whereas at high frequencies they flow near the surface (i.e. skin effect) and can, under limiting conditions, cancel the primary field. This latter effect has been referred to as 'eddy current saturation' and is most likely in the presence of highly conductive material.

The value of a 'low' or a 'high' frequency depends on the electrical conductivity (1/resistivity) of the material under investigation and the dimension of the problem in terms of the separation of the transmitting and receiving coils.

The low-frequency limit in electromagnetic geophysical surveying has become popular when the low induction-number criterion is also met (McNeil 1979). In this limit, the skin depth is far greater than the coil separation and the mutual induction of the secondary currents can be neglected. The technique has now become the method of choice for rapid, reconnaissance-style resistivity surveys for engineering and environmental studies.

Measurements of the secondary magnetic field, at low induction number, provide values that are inversely proportional to the electrical resistivity of the material being investigated. As such measurements could be made with small coil pairs at the surface of cylindrical rock cores, this technique was selected to be developed for non-contacting imaging of 'whole-round' core.

For a horizontally layered Earth, in the low induction number approximation, each 'filament' of induced current is considered to be independent of the other induced currents. This current flow is controlled by the primary field and the electrical conductance of the 'filament' in which it is flowing, and is unaffected by the electrical conductivity of the surrounding material. Response curves can be calculated, showing the relative contribution to the secondary field of different layers, as shown in Fig. 1. These curves enable the coil spacing and coil coupling to be optimised for the depth range of interest and for the theoretical response to be calculated. It is also possible to undertake localized electromagnetic 'soundings' using coil arrangements with gradually increasing depths of investigation, which can be interpreted in terms of layer thickness and resistivities. Figure 1 also demonstrates very different responses to near surface layers when the dipoles are changed from a vertical to a horizontal orientation.

For the laboratory situation where imaging sedimentalogical structures within cylindrical cores is required, a reasonable premise is that the dimensions of the transducer coils will be a compromise between very small, to identify sedimentalogical structures, and being large enough to ensure the secondary fields are measurable to the desired degree of accuracy.

Design of sensor coils

As described above, for non-contacting, low-induction-number, resistivity measurements, the primary magnetic field produced by the transmitter coil induces electric current in the core. These currents, in turn, generate a secondary magnetic field, which is inversely proportional to the resistivity of the core. The measurement of this secondary field is complicated because it is typically 107 smaller than the primary one. In order to measure the secondary field accurately we found it necessary to remove the primary field and amplify the residual field (secondary) to measurable levels. This is a complicated procedure, as at each stage it is necessary to avoid degradation of the minute secondary signal, while at the same time maximising its amplitude.

The measurement sensors are inductive coils. Experimentation showed the need for these coils to be specially designed, in order to exclude extraneous signals and to optimise the secondary signal. Sensor coils were designed and developed to be sensitive to changes in resistivity occurring over lateral distances of 10–20 mm over the cylindrical surface of whole-core. The coils were fixed at a separation of 42 mm for the experiments described below. In the first instance, the coil separation was selected to

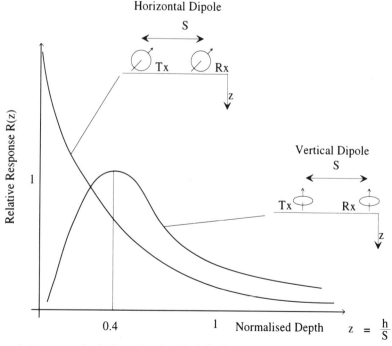

Fig. 1. Theoretical responses for horizontal and vertical dipole arrangements.

provide a substantial depth of investigation into the core to ensure edge effects did not dominate the performance. Consequently there is substantial potential for increasing the resolution by reducing the coil separation.

It is essential that the measurement provides an accurate and quantitative assessment of the resistivity of the core. In order to confirm the performance of the technology, measurements were made on the surface of a cylindrical Plexiglas tube, of internal diameter 100 mm, filled with a fluid of known electrical resistivity. Figure 2 demonstrates the performance of both

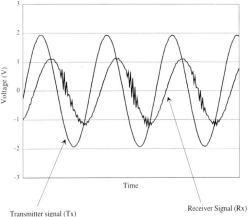

Transmitter signal (Tx)   Receiver Signal (Rx)

Fig. 3. Transmitter and receiver signals showing a 90° phase shift after processing to remove the primary magnetic field. The receiver signal is amplified relative to the transmitter signal.

horizontal and vertical dipoles using fluids having resistivities in the range 0.25–10 Ωm. The responses can be seen to be directly proportional to the electrical conductivity of fluid inside the core tube, confirming the measurement system is performing as expected. These data exhibit a straight line relationship between the measurement of the secondary field and the fluid conductivity, and were used to calibrate the output in terms of the resistivity of cylindrical core. Examples of the signals seen at the Transmitter (Tx) coil (primary field) and the Receiver (Rx) coil (secondary field) are shown in Fig. 3, illustrating the 90° phase difference between them. The Rx signal has been processed to remove the primary field generated by the Tx, and can be seen to have the greater noise. This is consistent with the Rx signal being extremely small compared to the Tx signal.

Experimental results

The performance of the sensor system to bed boundaries was studied using insulating 'targets' placed within the Plexiglas, fluid-filled, cylindrical tube described above. The results shown in Fig. 4 demonstrate that bed boundaries corresponding to changes in rock resistivity can be resolved. This was seen to be a crucial initial step towards the development of an imaging capability. Figure 4 shows the coil response to three horizontal resistive layers of differing thickness. The results resolve the interfaces of the layers

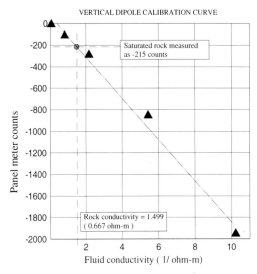

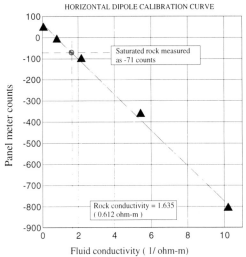

Fig. 2. Resistivity calibration curves for horizontal and vertical dipole arrangements.

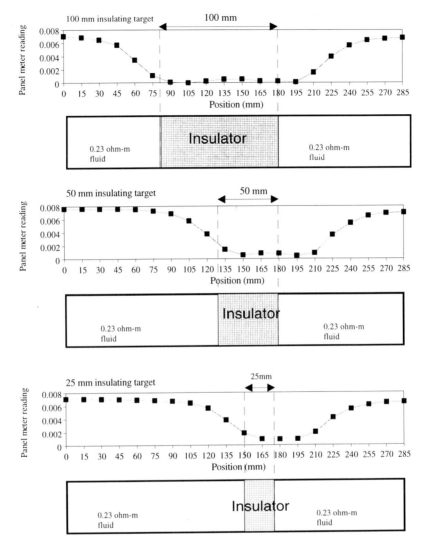

Fig. 4. Response to horizontal resistive layers of differing thicknesses. Dotted lines show where the leading and trailing edges pass the first (Rx) coil.

even when the layer thickness is less than the coil separation. Note the results are plotted as a log along the length of the core.

Multiple data sets were acquired to demonstrate the use of non-contacting resistivity methods in producing resistivity images. Figure 5 demonstrates the application of the technology in image acquisition mode. In order to build up an image, the cylindrical sample was measured at 15 mm intervals along its length and at 12 azimuthal positions around its circumference for each horizontal move. The data relate to a slab of low-porosity rock within a conductive fluid.

Figure 5 shows the coil response to a resistive layer with one horizontal interface and one dipping interface. The data are presented as an unrolled image, in addition to individual logs corresponding to maximum and minimum thickness (long core axis and short core axis), with the dark areas representing low conductivity (high resistivity) and the light areas representing high conductivity (low resistivity).

A comparison of Figs 4 and 5 shows that while the thickness of each layer is overestimated by a distance approximately equal to the coil separation, the dipping interface in Fig. 5

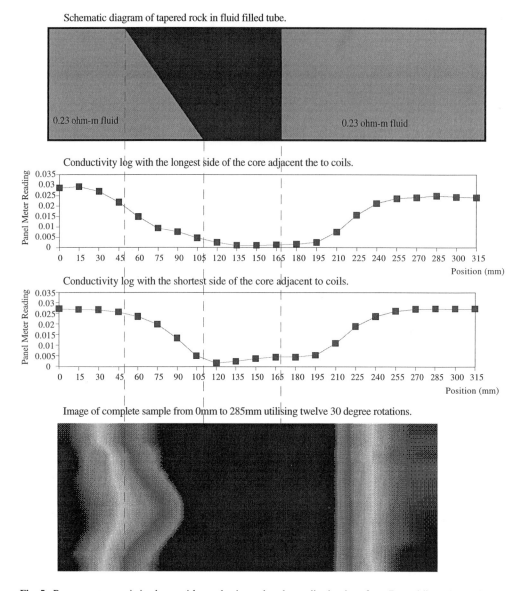

Schematic diagram of tapered rock in fluid filled tube.

0.23 ohm-m fluid

0.23 ohm-m fluid

Conductivity log with the longest side of the core adjacent the to coils.

Conductivity log with the shortest side of the core adjacent to coils.

Image of complete sample from 0mm to 285mm utilising twelve 30 degree rotations.

Fig. 5. Response to a resistive layer with one horizontal and one dipping interface. Dotted lines show where the leading and trailing edges pass the first (Rx) coil.

has been spatially, faithfully reproduced as a sinusoid in the unrolled image. This latter observation suggests the lateral resolution and depth
of investigation of the coil array are suitable for imaging sedimentalogical structure within whole-round core, even in the presence of invasion of the core by drilling mud and filtrate, because the dipping structure has been identified even though a relatively large depth of

investigation had been employed (comparable to the coil separation of 42 mm).

This approach was extended to a dipping layer far thinner than the dimensions of the coil array. The experimental set up can be seen in Fig. 6 where the thin dipping layer is constructed from two sandstone cores, each with complimentary, parallel dipping-interfaces. The separation of the two dipping interfaces was fixed at 25 mm, simulating a thin dipping conductive

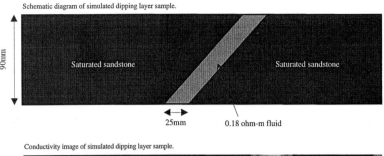

Schematic diagram of simulated dipping layer sample.

90mm

Saturated sandstone   Saturated sandstone

25mm   0.18 ohm-m fluid

Conductivity image of simulated dipping layer sample.

Fig. 6. Response to a dipping conductive layer.

layer, which could, for example, represent a narrow fluid-filled fracture. The resulting image is shown in Fig. 6 (bottom) with the dark areas representing low conductivity (high resistivity) with the converse applying to the light areas. The thin conducting layer is clearly discernible.

The use of different coil orientations was investigated as a means of assessing the resistivity structure of the core in three dimensions. This is of great importance because it would enable the assessment of, and compensation for, the effects of invasion on the core sample. A simple test of the response of the coil system in different orientations was carried out to demonstrate different depths of investigation that could be achieved.

Figure 7 shows the coil response for vertical and horizontal dipoles around a heterogeneous core. A resistive horizontal layer contains an azimuthal section of conductive material. This model is used to investigate the use of horizontal and vertical coil orientations in sampling different depths into the core. On the resistivity–position plot at the cut face both coil systems appear to be equally influenced by the near-field conductive fluid. In comparison, for the uncut face the vertical dipole continues to be influenced by the conductive fluid on the far side of the core

whilst the horizontal dipole measures an apparent resistivity which is closer to that of the core.

With reference to Fig. 8 it can be seen that the area beneath the relative response curves for the horizontal and vertical dipoles between 10 and 35 mm depth (conductive fluid) could be the same fraction of the corresponding areas beneath the response curves for depths greater than 10 mm (the height of the Tx and Rx coils above the fluid/rock). This provides an explanation of the equivalence of the resistivity values in Fig. 7, for the two dipole-orientations, when the coil array is adjacent to the cut face. It can also be seen in Fig. 8 that the response curve for the horizontal dipole decreases, in a fractional sense, far more quickly than the response curve for the vertical dipole. This is consistent with the horizontal dipole having a shallower depth of investigation than the vertical one, explaining the measurement response, described above, when the coil array is adjacent to the uncut face.

Conclusions

The results have demonstrated the potential of using different non-contacting inductive coil arrays to investigate the resistivity structure of

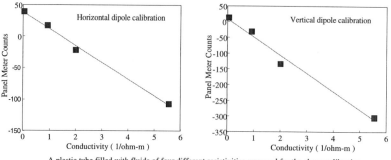

A plastic tube filled with fluids of four different resistivities was used for the above calibration.

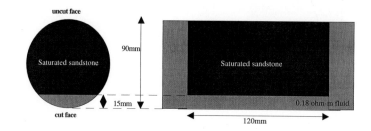

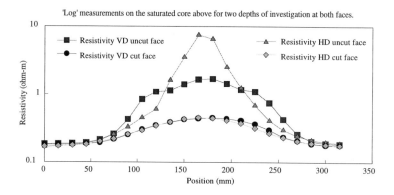

Fig. 7. Depth of investigation experiment using a resistive layer with a cut face.

whole-round core. It is readily apparent, however, that further, more detailed, investigation of the effects of coil separation, orientation and position would be beneficial. The results demonstrate the compatibility of our experimental investigations with theory and thus provide a basis for developing a full three dimensional capability.

The results provide an experimental basis to develop 2D and 3D portable, non-radiation-dependent scanning technology for application to the bulk of the core retrieved during drilling operations, both offshore and in laboratories.

Resistivity assessments at a scale of 10–20 mm are anticipated, which will offer the possibility of whole-core characterization routinely to the industry for the first time. These data could be used to upscale core data to lower resolution, but deeper penetrating, resistivity logging data will, in addition, provide quantitative calibration of down hole conductance images of the borehole wall. This would lead to an understanding of heterogeneities of direct relevance to other logging measurements at the scale of significant sedimentary features which would aid scaling up data from the borehole to the reservoir scale.

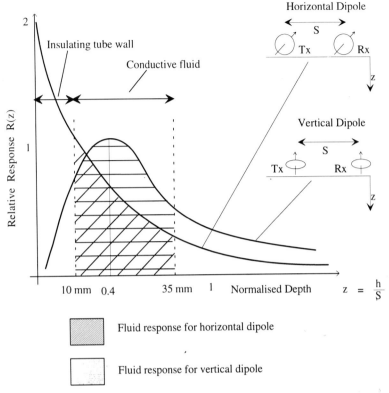

Fig. 8. Response curves for horizontal and vertical dipoles showing the likely response of the fluid phase introduced by the cut face.

The research reported here was part supported by the Oil and Gas Projects and Supplies Office of the UK Department of Trade and Industry. This paper is published with the permission of the Director of the British Geological Survey (NERC).

References

DOLL, H. G. 1949. Introduction to induction logging and application to logging of wells drilled with oil base mud. *Transactions of the American Institute of Mining and Metallurgical Engineers*, **186**, 148–162.

HMSO 1991. *Study of research and development requirements in core analysis*. Geoscience report to the Department of Energy (OSO). Oct 1991 HMSO.

GERLAND, S., RICHTER, M. VILLINGER, H. & KUHN, G. 1993. Non-destructive porosity determination of Antarctic marine sediments derived from resistivity measurements with an inductive method. *Marine Geophysical Researches*, **15**, 201–218.

HARVEY, P. K., LOVELL, M. A., JACKSON, P. D., ASHU, A. P., WILLIAMSON, G., SMITH, A. S., BALL, J. K. & FLINT, R. C. 1995. Electrical Resistivity Core Imaging III: Characterisation of an Aeolian Sandstone. *Scientific Drilling*, **5**, 164–176.

HUANG, M., LUI, C. & SHEN, L. C. 1993. Laboratory resistivity measurement using a contactless coil-type probe. *SPWLA 34th Annual Logging Symposium*, paper JJ, June 13–16 1993.

——, ——, —— & SHATTUK, D. 1995. Monitoring soil contamination using a contactless conductivity probe. *Geophysical Prospecting*, **43**, 759–778.

JACKSON, P. D., LOVELL, M. A., HARVEY, P. K., BALL, J. K., WILLIAMS, C., FLINT, R. C., GUNN, D. A., ASHU, A. P. & MELDRUM, P. I. 1995. Electrical Resistivity Core Imaging I: A new technology for high resolution investigation of petrophysical properties. *Scientific Drilling*, **5**, 139–151.

KELLER, G. V. & FRISCHKNECHT, F. C. 1966. *Electrical methods in geophysical prospecting*. Pergamon Press, Oxford.

LOVELL, M. A., HARVEY, P. K., JACKSON, P. D., FLINT, R. C., GUNN, D. A., WILLIAMSON, G., BALL, J. K., ASHU, A. P. & WILLIAMS, C. 1995. Electrical Resistivity Core Imaging II: investigation of fabric and fluid flow characteristics. *Scientific Drilling*, **5**, 153–164.

MCNEIL, J. D. 1979. *Interpretive aids for use with electromagnetic (non-contacting) ground resistivity mapping.* Forty-First European Association of

Exploration Geophysicists meeting, Hamburg, June 1979.

MORAN, J. H. & KUNTZ, K. S. 1962. Basic theory of induction logging and application to study two coil sondes. *Geophysics*, **27**, 829–858.

TELFORD, W. M., GELDHART, L. P. SHERIFF, R. E. & KEYS, D. A. 1976. *Applied Geophysics.* Cambridge University Press, London.

One-man-operable probe permeameters

D. J. PROSSER[1], A. HURST[2] & M. R. WILSON[3]

[1] Present address: Z & S (Asia) Ltd, First Floor, 46 Ord Street,
West Perth, 6005 WA, Australia

[2] University of Aberdeen, Department of Geology and Petroleum Geology, King's College,
Aberdeen AB24 3UE, UK

[3] Department of Engineering, University of Aberdeen, King's College,
Aberdeen AB24 3UE, UK

Abstract: A steady-state probe permeameter is designed that is robust, insensitive to moisture and can be operated using a low-pressure gas source. Consequently, substantial reduction in weight and bulk are possible and a truly one-man-operable, lightweight (<5 kg) instrument is constructed. Tests on natural samples using a prototype instrument reveal a similar level of precision to other permeameters over a wide range of permeability.

Probe permeametry has become a well-established experimental technique for the characterization of core from hydrocarbon reservoirs (Halvorsen & Hurst 1990), their outcrop analogues (Goggin et al. 1988b) and other porous media (see references in Hurst & Goggin 1995). When working on outcrops, in core stores or any remote, harsh environment, there is a need for equipment to be durable and safe. Although several portable probe permeameters are commercially available, field testing by a variety of users has proved them to have severe limitations in terms of transportibility and durability. If used at the (hydrocarbon drilling) well-site, particularly if airborne transportation is involved, there may be safety problems related to corrosive fluids in the power source.

Without dwelling on why many portable probe permeameters have ended up being relatively bulky, heavy and sensitive to environmental conditions, particularly moisture (splashes) and impact (bumps), it seems desirable to have an instrument design that is truly one-man portable, one-man operable and not sensitive to splashes and bumps. For these reasons design and testing of a new concept, minimized probe permeameter (MPP) was undertaken with a view to constructing a cheap, durable instrument capable of a variety of applications.

Requirements

To retain simplicity of construction a steady-state measurement system is preferred (Hurst et al. 1995) over an unsteady-state system (Jones 1994). Although unsteady-state measurements probably provide more rapid estimates on low-permeability core plugs held in Hassler-sleeves, there is some uncertainty associated to their application to probe measurements on more permeable media. The substantial bulk of pressure decay apparatus needed for unsteady-state measurements (Jones 1994) currently precludes design of a portable unsteady-state probe permeameter. Primarily the MPP must be capable of making estimates of gas flow-rate (from which permeability can be calculated) that are at least as good as those made by similar probe permeameters and comparable to Hassler-sleeve measurements on core plugs (Sutherland et al. 1993). Permeability estimates should be possible over a range of at least 0.1 mD to 10 D.

If the MPP is to be truly one-man operable it must be lightweight (<5 kg) and, preferably, be a single self-contained unit containing all metering, flow elements and power sources. Equally, the self-contained unit should be durable and tolerate splashes and bumps both during transportation and use. To maintain optimal operational flexibilty the unit should be able to operate from a variety of gas sources (pressurized gas cylinders, compressors) over a range of input pressures.

Given the potential harshness of some working environments (well-site, desert), and the desirability of not having to arrange special freighting and packaging every time the instrument is used, compactness and absence of any potentially corrosive or explosive (electrical) components is important. Ideally, the instrument should be sufficiently cheap to avoid making complete loss and subsequent replacement major issues.

The design goals for MPP are compared to some critical specifications for existing portable probe permeameters (Table 1). These data are

From Lovell, M. A. & Harvey, P. K. (eds), 1997, *Developments in Petrophysics*, Geological Society Special Publication No. 122, pp. 11–18.

Table 1. *Comparative data from existing commercial portable probe permeameters and MPP*

	EPS	TEMCO	MPP
Weight (kg)	27	15	<5
Bulk (cm)	$100 \times 50 \times 30$	$28 \times 48 \times 33$	$10 \times 12 \times 28$
Gas source	N_2 bottle	N_2 bottle	Flexible
Battery	Dry cell	Dry cell	C cell (flashlight)
	($7 \times 10 \times 8$ cm)	($7 \times 10 \times 8$ cm)	
Measurements/unit	<100	<100	No limit
Ruggedness			
(ability to tolerate)			
Splash	Possible	No	Yes
Bump	Doubtful	No	Yes
Air freight	No	No	Yes

Data for MPP are from the prototype model. Design specification for the MPP will further reduce weight and bulk.

not intended to be critical of the quality of existing instruments, but rather to draw attention to the potential for improving their one-man operability.

Practical considerations

Theoretical considerations and limitations of steady-state probe permeametry are presented elsewhere (Goggin *et al.* 1988a; Daltaban *et al.* 1989; Sutherland *et al.* 1993; Hurst 1995) are not the direct concern of this paper. Here, we simply wish to consider the basic components of a probe permeameter and demonstrate how they can be efficiently combined to produce an optimal design concept. Steady-state probe permeameters require the following components, a gas supply, a

pressure regulator, a mass flow controller, a pressure transducer and sensor, an electrical power supply, and a probe. These may be configured to include a range of mass flow rates, to make possible measurement of a wide range of permeability (Fig. 1).

Gas supply

The main requirement of the gas supply is that that it is dry and will not react with the sample. Nitrogen is widely acknowledged (Hurst & Goggin 1995 and references therein) as the most suitable flowing medium. However, use of commercially available nitrogen in pressurized cylinders leads to a trade-off between portability, small volume cylinders that do not fuel many

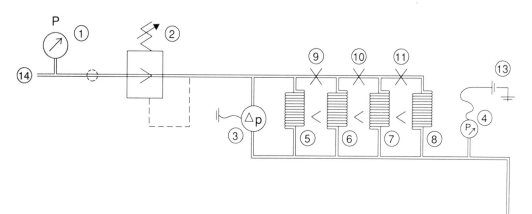

Fig. 1. Layout of probe permeameter components. 1, pressure gauge at gas source; 2, upstream pressure regulator; 3, micromanometer for measurement of ΔP across selected LFEs; 4, pressure transducer and indicator for measuring pressure within the flow element(s) downstream of the LFEs; 5–8, laminar flow elements (LFEs); 9–11, switches to control gas entry into LFEs; 12, probe assembly; 13, power supply; 14, gas source.

measurements, and durability, large volume cylinders that are heavy and awkward to transport manually. In many applications, air drawn from the atmosphere is the only practical option. At 273 K there is only a 2% difference in viscosity between air (17.05×10^{-6} kg ms^{-1}) and nitrogen (16.7×10^{-6} kg ms^{-1}); this difference has negligible effect on flow rate.

Of greater concern may be the moisture content of air drawn from the atmosphere that could dampen the porous media during measurement and so invalidate the single-phase flow status of measurements (Daltaban *et al.* 1991). Use of a chemical drier between the gas source and the porous media ensures operational confidence. However, expansion of air in a porous medium is, by consideration of adiabatic energy, at constant enthalpy. Inspection of a state diagram for H_2O shows that, for water vapour, a fall in pressure at constant enthalpy represents a movement away from saturation, i.e. implying that there is no risk of condensation in the porous media.

Pressure regulator

Commercially marketed pressure regulators are readily available. Their main drawback is that they tend to be large ($c.\,400$ cm^3) and heavy ($c.\,0.5$ kg).

Mass flow control

This is achieved by selecting a flow path through a choice of different resistors (Fig. 1). The lower the permeability the smaller the mass flow required. Flow rate can be measured in a variety of ways; for example, laminar flow elements and rotameters. A common problem with flow meters is their sensitivity to disturbance, 'bumps', often requiring them to be recalibrated before reliable meaurements can be made. Rotameters, although cheap and simple, are bulky and require operation from a stable, horizontal surface.

Pressure transducer and sensor

These are required to monitor pressure build-up and the subsequent steady-state achieved. They are standard components of permeameters and need operating ranges appropriate to the range of permeability to be investigated. To avoid any loss during pressure measurement, the pressure transducer should ideally be placed near to the point of gas inflow to the porous media.

Electrical power supply

Power is required for the operation of the transducer and all digital functions.

Probe

Probe design is often a function of application or user preference. Ideally, probes should be robust and lightweight. They need not be independent (attatched by an umbilical) from the main instrument.

Practical solution

To avoid dependence on a high-pressure gas cylinder supply the instrument system was designed to operate at low pressures (<5 bar). Obviation of a high-pressure gas source cuts down weight and essentially eliminates the need for motorized transport to carry the gas supply. To provide a low-pressure gas source a gas container is required that is lightweight and robust. Polyproylene is chosen as it is widely available in container form and has excellent physical properties. In our experimental model, a container based on a garden rose spray was used. To check for safety when using the container at 5 bar the hoop stress was calculated and found to be within the permitted tensile yield stress.

Mass flow control is achieved by construction of four laminar flow pathways. In our laboratory prototype these are simply tubing, the lengths of which have been calculated (Appendix) to ensure laminar flow at 2, 20, 200 and 2000 cm^3 per minute. For convenience, commercially available tubing was used, thus tubing diameter reflects only availability. In our final design laminar flow pathways will be engineered into a single (aluminium) block approximately $100 \times 60 \times 30$ mm. Pressure drop (ΔP) across the laminar flow elements is measured with a micromanometer by which a normalized gas throughput is calculated and then calibrated to give permeability. This design is compact and eliminates the need for incorporating any potentially 'bump-sensitive' mass flow meter components.

A PC board is included in the design that contains all the necessary electronics to measure and display pressure in the flow system downstream of the laminar flow pathways, ΔP across the laminar flow pathway, temperature and the interpreted permeability. The PC board monitors output from both the pressure transducer for the flow system and micromanometer, and records ΔP across selected flow pathways so that steady-state conditions can be identified. All data are recorded automatically, processed and stored once steady-state conditions are achieved. Downloading of data in ASCII format is then possible.

Instrument design

Our intention has always been to produce an instrument that is sufficiently lightweight and compact to allow one-man, single-hand operation. In concept, something similar in size, weight and appearance to a hand-operatable DIY drill. Clearly, gas source cannot be included in this design, thus, we leave the operator with the choice of selecting their own gas reservoir. As will be demonstrated, the results using a low pressure, polypropylene air reservoir compare favourably with other permeability measurements.

A layout for the new design (Fig. 2) reveals compactness and flexibility. Although our laboratory prototype is both heavier and more bulky than our design specification, this is mainly a consequence of having to use available components and materials rather than those stipulated in the design. Nevertheless, the prototype weighs only approximately 4.5 kg and has dimensions of $10 \times 12 \times 28$ cm. By making a compact and robust design, the instrument is easily enclosed in a waterproof casing and does not need to be stationed on a horizontal surface when taking measurements. The main constraint to bulk is the size of the gas regulator. We are confident that more compact pressure regulators are available.

An essential feature of all flow pathways in a probe permeameter is to ensure that laminar flow is maintained. Prior to construction of the prototype, calculations were carried out to ensure that turbulence would not be generated by the flow pathway design. In the prototype, successive opening of parallel flow pathways was adopted. For each flow rate (2, 20, 200 and 2000 cm^3 per min), and the tube dimensions of the available capillary tube (Table A1), the Reynolds number (Re) was calculated,

$$Re = \frac{\rho v d}{\mu}$$

where ρ = density of air (1.2 kg m^{-3}), v = velocity of air in the tube (m s^{-1}), d = diameter of the tube (m) and μ = the dynamic viscosity of air (18.12×10^{-6} kg ms^{-1}). A Reynolds number greater than 2000 implies turbulent flow. With laminar flow elements configured in series or in parallel, similar conditions of laminar flow are anticipated (Fig. 3).

Incorporation of a PC board into the body of the instrument makes single button control of

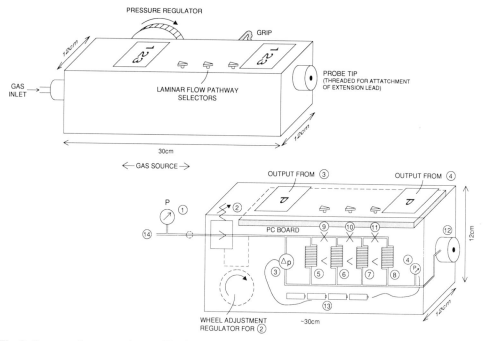

Fig. 2. Prototype instrument layout. Numbers as on Fig. 1.

a

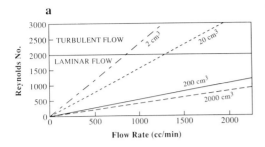

b

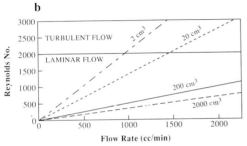

Fig. 3. Calculated Reynolds numbers for the laminar flow elements (LFEs): (**a**) LFEs open in series; (**b**) LFEs open in parallel.

the entire operation possible. In this way the instrument is very user-friendly; little operator skill is required before measurements can be made. Equally, all digital metering is clearly visible to the operator when holding the instrument.

Operation

In common with other probe permeameters, measurements are made by pressing the probe against a sample surface. During analysis of porous media the MPP measures the pressure drop across an operator-selected laminar flow pathway and the injection pressure at a location downstream of the flow pathways. Using the Poiseuille equation,

$$Q = \frac{\pi r^4 \Delta P}{8 \mu L}$$

where q = flow rate (m^3 s^{-1}), r = radius of capillary tube (m), ΔP = pressure difference (Pa), μ = viscosity of gas (Pa s) and L = length of capillary tube (m), the pressure drop along the flow system can be calculated for standard flow and operating conditions. For example, by rearranging the Poiseuille equation and assuming a 5 PSI injection pressure at 20°C,

the 2000 cm/min laminar flow pathway ($L = 0.357$ m, $r = 0.001494$ m) and using nitrogen gas one obtains,

$$\Delta P = \frac{((200/60) \times 10^{-6}) 8 (1.74 \times 10^{-5}) 0.357}{\pi (1.494 \times 10^{-3}) 4}$$

which gives a $\Delta P = 104.6$ Pa (10.5 mm H$_2$O).

From the measured pressure data the normalized gas flow rate through the sample (F_{norm} in cm^3/min) can be calculated,

$$F_{norm} = \Delta P (LFE/P_{cal}) \times (5/P_{inj})$$

where ΔP = pressure drop (mm H$_2$O) along the selected laminar flow path, LFE = maximum flow rate of the selected flow path, P_{cal} = calibrated pressure drop corresponding to the maximum flow rate of the selected flow path (mm H$_2$O) and P_{inj} = measured injection pressure (psi). Thus, by selecting an appropriate laminar flow path and measuring ΔP across the selected flow path while flowing a gas through a set of homogeneous core samples of known permeability, a calibration curve can be created for ΔP or normalized flow rate (F_{norm}) against permeability (Sutherland et al. 1993).

Experimental results

A series of preliminary calibrations were carried out on eight standard 1 inch and 1.5 inch diameter core plugs. All plugs were natural sandstone and have visible sedimentary heterogeneity. Although not 'homogeneous' standard materials they are representative of a range of reservoir characteristics and are comparable to data used for previous calibration of probe permeameters (Robertson & McPhee 1990). Despite concerns about plug sample heterogeneity, good calibration is obtained between Hassler sleeve measurements of permeability and probe permeameter flow rates over approximately three orders of magnitude permeability (Fig. 3). If standard homogeneous core plugs were used (see Halvorsen & Hurst 1990) we believe that better correlation will be achieved.

A second series of measurements were taken on a flat slab of Penrith Sandstone that has previously been used to examine the relationship between sedimentary heterogeneity and permeability (Fig. 4a). In this example, the permeability data illustrate that the probe permeameter is sensitive to subtle changes in permeability within a 300–800 mD range that corresponds to observed sedimentary heterogeneity (Fig. 4b).

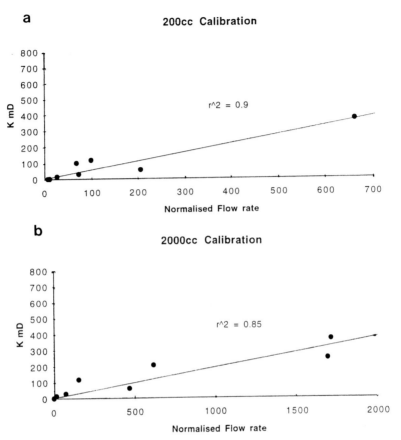

Fig. 4 Calibration curves for the MPP: (**a**) 200 cm³ flow element, (**b**) 2000 cm³ flow element. Data are from core plugs of known permeability

Discussion

Although the present prototype MPP is larger and heavier than the original specification, we are confident that components designed specifically for the MPP will allow the the original design goals to be met. In particular, avoiding having to use bulky 'off-the-shelf' manometers and pressure regulators, and the incorporation of a dedicated PC board and engineering flow elements, will greatly increase compactness and robustness and, decrease bulk.

In all our tests we have been sensitive to the possibility that the low pressure operation may not achieve satisfactory results, despite the favourable results of our pre-construction theoretical study. In all laboratory tests, including those reported in Figs 4 and 5 operation of MPP at pressure below 5 bar, has been problem-free. When the hand-pumped gas reservoir is used care is needed to ensure that the gas pressure does not fall too low. This situation is remedied by a short period of renewed pumping. Difficulty in attaining steady-state conditions, when no problems with probe tip seal are evident, appear to be caused by letting the pressure of the gas reservoir fall too low.

Results of the preliminary calibration (Fig. 4) indicate that the MPP provides estimates of permeability that are comparable to Hassler sleeve measurements over a range of permeability similar to that encountered in hydrocarbon reservoirs. By performing calibration on a set of homogeneous core plugs (Halvorsen & Hurst 1990; Hurst *et al.* 1995) we believe that the calibration between core plugs and flow rate will be improved. At present the level of correlation (R^2) is similar to that obtained with core plugs in similar calibration studies (Robertson & McPhee 1990; Sutherland *et al.* 1993). The investigation of the Penrith Sandstone slab (Fig. 5) demonstrates that MPP is senstitive to

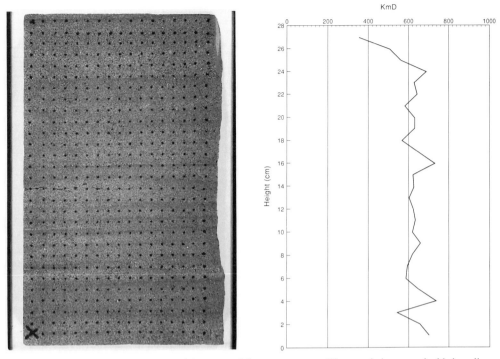

Fig. 5. (a) Slab of Penrith Sandstone used for permeabilty measurement. The sample is a cross-bedded medium-grained, aeolian dune sandstone. Points on the sample surface indicate positions where probe permeameter measurements have been taken. **(b)** Vertical log of permeability taken along the left-hand vertical transect of sample points shown in (a); measurements were taken at 5 mm intervals.

known variations in permeability that correspond to primary sedimentological features.

Conclusions

By considering the basic requirements of a portable probe permeameter a new design concept (MPP) is developed and patented that, even at prototype stage, provides a system that is six times lighter and at least ten times smaller than currently available systems. Because of the re-design the MPP is robust and generally not sensitive to conditions met in hostile environments (outcrop, well-site). MPP is self-contained and designed to operate from a wide range of gas sources. Preliminary testing of MPP shows that estimates of permeability are obtained that are comparable to those from other permeameters. Equally, permeability variations that are attributable to small-scale sedimentological variations are identifiable and quantifiable.

Table A1. *Dimensions of laminar flow pathways (LFEs)*

	L	OD	ID	ΔP_{air}	$\Delta P_{nitrogen}$
2 cm^3	0.265	0.159	0.508	9.1	9.4
20 cm^3	0.149	0.159	0.762	10.56	10.44
200 cm^3	0.962	0.318	2.159	10.07	10.46
2000 cm^3	0.357	0.476	2.997	10.07	10.56

L = length (m), OD = outer diameter (mm), ID = internal diameter (mm), ΔP_{air} and $\Delta P_{nitrogen}$ are the measured pressure drop for air and nitrogen at the given flow, 5 psi and 20°C (in mm H$_2$O).

Table A2. *Calculation of ΔP for all LFEs at an operating injection pressure of 5 psi at 20°C*

	Capillary radius (m)	Capillary length (m)	Flow Rate (m^3/min)	ΔP (Pa)	ΔP (mm H_2O)
Nitrogen					
2 cc LFE	0.000254	0.265	3.3333E-08	94.03	9.40327468
20 cc LFE	0.000381	0.149	3.3333E-07	104.44	0.4437022
200 cc LFE	0.0010795	0.962	3.3333E-06	104.63	10.4629127
2000 cc LFE	0.0014985	0.357	3.3333E-05	104.57	10.4570538
Air flow					
2 cc LFE	0.000254	0.265	3.333E-08	97.82	9.78156734
20 cc LFE	0.000381	0.149	3.3333E-07	108.64	10.8638512
200 cc LFE	0.0010795	0.962	3.3333E-06	108.84	10.8838344
2000 cc LFE	0.0014985	0.357	3.3333E-05	108.78	10.8777399

Absolute viscosity (Pa s) is 1.74×10^{-5} for air and 1.68×10^{-5} for nitrogen.

I. Morrison, C. Taylor, Tan Peng Kiat are acknowledged for their contributions to this work. R. T. Wilson is thanked for introducing us to the concepts of 'bumps' and 'splashes.'

Appendix 1: Calculation of laminar flow pathways

In the prototype system commercially available capillary tubing of 0.508 mm, 0.762 mm, 2.159 mm and 2.997 mm were used to construct laminar flow pathways (LFEs) for flow rates of 2 cm³/minute, 20 cm³/minute, 200 cm³/minute and 2000 cm³/minute respectively at a constant injection pressure of 5 psi. ΔP across the LFEs is measured by the micromanometer and used to to calculate a normalized throughput that can be calibrated to values of permeability (Fig. 3). Dimensions of the LFEs are summarized in Table A1. Using the Poiseuille equation values for ΔP are calculated for each LFE for air and nitrogen (Table A2).

References

DALTABAN, S., LEWIS, J. J. M. & ARCHER, J. S. 1989. Field minipermeameter measurements – their collection and interpretation. *In: Proceedings of the 5th European Symposium on Improved Oil Recovery*, Budapest, 25–27 April, 671–682.
—, WANG, J. S. & ARCHER, J. S. 1991. Understanding the physics of probe permeameter measurements through the use of the probe permeameter simulation program MIN-PER, *In: Minipermeametry in Reservoir Studies*. PSTI, Edinburgh, 27 June 1991.

GOGGIN, D. J., THRASHER, R. L. & LAKE, L. W. 1988a. A theoretical and experimental analysis of mini-permeameter response including gas slippage and high velocity flow effects. *In Situ*, **12**, 79–116.
—, CHANDLER, M. A., KOCUREK, G. & LAKE, L. W. 1988b. Patterns of permeability in eolian deposits: Page Sandstone (Jurassic), NE Arizona. *SPE Formation Evaluation*, **3**, 297–306.
HALVORSEN, C. & HURST, A. 1990. Principles, practice and applications of laboratory laboratory alipermeametry, *In*: WORTHINGTON, P. F. (ed.) *Advances in Core Evaluation, Accuracy and Precision in Reserves Estimation*. Gordon & Breach, Amsterdam, 521–549.
HURST, A. 1995. Probe permeametry – emergence of an old technology. *First Break*, **13**, 185–192.
— & GOGGIN, D. J. 1995. Probe permeametry: an overview and bibliography. *AAPG Bulletin*, **79**, 463–473.
—, HALVORSEN, C. & SIRING, E. 1995. A rationale for development of laboratory probe permeameters. *Log Analyst*, **36**, 10–20.
JONES, S. C. 1994. A new, fast, accurate pressure-decay probe permeameter. *SPE Formation Evaluation*, **9**, 193–199.
ROBERTSON, G. M. & MCPHEE, C. A. 1990. High resolution probe permeability: an aid to reservoir description, *In*: WORTHINGTON, P. F. (ed.) *Advances in Core Evaluation, Accuracy and Precision in Reserves Estimation* Gordon & Breach, Amsterdam, 495–520.
SUTHERLAND, W. J., HALVORSEN, C., HURST, A., MCPHEE, C. A., ROBERTSON, G., WHATTLER, P. R. & WORTHINGTON, P. F. 1993. Recommended practice for probe permeametry. *Marine and Petroleum Geology*, **10**, 309–317.

Network analogues of wettability at the pore scale

S. R. McDOUGALL, A. B. DIXIT & K. S. SORBIE

Department of Petroleum Engineering, Heriot-Watt University, Edinburgh EH14 4AS, UK

Abstract: This paper examines the macroscopic consequences of wettability variations at the pore scale. Important theoretical issues relating to distributed contact angles are examined and a process simulator capable of simulating the full flooding cycle characteristic of laboratory wettability tests is described. Simulations predict that final waterflood recoveries from weakly water wet (or even weakly oil wet) systems can often exceed those from strongly water-wet systems (even though initial imbibition rates may lead to the opposite conclusion). The underlying pore level physics has been explored by defining a capillarity surface, which incorporates the combined effects of both contact angle and pore dimension during the imbibition process. Finally, a regime-based framework has been developed which may go some way towards reconciling apparently contradictory wettability experiments.

Although the combined effects of permeability heterogeneity and gravity are extremely important in determining oil field recovery under waterdrive, the reservoir wettability state is an equally vital component that should not be overlooked. Numerous reports of experimental work relating to the role of non-uniform wettability in various aspects of oil recovery have been published, although a precise taxonomy of wettability is still lacking. Ageing temperature, brine pH/composition, crude oil composition, and connate water saturation have all been shown to play an important role in determining the ultimate wettability state of a system (Dubey & Waxman 1989; Hirasaki *et al.* 1990; Dubey & Doe 1991; Jadhunandan & Morrow 1991; Wolcott *et al.* 1991, 1993; Buckley 1993; Kovscek *et al.* 1993). Unfortunately, variations in laboratory procedure and experimental materials make generalized conclusions almost impossible and the precise wettability conditions prevailing at the end of a conditioning procedure are frequently difficult to deduce. Moreover, there are even serious doubts concerning the cross-correlation of the two main experimental methods used (Amott and USBM, Fig. 1) to determine core wettability. Indeed, until some clear theoretical interpretation of such matters is available, it is difficult to see how results from different laboratories can be adequately reconciled and understood.

To this end, the work presented here focuses upon the systematic investigation of wettability alteration at the pore scale and its subsequent impact upon waterflood displacement efficiency. Pore-scale models have been used to study a range of non-uniformly wet systems, their associated capillary pressure curves, spontaneous water imbibition characteristics and Amott indices. The models are capable of simulating the full flooding cycle characteristic

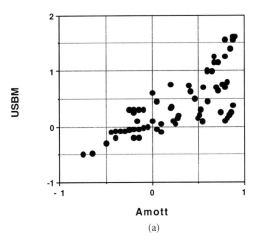

(a)

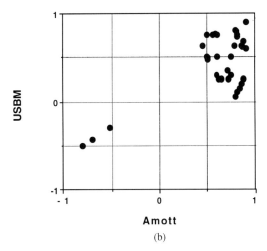

(b)

Fig. 1. Comparison of USBM and Amott indices. (a) J. S. Buckley (pers. comm.); (b) After Donaldson *et al.* (1969).

From Lovell, M. A. & Harvey, P. K. (eds), 1997, *Developments in Petrophysics*, Geological Society Special Publication No. 122, pp. 19–35.

of laboratory wettability tests (primary drainage, 'ageing', water imbibition, water drive, oil
imbibition, oil drive).

The first part of the cycle (primary drainage of
water by oil from a water-wet system) is
followed by an 'ageing' process, whereby wettability alterations occur at the pore scale. In a

physical system, wettability alteration depends
upon a variety of factors such as; ageing
temperature, brine composition, crude-oil composition and initial water saturation. The premise here, however, is that the net result of these
parameters is to produce a particular distribution of contact angles within the system.

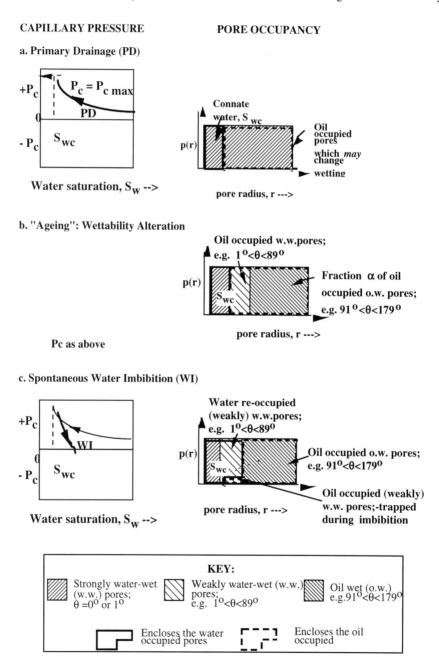

Fig. 2. Schematic sequence of network wettability conditioning simulations (mixed-wet case illustrated).

Consequently, the simulated ageing process produces oil-wet pores with contact angles in the range $\theta_{o1} < \theta_o < \theta_{o2}$ and water-wet pores with $\theta_{w1} < \theta_w < \theta_{w2}$. Of course, the precise range operating during an experiment depends upon complex interactions amongst the factors mentioned above. Thus, instead of quantifying the effects of contact-angle range for one particular case, the model has been used to examine the sensitivities of a cycle of displacements running under a number of different wettability scenarios. The results have been illuminating and have led to the concept of a capillarity surface, which can be used to predict

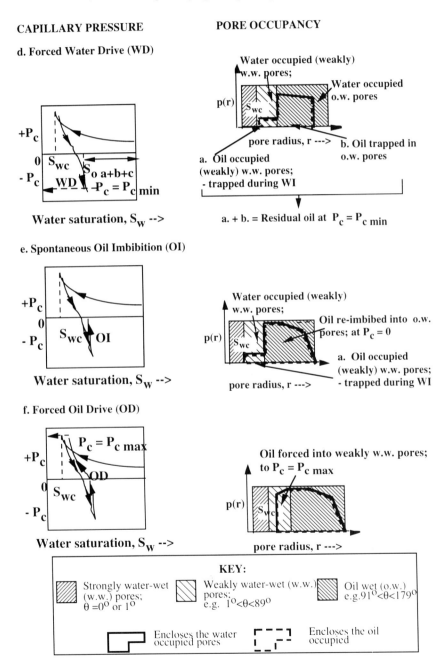

Fig 2. (*continued*).

phase distributions for any contact angle range. A number of contradictory experimental observations have been successfully reproduced and interpreted in light of this new theory.

Flow simulations in systems of non-uniform wettability

Generation of model and the flooding cycle

A 3D pore-scale model of average co-ordination number four ($z = 4$) has been used for the simulations (the co-ordination number of a network refers to the average number of pore elements meeting at a node (junction)). The overall network size was $15 \times 15 \times 15$, which gave a model containing approximately 7000 pores. For simplicity, a uniform pore-size distribution has been assumed: radii in the range $0.1–100\,\mu m$ are assigned to pore elements by using a random number generator. All pores are assumed to be of equal length ($100\,\mu m$) and have a cylindrical geometry. Note, however, that these are *not* limitations of the model, as any other pore size distribution and/or volume function can be included in order to mimic alternative geometries. Such sensitivities have been carried out and overall trends are very similar to those reported later.

A five-flood cycle has been performed on all networks and an 'ageing' simulation has been incorporated after primary drainage (Fig. 2a–f). This cycle simulates standard laboratory procedure by representing the following displacements: primary oil drainage, ageing, water imbibition, forced water drive, oil imbibition, and forced oil drive. No *a priori* assumptions are made concerning the distribution of oil-wet pores in the system; wettability alteration is considered only after primary oil drainage. Each stage of the process will now be described in more detail.

Primary drainage

Each primary drainage simulation begins with a strongly water-wet ($q = 0°$) network at 100% water saturation (Fig. 2a). Oil is initially introduced at the inlet face at negligible pressure. Now, oil will displace water in a water-wet pore only if the capillary pressure is positive. Therefore, one can anticipate practically no displacement of water by oil at near-zero capillary pressure (apart from some possible surface effect due to pore geometry). The pressure in the oil is then increased, leading to

an increase in capillary pressure. If P_{c1} is the capillary pressure prevailing at a particular instant, then oil will invade any pore of entry radius r if the following conditions prevail:

(1) the pore entry radius satisfies the condition

$$\left(0 < \frac{2\sigma\cos(\theta)}{r} \le P_{c1}\right);$$

(2) a continuum of oil from the inlet is present;
(3) a continuous path of water-filled pores is available from the target pore to the outlet (thereby allowing the wetting phase to escape).

Note that a capillary cut-off is implemented at the end of primary drainage, as failure to do so would ultimately yield complete recovery via film-flow in a strongly water-wet system.

It is evident from the above criteria that if the contact angle is the same in each pore, then the largest pore on the inlet face of the network is the first to be invaded. As the capillary pressure is further increased, oil invades successively smaller and smaller pores. However, if at any stage during the displacement, a particular water-filled pore loses its hydraulic continuity to the outlet, then water in that pore is considered to be trapped and cannot be displaced. Above a certain capillary pressure (P_{cmax}), no further displacement of water by oil is observed and the remaining water constitutes the irreducible water saturation (S_{wi}). This process is essentially 'invasion percolation' (Wilkinson & Willemson 1983) with wetting phase film drainage to a limiting capillary pressure.

One of the many advantages of using well-defined pore-scale models is that percolation theory can often be used to predict the onset of hydraulic discontinuity. Moreover, the number fraction of pores occupied by irreducible water at the end of primary drainage can also be deduced. This fraction is equal to the percolation threshold (PT) of the porous medium, and is related to the co-ordination number (z) and dimensionality (d) of the pore network by:

$$z(PT) = \frac{d}{d-1}. \tag{1}$$

The percolation threshold of the present model ($d = 3$ and $z = 4$) is therefore 0.375. This suggests that for a large 3D model of co-ordination number four, irreducible water will be present in 37.5% pores of the network.

The capillary pressure curve for primary drainage and the associated pore occupancy pattern are shown schematically in Fig. 2a. Note that, at this stage, the oil-occupied pores remain strongly water wet.

Wettability alteration

The premise used throughout this study is that wettability alterations that have occurred in actual reservoirs, as well as those associated with aged-core experiments, are non-uniform; i.e. that contact angles are not identical for all water-wet or oil-wet pores. This is based upon experimental evidence which indicates that various components of the crude oil interact differently with various mineral substrates in the rock e.g. quartz, feldspars, clays etc. (Jerauld & Rathmell 1994).

The degree of adsorption of polar components from crude oil onto a rock surface depends upon many factors, some of which have been outlined earlier. All of these factors affect the distribution of contact angles. The results presented later in this section will demonstrate the importance of this contact angle distribution during the flooding cycle. Indeed, Jerauld & Rathmell (1994) have observed distributions of contact angle in displacement experiments using etched micromodels aged in crude oil. They have also noted that the weak capillary forces often operating in mixed-wet systems are mainly related to variations in contact angle. Although

no experimental techniques are currently available to measure such contact angle distributions in natural porous media, the pore-scale simulator presented here provides an ideal theoretical tool with which to investigate and predict the expected consequences of various wettability/contact angle variations.

The two models of non-uniform wettability that have been considered are: (1) a mixed-wet system, where a certain fraction (a) of the largest pores containing oil after primary drainage become oil wet (Fig. 3a); and (2) a fractionally-wet system, where a random fraction of the pores containing oil become oil wet (Fig. 3b). For the sake of brevity, only mixed-wet cases will be discussed here, although similar conclusions also apply to fractionally wet media. In an earlier study (McDougall & Sorbie 1993), water-wet pores were simply assigned a zero contact angle, whilst all oil-wet pores had $\theta = 180°$. The new model, however, treats wettability alteration in a much more realistic manner (Fig. 2b). Contact angles (measured through the water phase) are now distributed randomly between 0 and 89° in the water-wet pores and between 91 and 180° in the oil wet pores. Pores containing irreducible water, however, remain strongly water wet ($\theta = 0°$). A range of different wettability conditions can be considered simply by allowing a different fraction (α) of oil-filled pores to become oil wet, whilst $(1 - \alpha)$ of the oil-filled pores remain water wet (although their

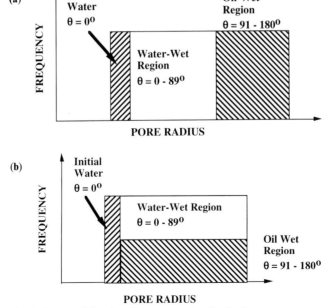

Fig. 3. Schematics of mixed wet and fractionally wet pore-size distributions.

contact angles are now distributed). Note that a is the fraction of oil-filled pores that have become oil wet; the actual oil-wet fraction (α') of the total number of pores in the network is consequently always less than α. The pore occupancy pattern after ageing with the crude oil is shown schematically in Fig. 2b.

Once ageing has occurred, the pore-scale wettability distribution remains invariant throughout the subsequent simulation (although, in principle, the simulator could also incorporate further wettability changes, due, for example, to desorption of oil components during other cycles of the flood).

The treatment of film-flow after wettability alteration

A realistic partial film-flow mechanism has been implemented during all four stages of the cycle following ageing. Film-flow affects not only the imbibition of a phase into the network but also its escape from it. The details of the partial film-flow mechanism are as follows.

Both water-wet and oil-wet pores are subdivided into two categories, depending upon their contact angle: (i) it is assumed that water can invade by film flow only in pores with contact angles in the range $0-\theta_{cw}$ (strongly water wet), whilst oil can only escape through films in pores having contact angles between θ_{co} and $180°$ (strongly oil wet). θ_{cw} and θ_{co} are critical cut-off values; (ii) a continuous path of water-*filled* pores is essential for the remaining weakly water-wet pores $(\theta_{cw} < \theta < 89°)$ to become available for invasion. Similarly, oil-wet pores with $91° < \theta < \theta_{co}$ require a continuum of oil to the outlet for escape. Analogous arguments hold when oil is the invading phase. In the present study, the values of θ_{cw} and θ_{co} have been taken as $45°$ and $135°$, respectively. The partial film-flow mechanism is probably the most likely mechanism operating in natural porous media.

Water imbibition

After the ageing process, water is introduced at the inlet face of the network and allowed to imbibe via the partial film-flow mechanism outlined above. The process is controlled by gradually reducing the capillary pressure from the highest value attained during primary drainage (P_{cmax}) to zero. The schematic is given in Fig. 2c.

Let P_{cwl} be any intermediate capillary pressure during water imbibition. The criteria for water imbibition into a given pore are:

(1) the pore should be water-wet in nature $(\theta = 0-89°)$;
(2) the partial film-flow criteria outlined above must be satisfied;
(3) the condition $(P_{cwi} \leq (2\sigma\cos(\theta)/r) \leq P_{cmax})$; must be met;
(4) a continuum of oil from the specified pore to the outlet should be available if oil is to escape (note that oil-wet pores still contain oil at this stage, and so the escape of oil from a water-wet pore via neighbouring oil-wet films need not be considered).

Forced water drive

It is evident from Laplace's equation that a negative capillary pressure is required for water to invade an oil-wet pore. Hence, once water imbibition is over, a forced water drive must be carried out for further recovery to take place. This is achieved by over-pressuring the water: the process is shown schematically in Fig. 2d. Forced water drive is primarily a drainage process, as water (the non-wetting phase) is now displacing oil (the wetting phase) from oil-wet pores. Thus, the criteria for the displacement of oil by water during this stage of the cycle are similar to those governing primary drainage. Consequently, forced water drive will be termed 'water drainage' here.

Let P_{cwd} be any intermediate capillary pressure during water drainage. The criteria for water invasion of a given pore are:

(1) the pore should be oil-wet $(\theta = 91-180°)$;
(2) the pore entry radius should satisfy the condition $(P_{cwd} \leq (2\sigma\cos(\theta)/r) \leq 0)$; (remember that P_{cwd} is now negative);
(3) a continuum of water from the inlet to the target pore should be present;
(4) (i) for strongly oil-wet pores $(\theta = 135-180°)$, a continuum of oil *or* a continuous oil-wet cluster to the outlet is all that is required for escape; (ii) for weakly oil-wet pores $(\theta = 91-135°)$, an uninterrupted path of oil-*filled* pores is required.

During this stage of the flooding cycle (forced water drive), the capillary pressure is decreased to successively larger negative values until no further displacement of oil by water is observed. The trapped oil remaining in the system constitutes the residual oil saturation (S_{or}).

It is clear from the earlier discussion that water cannot spontaneously imbibe into a water-wet pore if it is encircled by oil-wet pores. However, during the forced water drive, some of these oil-wet pores may indeed become invaded by water. This opens up the possibility of imbibition into previously shielded water-wet pores. The simulator described here accounts for such eventualities.

Oil imbibition

At the end of forced water drive, oil is introduced at the inlet face of the network and spontaneous imbibition of oil into oil-wet pores takes place. This process is analogous to water imbibition in water-wet pores. The capillary pressure is progressively increased, from the minimum reached at the end of forced water drive (a negative value) to zero. The criteria for oil imbibition are similar to those governing water imbibition. The associated capillary pressure curve and pore occupancy pattern for this stage are shown schematically in Fig. 2e.

Forced oil drive

The final flood of the cycle is the forced oil drive. Oil is forced into water-wet pores by slowly increasing the capillary pressure. The displacement criteria for the forced oil drive are similar to those for forced water drive. The schematic is shown in Fig. 2f.

Advancing–receding contact angles

It is well known that, unless very pure fluids are used, advancing contact angles (contact angles when the wetting phase displaces the non-wetting phase) are generally higher than receding contact angles. Moreover, data from the literature (Morrow & McCaffery 1978) show that the difference between an advancing and receding contact angle increases as the advancing angle approaches 90°. This phenomenon has been incorporated here: the distributed contact angles after ageing are taken to be the advancing angles and the receding angles are reduced by 20%. Hence, weakly wet pores are affected more than strongly wet pores.

Results and discussion

Although a wide range of sensitivities has been performed during this work (using different film-flow mechanisms, different co-ordination numbers, mixed- and fractionally wet systems), discussion will be restricted here to the role played by distributed contact angles during oil/

water displacements. Results from systems containing pores with distributed contact angles (i.e water-wet pores with $0° < \theta < 90°$ and oil-wet pores with $90° < \theta < 180°$), will be compared with those from systems containing pores with single-valued contact angles (i.e all water-wet pores having $\theta = 0°$ and *all* oil-wet pores having $\theta = 180°$, say).

Capillary pressure curves

Single-valued contact angles. Simulated capillary pressure curves for a variety of networks with single-valued contact angles are presented in Fig. 4. Note that the fraction of the total number of pores which are made oil wet (α') is presented, since this gives a better idea of how close to the percolation threshold the oil-wet cluster is. Consider first the strongly water-wet network ($\alpha = \alpha' = 0$, Fig. 4a). Following the imbibition curve downwards from the end of primary drainage, it is found that about 30% of the total pore volume is occupied by water after water imbibition. The third and fourth parts of the flooding cycle (forced water drive and oil imbibition), produce curves that are completely vertical: this is due to the absence of any oil-wet pores in the network The final stage of the procedure, oil-drive, is characterized by its hysteresis with the water imbibition curve.

An immediate observation from Fig. 4 is that the amount of water imbibed decreases with an increase in a (or α'). This is due to: (i) a reduction in the number of water-wet pores in the network, which cuts down the number of possible imbibition paths, and (ii) the poor connectivity of the system ($z = 4$), which means that the number of isolated water-wet clusters present after ageing also increases with α. These isolated water-wet clusters are effectively shielded from the invading water by oil-wet pores during imbibition.

The water saturation at the end of water drive is found to remain almost constant up to $\alpha' > 0.375$, because no connected path of oil-wet pores is yet available to transfer oil via film flow. Only after $\alpha' > 0.375$ should any difference be seen, as this is the percolation threshold for oil-wet pores. This has been borne out by simulation: for $\alpha' = 0.468$ and 0.625 ($\alpha = 0.75$ and $\alpha = 1.0$), the water saturations after forced water drive are found to be much higher than those with $\alpha' = 0.155$ and 0.313 ($\alpha = 0.25$ and $\alpha = 0.5$).

A continuous oil-wet cluster from the surface is also essential for oil imbibition during stage four of the cycle: no imbibition of oil is observed for $\alpha' = 0.155$ and 0.313 as expected. For $\alpha' > 0.375$, however, the extent of oil imbition increases rapidly, as the oil-wet pores aggregate to form larger spanning clusters.

When $\alpha = 1$ ($\alpha' = 0.625$), no imbibition of water takes place at all because all of the water-wet pores are associated with disconnected irreducible water (Fig. 4c): no oil drainage occurs for the same reason. Note also, that the end-of-cycle water saturation is slightly higher than the connate water saturation, which implies that water ultimately remains trapped in both oil-wet and water-wet pores. These observations are consistent with experimental results reported by Salathiel (1973).

Distributed contact angles. Having outlined the typical behaviour of single-valued contact angle systems, compare now the results obtained from systems with distributed contact angles. The corresponding capillary pressure curves are shown in Fig. 5. The extent of water imbibition differs significantly from that obtained from single-valued contact angle systems. Compare the results for $\alpha = \alpha' = 0$ (note that although $\alpha = 0$ means that no oil-wet pores exist, weakly water-wet pores ($0° < \theta < 89°$) are present). The extent of spontaneous water imbibition increases by c. 55% to 0.384 PV when the contact angles are distributed between $0°$ and $89°$. This compares with only 0.248 PV imbibed with single-valued contact angles ($\theta = 0°$). Moreover, for $\alpha' < 0.375$ (percolation threshold), the water saturations at the end of the forced water drive decrease significantly with α' when contact angles are distributed. Remember that this effect was absent after waterflooding systems characterized by single-valued contact angles. These results clearly indicate that a distribution of contact angles at the pore-scale plays a crucial role in determining recovery in systems of non-uniform wettability. In fact, it will now be shown that it is the actual range of contact angles that is of primary importance.

Table 1 shows water saturations and percentage oil in place recovered after water imbibition for various contact angle ranges in a 100% water-wet system. Notice that the extent of water imbibition with distributed contact angles up to $30°$ is almost identical with that resulting from single-valued contact angles. Even with $\theta = 0–60°$, the water saturation is only marginally higher than that obtained with all $\theta = 0°$. Only when contact angles are assigned from the range $0–89°$ is water imbibition substantially enhanced, implying something special about the range $60–89°$. This has been confirmed by distributing the contact angles within the restricted ranges $45–89°$ and $60–89°$ (see Table 1). This has major implications for laboratory wettability tests: these results show that a sample containing pores with a range of contact angles can imbibe up to 50% more water than one that is strongly water wet ($\theta = 0°$).

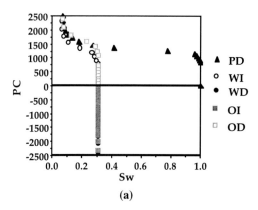

(a)

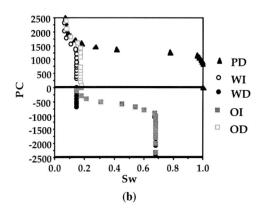

(b)

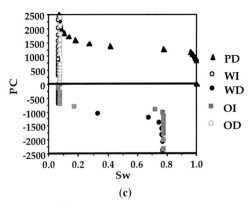

(c)

Fig. 4. Capillary pressure curves from three systems with single-valued contact angles: **(a)** $\alpha = \alpha' = 0$; **(b)** $\alpha = 0.75$, $\alpha' = 0.468$; **(c)** $\alpha = 1.0$, $\alpha' = 0.625$.

What is even more surprising, is that this is even true for a system that would appear to be neutrally wet on the basis of contact angle measurements ($\theta = 60$–$89°$). So, what are the reasons for this intriguing behaviour?

Contact-angle distributions and capillarity surfaces

Consider the simple example of five oil-filled cylindrical capillary tubes of differing radii. When the oil/water contact angle is the same

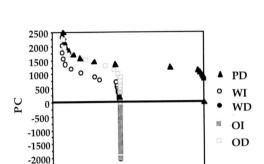

(a)

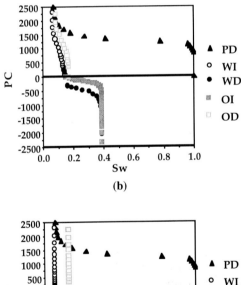

(b)

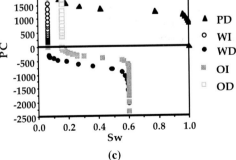

(c)

Fig. 5. Capillary pressure curves from three systems with distributed contact angles: (**a**) $\alpha = \alpha' = 0$; (**b**) $\alpha = 0.75$, $\alpha' = 0.468$; (**c**) $\alpha = 1.0$, $\alpha' = 0.625$.

Table 1. *Effect of contact angle range upon imbibition and oil recovery*

Contact angle range	S_w after imbibition	%OIP produced after imbibition
0–0°	0.31	26.13
0–15°	0.31	26.13
0–30°	0.299	25.22
0–45°	0.327	28.21
0–60°	0.337	29.28
0–89°	0.446	40.9
45–89°	0.408	36.85
60–89°	0.411	37.17

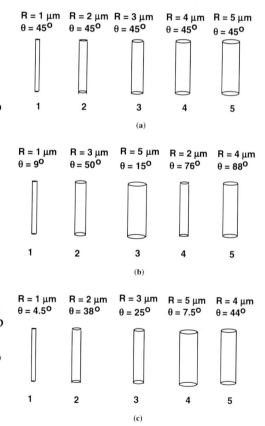

Fig. 6. Tube-filling sequences during imbibition for five tubes of differing radii: (**a**) all contact angles $= 45°$; (**b**) contact angles distributed from 0–90°; (**c**) contact angles half those in (**b**).

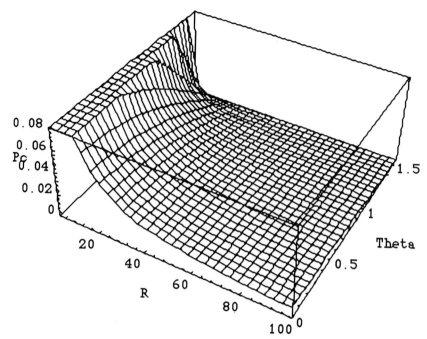

Fig. 7. Schematic example of a capillarity surface, $P_c(r, \theta)$. N.B. θ is in radians.

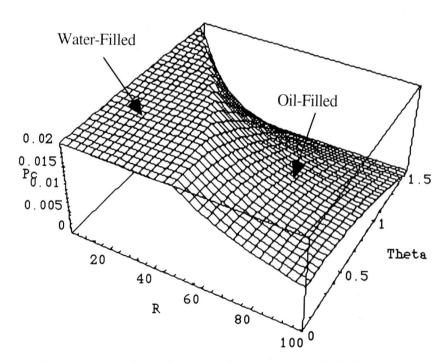

Fig. 8. Capillarity surface, $P_c(r, \theta)$, showing water and oil pore occupancy for imbibition.

in each tube, the imbibition filling sequence is entirely governed by pore radius through Laplace's equation: the smallest tube fills first, and so on (Fig. 6a). With distributed contact angles, however, this need no longer be the case, as shown in Fig. 6b; the filling sequence can be altered dramatically, because some of the smaller pores may now possess large contact angles and can be filled only at very low capillary pressures. This is rather intuitive and not too surprising, but the next example highlights a less-obvious effect. What happens to the filling sequence if the contact angle of each tube is simply halved; i.e. if the set of contact angles is changed from $\{9°, 15°, 50°, 76°, 88°\}$ to $\{4.5°, 7.5°, 25°, 38°, 44°\}$? On the face of it, the filling sequence should be the same as before, as all angles have simply been reduced by a factor of 2 and their rank order remains the same. The new filling sequence, shown in Fig. 6c, demonstrates, however, that this need not be the case. This rather surprising result is due to the non-linearity of the cosine function and its implicit non-linear effect upon capillary pressure. The combined effects of contact angle and capillary entry radius can perhaps best be understood by defining a capillarity surface. This is a three-dimensional plot of capillary pressure as a function of both contact angle and capillary entry radius; i.e the surface:

$$P_c(r, \theta) = \frac{2\sigma \cos \theta}{r}. \qquad (2)$$

A typical example is shown in Fig. 7 – the nonlinearity is self-evident. An important consequence of such a plot is that approximate phase distributions can be predicted for any array of capillary elements (including interconnected networks) simply by examining the associated surface in the light of percolation theory. This is best illustrated with reference to a fully connected water-wet network $(z = 6)$, which remains free of additional topological complications.

By taking a number of sequential horizontal cross-sections through the capillarity surface (i.e. planes of constant P_c), the progress of the imbibition floods can be studied as capillary pressure decreases (note that drainage floods can also be analysed, but accessibility has to be taken into account). Pores lying on the flat plateau of the sliced surface must have capillary entry pressures greater than that corresponding to the horizontal cut-off value and, if accessible, must be filled with water. Those pores not lying on the flat plateau must therefore be filled with oil as shown in Fig. 8. Capillary surfaces for

any range of contact angles or capillary entry radii can easily be plotted and sensitivities analysed. Such an analysis will now be conducted and used to examine two issues arising from the simulation results described earlier.

The first issue relates to the effect of contact angle range upon the early stages of water imbibition To investigate this, four ranges are considered and a constant capillary pressure cut-off is used to slice all four capillarity surfaces at the same point during the displacement. The results are given in Fig. 9. They clearly show that the amount of water initially imbibed reduces considerably once the contact angle range incorporates weakly water-wet pores $(60° < \theta < 90°)$.

The second, more important, issue arising from the simulation studies, concerns the reduction seen in residual oil once weakly water-wet pores are incorporated into the system. By examining the capillarity surface in light of percolation theory, the reasons for this behaviour become clear. For a well-connected network with $z = 6$, the percolation threshold is 0.25, so at the end of water imbibition, about 75% of the pores will contain water. Hence, the residual oil saturation can be estimated for all four cases by slicing the corresponding capillarity surface at a capillary pressure value that leaves 75% of the pores lying in the plateau region. The unshaded regions in Fig. 10 therefore show the distributions of oil-filled pores at the end of water imbibition. Although the unshaded regions all have the same area (approximately 25% of the square), their shapes differ: it can therefore be concluded from the figure that residual oil should be maximum when the contact angles are all 0° and minimum when they range from 0° to 90°.

The two sets of sensitivities just described mean that the capillary pressure curves from the four systems must actually cross one another. The implication for petrophysical practice is that wettability-imbibition trends depend entirely upon the capillary pressure cut-off employed in the laboratory. Moreover, the results presented here demonstrate that the contact angle range plays a major role in determining the form of capillary pressure curves and residual saturations.

The theory outlined above can be verified by comparing the pore occupancy histograms during water imbibition from single-valued and distributed contact angle systems with $\alpha = 0$ (Fig. 11). The variation in trapped oil distribution closely matches that predicted by the capillarity surface.

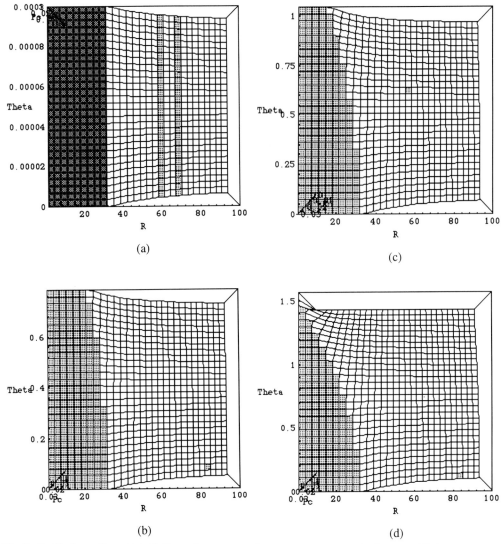

Fig. 9. Capillarity surfaces, $P_c(r, \theta)$, for various ranges of water-wet contact angle; the water-filled pores are shown shaded for each contact angle range **(a)** $\theta = 0°$, **(b)** $\theta = 0{-}45°$.

Oil production during water imbibition

The water imbibition capillary pressure curves as functions of produced OIP (oil in place) from single-valued and distributed contact angle networks are given in Fig. 12. The imbibition curve for the single-valued water-wet system ($\theta = 0°$ for all pores in the network) is also given in Fig. 12b for reference. With single-valued contact angles, imbibition is maximum in a 100% water-wet system ($\alpha = 0$) and diminishes monotonically with an increase in the fraction of oil-wet pores. In the case of distributed contact angles, however, oil production for $\alpha = 0$ and $\alpha = 0.25$ is actually higher than that obtained from the strongly water-wet system. For larger a production again decreases.

These results, when viewed in light of the earlier theoretical discussion, may go some way towards a clearer interpretation of many contradictory experimental results of the past. Jadhunandan (1990) has reported similar experimental trends from cores aged using various brines at 26°C (Fig. 13). For a brine composition of 4% NaCl and 0.2% CaCl$_2$, for example, the oil produced by water imbibition was found to be

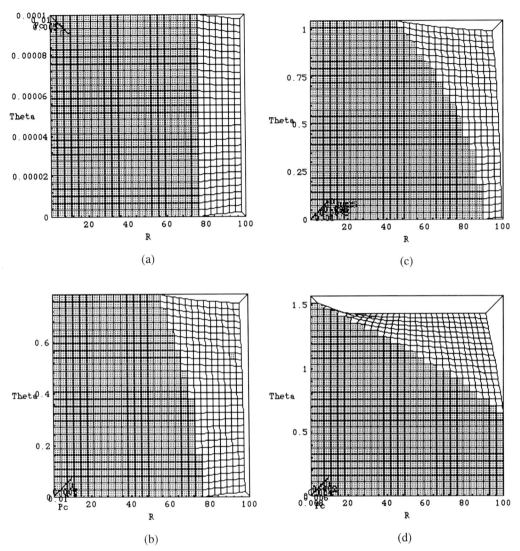

Fig. 10. Distribution of water- and oil-filled pores at residual oil; the water-filled pores are shown shaded for each contact angle range; (**a**) $\theta = 0°$, (**b**) $\theta = 0\text{--}45°$, (**c**) $\theta = 0\text{--}60°$ (**d**) $\theta = 0\text{--}90°$.

higher than that observed in a strongly water-wet core: no explanation was reported. The work reported here would suggest that the increase in water imbibition in the aged core is probably due to a distribution of contact angles at the pore scale; such anomalous behaviour cannot be explained by assuming single valued contact angles. Other experimental examples are also consistent with the theory developed here. Such experimental evidence provides useful support for the physics underpinning the pore-scale simulator.

A regime-based framework for the interpretation of wettability experiments

The waterflood recovery trends from a range of mixed-wet systems (i.e. differing α') terminated at various end-point capillary pressures are shown in Fig. 14 (note the importance of end-point capillary pressure here). Point A represents recovery from the 100% water-wet network with fixed contact angle ($\theta = 0°$), whereas all other points are derived from systems containing distributed angles (0–89°). Point B,

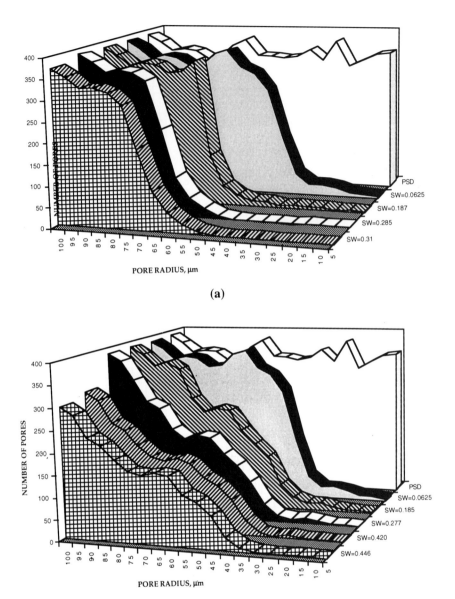

Fig. 11. Pore occupancy during spontaneous water imbibition for 100% water-wet system ($z = 4$); (**a**) single-valued contact angle; (**b**) distributed contact angles.

for example, represents the 100% weakly water-wet case ($\alpha' = 0$, $0 < \theta < 89°$). In order to apply the results presented so far to laboratory data, however, it is important to know where a given experiment lies with respect to Fig. 14; i.e. what was the end-point capillary pressure and what a' value has been achieved after the core conditioning process?

Although it is difficult to assess the precise fraction of water-wet and oil-wet pores within a sample experimentally, Amott wettability indices can be used to determine whether or not connected water-wet or oil-wet pathways exist. For example, a positive I_w value indicates that the fraction of strongly water-wet pores within a sample exceeds the percolation threshold

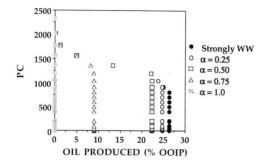

(a)

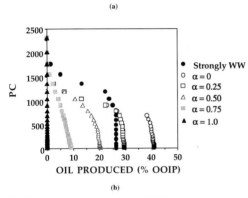

(b)

Fig. 12. Spontaneous water imbibition curves of mixed-wet systems; (**a**) single-valued contact angles (**b**) distributed contact angles.

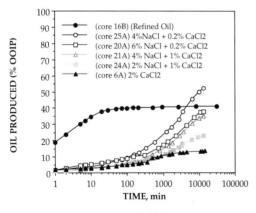

Fig. 13. Spontaneous water imbibition curves from cores aged with various brines. The refined oil case is strongly water wet (from Jadhunandan 1990).

$(\alpha' > \alpha'_{cr})$; similar reasoning holds for strongly oil-wet pores. Conversely, an Amott index of zero precludes the existence of a spanning cluster composed of strongly wetted pores $(\alpha' > \alpha'_{cr})$. Amott water and oil indices (I_w and I_o) corresponding to the simulations of Fig. 14 are given in Table 2.

By using the models described in this paper, the waterflood recovery trends from systems which cover strongly water-wet through to strongly oil-wet conditions can now be summarized in terms of various 'regimes' as follows.

Regime IA: from strongly water wet ($\theta = 0°$, point A) to weakly water wet ($\theta = 0-89°$, point B). $\alpha' = 0$

Recovery increases due to the effect of distributed contact angles at the pore-scale. I_w remains unity and $I_o = 0$. The end point capillary pressure has no effect upon recovery in this regime.

Regime IB: from weakly water wet ($\theta = 0-89°$, point B) to mixed wet, but $\alpha' < \alpha'_{cr}$

Recovery will be less than that derived from the weakly water-wet system. I_w decreases with increasing α' and I_o remains zero, as no oil-wet spanning cluster yet exists. Again, end-point capillary pressure has little effect upon recovery in this regime.

Regime II: from mixed wet with $\alpha' < \alpha'_{cr}$ to mixed wet with $\alpha' < (1 - \alpha'_{cr})$

In this regime, recovery can increase rapidly and may reach a local maximum depending upon the end point capillary pressure. Increased recovery is due to the availability of a spanning oil-wet cluster capable of transporting oil via film-flow to the outlet. I_w values decrease and I_o values increase as α' increases (both are non-zero). Notice how the end-point capillary pressure plays a major role in determining recovery in this regime.

Regime III: from mixed wet with $\alpha' < (1 - \alpha'_{cr})$ to $\alpha' = 1$

This regime has not been examined here, as the presence of connate water excludes the possibility of having $\alpha' > (1 - \alpha'_{cr})$. The regime is characterised by $I_w \to 0$ and $I_o \to 0$ and would yield optimum recovery if a sufficiently large negative capillary pressure could be reached. In practice, however, this is unlikely to be achieved.

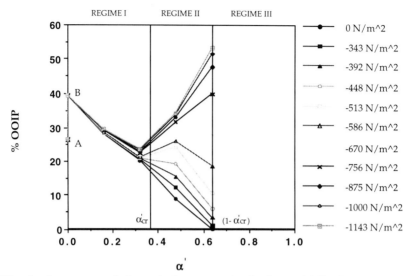

Fig. 14. Waterflood recovery trends from mixed-wet networks allowing partial film flow.

Table 2. *Amott indices from mixed-wet simulations*

% Oil-wet pores (α')	Water index (I_w)	Oil index (I_o)
0	1	0
0.17	0.96	0
0.32	0.86	0
0.48	0.27	0.7
0.64	1	0

The application of this regime-based framework to the interpretation of waterflood experiments is currently underway.

Conclusions

The work presented in this paper forms the theoretical basis from which a regime-based framework for the interpretation of wettability experiments has emerged. Capillary pressure curves from both mixed and fractionally-wet systems have been analysed using a 3-D pore-scale simulator which takes into account many pore level flow processes: non-uniform wettability alterations, partial film flow, trapping of wetting and non-wetting phases, variations in advancing and receding contact angles, etc. More specific conclusions are the following:

(1) It has been proposed that the ageing of cores produces pores with a distribution of contact angles; i.e. oil-wet pores in the range $\theta_{o1} < \theta_o < \theta_{o2}$ and water-wet pores with $\theta_{w1} < \theta_w < \theta_{w2}$.

(2) Simulations predict that final waterflood recoveries from weakly water-wet (or even weakly oil-wet) systems can often exceed those from strongly water-wet systems (even though initial imbibition rates may lead to the opposite conclusion).

(3) The underlying pore level physics has been explored by defining a capillarity surface. which incorporates the combined effects of both contact angle and pore dimension during the imbibition process.

(4) The development of spanning oil-wet and water-wet clusters in mixed-wet systems is strongly influenced by connectivity via the associated percolation threshold. These, in turn, affect oil/water wettability indices. More work is required in this area in order to obtain modified wettability indices which acknowledge the important role played by the percolation threshold.

(5) Theoretical models derived in this paper, in conjunction with pore-scale simulation, have led to the development of a regime-based framework for the interpretation of wettability experiments.

The authors would like to express their gratitude to the following organizations for their invaluable financial support: BP Exploration Ltd, The UK Department of Trade and Industry, Exxon Production Research Co., Shell UK Ltd, and Statoil. Useful discussions with C. Brown, P. Salino, D. Stern, P.-E. Øren and J. Buckley are also gratefully acknowledged.

References

BUCKLEY, J. S. 1993. Asphaltene precipitation and crude oil wetting. Society of Petroleum Engineers Paper 26675 presented at 68th Annual Technical Conference and Exhibition, Oct. 3–6 Houston.

DONALDSON, E. C., THOMAS, R. D. & LORENZ, P. B. 1969. Wettability Determination and Its Effect on Recovery Efficiency. *Society of Petroleum Engineers Journal*, March, 13.

DUBEY, S. T. & DOE, P. H. 1991. Base number and wetting properties of crude oils. Society of Petroleum Engineers Paper 22598 presented at 66th Annual Technical Conference and Exhibition, 6–9 Oct., Dallas.

—— & WAXMAN, M. H. 1989. Asphaltene adsorption and desorption from mineral surface. Society of Petroleum Engineers Papers 18462 presented at SPE International Symposium on Oil field Chemistry, 9–10 Feb., Houston.

HIRASAKI, G. J., ROHAN, J. A., DUBEY, S. T. & NIKO, H. 1990. Wettability evaluation during restored-state core analysis. Society of Petroleum Engineers Paper 20506 presented at 65th Annual Technical Conference and Exhibition, 23–26 Sep., Dallas.

JADHUNANDAN, P. P. 1990. *Effects of Brine Composition, Crude Oil and Aging Conditions on Wettability and Oil Recovery*. PhD Thesis, New Mexico Institute of Mining and Technology, New Mexico.

—— & MORROW, N. R. 1991. Effect of wettability on waterflood recovery for crude-oil/brine/rock systems. Society of Petroleum Engineers Paper 22597 presented at 66th Annual Technical Conference and Exhibition, 6–9 Oct., Dallas.

JERAULD, G. R. & RATHMELL, J. J. 1994. Wettability and relative permeability of Prudhoe Bay: a case study in mixed wet reservoirs. Society of Petroleum Engineers Paper 28576 presented at 69th Annual Technical Conference and Exhibition, 25–28 Sept. New Orleans.

KOVSCEK, A. R., WONG, H. & RADKE, C. J. 1993. A pore-level scenario for the development of mixed wettability in oil reservoirs. *American Institute of Chemical Engineers Journal*, **39**, 1072.

McDOUGALL, S. R. & SORBIE, K. S. 1993. The predictions of waterflood performance in mixed-wet systems from pore scale modelling and simulation. Society of Petroleum Engineers Paper 25271 presented at 12th SPE Symposium on Reservoir Simulation, 28 Feb.–3March, New Orleans.

MORROW, N. R. & McCAFFERY, F. G. 1978. Displacement studies in uniformly wetted porous media. *In*: PADDAY, G. F. (ed.) *Wetting, Spreading and Adhesion*. Academic Press, New York.

SALATHIEL, R. A. 1973. Oil recovery by surface film drainage in mixed wettability rocks, *Journal of Petroleum Technology*, Oct., 1216.

WILKINSON, D. & WILLEMSEN, J. F. 1983. Invasion percolation: a new form of percolation. *Journal of Physics A*, **16**, 3365.

WOLCOTT, J. M., GROVES, F. R. JR., TRUJILLO, D. E. & LEE, H. G. 1991. Investigation of crude-oil/mineral interactions: factor influencing wettability alteration. Society of Petroleum Engineers Paper 21042 presented in International Symposium of Oil Field Chemistry, 20–22 Feb., Anaheim, California.

——, —— & LEE, H. G. 1993. Investigation of crude-oil/mineral interactions: influence of oil chemistry on wettability alteration. Society of Petroleum Engineers Paper 25194 presented in International Symposium of Oil Field Chemistry, 2–5 March, New Orleans.

Pore-structure visualization in microdioritic enclaves

S. PUGLIESE & N. PETFORD

School of Geological Sciences, Kingston University, Penrhyn Road,
Kingston upon Thames KT1 2EE, UK
(e-mail: S.Pugliese@kingston.ac.uk)

Abstract: We present a simple methodology for estimating key petrophysical characteristics of partially molten (igneous) porous media using microgranular (microdiritic) enclaves from the Ross of Mull granite, Scotland, as examples. A number of enclaves have been infiltrated by granitic melt while still partially molten, with the infiltration (porosity) network now preserved as frozen-in granitic melt. By serial sectioning through individual enclaves, we show that the preserved infiltration channel network is interconnected in three dimensions. Using image analysis techniques, we have estimated the porosity (ϕ) of individual enclave sections and obtained the variation in porosity with depth. Simple representation of the pore-channel network on a branch and node chart allows useful petrophysical characteristics including the connectivity, genus and tortuosity of the network to be estimated. Computer-enhanced reconstructions of the pore network in three dimensions are shown that provide a powerful way of visualizing complex geometries in porous media.

It is now widely accepted that microgranular enclaves in granitic plutons result from the incomplete mixing of hot mafic to intermediate (45–55 wt% SiO_2) melts intruded into cooler granitic magma (e.g. Blake *et al.* 1965; Wiebe 1974, 1993; Castro *et al.* 1990*a*; Pitcher 1991; Chapman & Rhodes 1992). The importance of this process in controlling the final composition of calc-alkaline granitoid rocks depends critically on the ability of enclave and host granite magmas to interact while still partially molten. So far, three exchange mechanisms involving thermal, mechanical and chemical transfer have been recognized (Barbarin & Didier 1992). Despite continued uncertainty over the exact role played by each process during granitoid petrogenesis (see Didier & Barbarin 1991 for a review), most microgranular enclaves in exposed plutons exhibit some degree of interaction with their surrounding granite.

The ability of magmas of different chemical composition to mix and homogenize is principally governed by the initial physical properties (especially heat contents and mass fractions) of the magmas, as well as their rheological behaviour during cooling (Vernon 1984; Sparks & Marshall 1986; Fernandez & Gasquet 1994). Recent studies on microgranular enclaves have therefore focused on either diffusion (chemical and thermal) or mechanical incorporation as the principal exchange medium between magmas (Sparks & Marshall 1986; Baker 1992; Lesher 1994). While factors such as grain size and shape and degree of enclave crystallinity (porosity) have been shown to affect enclave rheology (Fernandez & Gasquet 1994), they also play a

fundamental role in controlling interstitial melt movement within crystallizing igneous systems (e.g. Marsh 1981; Tait *et al.* 1984; Kerr & Tait 1986; Hunter 1987; Cushman 1990; Petford 1993). To date, little attempt has been made to study the porosity and associated permeability of microgranular enclaves, despite the fact that the presence of a continuous fluid phase within an enclave will greatly influence fluid flow and diffusive mass transfer processes between enclave and host granite, and thus the final chemical composition of both magmas.

In this contribution, we outline a simple methodology for examining and quantifying key petrophysical properties (connectivity, tortuosity and two/three-dimensional pore geometry) of microgranular enclaves permeated by granitic melt during their crystallization. The methodology provides a first step towards quantifying the complex geometrical structures that result from melt infiltration in igneous systems, and is applicable to a wide range of other porous geological media.

Enclave samples

Regional geology and lithologies

The Ross of Mull granite lies at the western end of the Island of Mull, Scotland. The granite intruded Moine metasediments during late Caledonian times (Holdsworth *et al.* 1987) yielding a Rb–Sr whole rock age of 414 ± 3 Ma (Halliday *et al.* 1979). The pluton was emplaced concordantly along the western limb of the

From Lovell, M. A. & Harvey, P. K. (eds), 1997, *Developments in Petrophysics*, Geological Society Special Publication No. 122, pp. 37–46.

Assapol synform as a horizontal sheet approximately 3.5 km thick, dipping 30°E (Beckinsale & Obradovich 1973). A N–S trending foliation is developed within the granite which becomes accentuated along locally developed shear zones. Within the southern and southwestern portions of the pluton, numerous outcrops of Moine xenoliths and microdioritic enclaves are present, the latter showing the greatest concentration in topographic lows. Most of the microdioritic enclaves range in size from less than 5 cm to several metres in diameter and are either ovoid or ellipsoidal in shape, with aspect ratios between 2:1 and 50:1.

Mineralogy and textures

The host granite has been subdivided on textural criteria by R. H. Hunter & J. R. Reavy (pers. comm.) into two end-member types, the Fionnophort equigranular facies and the porphyritic alkali-feldspar facies. Grain sizes for the main rock forming minerals vary from 0.2 mm to >10 mm. Mineralogically the Fionnophort equigranular facies can be distinguished from the porphyritic alkali-feldspar facies by the virtual absence of plagioclase (maximum 5% modal) and increased amounts of quartz and alkali feldspar.

The microgranular enclaves display a wide range of textural and mineralogical features which we have used to classify them into (1) megacrystic and (2) megacryst-free varieties. The megacryst-free enclaves are composed largely of amphibole ($c.$ 30%), biotite ($c.$ 20%), albite twinned interstitial plagioclase ($c.$ 45%) and show recrystallization textures. The megacrystic enclaves can be subdivided into two groups depending on the megacryst assemblage: (a) plagioclase + quartz and (b) plagioclase + hornblende ± quartz ± alkali feldspar. Mantling of the megacrysts by coronas of plagioclase or amphibole ± biotite is common.

These observations, coupled with the presence of crenulate and cauliform contacts (e.g. Castro *et al.* 1990*b*), provide evidence for magmatic interactions between enclaves and host granite. In exceptional cases, a pervasive veining of the enclaves by the host granite (Fig. 1) has led to the development of channel-like morphologies referred to by Petford *et al.* (1996) as melt channels. In the field the channels range in shape from polygonal structures (akin to desiccation cracks) to straight felsic strands (Fig. 1). In thin section the channels are composed of plagioclase feldspar (80%), alkali feldspar (10%) and quartz (5%) and form mesoscopically sharp boundaries with the enclosing enclave matrix. Where megacrystic euhedral amphibole is present, mantling

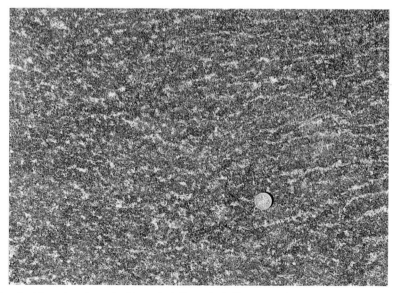

Fig. 1. Microdioritic enclave from the Ross of Mull granite infiltrated during crystallization by melt from the surrounding host granite (also crystallizing). The geometry of the infiltrating melt is largely polygonal but becomes linear towards the right hand side of the plate. Coin is approximately 2 cm.

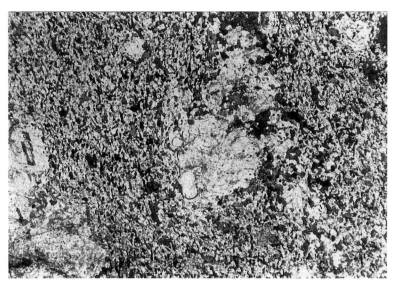

Fig. 2. Photomicrograph of a melt channel, occupied by a plagioclase feldspar megacryst. The enclave matrix shows a magmatic foliation running north–south. Field of view is 4.5 cm.

of the crystals by plagioclase is common. Euhedral titanite crystals (up to 10 mm) are a frequent accessory phase found within the melt channels, where they are often associated with subordinate amounts of secondary calcite. A well-developed foliation, defined by aligned plagioclase laths, is common within the enclave matrix and wraps around the melt channels (Fig. 2).

Enclave pore structure

2D porosity

Porosity and pore structure exert a fundamental control on the movement of fluids through a porous medium (Bear 1972). Porosity (ϕ) is defined as the fraction of the bulk volume of porous sample (U_b) that is occupied by pore (void) space, U_v, so that:

$$\phi = U_v/U_b = (U_b - U_s)/U_b \qquad (1)$$

where U_s is the volume of solids within U_b (Bear 1972). In sedimentary systems, the void space is normally filled by either air, cement or hydrocarbons. In enclaves, and other igneous systems where melt is dispersed within crystalline matrix, porosity is defined by the melt fraction (e.g. Turcotte 1982; McKenzie 1984). Remnant porosity (melt fraction) within the Ross of Mull enclaves is preserved as frozen-in melt channels (Petford *et al.* 1996).

The porosity of individual enclave slices was estimated by scanning each rock slice into a PC as a grey-scale image (Fig. 3). Each captured image was then analysed using standard image analysis software that allowed us to isolate that area of rock occupied by melt channels (shown black in Fig. 3) from the enclave matrix, and thus estimate the porosity for each (2D) serial slice. Although we identified three types of pore space (effective, isolated and dead end), within the enclaves, only the effective porosity (ϕ_e) gives the medium its ability to transport material. By serial sectioning the enclave at 2.5 mm intervals (the minimum spacing achievable using a standard rock saw during this study), it was possible to trace individual melt channels from successive slices through the enclave. It is these channels that contribute to the effective porosity of the enclave.

The obtained variation in porosity with depth as estimated from individual rock slices (see Fig. 3) is shown in Fig. 4. Estimated porosities range from a minimum of 11% to a maximum of 23%. Two regions of high porosity occur approximately half way through the enclave, where melt channels combine to isolate areas of the enclave matrix (e.g. Fig. 3 section 5U). These 'melt envelopes' form an important part of the three-dimensional structure of the porosity network and represent areas where flow rates will be lowest.

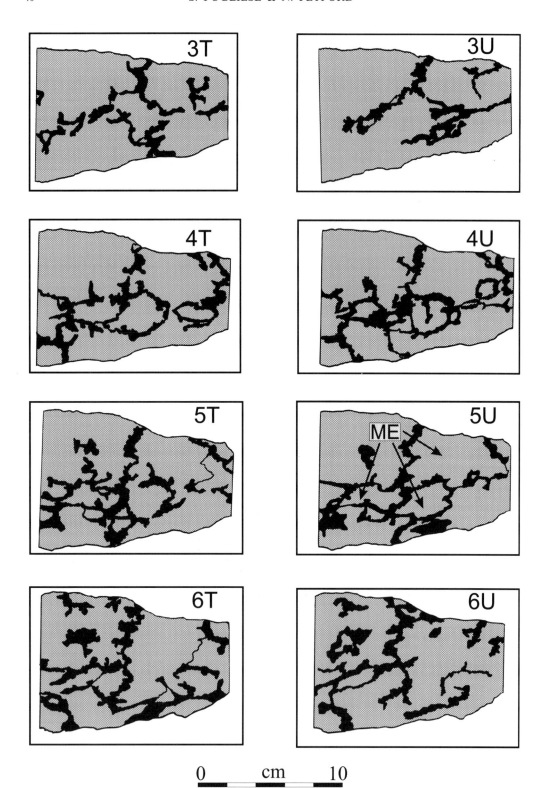

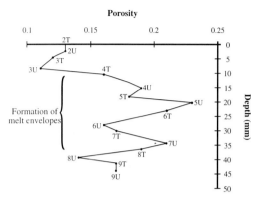

Fig. 4. Variation in remnant magmatic porosity with depth through the microdioritic enclave constructed from serial sections shown in Fig. 3, measured at 2.5 mm increments. Maximum porosity (23%) is found approximately halfway through the enclave where polygons connect to form 'melt envelopes' (e.g. Fig 3, 5U).

Genus and tortuosity

Although melt channels are clearly linked in two dimensions (Fig. 3), a full description of their geometry can only be obtained by reconstructing the melt channels in three dimensions. One way of achieving this is to construct a branch and node chart (Dullien 1992) from individual serial sections. Branch and node charts allow a general qualitative and quantitative evaluation of porous material to be made from an estimate of their connectivity (C). Connectivity is related to the genus (G), a topological parameter that describes the interconnectedness of structures from:

$$G = C = b - n + N \qquad (2)$$

where b is the number of branches, n is the number of nodes and N is the number of separate networks (Dullien 1992). Figure 5 is a branch and node chart made from combining the 2D images shown in Fig. 3. The chart shows the pore network contains 21 branches ($b = 21$), 21 nodes ($n = 21$) and three separate channel networks. It thus follows from equation (2) that $C = G = 3$.

Figure 5 further shows that the melt channels are not simple linear features but instead possess a tortuosity. Tortuosity (Υ) plays an important

role in controlling the efficiency of material transport in porous media by constraining flow rates (Bear 1972) and is defined as:

$$\Upsilon = \frac{L}{L_0} \qquad (3)$$

where L is the average path length of the flow and L_0 is the sample length in the direction of macroscopic flow (Dullien 1992). Using the branch and node chart, it is possible to estimate the tortuosity for each of the melt channels in the enclave. For the main three channel networks within the enclave, Υ lies approximately between 1 and 2 (see Fig. 5).

A further constraint on the degree of material transport through the enclave will be to what extent individual melt channels are continuous throughout the length of the enclave. A number of terminating channels, referred to in Fig. 5 as dead-end pores were observed in the enclave, and may act as a sites for *in-situ* crystal growth (Petford *et al.* 1996). We stress that factors such as tortuosity and dead-end pores can only be revealed by visualizing the geometry of the pore network in 3D.

Permeability

The ability of a liquid or gas to flow through a porous medium is defined as the permeability and is determined by the porosity, pore geometry and physicochemical properties of the fluid (Bear 1972). However, as the permeability in the same porous sample may not be constant, and will change with the physical properties of the permeating fluid and the mechanism of permeation, a parameter termed the specific (intrinsic) permeability (k) is introduced. This parameter is independent of both fluid properties and flow mechanisms and so its value is uniquely determined by the pore structure (e.g. Dullien 1992). In generalized form, the specific permeability is expressed as:

$$k = \frac{a^2 \phi^n}{b} \qquad (4)$$

where a is the grain size (diameter) of the matrix through which flow occurs, ϕ is the porosity of the system and n and b are dimensionless parameters (e.g. McKenzie 1984). For low

Fig. 3. Image-enhanced serial slices through a microdioritic enclave infiltrated by granitic melt. The letters T and U after the numbers refer to the top and underside of the slice respectively. The black areas are frozen-in granitic melt and define the remnant magmatic porosity of the enclave. Where the porosity interconnects, melt channels are formed. Note the crude polygonal geometry of the porosity, and the tendency of larger polygons to isolate areas of the enclave matrix to form 'melt envelopes' (ME, slice 5U).

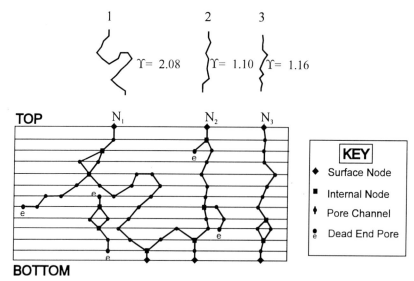

Fig. 5. Branch and node chart representation of the connectivity of the mesoscopic network of interconnecting melt channels shown in Figs 3 and 4. The chart is defined by three networks and has a genus (G) of 3. The generalised tortuosity (Υ) of each network is shown above the chart for comparison. Values range from 1.10 to 2.08.

porosity systems ($\phi \ll 1$), n usually lies between 2 and 3 while b is partly related to the tortuosity of the system.

Porosity estimates from Fig. 4 were used to obtain permeability ranges for the enclave shown graphically in Fig. 3. Ten permeability models taken from the literature based on conduit flow and phenomenological models were applied. Parameters such as porosity and pore throat radius were kept constant ($\phi = 0.17$,

Table 1. *Log values for both specific permeabilities and flow rates for various phenomenolgical and conduit porous media flow models*

Models	Log specific k (m^2)	Darcian flow velocity (m s^{-1})
Conduit flow		
Carman–Kozeny		
Spherical grains	−8.85	4.0×10^{-10}
Circular cylindrical grains	−8.29	1.5×10^{-9}
(flow parallel to cylinder axis)		
Circular cylindrical grains	−8.54	8.5×10^{-10}
(flow perpendicular to cylinder axis)		
Blake–Kozeny	−8.77	5.0×10^{-10}
Capillaries		
Bundles of identical capillaries (1D)	−6.72	5.6×10^{-8}
Pseudo3D ($\Upsilon = 1.10$)	−6.76	5.1×10^{-8}
Pseudo3D ($\Upsilon = 1.16$)	−6.78	4.8×10^{-8}
Pseudo3D ($\Upsilon = 2.08$)	−7.04	2.7×10^{-8}
Phenomenological		
Beds of particles		
Rumpf & Gupte	−9.42	1.1×10^{-10}
Fibrous beds		
Davies	−7.65	6.6×10^{-10}
Chen	−	−

Variables governing permeability were kept constant at ϕ (porosity) = 0.17, r (grain size) = 0.003 m, μ (granitic melt viscosity) = 10^4 Pa s. Υ defines the tortuosity of the channels.

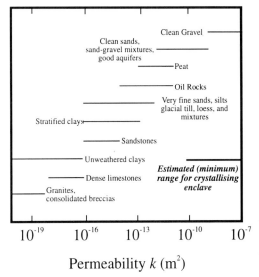

Clean Gravel

Clean sands, sand-gravel mixtures, good aquifers

Peat

Oil Rocks

Very fine sands, silts glacial till, loess, and mixtures

Stratified clays

Sandstones

Unweathered clays

Estimated (minimum) range for crystallising enclave

Dense limestones

Granites, consolidated breccias

10^{-19} 10^{-16} 10^{-13} 10^{-10} 10^{-7}

Permeability k (m^2)

Fig. 6. Estimated range in specific magmatic permeabilities of Ross of Mull enclaves compared with values typical of common geological materials. Magmatic permeabilities are high and are comparable with clean sands and gravel. During active melt infiltration prior to freezing, magmatic permeabilities may have been even higher. Values and model variables are given in Table 1 (after Lerman 1979).

$r = 0.003$ m) in order to compare the variance between each model. The results are shown in Table 1. Permeabilities range from a maximum of 1.9×10^{-7} m^2 (log -6.7 m^2) to a minimum of 3.8×10^{-10} m^2 (log -9.4 m^2). Although a number of different permeability models have been applied, the spread of values is tightly constrained (Fig. 6), with the mean lying at log -8 m^2.

Our estimates should, however, be treated with caution as they represent the final stage in a complex infiltration process governed ultimately by cooling rates within the enclave. For example, crystallization within the enclave will lead to a reduction in the permeability and a subsequent increase in tortuosity with time. It is thus likely that the estimates for the porosity and hence permeability are minimum values that reflect conditions within the enclave at the point of freezing. Actual magmatic values may well have been much greater.

3D pore-structure visualisation

We are now in a position to consider in more detail the three dimensional pore structure of the enclaves. Pore structure visualization was achieved by firstly digitizing each individual slice

as in Fig. 3 and then reducing the image size by 25%. Each new image has a resolution of 75 dots per inch (dpi) and is approximately 300 kilobytes in size. Individual images were then overlaid sequentially and read into 3D Studio, a 3D modelling package running on a Pentium PC, in order to reconstruct and visualize the pore geometry in three dimensions.

Figure 7 shows the resultant enclave pore structure in three dimensions. The enclave matrix is false coloured in light grey with the melt channel network highlighted in dark grey. By joining melt channels in two dimensions between successive slices that (a) directly overlap and (b) lie within a distance equal to or less than the diameters of the melt channels, it was possible to reconstruct the 3D pore geometry of the melt channels in isolation of the enclave matrix (Fig. 8).

Figure 8 has been vertically exaggerated by a factor of two in order to enhance the finer details of the pore structure. Several features identified in two dimensions can be seen in the 3D reconstruction that are not apparent in the simple branch and node chart in Fig. 5. For example, dominant channels can be traced throughout the length of the enclave with smaller subsidiary channels branching off them. Variations in the cross sectional area of channels (serial type pore non-uniformities) occur in both main and secondary channels, while dead-end pores and isolated pockets are clearly visible. Although the resolution is good, the angularity of the melt channels in Fig. 8 is an artefact of the modelling package, and in reality they are much smoother. The importance of this kind of visualization is that it provides a first step towards measuring and quantifying variables such as channel connectivity, effective porosity and tortuosity that control rates of fluid flow and hence advective and diffusive mass transfer processes within the system.

Summary

Simple visualization methods have been used to examine the internal pore structure and petrophysical properties of microgranular porous enclaves from two-dimensional serial sections. Enclave porosity defined as frozen granitic melt within the enclave, has been estimated from serial sections taken through the enclave. Average enclave porosity is 14%, with highest porosities of 23% occurring where individual melt channels coalesce into envelope-like structures. Using estimates of 2D porosity, a number of permeability models for the infiltration of

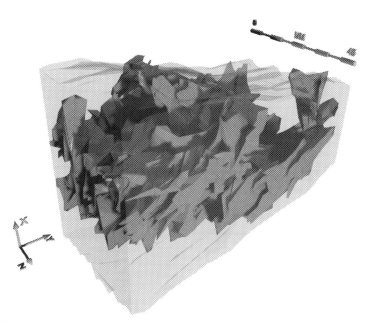

Fig. 7. Grey-scale computer-generated image of a microdioritic enclave reconstructed from 18 digitized serial sections. The light grey areas represent the enclave matrix, while the dark grey areas define the melt channels.

granitic melt into the enclaves have been applied. Values range from $c.\log -9\,\mathrm{m}^2$ to $\log -7\,\mathrm{m}^2$. Reconstructions of the melt channel network using branch and node charts provide a simple way of representing the geometry of the 3D pore structure and estimating the connectivity and tortuosity of the pore network. More sophisticated computer imaging and visualization of the pore network from serial slices reveal the true geometry of the system and suggest that

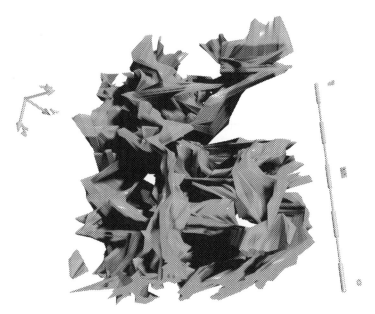

Fig. 8. Computer-generated three-dimensional image of enclave pore structure defined by granitic melt channels (dark grey). The pore-channel network has been isolated from the enclave matrix and exaggerated vertically to highlight the fine detail contained within the pore geometry.

different fluid flow models should be considered when estimating petrophysical properties of porous rocks. Future work will aim at quantifying more fully the petrophysical characteristics of igneous systems and other geological porous media using the methodology and visualization techniques described here.

T. Fletcher is thanked for introducing S.P. to 3D Studio and for help in the construction of Figs 7 & 8. Two anonymous reviewers are also thanked for their constructive comments. This work forms part of an ongoing PhD study by S.P. at Kingston University, who are acknowledged for financial and technical assistance.

References

BAKER, D. R. 1992. Tracer diffusion of network formers and multicomponent diffusion in dacitic and rhyolitic melts. *Geochimica et Cosmochimica Acta*, **56**, 617–631.

BARBARIN, B. & DIDIER, J. 1992. Genesis and evolution of mafic microgranular enclaves through various types of interaction between coexisting felsic and mafic magmas. *Transactions of the Royal Society of Edinburgh: Earth Sciences*, **83**, 145–153.

BEAR, J. 1972. *Dynamics of Fluids in Porous Media*. Elsevier, New York.

BECKINSALE, R. D. & OBRADOVICH, J. D. 1973. Potassium–Argon ages for minerals from the Ross of Mull, Argyllshire, Scotland. *Scottish Journal of Geology*, **9**, 147–156.

BLAKE, H. R., ELWELL, R. W. D., GIBSON, I. L., SKELHORN, R. R. & WALKER, G. P. L. 1965. Some relationships resulting from the intimate association of acid and basic magmas. *Quarterly Journal of the Geological Society of London*, **121**, 31–49.

CASTRO, A., DE LA ROSA, J. D. & STEPHENS, W. E. 1990*a*. Magma mixing in the subvolcanic environment: petrology of the Gerena interaction zone near Seville, Spain. *Contributions to Mineralogy and Petrology*, **205**, 9–26.

——, MORENO-VENTAS, I. & DE LA ROSA, J. D. 1990*b*. Microgranular enclaves as indicators of hybridization processes in granitoid rocks, Hercynian belt, Spain. *Geological Journal*, **25**, 391–404.

CHAPMAN, M. & RHODES, J. M. 1992. Composite layering in the Isle au Haut Igneous Complex, Maine: evidence for periodic invasion of a mafic magma in to an evolving magma reservoir. *Journal of Volcanology and Geothermal Research*, **51**, 41–60.

CUSHMAN, J. H. 1990. *Dynamics of Fluids in Hierarchical Porous Media*. Academic Press, London.

DIDIER, J. & BARBARIN, B. 1991. *Enclaves and Granite Petrology*. Developments in Petrology **13**. Elsevier, Amsterdam.

DULLIEN, F. A. L. 1992. *Porous Media – Fluid Transport and Pore Structure*, 2nd Edition. Academic Press.

FERNANDEZ, A. N. & GASQUET, D. R. 1994. Relative rheological evolution of chemically contrasted coeval magmas: example of the Tichka plutonic complex (Morocco). *Contributions to Mineralogy and Petrology*, **116**, 316–326.

HALLIDAY, A. N., AFTALION, M., VAN BREEMEN, O. & JOCELYN, J. 1979. Petrogenetic significance of Rb–Sr and U–Pb isotopic systems in the 400 Ma old British Isles granitoids and their hosts. *In*: HARRIS, A. L., HOLLAND, C. H. & LEAKE, B. E. (eds) *The Caledonides of the British Isles – Reviewed*. Geological Society, London, Special Publications, **8**, 653–661.

HOLDSWORTH, R. E., HARRIS, A. L. & ROBERTS, A. M. 1987. The stratigraphy, structure and regional significance of the Moine rocks of Mull, Argyllshire, W. Scotland. *Geological Journal*, **22**, 83–107.

HUNTER, R. A. 1987. Textural equilibration in layered igneous rocks. *In*: PARSONS, I. (ed.) *Origins of Igneous Layering*. Kluwer Academic Publishers, Dordrecht, 473–503.

KERR, R. C. & TAIT, S. R. 1986. Crystallisation and compositional convection in a porous medium with application to layered igneous intrusions. *Journal of Geophysical Research*, **91**, 3591–3608.

LERMAN, A. 1979. *Geochemical Processes: Water and Sediment Environments*. Wiley, New York.

LESHER, C. E. 1994. Kinetics of Sr and Nd exchange in silicate liquids: Theory, experiment and application to uphill diffusion, isotopic equilibrium and irreversible mixing of magmas. *Journal of Geophysical Research*, **99**, 9585–9604.

MARSH, B. D. 1981. On the crystallinity, probability of occurrence and rheology of lava and magma. *Contributions to Mineralogy and Petrology*, **78**, 85–98.

MCKENZIE, D. P. 1984. The generation and compaction of partially molten rock. *Journal of Petrology*, **25**, 713–765.

PETFORD, N. 1993. Porous media flow in granitoid magmas: an assessment. *In*: STONE, D. B. & RUNCORN, S. K. (eds) *Flow and Creep in the Solar System: Observations, Modelling and Theory*. Kluwer Academic Publishers, Netherlands, 261–286.

——, PATERSON, B. A., MCCAFFREY, K. J. W. & PUGLIESE, S. 1996. Melt infiltration and advection in microdioritic enclaves. *European Journal of Mineralogy*, **8**, 405–412.

PITCHER, W. S. 1991. Synplutonic dykes and mafic enclaves. *In*: DIDIER, J. & BARBARIN, B. (eds) *Enclaves and Granite Petrology*, Elsevier, Amsterdam, 383–391.

SPARKS, R. S. J. & MARSHALL, L. A. 1986. Thermal and mechanical constraints on mixing between mafic and silicic magmas. *Journal of Volcanology and Geothermal Research*, **29**, 99–124.

TAIT, S. R., HUPPERT, H. E. & SPARKS, R. S. J. 1984. The role of compositional convection in the formation of adcumulus rocks. *Lithos*, **17**, 139–146.

TURCOTTE, D. L. 1982. Magma migration. *Annual Review of Earth and Planetary Science*, **10**, 397–408.

VERNON, R. H. 1984. Microgranitoid enclaves in granites: globules of hybrid magma quenched in a plutonic environment. *Nature*, **309**, 438–439.

WIEBE, R. A. 1974. Coexisting intermediate and basic magmas, Ingonish, Cape Breton Island. *Journal of Geology*, **82**, 74–87.

WIEBE, R. A. 1993. The Pleasant Bay layered gabbro-diorite, Coastal Maine: ponding and crystallisation of basaltic injections into a silicic magma chamber. *Journal of Petrology*, **34**, 461–489.

Pore-size data in petrophysics: a perspective on the measurement of pore geometry

PAUL B. BASAN[1], BEN D. LOWDEN[1], PETER R. WHATTLER[2]
& JOHN J. ATTARD[3]

[1] *Applied Reservoir Technology Ltd, Moat Farm, Church Road, Milden,
Ipswich IP7 7AF, UK*
[2] *Enterprise Oil plc, Grand Buildings, Trafalgar Square, London WC2N 5EJ, UK*
[3] *SINTEF Unimed, N-7034, Trondheim, Norway*

Abstract: Mercury injection capillary pressure (MICP), backscattered electron images
(BSEI) and nuclear magnetic resonance (NMR) provide pore-geometry parameters useful
for understanding variations in rock properties. MICP pore size is an area-equivalent
diameter of the throats connecting the pore system. The MICP distribution contains a point
(the turning point) that reveals where mercury first encounters the permeable network. This
point identifies both critical pore size and connected porosity. BSEI measure pore area and
diameter. Both parameters are unrelated to any specific morphological element. NMR pore
size is a derivative of the pore-surface:volume ratio. Like BSEI, NMR pore size is unrelated
to morphological elements. Like MICP, NMR provides a pore-size distribution that
represents the entire sample volume.

All three techniques provide a data distribution suitable for cross-correlation. Matching
distributions determines the character of the permeable pore system. NMR and MICP
distributions compare through the entire time and pressure spectra, matching best in the
central part of the distributions. The match suggests pore-throats merge with the pore
channels in the permeable part of the system. BSEI distributions match only with the low-
pressure, slow-time parts of the other distributions. However, image porosity has a stronger
correlation with permeability than any of the other parameters, suggesting that the BSEI
distribution is near the turning point.

The need to characterise pore structures, beyond
the simple visual descriptions provided by thin
section and secondary electron microscopy
(SEM), has taken on renewed importance over
the past several years. This requirement evolves
from the need to have quantitative data at all
scales for both reservoir evaluation, in the short
term, and reservoir simulation in the long term.
New tools, like the high-resolution borehole
imaging and the nuclear magnetic resonance
(NMR) logs, have the potential not only to
impact the quality of reservoir simulations, but
also to offer important information about what
regulates reservoir performance. In many
respects, however, technical developments in
detection tools has advanced faster than our
understanding of how to use the data.

Research journals contain a large body of
information about the physical and mathematical characteristics of ideal pore systems (e.g.
Graton & Fraser 1935; Wong *et al.* 1984; Koplik
et al. 1984; Wong 1994). Other investigators
used many of these ideas to develop predictive
models for hydraulic (i.e. permeability) and
electric conductivity (e.g. Archie 1942, 1947;
Swanson 1981; Wong *et al.* 1984; Katz &
Thompson 1986, 1987; Thompson *et al.* 1987;

Herrick 1988; Herrick & Kennedy 1994). Only a
few of these articles, however, use actual rock
data to test the models, and fewer still offer
explanations of the geometric parameters used
in the models.

Most sedimentological investigations of pore
geometry, on the other hand, rely solely on rock
data. These investigations use visualization
techniques that treat pore geometry only in a
casual way. Visualization depends on organizing
petrographic descriptions into category types,
like primary and secondary pores (Schmidt &
McDonald 1983), that detract from the characterisation of pore structure. A few researchers
used petrographic image analysis (PIA) to
establish the relationship between pore geometry
and rock properties. These studies attempted to
overcome the limitations inherent in visualization techniques by classifying pore types based
on measurement parameters (Ehrlich *et al.* 1984,
1991).

Most serious work on determining how pore
geometry affects rock properties, cited above,
uses either mercury injection capillary pressure
(MICP) and/or electrical properties to define the
pore system. More recently, the renewed interest
in nuclear magnetic resonance (NMR) showed

From Lovell, M. A. & Harvey, P. K. (eds), 1997, *Developments in Petrophysics*, Geological Society Special
Publication No. 122, pp. 47–67.

yet another technique for establishing the link between pore size and rock properties (Kenyon *et al.* 1989; Howard and Kenyon 1992).

One characteristic that all these previous studies have in common is that the technical conclusions came from samples having limited sedimentological and petrophysical variation. Consequently, one of the objectives for our article is to show that mercury injection capillary pressure (MICP), image analysis of backscattered electron images (BSEI), and nuclear magnetic resonance (NMR), from a large and varied data base, provide accurate pore-geometry data for characterizing the permeable system in sandstone formations. Our other objectives are to show that: (1) used individually each laboratory measurement has a place in reservoir evaluation because each portrays the pore system in a slightly different way; (2) MICP, BSEI and NMR measurements differ, especially in the way they characterize the volume of the system; and (3) gaining a better understanding of pore geometry is a prerequisite for improving reservoir interpretations, especially where imaging and NMR logs are involved.

Pore-geometry in sandstone formations

Darcy's law and Archie's law at present are the most familiar and widely accepted rules that embody a pore-geometry theory. Darcy suggested that in a simple pore system, like a straight pipe, the length of the pipe and its cross-sectional area control both the rate and volume of flow. Archie (1947) pointed out that porosity (pore volume) controls resistivity. These laws imply, therefore, that cross-sectional area, length, and volume are fundamental parameters for characterizing pore geometry.

Defining geometric parameters in a single pipe, or even a set of capillary tubes, is relatively easy because these systems have fixed dimensions. However defining a three-dimensionally interconnected pore system, controlled by a structure of randomly sized and packed solid particles, is more difficult. The depositional and diagenetic activities governing packing styles, grain size and grain shape in sedimentary rocks are too complex for the application of simple models. Even porosity (pore volume), while easily estimated in a laboratory is not a straightforward variable. At present the distinction between total and effective porosity, and their role in petrophysical interpretation, is still controversial and ill-defined.

If the complications of evaluating microscopic pore structures frequently modified by several stages of diagenesis are not enough, consider that our main interest is the geometry of pore channels, and not strictly the pores or pore types visible on thin section or SEM photomicrographs. Dullien (1979) pointed out that, with few exceptions, rocks do not have pores in the strictest sense. This distinction may at first seem obvious, but understanding pore geometry requires linking what we can see in two-dimensional images with how this system controls hydraulic and electrical properties.

Fluid transport through a sedimentary rock depends on both the three-dimensional continuity of channel pathways and the size of the limiting channel cross section. Cutting a section through an open channel bounded by particles gives the illusion that the system consists of solids and pores. The reality of the situation is that each opening, called porosity elements or porels by Ehrlich *et al.* (1991), represents different parts of the permeable channels. We introduce the term pore-form here because, as discussed later, pore size and porosity are not related. A pore-form is simply the opening coloured blue in a thin section, or fixed more precisely by segmenting the feature using a well-defined scale (e.g. grey scale).

Our objective for introducing a new term is to focus attention on the channel-like structure of sandstone pore systems, and away from the idea of isolated void space. Although isolating pores and assigning an origin to these features furnishes useful petrographic information, this process is inappropriate for extracting petrophysical information. For example, the concept of pore throats and pore bodies partitions the system into elements that, while useful in a practical sense, are artificial in a three-dimensional sense. Similarly, the idea of primary and secondary pores, frequently used by sedimentologists, has no meaning in three-dimensions. We do not propose the abandonment of these terms because they serve a function in the visualization of reservoir properties. However, we do propose that petrophysicists, geologists and engineers embrace an understanding of pore geometry that goes beyond these two-dimensional concepts.

This article focuses primarily on pore length and cross-sectional area. We will illustrate that pore sizes, in a generic sense, obtained directly from BSEI analysis and NMR relaxation time, and indirectly from MICP calculations, characterize the three-dimensional pore structure. We will also show that measuring the area of pore-forms on a backscattered image usually has a stronger correlation to permeability than any other parameter evaluated in this article. If we define effective porosity as the amount that contributes to permeability, then this measure of cross-sectional area fits the description.

Pore area is a well-defined and understood term. The term pore length, on the other hand, is not so well defined. Both pore area and pore length are measures of pore size. Physicists use length in a generic sense to signify a dimension, usually in units of microns. Katz & Thompson (1986, 1987) defined a term l_c (critical length) as the pore size where mercury first encounters the interconnected pore system. Clearly, this length is a diameter (or radius) related to mercury injection pressure. Later, as we discuss the various techniques for measuring pore size, it will become clear that each measures a different pore length, and in a different manner. Consequently, while length is well defined in terms of scale and dimension, in most cases the location of the measurement is poorly defined. The commonality among the different measurements is that all attempt to collect a distribution of pore sizes to characterise the same microscopic stricture.

Acquiring some measure of pore geometry

Traditional petrographic techniques

Petrography furnishes useful mineralogical and categorical information for supporting reservoir characterization. However, these data have limited value in quantitative petrophysical evaluations because, previously, the technology to see pores has outstripped our ability to measure them. We know, for example, that a random section through the resin-filled rock captures a distribution of cross-sections through the three-dimensional pore channels. Petrographic techniques, by necessity, treat pore-forms as discrete features, which ignores the three-dimensional information inherent in the section.

Petrographic point-count is an example of isolating the pore-form from its three-dimensional configuration. A point in random space is a semiquantitative technique that functions as a volume estimator (Gundersen *et al.* 1988). Consequently, thin section analysis should provide an estimate equivalent to core porosity. The fact that point-count data seldom has a significant correlation either to porosity or permeability is a function of both the low resolution of the technique and the non-statistical representativeness of the data. More important, beyond attempting to measure the volume of mineral particles and pore space, obtaining meaningful information on pore size,

or for that matter grain size, is not only tedious but also fraught with problems (e.g. defining the boundaries of pores).

Backscattered electron imaging (BSEI) and image analysis

Using computers to analyse rock images solved many of the statistical and logistical problems associated with petrographic observation. Image processing removes the bias inherent in visual examination by investigating a statistically significant number of pores, without considering either the origin of the feature or its position in three-dimensional space (i.e. the measurement does not take into account that the feature is a primary or secondary pore or a pore-body or pore-throat).

Additionally, image analysis techniques make measurements unobtainable from routine petrography. For example, image analysis software usually defines (segments) pore-forms automatically by using a precise and accurate calibration standard. After segmenting the pore-forms, the software counts the number of pores and measures their perimeter and area (Fig. 1). Simultaneously, image analysis programs estimate pore diameters by measuring vector lengths from different positions on each pore-form. The processing power of computers makes these estimates on thousands of pore-forms in minutes. An acquisition set for a sandstone analysis typically contains information on

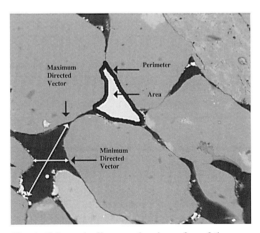

Fig. 1. Schematic diagram showing a few of the different measurements made by BSEI image analysis software, including an approximation of two vector measurements. The processing program makes vector (chord lengths) measurements every 5.625° (64 vectors). The white dot approximates the centroid.

10,000 pore-forms, as compared with a petrographic study typically containing < 500 grain and pore counts.

Analysis of BSE images is the only practical way to obtain a measure of pore area (called image porosity for convenience), a parameter critical to the evaluation of rock conductivity (Eq. 1). An important point is that image porosity seldom matches core porosity (Ehrlich *et al.* 1991; Whattler *et al.* 1995). This situation does not negate the importance of either measurement, but instead largely reflects the way core and image analyses resolve pore space. Core analysis uses gas molecules to estimate volume. BSEI, on the other hand, uses pixels having dimensions of square-microns. The image porosity shown in this article has an imposed lower resolution of $9 \, \mu m^2$ (a pore diameter of approximately $3 \, \mu m$). Another way to view the difference between core and image porosity is to think of the pore area as equivalent to the volume associated with pore space $>3 \, \mu m$. As discussed below, imposing a known limit on the minimum pore size focuses on the part of the system most involved in conducting fluids.

$$\text{Image } \phi = \frac{\text{Segmented pore pixels}}{\text{Total image pixels}}. \quad (1)$$

Only image analysis, among the different techniques for deriving pore size, makes pore-vector measurements. Vectors define more than one pore length, normally the longest, shortest and mean vectors for each pore. Consequently, one of the advantages of BSEI image analysis is that it defines the *length* term more precisely than other pore-size techniques. The BSEI length parameter used to illustrate the data in this article is the shortest, average diameter from the distribution of minimum vectors (Fig. 1). The distribution of vector measurements provides the data to estimate central tendency (the average) in different ways.

Mercury injection capillary pressure (MICP)

Mercury injection explores the distribution of pressure-volume relationships regulating the entry of a non-wetting phase (mercury) into a pore system. Capillary forces inhibit entry of the mercury into the pore volume, and therefore mercury intrusion only progresses after applying an ever increasing pressure to overcome these forces. The accepted view of this procedure is that MICP detects a sequence of pore throats.

Transforming pressure steps to pore size, using the Washburn equation, models the pore system as a series of interconnected, cylindrical tubes (Eq. 2). The MICP pore length therefore is the area equivalent diameter of a circle, and consequently provides no information about either the vector length or the aspect of the pore system.

$$\text{Pore diameter} = \frac{4\gamma c - \text{Cos}\,\theta}{P_c} \quad (2)$$

where: $c =$ Washburn constant; $P_c =$ Capillary pressure; $\gamma =$ surface tension constant at 480 dynes/cm; $\theta =$ contact angle through the wetting fluid ($140°$).

MICP furnishes two estimates of the throat size associated with the pore channels. Traditionally, we use the median point on the curve because the distribution is strongly skewed. The other estimate is the pore-size equivalent to the pressure at the inflexion point on a closure-corrected MICP distribution curve (e.g. Swanson 1981; Katz & Thompson 1986) (Eq. 3, Fig. 2).

$$\frac{S_b}{P_{c(A)}} \quad (3)$$

where: $P_c =$ capillary pressure; $S_b =$ bulk mercury saturation; $A =$ the pressure at the turning point.

We will refer to this reference point as the turning point, for convenience, throughout this article (see Swanson 1981 for the procedure of obtaining the turning point).

The difference between these two estimates is that the median statistic reflects the influence exerted by the tails of the distribution. The threshold size reflects the point at which mercury first joins the connected system, and therefore ignores the throats below the turning point (Swanson 1981; Katz & Thompson 1986).

Recall that BSEI, by comparison, examines pore size without regard to their origin or position within the pore system. These individual measurement strategies point out some important differences in BSEI and MICP data. First, any statistical measure of the MICP distribution reflects the pressure and pore size influenced by the total intrusion distribution, including the smallest, hydraulically ineffective parts. BSEI distributions do not contain the small pores associated with the high pressure part of MICP distribution.

Second, BSEI pore size does not cover the same dynamic range or volume as the MICP experiment. Therefore, deriving an average from these different distributions will obviously yield a different estimate of central tendency.

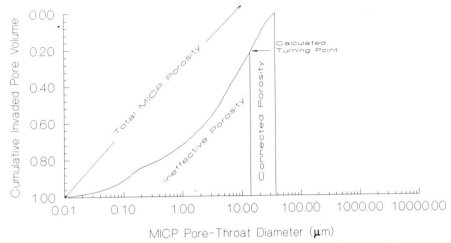

Fig. 2. Diagram of an MICP cumulative distribution showing the distribution of porosity elements contained within the total volume. The turning point, calculated from the MICP data using Equation 3 marks the threshold pore size (see Swanson 1981 for method of obtaining the turning point).

Third, MICP primarily measures the entry-throat diameter, while missing most of the volume behind the throat. BSEI simply measures the sizes of pore-forms within the imposed limits. This characteristic of the BSEI measurements means the size reflects all the parts of the system contained in the image, without regard to some morphological feature. As illustrated later, the only time MICP and BSEI would have a similar average pore size is where channels have nearly equal dimensions at all cross sections, and the cross section is $\geq 9\,\mu m^2$.

The distinction between the pore size estimates, from MICP and BSEI, that characterize the permeable part of the pore system is both important and informative, and therefore requires further explanation. Earlier we discussed the turning point on the MICP distribution. Defining the turning point requires having a distribution with enough dynamic range to incorporate the point. Therefore, finding an average BSEI diameter to represent the entire pore system is complicated by the limited range of the data. Any estimate of central tendency is biased by the absence of the pore sizes $<3\,\mu m$. Later, we will show that the smallest pores in BSEI distributions are close to the turning point value. Nevertheless, calculating an average produces a pore-size parameter that only characterises the geometry of high permeability samples.

Nuclear magnetic resonance (NMR)

Today NMR technology is catching the interest of a range of petroleum scientists. This expansion of interest parallels the arrival of new wireline tools and laboratory instruments that are increasing the industry's awareness of NMR data, and the importance of pore geometry in petrophysics. Numerous articles detail NMR theory. We refer the reader to the publications of Kenyon (1992), Kenyon et al. (1988, 1989), Howard et al. (1990), Howard & Kenyon (1992), and Kleinberg et al. (1993) for this information. Our purpose here is simply to explain why NMR plays an important role in the description of pore geometry.

Brownstein & Tarr (1979) showed that NMR relaxation rate is proportional to the surface area-to-volume ratio of pores in tissue. The early works of Brown & Gamson (1960), Seevers (1966), Timur (1969) and Loren & Robinson (1970) revealed the potential of NMR as both a core analysis and wireline methodology. Simply stated, the NMR signal depends on: (1) the type and quantity of fluid saturating the porous medium (saturation and porosity); (2) the physical and chemical environment with the pore structure (pore size, wettability, and other information).

Relaxation times, symbolized as T_1 for longitudinal relaxation-time and T_2 for transverse relaxation-time, are the most frequently used parameters. T_1 is directly proportional to pore size; T_2 relaxation rate also is proportional to pore size, although the parameter is affected by diffusion (Kenyon 1992). Kleinberg et al. (1993) showed, however, that T_2, acquired at low frequency and short pulse spacing, provides virtually the same pore size information as T_1. Most NMR wireline logging projects collect T_2

52   P. B. BASAN *ET AL.*

data, and therefore our work concentrates mostly on this time constant.

The dependence of relaxation time on pore surface is important for relating NMR parameters to the petrophysical properties of reservoir rocks. Two important points to remember are: (1) the interactions of fluid molecules with the pore surface, and any paramagnetic centres at the pore surface, control relaxation time in most rock-pore systems (Kleinberg *et al.* 1994). (2) the relationship of relaxation time to pore surface area and pore volume suggests that the rate of proton decay is proportional to a pore length (Eq. 4).

$$\frac{1}{\text{Relaxation time}} = \frac{\text{Surface}(l^2)}{\text{Volume}(l^3)} = \frac{1}{\text{Length}} \quad (4)$$

The equations,

$$\text{Length} = T_{1,2} \times \rho_{1,2} \quad (5)$$

$$\rho_{1,2} = \frac{1}{T_{1,2}} \times \frac{S(\text{surface})}{V(\text{volume})} \quad (6)$$

where $\rho_{1,2}$ is the pore-surface strength for transforming relaxation time to pore size. Obviously ρ is an important parameter where the objective is to transform relaxation time into pore size. Obtaining ρ, however, depends on some estimate of the pore surface-to-volume ratio. Determining how to measure the surface-to-volume ratio presents a difficult problem. The S/V term clearly has the dimensions of length (Eq. 4). The key question is which length relates relaxation time to pore size.

At present no method for calculating ρ is universally accepted. The primary problem is each pore-size estimator investigates either a different aspect of the pore system or resolves pore size at a different level. Consequently, each technique produces a non-unique estimate of ρ. Gas-absorption techniques use gas molecules to investigate surface area, and therefore furnish data more sensitive for estimating ρ than either MICP or BSEI. However, Kleinberg *et al.* (1995) found that the surface area obtained from gas absorption produced calculations of ρ an order-of-magnitude greater than NMR diffusion measurements (Fig. 3). Notice that capillary pressure (MICP) produces values for ρ that best fit the pulse-field gradient NMR data. Pore-size determinations from optical thin section images, on the other hand, produce estimates of ρ about an order-of-magnitude less sensitive than the NMR diffusion measurement, and two to three orders-of-magnitude greater than gas absorption (e.g. Kenyon *et al.* 1989).

The pore length investigated by NMR, like MICP, is obscure because the experiment furnishes no intrinsic reference point for

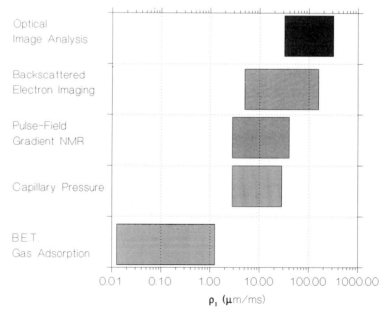

Fig. 3. Schematic diagram showing the relative sensitivity of different pore size (length) measurements on the calculation of the pore-scaling parameter ρ. This diagram was modified from one provided by Robert Kleinberg, Schlumberger-Doll Research. All scales are relative.

making the conversion from time to size. However, NMR relaxation time, like BSEI, occurs without regard to the morphology or origin of the pore channels. The length therefore is a measure of some average channel size as opposed to some specific feature (e.g. pore throat) in the system. Nevertheless, remember that the low dynamic range of BSEI pore diameters affects the calculation of central tendency. Consequently, the calculation of ρ from the median MICP diameter or the threshold diameter will have a greater dynamic range than one calculated from the average BSEI diameter (Fig. 3). An important point is that any meaningful estimate of ρ should come from matching the entire distribution of pore sizes in a sample against the entire NMR relaxation-time distribution. Matching a single point to obtain ρ only has meaning, for our purposes, if it estimates the minimum pore size connecting the permeable system (i.e. a hydraulic ρ).

Evaluating pore geometry from laboratory measurements

The industry has a great deal of reservoir information based on standard porosity and permeability measurements. Consequently, one way of testing new measurement techniques is to determine how they compare with the porosity–permeability model. Reservoir scientists already recognize that pore-throat size relates to permeability. The objective of this section is to show that MICP is not the only pore-size estimator that offers valuable reservoir information. NMR and BSEI provide easily acquired pore-size data that extend our understanding of rock

properties. Additionally, present-day NMR technology is moving pore-size information from the laboratory to the borehole, and opening up opportunities for reservoir evaluation previously unavailable to geologists, engineers and petrophysicists. Using NMR wireline data effectively requires understanding both how to calibrate the wireline signal (time and amplitude) and how to use the data. MICP and BSEI therefore have a role in transforming pore size from laboratory experiment to borehole interpretation.

Porosity and permeability

Both porosity and permeability play an important role in reservoir description because the former describes reservoir storage capability, while the latter describes the ability of the rock formation to transmit fluids. However, cross plots of these reservoir variables show that porosity seldom has a statistical correlation to permeability significant enough to develop predictive models. For example, the data for Fig. 4 is a collection of samples from a wide range of rock types, age groups and diagenetic styles. The simple linear regression for the 830 data points has a correlation coefficient of $r = 0.689$. A subset of the data base, containing 180 samples, has a similar correlation coefficient of $r = 0.682$ (Fig. 5). We will use this subset to illustrate our points because the data base contains complementary pore geometry measurements from MICP, BSEI and NMR.

Analytically, correlations between permeability and porosity relate a variable having dimensions of length-squared (area) to a variable

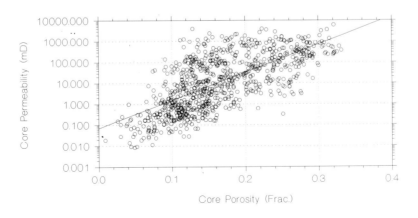

Fig. 4. Correlation of permeability to porosity from the ART North Sea data base. Many of the data points reside in the porosity band between 10% and 20%, spanning nearly six orders of permeability.

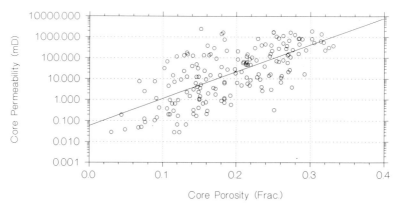

Fig. 5. Correlation of permeability to porosity for a subset of samples from the ART North Sea data base. The subset captures the variability contained in the parent data base. All samples also have NMR, MICP and BSEI measurements. The correlation of porosity and permeability in this subset forms the basis for comparing other correlations shown in this article.

having no dimensions; conceptually, the correlation attempts to use total volume to predict the effective cross-sectional area for conducting fluids. However, total porosity is a static property related only to storage capacity, and not to the dynamic, flowing capacity of the pore system (e.g. Perez-Rosales 1976; Katz & Thompson 1986; Herrick & Kennedy 1994).

Total porosity is largely independent of pore channel proportions, and therefore only correlates well with permeability where the smallest effective part of the pore channel approaches the size of the largest part of the pore channel. In two-dimensional terms, a situation where the pore-throat approaches the size of the pore body. Such conditions suggest that the cross-sectional area of connected channels reflect the total porosity of the pore system (i.e. area-=volume as suggested by DeLesse 1847). This relationship is never perfect in natural systems. Nevertheless, good correlations between core porosity and permeability occur in well-ordered systems at both ends of the spectrum (i.e. both high and low permeability).

Kleinberg *et al.* (1995) pointed out that porosity and permeability correlations work well in predictive models where the system has a fixed multiplicative relationship between the channel proper and the part of the channel that connects the system (i.e. the narrow connectors are in fixed proportion, throughout the system, to the enlarged part of the channel defined by grain packing). Although granular systems are generally well ordered, permeability and porosity correlations often have a spread ranging as much as six orders of magnitude. This spread of

data illustrates that pore systems having similar volumes differ in their ability to transmit fluids (Figs 4 & 5). In other words, while the pore systems in a sequence of reservoir sandstones may fit one or the other conditions discussed above, each unit (facies or flow unit) frequently has a different multiplicative relationships between the largest and smallest part of the pore system. Herrick & Kennedy (1994), working on electrical conductivity, viewed this spread as variations in transport efficiency. They defined a geometrical parameter *E* and explained that it contains the effects of tortuosity, pore-throat size, connectivity, and other characteristics that affect electrical conductance.

Pore size and permeability

Mercury injection capillary pressure pore-throat size

High pressure mercury intrusion (up to 60 000 psi) fills most of the pore volume over a sequence of 117 pressure steps. Transforming the pressure steps to pore size (Eq. 2) produces a distribution that portrays the pore system as a series of connected throats. The link between the pore throat and permeability is legendary in petrophysics, especially the idea that the smallest throat controls permeability. Numerous small entry ports appear at the high pressure part of the MICP distribution, and below the threshold size, or turning point, defined by Swanson (1981) and Katz & Thompson (1986) (Eq. 3). Realistically, therefore, it is not the smallest pore-throat that regulates permeability, but instead some critical size that connects the permeable volume (Kamath 1992).

Examining a cross plot of permeability to median pore-throat size shows the strong linear correlation between these two variables ($r = 0.882$) (Fig. 6). The threshold pore-throat size obtained from equation 2 has a lower correlation coefficient ($r = 0.837$) than the median pore-throat size (Fig. 7). The median diameter estimates the central tendency of the entire distribution including the smallest, albeit hydraulically ineffective pore-throats, whereas the threshold diameter is not an estimate of central tendency. Instead, this parameter is the diameter equivalent to the turning point on the distribution and corresponds to the channel cross-section where mercury enters the pore

system with the greatest efficiency. Therefore, the threshold diameter is independent of high-pressure tails (Fig. 2).

The decrease in the correlation coefficient is surprising because the Swanson parameter, from which the threshold diameter is derived, usually provides an excellent estimation of permeability. One possibility is that the Swanson parameter is not a perfect way of finding the true turning point. Another part of the problem is that the toxicity of mercury usually prevents using the identical sample for both laboratory measure-ments. Most of the samples used for MICP are offcuts from the porosity–permeability plugs. The far outliers in the cross plot therefore probably reflect sample differences.

In spite of the sample differences, both correlations are a significant improvement over

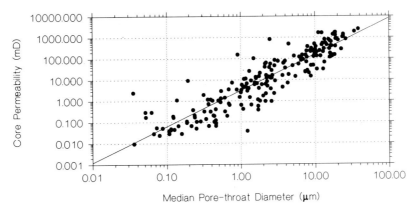

Fig. 6. Correlation of permeability to median mercury pore-throat diameter. The conversion from pressure steps to pore size was derived using equation 1.

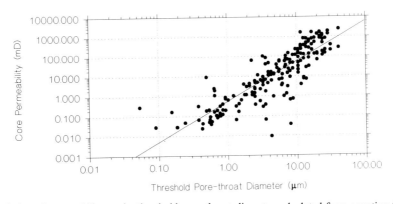

Fig. 7. Correlation of permeability to the threshold pore-throat diameter calculated from equation 2. The threshold pore diameter is independent of high-pressure tails of the distributions, and therefore is not a true statistical measure of central tendency. The diameters represent a single pressure and volume point (the turning point) on the distribution. The outliers in this plot largely reflect the minor differences that occur in the core plugs used for the two experiments.

the porosity–permeability correlation, but still shows a spread of pore sizes relative to permeability. Comparing the porosity to permeability relationship with the pore size to permeability relationships offers some interesting observations (see Figs 5, 6 & 7). Notice that maximum vertical spread of porosity–permeability data, of nearly six-orders of magnitude, occurs in the mid-range porosity values (10–20%). This association shows typical contrast between the total amount of pore space and the ability of the pore system to transmit fluids through the space.

The permeability to pore-throat cross plots have a two to three order-of-magnitude data spread largely between 1 μm and 10 μm (i.e. different pore systems have different efficiencies). This spread diminishes at larger pore-throat sizes, suggesting that the pore-throat size essentially merges with the other parts of channel, and therefore approaches a system having the optimum efficiency.

Backscattered electron image pore size

BSE images record pore size without regard to any specific feature in the system (e.g. pore-throat or pore-body). Remember the differences in BSEI and MICP discussed above, including that BSEI provides two pore-size estimators: area and diameter. Let us ignore the distinction between pore-throat and pore-body, and simply focus on the fact the BSEI technique has a pore-size resolution threshold of 3 μm. The difference in resolution means that the BSEI data does not cover the same dynamic range as MICP data. Figure 8 illustrates that channel diameters in low permeability pore structures overlap only at smallest BSEI sizes and largest MICP sizes. The tight sample has 12% MICP porosity contained in channels having pore-throats generally less 1 μm in diameter (Fig. 8 top diagram). Image porosity is only 1% in this sample, suggesting that only a small proportion of the channels are

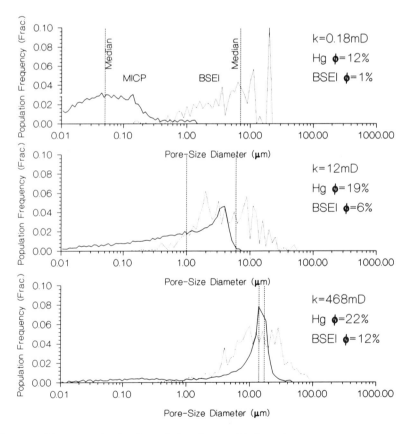

Fig. 8. Frequency histograms showing the variations between MICP and BSEI pore-size estimates from samples having a range of permeability. Notice that the porosity and dynamic range of pore sizes from BSEI is significantly smaller than the MICP distribution, especially at low permeability where the channels have the smallest dimensions. The central tendency of the distributions nearly overlap at high permeability.

>3 μm. The pore-size distributions progressively overlap where samples have higher levels of permeability, and the differences in porosity values become smaller.

An important point illustrated by Fig. 8 is that the central tendency of the distributions become increasingly similar as permeability increases (Fig. 8 bottom diagram). The decrease in the volume of pore-throats <1 μm, and the increase in image porosity, causes the distributions to merge toward high permeability. The upper diagram in Fig. 8 has a median pore-throat diameter of 0.05 μm, whereas the BSEI distribution has an average diameter of 7 μm. The pore system in the high permeability example has a MICP pore-throat diameter of 14 μm, while the BSEI diameter is 17 μm. This sequence of distributions suggests that the contrast between the largest and smallest parts of the pore channels (i.e. the throats and bodies) diminishes as permeability increases. Another way to interpret this overlap is that pore-throat size approaches the general channel size.

The cross plot of the average, minimum pore-size vector clearly shows the BSEI data has less scatter around the regression line than MICP data because the parameter has a narrow range compared to permeability (about two orders of magnitude as compared to three-orders of magnitude for the MICP data) (Fig. 9). The regression has a linear correlation of $r = 0.803$, which expresses both the linearity and the absence of pore sizes below 3 μm (compare Figs. 6, 7 & 9). The ramification of the converging median values, shown in Fig. 8, and the comparison of permeability to minimum pore size shown in Fig. 9, suggests that the BSEI and MICP average diameter start to merge between 10 mD and 100 mD.

An important observation is that the BSEI data is more similar to the MICP threshold diameter than to the median pore-throat diameter because the differences in dynamic range are smaller (see Figs 7 & 9). Picking a threshold pressure/pore size point also eliminates the volume associated with the tails of the MICP distribution (Fig. 2). Consequently, the part of the MICP distribution represented by the volume greater than the turning point is now something other than total porosity (i.e. effective or connected porosity by some definition). MICP porosity usually is equal to core porosity. The connected MICP porosity (or the porosity associated with pore-throats larger than the critical size) now becomes more equivalent to image porosity (Fig. 10). The cross plot of the two porosity variables contains scatter, but the values have a similar dynamic range, and unquestionable linearity ($r = 0.786$). Part of the scatter is attributable to the difficulty of calculating the precise turning point on an MICP distribution and also to the differences in the sample pieces used in the two experiments.

The correlation of MICP connected porosity to permeability has a coefficient of $r = 0.843$ (Fig. 11). This correlation is a significant improvement over the base-line porosity and permeability relationships used here for comparison (Fig. 5). Image porosity for the same set of data points has an even more impressive correlation with permeability ($r = 0.926$) (cf. Figs. 5 11, 12).

The improvement in the correlation to permeability using image porosity has interesting ramifications. The implication is that pore sizes >3 μm characterise the critical pore size in a large number of sandstone formations. In fact, the correlation implies that criteria established

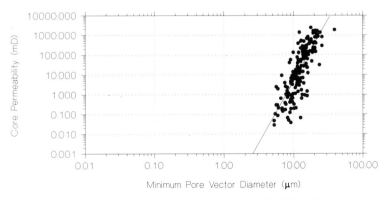

Fig. 9. Correlation of permeability to image pore size from the ART North Sea data base. Image analysis furnishes pore size estimates at different vectors through the pore space. This distribution represents the median of the distribution from the smallest (minimum) directed vector.

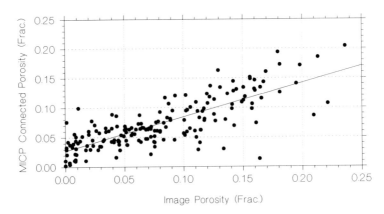

Fig. 10. Correlation of MICP connected porosity to BSEI porosity showing that the data have the same dynamic range. MICP and BSEI measurements are made on different sample pieces, although most come from the same core plug. Sample differences account for most of the scatter.

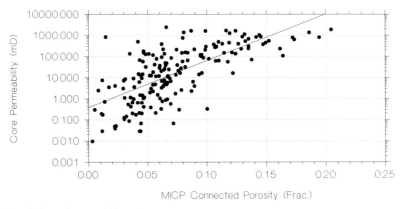

Fig. 11. Correlation of permeability with MICP connected porosity. The correlation is significantly better than the porosity–permeability correlation (cf. Fig. 5), but still contains a great deal of scatter. The amount of the scatter is due, in part, to differences in samples, and in part to difficulties in calculating the precise turning point.

for obtaining image porosity actually matches the concept of a critical pore size better than those for MICP established by Swanson (1981) and Katz & Thompson (1986). In other words, the turning point in many sandstone formations approaches the pressure point associated with a channel area of approximately $9\,\mu m^2$.

Additionally, the strong correlation implies that the turning-point pore size is around $3\,\mu m$ in many sandstones, which illustrates why an average BSEI pore size only correlates well with high permeability values. Image porosity incorporates the area of all pore-forms greater than the imposed resolution threshold. Extracting an average from the distribution produces a value higher than the critical point, especially at low permeability (see Fig. 8).

Nuclear magnetic resonance relaxation time

Many authors recognized that MICP distributions have a character similar to the NMR relaxation-time distribution (e.g. Kenyon *et al.* 1989; Howard *et al.* 1990; Marschall *et al.* 1995). Kleinberg *et al.* (1995) discovered that calculating ρ from MICP data produces a general relationship for a significant number of samples (Eq. 7).

$$P_c T_n = 10\,\mathrm{psi}\,\mathrm{s}^{-1}$$
$$= 6.8 \times 10^4\,\mathrm{Pa}\,\mathrm{s}^{-1} \Rightarrow \rho_n = 5\,\mu\mathrm{m}\,\mathrm{s}^{-1}. \quad (7)$$

He derived the relationship by modifying the Washburn equation so that:

$$P_c = \frac{2\gamma\,\mathrm{Cos}\,\theta}{r} \quad \text{and} \quad \frac{1}{T_n} = \rho_n\frac{S}{V} = \rho_n\frac{2}{r},$$

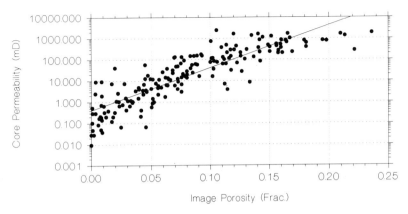

Fig. 12. Correlation of permeability with BSEI porosity. Image porosity, which is actually a pore size (pore area), usually has a consistently stronger correlation with permeability than any other variable. The result implies that the imposed pore size cutoff used for image porosity approaches the turning-point volume.

and therefore

$$\rho_n = \frac{\gamma \cos \theta}{P_c T_n} \tag{8}$$

where: P_c = capillary pressure; r = pore radius; c = surface tension constant at $0.4855\,\mathrm{Pa\,m^{-1}}$; $\theta = 140°$; ρ_n = the pore scaling parameter for either T_1 or T_2; S/V = the pore surface-to-volume ratio; T_n = the relaxation-time parameter T_1 or T_2.

Kleinberg *et al.* (1994) suggested that the critical length, in terms of NMR relaxation rate, is the shortest distance required for the protons to reach the surface. Many individuals working on the relationship between pore size and relaxation time suggest that the recorded length only characterises the pore-body (e.g. Kleinberg *et al.* 1995). The debate on what part of the system NMR measures is beyond the scope of this discussion. However, our opinion is that relaxation time is not specific to any special part of the pore system. We accept that proton relaxation is fast where fluid resides in very small capillaries or associated with clay-bound water. On the other hand, the channel pathways connecting the permeable part of the pore system obviously have throats approaching the size of the openings bounded by framework particles. Relaxation time, therefore, must be a measure of apparent, average pore size. The problem is not that NMR relaxation time fails to record pore-throats, but instead, as in the case in all other measurement techniques, we have no unambiguous way to separate the different elements in a natural, three-dimensional network.

The idea that NMR simply investigates the pore system without bias to any special part of

the channel morphology suggests that relaxation time corresponds better to BSEI pore sizes than to MICP. However, we pointed out that the NMR distribution more closely resembles the MICP distribution because both have similar volumes of investigation. Our work shows that BSEI and NMR data are closely aligned through the upper part of the distribution. BSEI does not record the part of the pore system characterised by fastest relaxation times. Attempting to match the entire NMR distribution with the PIA or BSEI distribution perhaps created the impression that NMR only records pore-bodies. Certainly, using the average pore-size value would produce a ρ biased by the big pores. Although not yet tested in our data base, the evidence shown here strongly indicates that ρ_{image} should approach $\rho_{\text{hydraulic}}$.

NMR relaxation-time distributions account for virtually the same volume of pore space as core analysis and MICP, as shown by the close relationship among the porosity values (Fig. 13). Core and NMR porosity are more strongly correlated than core and MICP porosity ($r^2 = 0.963$ v. $r^2 = 0.893$). The scatter in the diagram shows variations associated mostly with the inability to use the same sample for both MICP and porosity. Nevertheless, similarities in the volume of investigation makes MICP and NMR distributions comparable through the entire range of time and pressure steps.

Marschall *et al.* (1995) and Marschall (1996) showed how to match a point on the MICP distribution with a point on the NMR distribution, and also how to find the turning point on these curves. Matching distributions provides

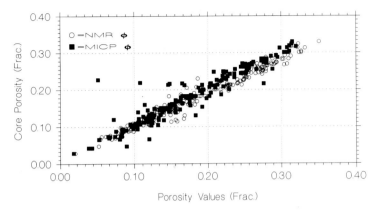

Fig. 13. Correlation of porosity values from core, NMR and MICP analyses. NMR porosity is slightly higher than core porosity. The scatter in the core and MICP porosity is probably due to sample differences.

useful information about the time–pore-size distribution in rock samples. However, using the information requires understanding the acquisition styles because NMR is a static measurement while MICP is a dynamic measurement. NMR records all the channel sizes that relax at a given time simultaneously, and therefore the distribution is ranked (i.e. the distribution contains no dynamic information). MICP, on the other hand, records information as the mercury reaches different parts of the sample. Consequently, while the shape of the distributions look similar, the volume of protons relaxing at a given time do not strictly correspond to the MICP distribution at a given pressure.

Our observations on numerous matches is that the central part of the distributions usually match well, but the quality of match declines at fast times – high pressures and at slow times-low

pressures (Marschall *et al.* 1995, figs 1 & 2). This kind of information suggests the distinction between the widest and narrowest part of the pore channels is less important in the effective part of the pore system, and also cautions trying to extract information from these two distributions without understanding how they differ and what the result means.

However, the fact that MICP and NMR distributions investigate similar volumes offers the opportunity to interrogate the NMR distribution in the same fashion as the MICP distribution. The standard statistical treatment of an NMR distribution is to obtain the geometric mean of the spectrum. The cross plot of permeability to the geometric mean T_2 relaxation-time parameter has linearity, but also a four order-of-magnitude spread in the data ($r = 0.827$) (Fig. 14). Nevertheless, the

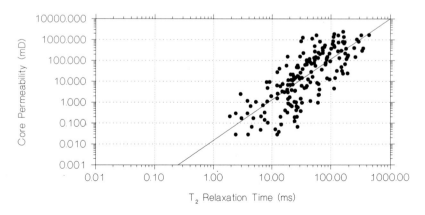

Fig. 14. Correlation of core permeability and geometric mean T_2 relaxation time. Units of time and permeability are different, but the pore-size attributes of relaxation time produces a correlation containing considerable linearity. The vertical spread of data is about four orders of magnitude.

correlation is still a significant improvement on the statistical correlation between our porosity and permeability baseline statistic. Remember that the geometric mean parameter, like the median MICP parameter, contains the influence of the fast (small pore) tails.

MICP and NMR distributions have a similar shape throughout the pressure and time spectra (Fig. 15). However, the similarity in shape does not signify that NMR measures any particular pore morphology. Image porosity, on the other hand, seldom matches MICP or NMR porosity. Nevertheless, where BSEI porosity approaches NMR and MICP porosity, the pore-size distribution not only has a similar shape, but nearly overlies the NMR curve (Fig. 15, top). Here, only a 2% difference separates image porosity from the NMR and mercury porosity.

The low permeability example illustrates the more common situation (Fig. 15, bottom). Usually image porosity is significantly different

from total porosity. Now, the adjustment to accommodate the difference in porosity produces a BSEI distribution that corresponds only to the lower pressures and slower relaxation times. These curves demonstrate the important link between pore size and pore volume. Unrestricted the distributions show all pores sizes represented within the total pore system. Restricting the measurement to a volume of known pore sizes focuses the distribution on the hydraulic part of the system.

The correlation of image porosity to permeability, as compared with the correlation of core porosity to permeability, indicates that the information from BSEI images more closely approximates the permeable volume. This observation also suggests that the porosity-restricted part of the pore-size distribution obtained from image analysis corresponds to the hydraulically effective part of the system. The implication is that the BSEI distribution furnishes a guide for

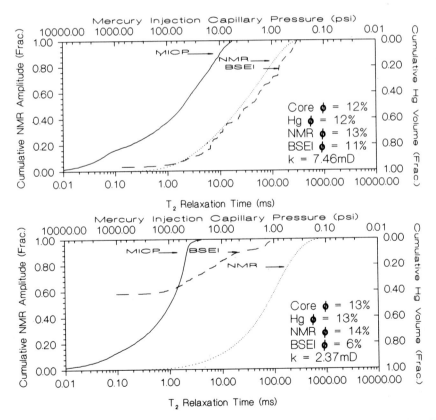

Fig. 15. Cumulative frequency diagrams showing the similarities and difference between unmatched MICP, NMR and BSEI distributions. Top diagram illustrates distributions having similar volumes. The bottom diagram is an example where BSEI porosity is different from the other estimates, and only compares with the lower pressure (MICP) and slower times (NMR). The BSEI distribution approximates the turning point.

finding the part of the NMR distribution that represents the porosity associated with the permeable channels.

Discussion

The geometry of pore structures has always interested petroleum scientists because ultimately this characteristic of rock formations regulates the storage capacity and deliverability of hydrocarbon reservoirs. Until recently, however, only a few laboratory measurements furnished quantitative information about pore geometry. Now, NMR technology not only offers laboratory measurements that characterise pore structure, but also new wireline tools that furnish a continuous record of this information. Consequently, understanding what each measurement records, and how each differs is not only informative, but also practical.

Informative because obtaining data from sources that measure pore geometry in different ways furnishes a method for both visualising and quantifying how changes in pore structure vary from layer-to-layer (Lowden 1996). Visualizing these changes provides models that lead to a better understanding of how to apply pore geometry data. Practical because wireline logs seldom record the full complement of information. A combination of borehole conditions and instrument limitations frequently causes the tool to miss the fastest and slowest times. NMR core analysis, combined with MICP and BSEI data, will help reconstruct the distribution in order to find the key parameters for interpreting the results (e.g. the T_2 cutoff, BVI, FFI).

Worthington (1995) stated that wireline permeability is one of the most sought-after and elusive parameters in reservoir geoscience. He discussed several approaches to permeability prediction most of which attempt to use aspects of pore geometry in some form. However, except for the NMR tool, wireline log response has no recognised link to the parameters that characterise variations in pore geometry. Even the NMR log response does not always have a clear link to pore geometry because the presence of gas and oil change the influence of the pore surface on relaxation rate (Akkurt et al. 1995). Nevertheless, while reservoir fluids may hide the true character of the pore structure, the NMR parameters are still more closely aligned to permeability than bulk properties like shale volume or grain density.

Today, NMR tools are collecting larger amounts of data, and new signal processing techniques are improving the quality of the data.

However, our understanding of what tools tell us about the formation still lags behind. Consequently, the major tasks facing the industry today are to acquire a better understanding of rock properties that create the log response, and to develop data applications that will improve interpretation strategies.

For example, even the simple form of the standard NMR permeability equations usually provide a better estimate of reservoir deliverability than equations based on porosity alone (see Figs 5, 16 & 17). The equations used to derive these results are from Coates et al. (1991a,b) and Kenyon et al. (1988) (Eqs 9, 10).

$$k = \left[\left(\frac{\text{FFI}}{BVI}\right) \times \left(\frac{\phi}{10}\right)^2\right]^2 \qquad (9)$$

where: k = Linear permeability; FFI = the free-fluid index (= bulk-volume movable); BVI = the bulk-volume irreducible; ϕ = NMR porosity in p.u.; 10 is a factor to scale results to units of permeability when ϕ is in p.u.

$$k = T_2^2 \times \phi^4 \qquad (10)$$

where: k = linear permeability; T_2 = the geometric mean of the T_2 distribution; ϕ = NMR porosity, fractional.

Both estimates of permeability provide similar results, and have correlation coefficients with core permeability of $r = 0.888$ and $r = 0.875$ respectively. These correlations are significantly better than the porosity-permeability comparison base line of $r = 0.682$, $r^2 = 0.462$. The primary reason that the NMR permeability equations work well is that the parameters partition the volume into parts that retain fluid and parts that transmit fluid (Fig. 2). This approach is similar to finding the turning point or critical pore size in the system, and therefore firmly based in concepts of pore geometry and its influence on permeability.

Straley et al. (1995, fig. 1), among others, showed that the NMR response is unquestionably proportional to pore size in rock formations. Of course, at present we do not know exactly which pore size corresponds to the NMR response. Consequently, the working assumption is that the NMR distribution contains a turning point that has the same significance as the MICP distribution.

Both standard NMR permeability equations use the term ϕ^4. Kenyon et al. (1988) suggested that the exponent essentially transforms the length expressed by relaxation time to the critical pore-throat size. In essence, their interpretation implies that initial estimates of the

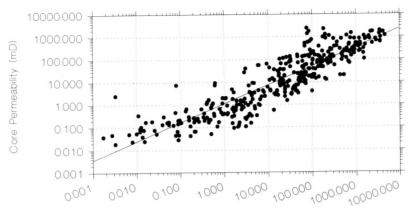

Fig. 16. Cross plot showing the correlation between core permeability and the permeability estimated by the equation $(\text{FFI}/\text{BVI} \times \text{NMR}_\phi^2/10)^2$.

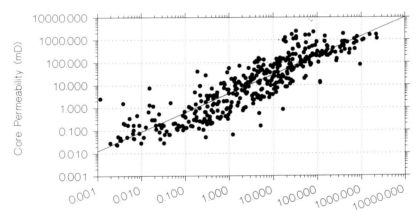

Fig. 17. Cross plot showing the correlation between core permeability and the permeability estimated by the equation $T_2^2 \times \phi^4$.

length term (either T_2 or FFI/BVI) only approximate the critical diameter or turning point. Apparently, the porosity term refines the length term relative to the connected volume.

Interestingly, ϕ^4 works well as a default refinement of the important length term in the equation. Refining this estimate further requires calibrating the distribution against other pore-size distributions. Marschall *et al.* (1995), and others, established that the Swanson parameter (Eq. 2) in many cases provides a way to link the turning point on the MICP distribution with a point on the NMR distribution. As we show in Figs 7 and 11, the Swanson parameter is not always perfectly correlated with permeability. Moreover, MICP is not always perfectly corre-

lated with NMR (Fig. 18). Although the two distributions cover a similar volume and investigate the same pore system, they still measure pore size in different ways. Earlier we mentioned that the central part of the NMR and MICP distributions match in virtually all situations, but that the volume associated with the extreme tails of the two distribution usually have degrees of mismatch (Fig. 18). The fact that MICP is a dynamic distribution while NMR is a static distribution accounts for most of the mismatch. However, the difference in the tails suggests that the turning point on the MICP distribution may not be directly equivalent to the turning point on the NMR distribution. This situation implies that simply deriving a scaling factor by matching

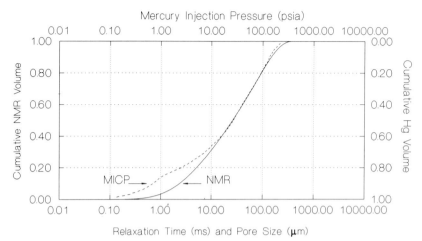

Fig. 18. Automated cross-correlation match between an NMR and MICP distribution. Notice that fast-times and high-pressure tails do not correspond. The distributions fit perfectly in the central part, but the curves again depart at slow times and low pressure.

the distributions is not enough. Lowden (1996) showed a cross-correlation technique that provides both a scaling factor and a correlation coefficient. The correlation coefficient provides an indicator of how well the two distributions match.

We also showed that image porosity has a more substantial correlation to permeability than any of the parameters illustrated in this article. Although we have not fully evaluated the image porosity distribution as a way to determine the turning point, we have determined that this parameter has the same dynamic range as MICP connected porosity (Fig. 11). Image porosity also has a strong correlation to free-fluid index (FFI) used in one of the permeability

equations (see Eq. 9). The correlation coefficient for these two variables is $r = 0.893$; image porosity is within $\pm 3\%$ of FFI obtained from laboratory centrifuge experiments (Fig. 19). This correlation suggests that BSEI, independent of other measures, locates the turning point at a reasonable accuracy, and that the BSEI turning point is essentially equivalent to the volume of fluid that is free within the pore system.

This result has practical applications because the free-fluid volume is often difficult to ascertain on NMR wireline logs where the reservoir contains hydrocarbons. BSEI on side-wall cores, and in theory even ditch cuttings, could provide a first-order estimate of this important parameter.

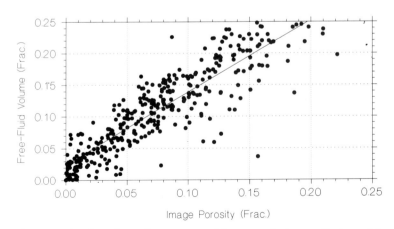

Fig. 19. Cross plot comparing the volume of water removed from samples by centrifuging and image porosity.

Summary and conclusions

Until recently, our understanding pore geometry had not progressed much further from the time Graton & Fraser (1935) first published the idea of a pore system containing duct elements and throat-planes. One purpose of this article was to share our ideas and findings with the industry because modern-day techniques require a better appreciation of pore geometry.

We showed that NMR, MICP and BSEI provide laboratory data for obtaining a visual and quantitative view of pore structure. Each provides data that extends our understanding of pore geometry. Although mercury intrudes the entire volume, as evidenced by the similarities between core and mercury porosity, the measurement focuses on pressure variations created by restrictions in the pore channels. These restrictions, or pore-throats, in the system regulate the way fluids flow through the rock. However, not all of these throats have a strong influence on flow. Instead, MICP volume distributions contain a point (the turning point) that shows where mercury first encounters the connected system (the connected porosity and threshold pore size).

Transforming pressure points to pore size assumes that the system has a tube shape, and therefore the measure is an area-equivalent diameter. An area-equivalent diameter does not take into account the aspect of a pore system composed of irregularly shaped channels. Additionally, because the data reflects pressure steps at restrictions, the measurement excludes much of the information about the average size of the entire pore system.

BSEI measures several different pore sizes two of which are interesting from the stand point of comparison against other measures of pore geometry and permeability. Computer analysis of digital images provides the cross-sectional area and pore-vector diameter for each pore-form captured on the image. Only BSEI provides a measurement oriented by a vector. However, the pore-forms are undifferentiated in terms of the duct elements defined by Graton & Fraser (1935) (i.e. BSEI does not differentiate between pore-throat and pore-body). Consequently, unlike MICP, the distribution of pore sizes obtained from BSEI contains elements representing the entire pore system.

Practical considerations limit the minimum size (area) measured by the image analysis software. The pore resolution of BSEI data used in this article is $9 \, \mu m^2$, or a diameter of $3 \, \mu m$. Restricting the amount of pore space produces a BSEI distribution that is not usually comparable to the entire spread of MICP pressure points or NMR relaxation times. Using the BSEI data as a basis of comparison with MICP and NMR requires volume-normalizing the distributions.

We showed that image porosity has a stronger correlation to permeability than either MICP pore size or connected porosity. Although we have not fully evaluated image porosity in light of this finding, it appears like the smallest pore in the BSEI data approximates the turning point in a large number of cases. Extracting an average pore size (in essence taking the average from the average pore size) produces a parameter that only fits permeability values $>10 \, mD$.

NMR relaxation time distributions record a pore size related to the pore surface:volume ratio. This estimator of pore size investigates all the fluid-filled channels without any special emphasis on the duct elements. Consequently, NMR measures pore geometry in a way not unlike BSEI. However, as evidenced by its similarity to core and MICP porosity, NMR samples a complete spectrum of pore-channel sizes. The shape of the NMR relaxation time distribution usually compares well with the MICP distribution, especially through the central parts. NMR produces a ranked distribution in the sense that all parts of the pore channels relaxing at a given time appear in the same place on the distribution, regardless of their position within the pore system. MICP, on the other hand, produces a dynamic distribution. This difference in distribution style usually accounts for the incomplete fit at fast times and low pressures (i.e. where MICP detects small throats connecting large channels).

The correspondence of these two measurements through the central part of the distribution is fundamental to understanding pore geometry. First, the central part of the distribution is above the turning point. Second, the match implies that the pore-throats merge with the general part of the pore system. One interpretation of this merging is that the size of pore-throats and pore-bodies approach equality in the permeable part of the system. We might also suggest that one way to distinguish the true turning point is to find where the distributions first match.

Listing all the individuals who contributed to this article is impossible because scientists from more than 30 oil companies aided our understanding of pore structure. However, we wish to make special acknowledgements to J. Howard, Phillips Petroleum Company, B. Kleinberg, Schlumberger-Doll Research, T. Pritchard, British Gas Research, and B. Moss, Moss Petrophysics, for evaluating our ideas and offering constructive advice.

R. Pratt, Applied Research Technology Ltd, generated the image analysis data used in this presentation. He also is responsible for writing the software programmes used for integrating mercury injection, image analysis and NMR data. S. Walker, Applied Reservoir Technology Ltd, helped to produce the final manuscript.

Finally, we acknowledge the cooperation of Enterprise Oil plc, British Gas plc, Phillips Petroleum Company, and Conoco Incorporated for releasing the information used here and in previous publications. The support of these companies was critical to the evolution of our thoughts on pore geometry.

References

AKKURT, R., VINEGER, H. J., TUTUNJIAN, P. N. & GUILLORY, A. J. 1995. NMR Logging in Natural Gas Reservoirs. SPLWA 36th Annual Symposium, Paper N.

ARCHIE, G. E. 1942. The electrical resistivity log as An aid in determining some reservoir characteristics. *AIME Petroleum Technology*, 54–62.

——1947. Electrical resistivity: an aid in core analysis interpretation. *American Association of Petroleum Geologists Bulletin*, **31**, 350–366.

BROWN, R. J. S. & GAMSON, B. W. 1960. Nuclear magnetism logging. *Journal of Petroleum Technology*, **219**, 199–201.

BROWNSTEIN, K. R. & TARR, C. E. 1979. Importance of classical diffusion in NMR studies of water in biological cells. *Physics Reviews A*, **19**, 2446–2453.

COATES, G. R., MILLER, M., GILLEN, M. & HENDERSON, G. 1991a. The MRIL in Conoco 33–1, an investigation of a new magnetic resonance imaging log. SPLWA 31st Annual Symposium.

——1991b. The magnetic resonance imaging log characterized by comparison with petrophysical properties and laboratory core data. *66th Annual Technology Conference and Exhibition, Dallas*, SPE Paper 22723, 627–635.

DELESSE, M. A. 1847. Procede mechanique pour determiner la composition des roches. *Comptes Rendus de l'Academie des Sciences. Paris*, **25**, 544.

DULLIEN, F. A. L. 1979. *Porous Media-Fluid Transport and Pore Structure*. Academic Press, New York.

EHRLICH, R., CRABTREE, S. J., HORKOSWITZ, K. O. & HORKOSWITZ, J. P. 1991. Petrography and reservoir physics 1: objective classification of reservoir porosity. *American Association of Petroleum Geologists Bulletin*, **75**, 1547–1562.

——1984. Petrographic image analysis, 1. analysis of reservoir pore complexes. *Journal of Sedimentary Petrology*, **54**, 1365–1378.

GRATON, L. C. & FRASER, H. J. 1935. Systematic packing of spheres, with particular relation to porosity and permeability. *Journal of Geology*, **43**, 785–909.

GUNDERSEN, H. J. G., BENDTSEN, T. F., *ET AL.* 1988. Some new simple and efficient stereological methods and their use in pathological research and diagnosis. *Acta Pathologica, Microbiologica et Immunologica Scandinavica*, **96**, 379–394.

HERRICK, D. C. 1988. Conductivity models, pore geometry and conduction mechanisms. SPWLA, 29th Annual Logging Symposium, Paper D.

—— & KENNEDY, W. D. 1994. Electrical efficiency: a pore geometric model for the electrical properties of rocks. SPLWA 34th Annual Symposium, Paper HH.

HOWARD, J. J. & KENYON, W. E. 1992. Determination of pore size distribution in sedimentary rocks by proton nuclear magnetic resonance. *Marine and Petroleum Geology*, **9**, 139–145.

——, —— & STRALEY, C. 1990. Proton magnetic resonance and pore size variations in reservoir sandstones. *65th Annual Technology Conference and Exhibition*, New Orleans, SPE Paper 20600, 733–741.

KAMATH, J. 1992. Evaluation of accuracy of estimating air permeability from mercury-injection data. *Society of Petroleum Engineers Formation Evaluation*, **7**, 304–310.

KATZ, A. J. & THOMPSON, A. H. 1986. Quantitative prediction of permeability in porous rock. *American Physical Society, Physical Reviews B*, **34**, 8179–8181.

—— & —— 1987. Prediction of rock electrical conductivity from mercury injection measurements. *Journal of Geophysical Research*, **92**, 599–607.

KENYON, W. E. 1992. Nuclear magnetic resonance as a petrophysical measurement. *Nuclear Geophysics*, **6**, 153–171

——, DAY, P. I., STRALEY, C. & WILLEMSEN, J. F. 1988. A three part study of NMR longitudinal relaxation properties of water-saturated sandstones. *Society Petroleum Engineers Formation Evaluation*, **3**, 622–636.

——1989. Pore-size distribution and NMR in microporous Cherty Sandstones. *SPWLA, 32nd Annual Logging Symposium* 1–24.

KLEINBERG, R. L., FAROOQUI, S. A. & HORSFIELD, M. A. 1993. T_2/T_2 ratio and frequency dependence of NMR relaxation in porous sedimentary rocks. *Journal of Colloid and Interface Science*, **158**, 195–198.

——1995 (in press). Utility of NMR T_2 distributions, connection with capillary pressure, clay effects, and determination of surface relaxivity parameter ρ_2. *Journal of Magnetic Resonance*, in press.

——1994. Mechanism of NMR relaxation of fluids in rock. *Journal of Magnetic Resonance, Series A*, **108**, 206–214.

KOPLIK, J., LIN, C. & VERMETTE, M. 1984. Conductivity and permeability from microgeometry. *Journal of Applied Physics*, **56**, 3127–3131.

LOREN, J. D. & ROBINSON, J. D. 1970. Relations between pore size, fluid, and matrix properties, and NML measurements. *Society of Petroleum Engineers Journal*, **10**, 268–278.

LOWDEN, B. D. 1996. NMR facies analysis. *In: Improving NMR Log Interpretations Using Core Data: Workshop*. SLWA 37th Annual Symposium, Paper 13.

MARSCHALL, J. S. 1996. NMR analog of the Swanson parameter (T_2Sb) provides reliable permeability estimator. *SCANews*, **10**, 3–4.

MARSCHALL, J. S., GARDNER, J. S., MARDON, D. & COATES, G. R. 1995. Method for correlating NMR relaxometry and mercury injection data. *Proceedings of the 1995 International Symposium of the Society of Core Analysts*, Paper SCA-9511.

PEREZ-ROSALES, C. 1982. On the relationship between formation resistivity factor and porosity. *Society of Petroleum Engineers Journal*, **22**, 531–536.

SCHMIDT, V. & McDONALD, D. A. 1983. *Secondary Reservoir Porosity in the Course of Sandstone Diagenesis*. AAPG Continuing Education Course Note Series, **12**.

SEEVERS, D. O. 1966. A nuclear magnetic method for determining the permeability of sandstones. SPWLA, 7th Annual Logging Symposium.

STRALEY, C., MORRISS, C. E., KENYON, W. E. & HOWARD, J. J. 1995. NMR in partially saturated rocks insights on free fluid index and comparison with borehole logs. *The Log Analyst*, **36**, 40–56.

SWANSON, B. F. 1981. A simple correlation between permeability and mercury capillary pressure. *Journal of Petroleum Technology*, **33**, 2498–2504.

THOMPSON, A. H., KATZ, A. J. & RASCHKE, R. A. 1987. Estimation of absolute permeability from capillary pressure measurements. *62nd Annual Technical Conference and Exhibition, Dallas, TX*, SPE Paper 16794, 475–481.

TIMUR, A. 1969. Pulsed nuclear magnetic resonance studies of porosity, movable fluid, and permeability. *Journal at Petroleum Technology*, **21**, 775–786.

WHATTLER, P. R., BASAN, P. B. & MOSS, B. P. 1995. Pore geometry and rock properties. *Proceedings of the 1995 International Symposium at Society Core Analysts*, Paper SCA-9521.

WONG, P.-Z. 1994. Flow in porous media: permeability and displacement patterns. *MRS Bulletin*, 32–38.

WONG, PO-ZEN, KOPLIK, J. & TOMANIC, J. P. 1984. *Conductivity and permeability of rocks*. American Physical Society, Review B, **30**, 6606–6614.

WORTHINGTON, P. F. 1995. Estimation of intergranular permeability from well logs. *Dialog*, **3**, 7–10.

Acoustic wave propagation and permeability in sandstones with systems of aligned cracks

M. S. KING, A. SHAKEEL & N. A. CHAUDHRY

*Department of Earth Resources Engineering, Royal School of Mines,
Imperial College, London SW7 2BP*

Abstract: A polyaxial (true triaxial) stress-loading system has been developed for testing cubic specimens of rock, with the capability of varying each of the principal stresses independently. Facilities are incorporated for measuring all nine components of P and polarized S-wave velocities and attenuation in the principal directions, and for fluid permeability in the 1-direction. Tests have been performed on specimens of five sandstones, in which systems of aligned cracks have been introduced by increasing σ_1 and σ_2 together to near failure, while maintaining σ_3 constant at some low level. Results indicate that there are excellent correlations between permeability in the plane of the aligned cracks as they are closed under hydrostatic stress and both velocities and attenuation of (1) P and S waves propagating in a direction normal to the plane of the cracks, and (2) S waves propagating in the plane of the cracks and polarized normal to the plane.

Discontinuities in the rock mass are one of the most prominent features of the Earth's upper crust, ranging in size from microcracks in igneous rocks to large-scale joints, fractures and faults in sedimentary rocks. At shallow depths they are often found to be aligned by tectonic stresses in a direction normal to the minimum principal field stress (Nur & Simmons 1969; Engelder 1982; Babuska & Pros 1984). Those cracks, joints and faults in the top 10–20 km of the crust are hypothesized to be liquid-filled, by either meteoric water or hydrocarbons (Crampin & Atkinson 1985).

The presence of interconnected discontinuities, their orientation and the fluids they contain significantly affect the elastic and transport properties of the rocks containing them (Walsh 1965*a, b, c*, 1981; King 1966; Gibson & Toksöz 1990). Where the interconnected discontinuities are randomly oriented and the state of stress is hydrostatic, the rock will generally exhibit isotropy in its elastic and transport properties (King 1970; Gangi 1978). An otherwise isotropic rock mass containing a system of interconnected, aligned cracks will, however, behave in an anisotropic manner in these properties (Lo *et al.* 1986).

If the principal stresses are altered on an isotropic rock containing randomly oriented cracks and initially subjected to a hydrostatic state of stress, the crack distribution no longer remains randomly oriented (Nur & Simmons 1969; Nur 1971). Those cracks with their normals in directions close to the new major principal stress will tend to be closed more than those with normals in directions close to the new

minor principal stress (Walsh 1981; Holt & Fjaer 1987; Zamora & Poirier 1990). The elastic and transport properties of the rock will then become anisotropic, with the degree of anisotropy depending on the magnitudes of the new principal stresses and on the shapes and interconnectivity of the original cracks.

Theoretical studies can provide a basic understanding of how the presence of cracks affects the physical properties of porous rocks. The establishment of theoretical relationships between discontinuity parameters and the elastic and transport properties of porous rocks makes it possible to map the pore and crack structure using geophysical data, with important implications for oil and gas reservoir characterization. The employment of such studies to charactize discontinuities in rocks has only comparatively recently received attention from geophysicists (Toksöz *et al.* 1976; Douma 1988; Schoenberg & Douma 1988).

Few experimental (rather than theoretical) studies have been reported, however, of the influence of sets of randomly oriented or aligned cracks and fractures on the elastic and transport properties of rocks. Sayers *et al.* (1990) describe a polyaxial-stress loading system for testing $50 \times 50 \times 50$ mm cubic rock specimens to a maximum compressive stress of 120 MPa in each of the three principal directions. Results and analyses of ultrasonic velocity tests demonstrating the effects of changes in state of stress leading to the anisotropic velocity behaviour of a dry sandstone are reported here and in a subsequent paper (Sayers & van Munster 1991). Wu *et al.* (1991) discuss similar behaviour for

From Lovell, M. A. & Harvey, P. K. (eds), 1997, *Developments in Petrophysics*, Geological Society Special Publication No. 122, pp. 69–85.

another dry sandstone and conclude that further research is required to characterize fully stress-induced anisotropy in porous rocks containing microcracks.

Xu & King (1992) report the results of conventional triaxial tests on cylindrical specimens of slate, in which cracks in the cleavage plane parallel to the specimen axis were introduced by loading the specimen axially in a servo-controlled compression testing machine while maintaining a small confining stress. The degree of anisotropy indicated by ultrasonic velocities in and fluid permeability of the slate both was shown to increase dramatically as orientated microcracks were introduced.

Smart (1995) describes a polyaxial stress-loading system based on a conventional Hoek triaxial test cell with cylindrical rock specimens. The intermediate and minor principal stresses in the radial direction are applied by a number of flexible tubes located between the cell wall and rock specimen jacket. These tubes are each maintained at different pressures chosen to provide a differential radial confining stress around the specimen. Although possessing several advantages over those using cubic rock specimens, this system is handicapped by being limited to a maximum stress difference of some 15 MPa between the intermediate and minor principal stresses.

Experimental test system

A polyaxial (true triaxial) stress loading system, based in part on that described by Sayers *et al.* (1990), has been developed for testing $51 \times 51 \times 51$ mm cubic rock specimens. Each of the three principal stresses may be varied independently in the range 0–115 MPa in the horizontal principal directions and to over 600 MPa in the vertical major principal direction. Facilities for maintaining the pore pressure in the range 0–5 MPa are also incorporated.

The system, described in a preliminary technical note by King *et al.* (1995) and in detail by Chaudhry (1995), consists of a loading frame in the form of an aluminium alloy ring within which two pairs of hydraulic rams (matched in

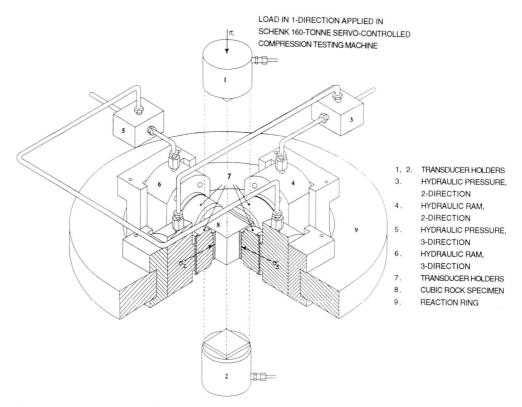

1, 2.	TRANSDUCER HOLDERS
3.	HYDRAULIC PRESSURE, 2-DIRECTION
4.	HYDRAULIC RAM, 2-DIRECTION
5.	HYDRAULIC PRESSURE, 3-DIRECTION
6.	HYDRAULIC RAM, 3-DIRECTION
7.	TRANSDUCER HOLDERS
8.	CUBIC ROCK SPECIMEN
9.	REACTION RING

LOAD IN 1-DIRECTION APPLIED IN SCHENK 160-TONNE SERVO-CONTROLLED COMPRESSION TESTING MACHINE

Fig. 1. Isometric sketch of loading system.

load-actuating pressure characteristics) and ultrasonic transducer holders are mounted to provide orthogonal stresses on the cubic rock specimen in the horizontal plane. The horizontal principal stresses may be servo-controlled using facilities associated with a Schenk compression testing machine. The loading frame is shown diagrammatically in Fig. 1.

The vertical major principal stress is provided by ultrasonic transducer holders mounted in a

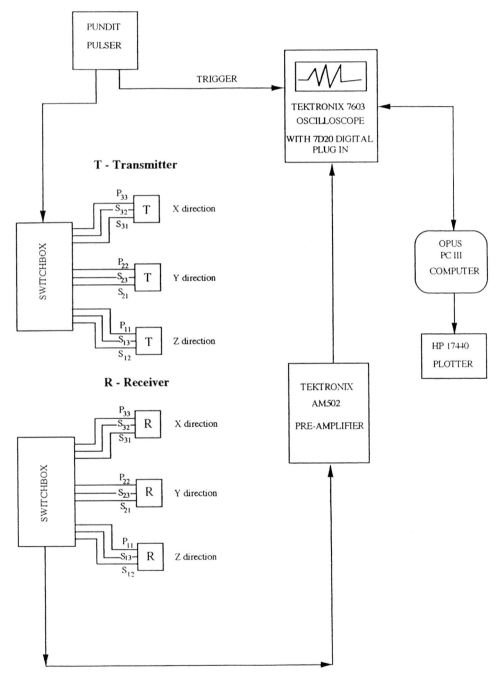

Fig. 2. Ultrasonic velocity and attenuation measurements system.

Schenk 160-tonne closed-loop servo-controlled compression testing machine. Stress is transmitted to each of the six faces of the cubic rock specimen through 5 mm thick faceplates matching approximately the elastic properties of the rocks being tested. Magnesium has been found a suitable material for use with porous sandstones. Deformation of the rock specimen is measured by pairs of LVDTs mounted in each of the three principal directions.

Each of the three pairs of transducer holders contains stacks of PZT piezo-electric transducers capable of producing or detecting pulses of compressional (P) or either of two shear (S) waves polarized at right angles propagating in one of the principal stress directions. The transducer holders are similar to those described by Tao & King (1990), with band widths in the range 450–800 kHz for P-wave and 350–750 kHz for S-wave pulses. A block diagram of the ultrasonic measurement system is shown in Fig. 2.

The system permits the measurement of the deformation and the elastic wave velocities and attenuation in each of the three principal directions as the rock specimen is subjected to a polyaxial state of stress. Employing the time-of-flight technique (King 1983) with digitized elapsed time data and correcting for specimen deformation, nine components of velocity are calculated: three compressional (V_{P11}, V_{P22} and V_{P33}) and six shear (V_{S12}, V_{S13}, V_{S21}, V_{S23}, V_{S31} and V_{S32}). The convention used here for velocities is that the first numeral subscript refers to the direction of propagation and the second to the direction of particle motion as indicated in Fig. 3. Both P- and S-wave velocities may be measured with an accuracy of $\pm 1\%$ and a precision of $\pm 0.5\%$. The redundancy in S-wave velocity measurements (theoretically $V_{Sij} = V_{Sji}$) provides the opportunity to confirm that the state of stress within the rock specimen is indeed homogeneous.

Attenuation measurements are made using the spectral ratios technique with a $51 \times 51 \times 5$ mm aluminium cube as a standard, in a manner similar to that described by Tao et al. (1995) including corrections for the effects of diffraction. Measurements of fluid permeability and, under certain circumstances, those of complex electrical conductivity are possible in the major principal stress direction.

Experimental results and discussion

Tests have been performed on dry specimens of several sandstones in which systems of aligned

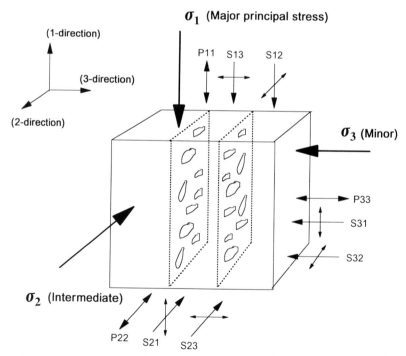

Fig. 3. Direction of propagation and polarization of the nine components of velocity with respect to the principal stress axes, and planes of aligned cracks produced.

cracks have been introduced by increasing the major (σ_1) and intermediate (σ_2) stresses in unison to near failure of the specimen, while maintaining the minor principal (σ_3) stress constant at some prescribed low level. The nine components of velocity and gas permeability in the 1-direction are measured during three separate stress cycles. The first involves measurements on the fresh, uncracked rock specimen during the application of an increasing

hydrostatic state of stress. The second cycle involves measurements while a system of aligned cracks is formed in the rock specimen (cracking cycle). The third cycle involves measurements during the application of a further increasing hydrostatic state of stress to close the cracks formed (crack-closing cycle).

Discussed here as being characteristic of the tests performed on several sandstones in this research programme will be those on Penrith

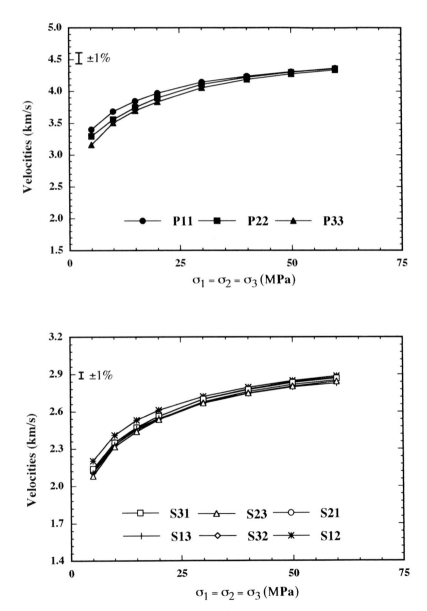

Fig. 4. P- and S-wave velocities as a function of initial hydrostatic stress cycle.

sandstone. This is a fine-to-medium-grained sandstone of lower Permian age having a low clay content (3%), an effective porosity of *c.* 13% and a permeability of *c.* 150 mD. For the first test cycle, Fig. 4 shows the three P- and six S-wave velocities for Penrith sandstone plotted as a function of hydrostatic stress ($\sigma_1 = \sigma_2 = \sigma_3$) to 60 MPa. It will be observed that the sandstone exhibits behaviour that is close to being isotropic, with both sets of P- and S-wave velocities increasing in magnitude with increasing stress and lying within the ±1% error bar, except at the lowest stress levels.

For the second (cracking) cycle, Fig. 5 shows the three P- and six S-wave velocities plotted as a function of major (σ_1) and intermediate (σ_2) principal stresses, with $\sigma_1 = \sigma_2$ up to a stress level of 100 MPa and σ_1 increasing with

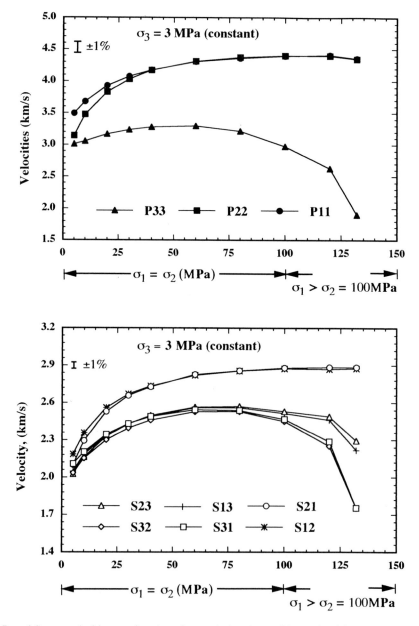

Fig. 5. P- and S-wave velocities as a function of stress during the cracking cycle with $\sigma_3 = 3$ MPa.

$\sigma_2 = 100$ MPa (equipment limitation) for σ_1 levels above 100 MPa. The minor principal stress (σ_3) was maintained throughout the test at $\sigma_3 = 3$ MPa. It is concluded from Fig. 5, with $V_{P11} \sim V_{P22}$ and $V_{S12} \sim V_{S21}$ all increasing monotonically, that the majority of the cracks formed are aligned in the 12-plane, perpendicular to the 3-axis. The S-wave velocities plotted in Fig. 5 indicate that the magnesium loading plates match the sandstone well in elastic properties up to stresses of $\sigma_1 = \sigma_2 = 100$ Mpa.

As σ_1 is increased above 100 MPa during the cracking cycle, it is observed that the magnitudes of $V_{S13} \sim V_{S23}$ are higher than and diverge from those of $V_{S32} \sim V_{S32}$. The reason for this behaviour lies probably in the inhomogeneous nature of the state of stress in the rock specimen at these high σ_1 and σ_2 stress levels. The propagation paths for V_{S13} and V_{S23} lie in the 12-plane at the centre of the specimen, where confinement caused by the loading platens leading to the inhibition of crack formation is

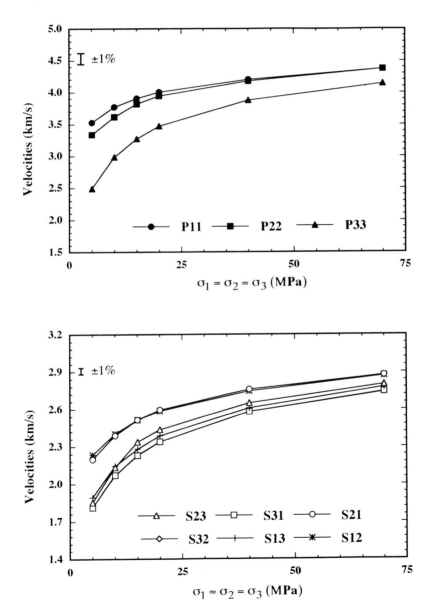

Fig. 6. P- and S-wave velocities as a function of stress during the crack-closing cycle.

a maximum. The propagation paths for V_{S31} and V_{S32}, on the other hand, are in the 3-direction and must pass through all the aligned cracks. This behaviour suggests that the aligned crack density towards the extremities of the specimen in the 3-direction is higher than in the centre for values of σ_1 greater than 100 MPa.

Figure 6 shows the three P- and six S-wave velocities plotted as a function of hydrostatic stress during the crack closing cycle. It is clear from this figure that the state of stress throughout the rock specimen is close to being homogeneous for the range of stresses shown (5–60 MPa). As the stress is increased, both sets of P- and S-wave velocities appear to be approaching asymptotic values that are only slightly lower in magnitude than those shown in Fig. 4 for the preliminary uncracked cycle.

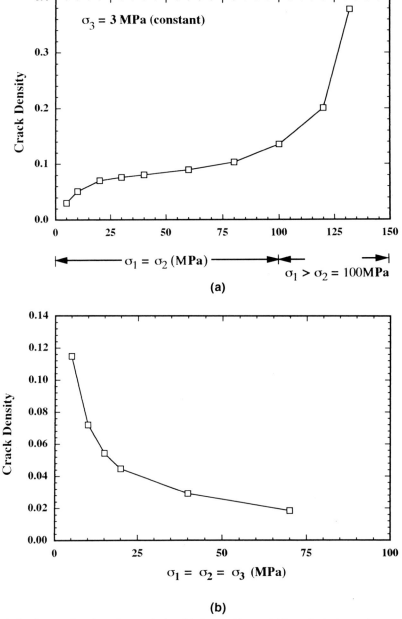

(b)

Fig. 7. Crack density as a function of stress during (**a**) the cracking and (**b**) crack-closing cycles.

Upon removal from the loading frame after completion of the tests, the specimens all showed signs of through-going fractures aligned close to normal to the 3-direction.

The nine components of velocity determined as a function of stress in the second (cracking) cycle reported above have been used to calculate the crack density for the aligned cracks introduced. The procedure, employing Nishizawa's (1982) theory first to model the velocity data, is described in detail by Chaudhry (1995). Chaudhry found excellent fits (within ±1% at all stress levels) in comparing the nine theoretically modelled and the laboratory measured velocities during both the cracking and crack closing cycles. The calculated crack density as a function of stress level during each of these cycles is shown in Fig. 7a for the cracking cycle and Fig. 7b for the crack closing cycle, with an average crack aspect ratio of 0.015.

Figure 8 shows the gas permeability in the 1-direction measured during the cracking cycle as a function of stress level, with σ_3 kept

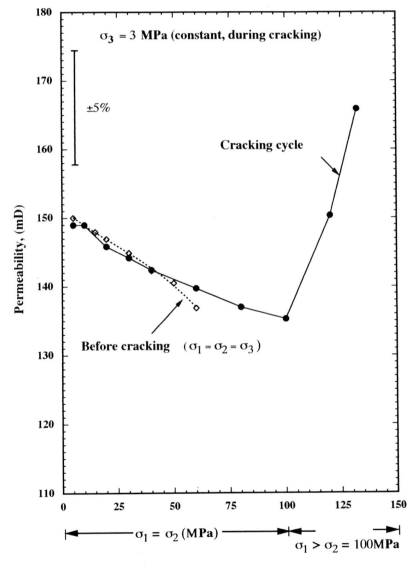

Fig. 8. Permeability in the 1-direction as a function of stress during the initial hydrostatic stress cycle and during the cracking cycle with $\sigma_3 = 3$ MPa.

constant at 3 MPa. Also shown on the plot is the permeability before cracking as a function of hydrostatic stress. It will be observed that the permeabilities before and during cracking are close in magnitude for hydrostatic stresses to 50 MPa before cracking and to $\sigma_1 = \sigma_2 = 50$ MPa during cracking. At higher stress levels during cracking there is first a tendency for the rate of decrease in permeability to slow down and then, at $\sigma_1 = \sigma_2 = 100$ MPa, for the permeability to increase sharply as the stress level is further increased. This sudden increase in permeability at $\sigma_1 = \sigma_2 = 100$ MPa probably represents the point at which aligned cracks

tend to coalesce to form continuous channels through the specimen in the 12-plane.

The gas permeability in the 1-direction as the system of aligned cracks is closed during the third cycle is shown in Fig. 9 as a function of hydrostatic stress to 70 MPa. The permeability at the higher stresses is some 15% lower than the permeability measured at the same stress level on the uncracked specimen. This indicates the possibility that some plugging of original flow channels in the uncracked rock has occurred during the cracking cycle.

Attenuation in the form of $1000/Q$, where Q is the quality factor calculated by the spectral

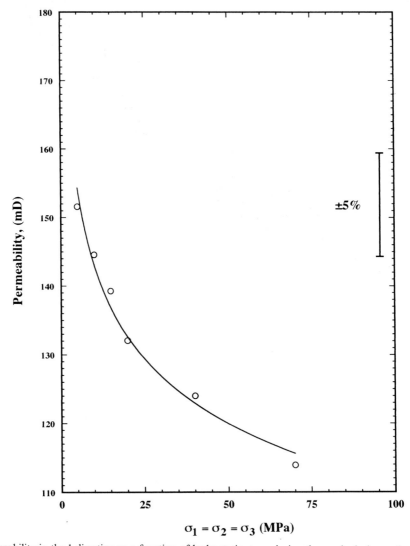

Fig. 9. Permeability in the 1-direction as a function of hydrostatic stress during the crack-closing cycle.

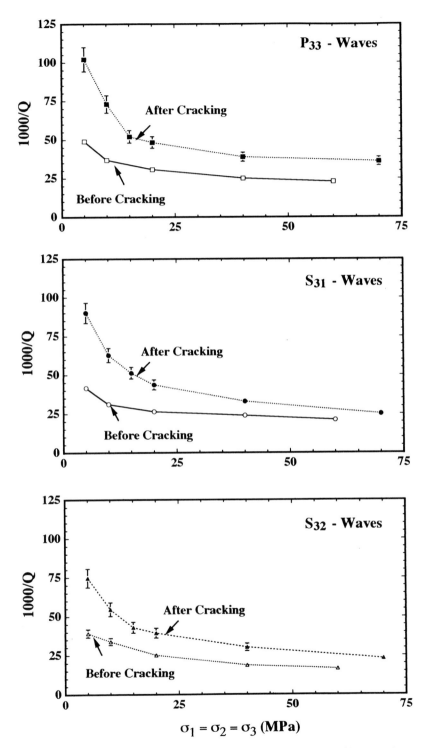

Fig. 10. Attenuation of P and S waves propagating in the 3-direction as a function of hydrostatic stress during the initial and crack-closing cycles.

80   M. S. KING *ET AL.*

ratios technique (Tao *et al.* 1995), for P and S waves propagating in the 3-direction as a function of hydrostatic stress are shown in Fig. 10. Attenuation in the rock is shown for the initial stress cycle (before cracking) and for the final cycle as the aligned cracks were closed. It is seen that there is a tendency for the attenuation at higher stress levels during the crack closing cycle to approach that observed during the initial uncracked rock cycle.

Attenuation during the cracking cycle is shown in Fig. 11 as a function of stress for P and S waves propagating in the 2- and 3-directions. It is seen that P and S waves propagating in the 3-direction all experience sharp increases in attenuation as σ_1 is increased above 100 MPa. In the 2-direction, however, only the S wave polarized in the 3-direction (S_{23}) experiences this sharp increase as σ_1 is increased above 100 MPa. The remaining P(P_{22}) and S(S_{21}) waves show only

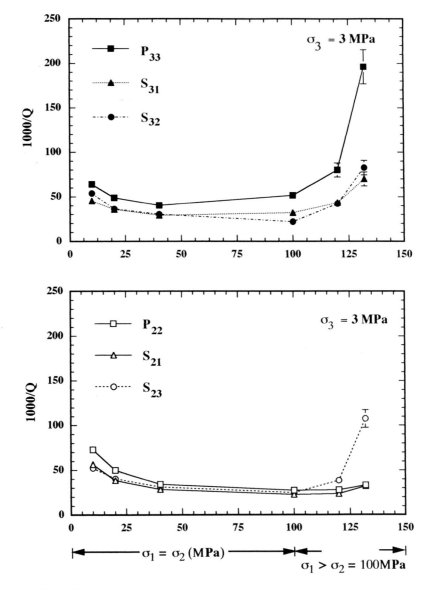

Fig. 11. Attenuation of P and S waves propagating in the 2- and 3-directions during the cracking cycle with $\sigma_3 = 3$ MPa.

very small increases in attenuation at the highest stress levels. The attenuation behaviour during the cracking cycle appears to mirror that of permeability, with large increases in permeability in the 12-plane being associated with large increases in attenuation of P and S waves propagating in the 3-direction.

Attenuation of $P(P_{33})$ and $S(S_{32})$ waves propagating in the 3-direction during the crack closing cycle are shown in Fig. 12 as a function of crack density (Fig. 12a) and permeability in the 1-direction (Fig. 12b). The attenuation–crack density relationship is seen to be almost linear. The attenuation–permeability relation-

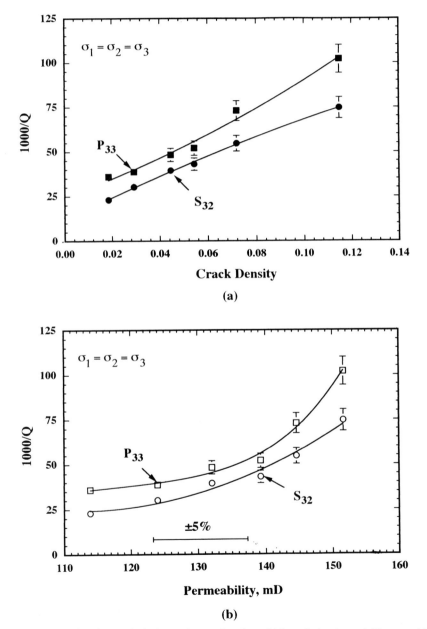

Fig. 12. Attenuation during the crack-closing cycle as a function of (**a**) crack density and (**b**) permeability in the 1-direction.

ship shows a decreasing sensitivity of attenuation to changes in permeability as the permeability decreases during the crack-closing cycle. There is an approximately 65% reduction in attenuation for both P and S waves propagating in the 3-direction associated with the 25% decrease in permeability caused by closure of the aligned cracks.

Figure 13 shows the permeability in the 1-direction (Fig. 13a) and S-wave velocity (V_{S31})

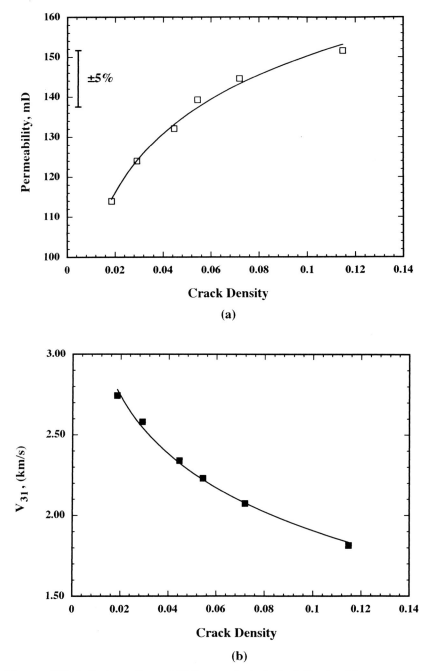

(a)

(b)

Fig. 13. (a) permeability in the 1-direction and (b) S-wave velocity in the 3-direction (V_{S31}) as a function of crack density.

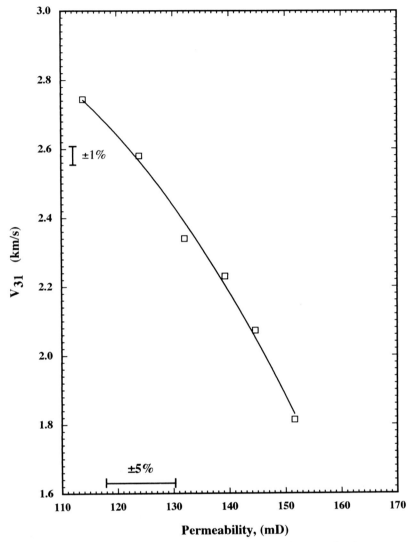

Fig. 14. S-wave velocity in 3-direction (V_{S31}) as a function of permeability in 1-direction.

in the 3-direction (Fig. 13b) as functions of crack density during the crack-closing cycle. It will be seen that the correlations between both parameters and the crack density are excellent. Figure 14 shows the S-wave velocity in the 3-direction (V_{S31}) as a function of permeability in the 1-direction, with V_{S31} increasing as the permeability decreases during the crack-closing cycle. It is seen that the correlation is excellent and that a significant increase in V_{S31} (51%) is associated with the decrease in permeability caused by closure of the aligned cracks.

Since all S waves either propagating or polarized in the 3-direction are approximately equal in magnitude during the crack-closing cycle, it is clear that the velocities of S waves propagating in the direction of the aligned cracks and polarized normal to them in the 3-direction may be related to changes in permeability as the cracks are closed. It should be noted at this point that the corresponding change in V_{P33} for P waves propagating in a direction normal to the plane of the cracks is 64%. Either the velocities or attenuation of P waves propagating perpendicular to the cracks or S waves propagating in the plane of the cracks and polarized perpendicular to them might therefore be used to predict changes in permeability of a rock containing aligned cracks and undergoing changes in stress.

Conclusions

(1) The experimental system has been successfully used to introduce sets of fractures and microcracks aligned in a plane normal to the minimum principal stress in cubic sandstone specimens. These aligned discontinuities are distributed in a reasonably homogeneous manner throughout the rock specimen, as indicated by the ultrasonic velocity measurements during crack closure.

(2) The results of the research programme to date suggest that, for sandstones with sets of aligned fractures and cracks, there are strong relationships existing between permeability in the plane of the aligned discontinuities and the velocities and attenuation of certain P and S waves as the aligned cracks are closed under pressure. Specifically, the P and S waves are those propagating normal to the plane and S waves propagating parallel to the discontinuity plane and polarized normal to it.

(3) The experimental system provides the opportunity to study the influence of changes in the intermediate principal stress on the mechanical (static and dynamic elastic, and failure properties) and transport (permeability and electrical conductivity) behaviour of porous sedimentary rocks. It is intended during the next stage of the research programme to study the behaviour of sedimentary rocks approaching failure when subjected to large differences between each of the principal effective stresses, as found adjacent to a deep borehole during drilling.

We wish to acknowledge with thanks the support provided by Shell Expro, British Gas, BP Exploration and AGIP for this research project. We also thank BP Research for donating the Schenk servo-controlled compression testing machine to the College.

References

BABUSKA, V. & PROS, Z. 1984. Velocity anisotropy in granodiorite and quartzite due to distribution of microcracks. *Geophysical Journal of the Royal Astronomical Society*, **76**, 113–119.

CHAUDHRY, N. A. 1995. *Effects of aligned discontinuities on the elastic and transport properties of reservoir rocks*. PhD thesis, Imperial College of Science, Technology and Medicine, University of London.

CRAMPIN, S. & ATKINSON, B. K. 1985. Microcracks in the Earth's crust. *First Break*, **3** (March), 16–20.

DOUMA, J. 1988. The effect of the aspect ratio on crack-induced anisotropy. *Geophysical Prospecting*, **36**, 614–632.

ENGELDER, T. 1982. Is there a genetic relationship between selected regional joints and contemporary stress within the lithosphere of north America? *Tectonics*, **1**, 161–177.

GANGI, A. F. 1978. Variation of whole and fractured porous rock permeability with confining pressure. *International Journal of Rock Mechanics & Mining Sciences & Geomechanics Abstracts*, **15**, 249–257.

GIBSON, R. L. & TOKSÖZ, M. N. 1990. Permeability estimation from velocity anisotropy in fractured rock. *Journal of Geophysical Research*, **95**, 15,643–15,655.

HOLT, R. M. & FJAER, E. 1987. Acoustic behaviour of sedimentary rocks during failure. *In*: KLEPPE, J., BERG, E. W., BULLER, A. T., HJELMELAND, O. & TORSAETER, O. (eds) *North Sea Oil and Gas Reservoirs*. Graham & Trotman, London. 311–316.

KING, M. S. 1966. Wave velocities in rocks as a function of changes in overburden pressure and pore fluid saturants. *Geophysics*, **31**, 50–73.

——1970. Static and dynamic elastic moduli of rocks under pressure. *In*: SOMERTON, W. H. (ed.) *Rock Mechanics – Theory and Practice, Proceedings of the 11th US Symposium on Rock Mechanics*. AIME, New York, 329–351.

——1983. Static and dynamic properties of rocks from the Canadian Shield. *International Journal of Rock Mechanics & Mining Sciences & Geomechanics Abstracts*, **20**, 237–241.

——, CHAUDHRY, N. A. & SHAKEEL, A. 1995. Experimental ultrasonic velocities and permeability for sandstones with aligned cracks. *International Journal of Rock Mechanics & Mining Sciences & Geomechanics Abstracts*, **32**, 155–163.

LO, T. W., COYNER, K. B. & TOKSÖZ, M. N. 1986. Experimental determination of elastic anisotropy of Berea sandstone, Chicopee shale, and Chelmsford granite. *Geophysics*, **51**, 164–171.

NISHIZAWA, O. 1982. Seismic velocity anisotropy in a medium containing oriented cracks – transversely isotropic case. *Journal of the Physics of the Earth*, **30**, 331–347.

NUR, A. 1971. Effects of stress on velocity anisotropy in rocks with cracks. *Journal of Geophysical Research*, **76**, 2022–2034.

—— & SIMMONS, G. 1969. Stress-induced velocity anisotropy in rock: an experimental study. *Journal of Geophysical Research*, **74**, 6667–6674.

SAYERS, C. M. & VAN MUNSTER, J. G. 1991. Microcrack-induced seismic anisotropy of sedimentary rocks. *Journal of Geophysical Research*, **96**, 16 529–16 533.

——, —— & KING, M. S. 1990. Stress-induced ultrasonic anisotropy in Berea sandstone. *International Journal of Rock Mechanics & Mining ciences & Geomechanics Abstracts*, **27**, 429–436.

SCHOENBERG, M. & DOUMA, J. 1988. Elastic wave propagation in media with parallel fractures and aligned cracks. *Geophysical Prospecting*, **36**, 571–590.

SMART, B. G. D. 1995. A true triaxial cell for testing cylindrical rock specimens. *International Journal of Rock Mechanics & Mining Sciences & Geomechanics Abstracts*, **32**, 269–275.

TAO, G. & KING, M. S. 1990. Shear-wave velocity and Q anisotropy in rocks: a laboratory study. *International Journal of Rock Mechanics & Mining Sciences & Geomechanics Abstracts*, **27**, 353–361.

——, —— & NABI-BIDHENDI, M. 1995. Ultrasonic wave propagation in dry and brine-saturated sandstones as a function of effective stress: laboratory measurements and modelling. *Geophysical Prospecting*, **43**, 299–327.

TOKSÖZ, M. N., CHENG, C. H. & TIMUR, A. 1976. Velocities of seismic waves in porous rocks. *Geophysics*, **41**, 621–645.

WALSH, J B. 1965a. The effect of cracks on the compressibility of rock. *Journal of Geophysical Research*, **70**, 381–389.

——1965b. The effect of cracks on the uniaxial compression of rocks. *Journal of Geophysical Research*, **70**, 399–411.

——1965c. The effect of cracks in rock on Poisson's ratio. *Journal of Geophysical Research*, **70**, 5249–5257.

——1981. Effect of pore pressure and confining pressure on fracture permeability. *International Journal of Rock Mechanics & Mining Sciences & Geomechanics Abstracts*, **18**, 429–435.

WU, B., KING, M. S. & HUDSON, J. A. 1991. Stress-induced ultrasonic wave velocity anisotropy in a sandstone. *International Journal of Rock Mechanics & Mining Sciences & Geomechanics Abstracts*, **28**, 101–107.

XU, S. & KING, M. S. 1992. Modelling the elastic and hydraulic properties of fractured rock. *Marine and Petroleum Geology*, **9**, 155–166.

ZAMORA, M. & POIRIER, J. P. 1990. Experimental study of acoustic anisotropy and birefringence in dry and saturated Fontainebleau sandstone. *Geophysics*, **55**, 1455–1465.

A simple but powerful model for simulating elastic wave velocities in clastic silicate rocks

SHIYU XU[1], JUST DOORENBOS[2], SUE RAIKES[3] & ROY WHITE[1]

[1] *Research School of Geological and Geophysical Sciences, Birkbeck and University Colleges, London WC1E 7HX, UK*
[2] *BP–Statoil Technology Alliance, Norway*
[3] *BP Exploration Technology Provision, Sunbury-on-Thames, Middlesex TW16 7LN, UK*

Abstract: We describe a practical velocity model for clastic silicate rocks developed from the Kuster and Toksöz, differential effective medium and Gassmann theories. The model divides the total pore space into two parts, one associated with sand grains and the other associated with clays (including bound water). The difference in pore geometry makes the clay fraction more compliant with increasing porosity than the sand fraction. The model accurately simulates the combined effect of lithology, porosity, clay content, water saturation and fluid type on laboratory and logging P- and S-wave velocities. Velocity dispersion is modelled by considering the relaxed and unrelaxed extremes of fluid flow. Of three possible schemes for predicting S-wave logs from other logs, prediction from the P-wave sonic and porosity logs is generally the most accurate; comparisons of blind test predictions and S-wave logs at two Norwegian North Sea wells demonstrate the accuracy and robustness of the method. The predictions were further improved once geological information was provided on the formations encountered in the well. Examples of application to the prediction of velocity dispersion and dry frame moduli, the detection of hydrocarbons from crossplotting the P-wave and S-wave sonic velocities, and sonic log editing confirm the flexibility, accuracy and reliability of the model. The model has proved invaluable in constructing accurate models of the seismic reflection response in the vicinity of wells.

Seismic techniques for direct hydrocarbon indication depend on seismic attribute anomalies which are related to changes in fluid content of sedimentary rocks. Interpretation and modelling of direct hydrocarbon indicators (DHI) require a comprehensive understanding of how these seismic attributes are affected by pore fluid content and other factors such as pressure, clay content, and frequency. In a recent review, Castagna *et al.* (1995) listed the following as key issues for DHI interpretation:

(1) pore fluid properties;
(2) S-wave velocity prediction;
(3) rock frame properties and fluid substitution;
(4) sonic log editing and quality control;
(5) velocity dispersion (relating laboratory, log and seismic measurements).

Various models have been proposed to tackle these issues (Castagna *et al.* 1985; Han *et al.* 1986; Mavko & Jizba 1991; Greenberg & Castagna 1992; Mukerji & Mavko 1994). The majority of such models are empirical or semi-empirical and each is generally designed for particular conditions. Theoretical models on the other hand often require dry rock frame moduli (e.g. Mukerji & Mavko 1994) and may involve specialized knowledge of other rock parameters even less accessible to routine measurements.

This restricts their application to exceptional circumstances. For practical application a model must be capable of relating elastic properties directly to basic petrophysical measurements, such as clay content. In order to cope with a wide range of practical problems, it is important to have a model that can provide a consistent and complete representation of the elastic behaviour of particular rock types. In this paper we consider the elastic properties of clastic silicate rocks.

Castagna *et al.* (1995) point out a number of unsolved problems in the five areas listed above. Thus the determination of frame bulk and shear moduli, needed for fluid substitution, remains problematical. The velocity dispersion predicted from Biot's equations is, in most cases, considerably lower than observed (Mavko & Jizba 1994). These problems need to be solved before a proper practice for fluid substitution can be established.

Xu & White (1995*a*) have shown that the effect of clay on elastic wave velocities can be divided into two parts:

● the mineralogy effect, since clay particles are normally softer than sand grains;
● the pore geometry effect, since clay particles are mostly flaky and form flatter pores than sands.

From Lovell, M. A. & Harvey, P. K. (eds), 1997, *Developments in Petrophysics*, Geological Society Special Publication No. 122, pp. 87–105.

They proposed a clay–sand mixture model which employs (1) the time-average model to simulate the effect of lithology and (2) the Kuster & Toksöz (1974) and differential effective medium (DEM) theories to model the effect of pore shape and (3) the Gassmann (1951) theory to model fluid relaxation. Its results simulate well-established relations between P-wave and S-wave velocities and porosity, clay content and fluid content in sand-shale systems. Unlike common applications of Gassmann's theory, this derived the elastic moduli of the dry rock frame from Kuster & Toksöz and effective medium theories before using Gassmann theory to simulate fluid relaxation. The predictions are thus derived from tabulated grain matrix parameters, together with values inferred from a 'pure shale' zone in log-based prediction, and this makes it easy to implement fluid substitution.

In previous studies Xu & White (1995*a*, *b*, 1996) have demonstrated the predictive ability of the model in the following areas:

- it simulates the branched dependence on porosity of shaly sands and sandy shales reported by Marion *et al.* (1992);
- it predicts that the first few percent of clay dominate its effect on P-wave and S-wave velocities, as reported by Blangy *et al.* (1993);
- the predicted P-wave velocities agree well with the laboratory measurements classified by Vernik (1994);
- it produces a modulus–porosity relationship for sandstones that can be approximated by a linear function while the relationship for shale is highly non-linear (Blangy *et al.* 1993);
- it provides robust prediction of S-wave logs from other logs, together with a self-checking scheme for the consistency of the prediction (Xu & White 1996);
- it simulates v_p/v_s from log measurements corresponding to varying shale volumes and v_p/v_s for laboratory sandstone samples as clay content changes from 0% to 50% (classified by Vernik 1994);
- velocity dispersion due to fluid flow mechanisms can be modelled; the results show that the high-frequency version simulates laboratory measurements while the low-frequency version simulates seismic and sonic measurements;
- fluid substitution can be readily carried out once the bulk modulus and density of the fluid mixture have been established.

These results show that the clay–sand mixture model does provide a comprehensive representation of the elastic behaviour of siliciclastic rocks.

Because it is a physical model, it has an in-built capability to simulate the combined effects of porosity, clay content, fluid content and frequency on P-wave and S-wave velocities that makes it an ideal tool for DHI modelling and interpretation.

In the following section the clay–sand mixture model is reviewed. We then show how it tackles issues 2 to 5 in the list from Castagna *et al.* (1995). Its applications to velocity dispersion, the evaluation of frame moduli, the effect of fluid content, and sonic log editing are illustrated. Two blind tests are described that demonstrate the ability of the model to predict S-wave logs accurately.

The clay–sand mixture model

The clay–sand mixture model (Xu & White 1995*a*) is an idealized physical model for velocities in siliciclastic rocks based the Kuster & Toksöz (1974), differential effective medium (Bruner 1976; Cheng & Toksöz 1979) and Gassmann (1951) theories. The model has two key features. The first is that it models the compliances of pores in the rock by assigning separate pore spaces to the sand and clay mineral fractions of the rock, each having different effective pore aspect ratios (ratios of short semi-axis to long semi-axis). The second is the use of Kuster & Toksöz (1974) and DEM theory to compute the elastic moduli of the dry frame. This overcomes a limitation of the KT theory to dilute concentrations of pores. Given the dry frame moduli, application of Gassmann's equations then gives the low-frequency velocity in the fluid-saturated rock. Figure 1 is a schematic diagram showing the sequence for calculating low-frequency P-wave and S-wave velocities.

The inputs required by the model are the densities and elastic moduli of the sand grains, clay particles and the pore fluid, the porosity and clay content of the rock frame. The model also requires two aspect ratios, one (α_s) for pores associated with sand grains and one (α_c) for pores associated with clay minerals (including bound water). Xu & White (1995*a*, fig. 1) give a flow chart showing how P-wave and S-wave transit times (strictly slownesses, reciprocals of velocities) are constructed from these inputs.

Shale volume replaces clay content in well-log applications. In these circumstances, the porosity used for velocity prediction becomes an effective porosity which excludes micro-pore spaces contained in a 'pure shale'. Xu & White (1996) describe the procedures adopted in dealing with shale volume when applying the model

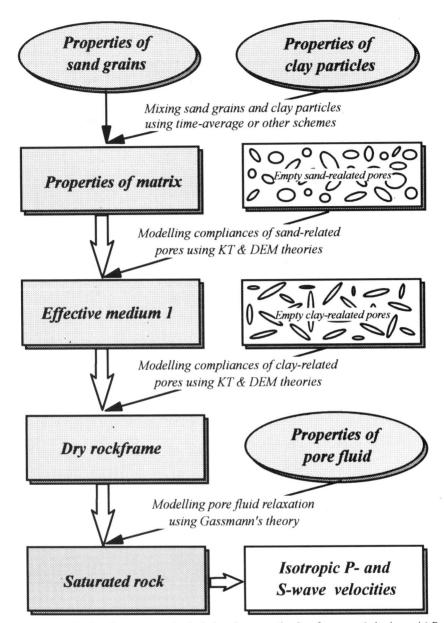

Fig. 1. Schematic showing the sequence of calculations in computing low-frequency (seismic, sonic) P-wave and S-wave velocities using the clay–sand mixture model.

to predict P-wave and S-wave logs. The density and moduli of the shale fraction are extracted from a 'pure shale' interval, guided by tabulated values for clay minerals, those for the sand grains are essentially values for quartz but can be varied within a small range according to the composition of the sandstone, and the parameters for the fluid fraction are taken from tables or laboratory measurements. Porosity is evaluated by standard logging methods and is checked against measurements from core where available.

The two aspect ratios are determined by a straightforward best-fit. There are two major difficulties in measuring aspect ratios directly: (1) we assume an idealized ellipsoidal pore shape which can hardly be expected in real rock, and (2) describing the hypothetical pore spaces associated with the sand and clay mineral fractions of the rock by just two aspect ratios is obviously another gross simplification. Although a model with two aspect ratio distributions may seem more plausible, the need to specify the parameters of these distributions makes this option impractical. In fact a single aspect ratio can mimic the elastic response of a uniform distribution of pores rather well (M. Sams, pers. comm.). The pragmatic view is that the aspect ratios simply characterise how the compliances of the two components respond to changes in porosity. It is an elastic rather than a geometrical concept that is quantified by aspect ratio and in practice there is no difficulty in estimating the two effective aspect ratios from the best fit between the predictions and observations. A post-hoc justification of the concept is that aspect ratio is a stable parameter: all results to date show that α_s is close to 0.1 and α_c of the order of 0.03.

To apply Kuster & Toksöz' theory (1974) to compute the elastic properties of the dry frame, we need the bulk and shear moduli of the mixture of sand grains and clay minerals. There are several schemes for combining moduli (Wang & Nur 1992) which can be used to approximate the effective elastic constants of the grain mixture. Xu & White (1996) present results showing that, apart from the Voigt and Reuss schemes which define upper and lower bounds, there is little to choose between the various schemes. The reason for this lies in the relatively low contrast between the clay mineral and sand grain moduli in comparison with that between sand grain (or clay mineral) and brine. We used time-average equations to compute P- and S-wave transit times of the grain mixture; the elastic bulk and shear moduli of the mixture are then calculated from its P- and S-wave transit times and density (Xu & White 1995a, 1996). This well-known scheme gives results close to other more complex recommended schemes and is a simple approximation to the low-frequency conditions that mainly concern us.

Finally Gassmann's equations are applied to model the effect of fluid motion on wave propagation.

Velocity dispersion

Mukerji & Mavko (1994) and Castagna *et al.* (1995) point out that the use of Gassmann's theory leads to a low-frequency approximation for P- and S-wave velocities. This is because the viscous drag of the pore fluid is ignored in Gassmann's model. Thus the model described so far is a low-frequency one since it simulates the effect of fluid flow using the Gassmann (1951) model. In applying Kuster & Toksöz (1974) theory to calculate the moduli of the dry rock frame, the frame is taken to be empty so that there can be no viscous drag and no consequent dependence on frequency.

To model wave velocities at high frequencies, Gassmann's theory is set aside and effective elastic moduli are calculated for the saturated rock by including the bulk modulus of the pore fluid inclusions when applying the Kuster & Toksöz and DEM theories (Xu & White 1995b, 1996). This procedure assumes that fluid flow is the prime cause of velocity dispersion in rocks and is based on the results of Mavko & Jizba (1991) and Mukerji & Mavko (1994) showing that theories, such as those of Hudson (1981) and Kuster & Toksöz (1974), which assume isolated pores are high-frequency approximations. Figure 2 shows the sequence of calculations adopted in computing high-frequency P-wave and S-wave velocities.

A numerical simulation was performed to investigate the effect of clay content and porosity on velocity dispersion. Parameters used in the simulation are listed in Table 1 (the same parameters are applied in the later simulations if not otherwise mentioned). Figure 3 shows the low- and high-frequency P-wave velocities predicted from the clay–sand mixture model as a function of porosity and clay content. The low-frequency velocities are always lower than the high-frequency velocities and the magnitude of the dispersion (difference between the two) increases with porosity and clay content. According to the model, one expects that the velocity dispersion for shales (especially soft shales) would be greater than that of clean sandstone. Klimentos & McCann (1990) measured P-wave attenuation coefficients (α) of 42 sandstone samples at 1 MHz and 40 MPa confining pressure. They observed a significant decrease in P-wave attenuation coefficient with increasing clay content. According to the Kramers–Krönig relation, this implies that velocity dispersion increases with clay content.

Indications of velocity dispersion between sonic log frequencies and laboratory frequencies can also be inferred by comparing predictions for laboratory and log measurements. Xu & White (1995a) employed the low-frequency version of the model to simulate laboratory data measured by Han *et al.* (1986) at 40 MPa confining pressure and 1 MPa pore pressure. The best fit aspect ratios were 0.14 for sand-related pores and 0.04 for clay-related pores. The former is noticeably higher than values found

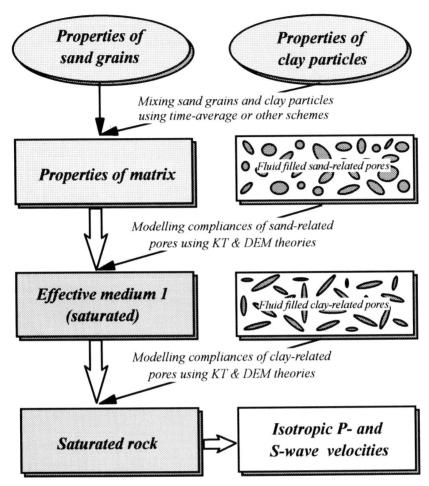

Fig. 2. Schematic showing the sequence of calculations in computing high-frequency (laboratory) P-wave and S-wave velocities using the clay–sand mixture model.

in well log modelling, for which α_s is very stable and typically 0.12 at most; α_c is less stable but appears to be 0.03 for well-compacted 'pure' shales. The discrepancy arises from the stiffening of the pores by the fluid at high frequencies. Application of the high-frequency version of the model to these data remedies this discrepancy (Fig. 4b). The low-frequency predictions using $\alpha_s = 0.12$ and $\alpha_c = 0.03$ clearly underestimate the laboratory measurements (Figure 4a).

These results indicate that fluid flow appears to be the main cause of velocity dispersion in fully-saturated rocks, and laboratory frequencies (normally hundreds of KHz to MHz) fall in the high-frequency range while logging frequencies (15–25 KHz) fall in the low-frequency

Table 1. *Transit times, density and aspect ratio of different rock types*

Lithology	T^p (μs m^{-1})	T^s (μs m^{-1})	Density (kg m^{-1})	Aspect ratio
Sandstone	171	256	2650	0.12
Shale	230	394	2600	0.03
Brine	617		1050	

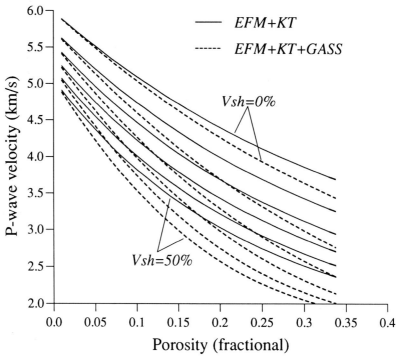

Fig. 3. Illustration of the dependence of P-wave velocity on porosity and shale volume at low and high frequencies. Dashed lines represent the P-wave velocities at low frequencies (from differential effective medium, Kuster and Toksöz and Gassmann theories) and the solid lines represent the P-wave velocities at high frequencies (from differential effective medium and Kuster & Toksöz theories).

range. Thus the low-frequency version of the model simulates seismic and certain log measurements while the high-frequency version simulates laboratory measurements.

Castagna *et al.* (1995) suggested the use of Biot's high-frequency equations for modelling laboratory measurements and Gassmann's equation for modelling seismic velocities. However it is increasingly evident that the velocity dispersion predicted from Biot's theory (1956*a, b*) is generally lower than measured velocity dispersion (Mavko & Jizba 1994). Mavko & Jizba (1994) and Mukerji & Mavko (1994) explain how local or 'squirt' flow usually dominates velocity dispersion, especially when 'soft' porosities are present, while Biot's theory can consider only the effect of the 'macro' or 'large scale' fluid flow on velocity dispersion. Because the clay–sand mixture model predicts velocity dispersion using the idea proposed by Mukerji & Mavko (1994), it predicts a higher

velocity dispersion than that predicted by Biot's theory. For a sandstone with 20% porosity and 10% clay content, the predicted velocity dispersion is about 10%.

Frame moduli

The values of the frame moduli of a shaly sandstone are vital parameters in any theory aiming to predict its elastic and anelastic properties. The potential softening effect of pore fluid on shales makes it difficult to predict in *in situ* conditions. Castagna *et al.* (1995) note that dry rock moduli determined in the laboratory may be inapplicable because of chemical interactions between pore fluids and clay minerals. It is common practice in fluid substitution modelling to estimate dry rock frame moduli from measured P-wave and S-wave velocities using Gassmann (low frequency) or Biot (high

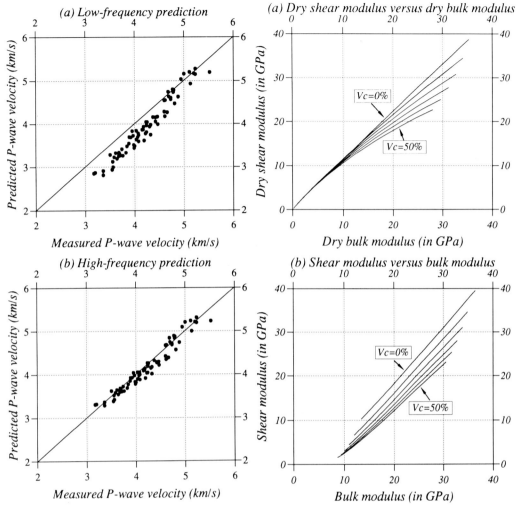

Fig. 4. Cross-plot of measured P-wave velocities and the velocities predicted from (**a**) the low-frequency version and (**b**) the high-frequency version of the clay–sand mixture model. The measurements were taken at 40 MPa confining pressure and 1 MPa pore pressure by Han *et al.* (1986).

Fig. 5. Illustration of the predicted relationships between shear and bulk moduli for (**a**) dry rock frame and (**b**) 100% brine saturation. Different lines represent different clay content.

frequency) theory. Estimation of frame moduli from P-wave and S-wave logs relies on the two critical conditions:

(1) a reliable S-wave log, in addition to P-wave and density logs, is available;
(2) there exists a reference rock in the vicinity of the reservoir which is fully brine-saturated and has characteristics very similar to the reservoir rock (e.g. same rock type, same porosity, same clay content).

If there is no S-wave log, one can assume a frame shear modulus equal to the frame bulk modulus. From laboratory measurements on clean quartz sandstone samples at high confining pressure, Murphy *et al.* (1993) found the relationship

$$K_d = 0.9 \times \mu_d$$

where μ_d and K_d are the shear and bulk moduli of the rock frame respectively. On the other hand, Castagna *et al.* (1993) published a number of linear frame moduli trends for sandstone, limestones, shales and dolomite and their results suggest that the bulk modulus for a sandstone frame is about equal to its shear modulus.

Castagna *et al.* (1995) use laboratory measurements by Han *et al.* (1986) to point out that the effect of clay content and weak cementation on frame moduli may be highly non-linear.

The clay–sand mixture model gets round this problem by predicting frame moduli from porosity, clay content and grain matrix properties. Xu & White (1996) compared the dry shear moduli predicted from the model with the laboratory measurements of Blangy *et al.* (1993) and found very good agreement between them. The model can also predict the relationship between frame bulk and shear moduli. Figure 5 shows the μ_d–K_d trends predicted in the cases of (a) dry and (b) 100% brine saturation. Several important points are suggested by the figure.

- For clean sandstones, the predicted K_d–μ_d relation is very close to that ($K_d = 0.9\mu_d$) observed by Murphy *et al.* (1993).
- For reservoir quality sandstones with about 10% clay content, the dry shear modulus is roughly equal to the bulk modulus. As clay content increases, the μ_d–K_d relationship becomes increasingly non-linear (cf. Castagna *et al.* 1995).
- In 100% brine-saturated rocks, the shear modulus can be approximated by a linear function of bulk modulus whose slope and intercept vary with clay content, indicating a significant role for both lithology and pore geometry.
- The relationship between the dry bulk modulus and dry shear modulus at high porosities (low values for the moduli) is less scattered than that for 100% brine-saturated rocks.

Shear-wave prediction

The interpretation of amplitude-versus-offset analyses and multicomponent seismic and VSP data requires S-wave velocity logs. Although the logging industry now offers a number of S-wave velocity tools, they have been run in relatively few wells and are still not used routinely. There is therefore a need to predict S-wave velocities from other logs whenever measured S-wave velocity logs are unavailable. Even when an S-wave log has been run, comparison with its prediction from other logs can be a useful quality control. Prediction can be used to fill gaps in S-wave logs run with, say, monopole sources or to enhance the resolution of logs from long spacing tools.

A number of empirical or semi-empirical laws have been published to predict S-wave logs from other logs. Xu & White (1996) discussed their advantages and disadvantages. Prediction of S-wave logs using the clay–sand mixture model can take one of three routes:

(1) from grain matrix parameters (lithology), porosity and clay content,
(2) from lithology, P-wave velocity and clay content, or
(3) from lithology, P-wave velocity and porosity.

If all the measurements were error-free, it would not matter which scheme was used. Applications of the model to laboratory measurements by Han *et al.* (1986) and two well log data sets indicate that schemes 2 and 3 are more robust than scheme 1 (Xu & White 1995*b*, 1996). This is not surprising since there is a very strong correlation between v_p and v_s: both are similarly affected by porosity, clay content and lithology. On top of this the sonic log is usually more reliable than estimates of porosity and shale volume for which errors, besides those in the measurements, may be introduced from the use of imperfect models and imperfect parameters.

Blind tests

Blind tests were made using the model on two well data sets from the Norwegian North Sea where good quality S-wave log data were recorded. A standard suite of well logs without S-wave logs was provided in each case. The S-wave logs, in addition to other related geological and petrophysical information, were later provided to check the accuracy of the S-wave predictions. Figures 6 and 7 show the logs provided at the two wells.

The geological setting of the field can be briefly described as follows. The main reservoirs are in the Jurassic Brent Formation (3495–3575 m at Well A and 2865–2915 m at Well B) and Dunlin Formation (3760–3815 m at Well A and 3070–3090 m at Well B). The Brent sandstones were deposited in a marginal marine to deltaic environment. The Etive and Ness formations at the top consist of a mixture of clean sands, conglomerates and shale layers. The Oseberg Formation within the Brent consists mainly of good-quality homogeneous sandstones, with alternating sandy shales. The Osberg sandstones were deposited by high-density turbidites and contain thin clay lenses. Some carbonate cementation is present. The Brent sands were originally almost shale free, and consist of mainly quartz, biogenic carbonate cement and some feldspar. Extensive feldspar dissolution has taken place. This process is the main source of the authigenic clay minerals. Below the Oseberg sands are the Drake mudstones and shales. The Dunlin sands were

deposited in a nearshore or estuarine environment, coarsening upwards.

The WDS log analysis package was used to evaluate porosity, shale volume and water

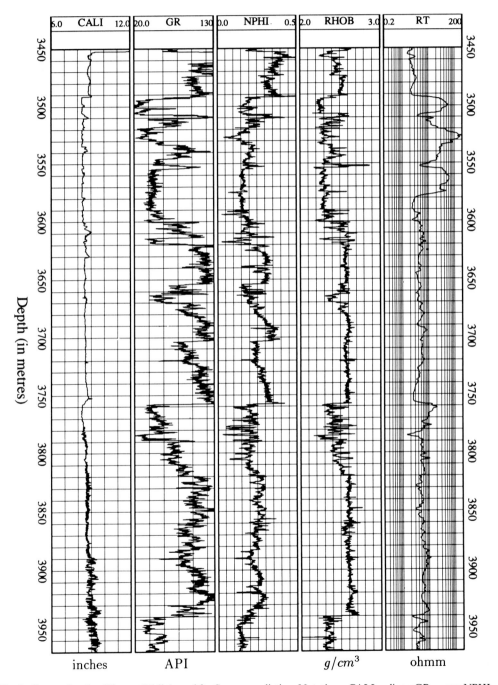

Fig. 6. Conventional well logs at Well A used for S-wave prediction. Notations: CALI, caliper; GR, γ-ray; NPHI, neutron porosity; RHOB, bulk density; RT, deep resistivity.

saturation from the logs. Laboratory measurements of porosities were used to make sure that the estimated porosities were sensible. Parameters used in the predictions are listed in Table 2. The S-wave logs from the blind tests were predicted from the P-wave and the estimated shale volume logs. Figures 8 and 9 show the results of the formation evaluation

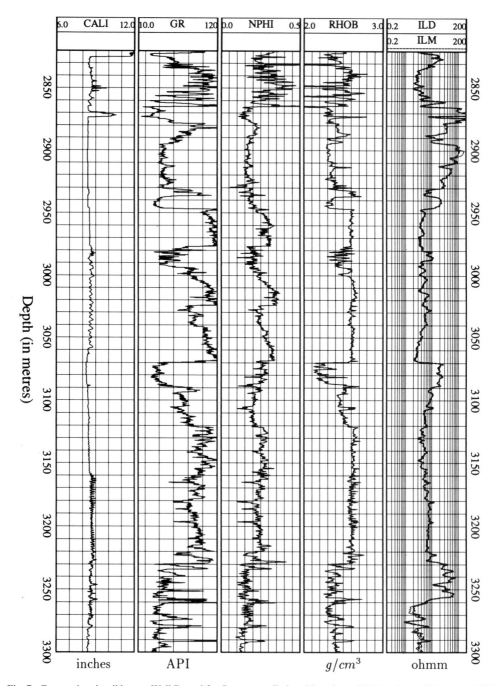

Fig. 7. Conventional well logs at Well B used for S-wave prediction. Notations: CALI, caliper; GR, γ-ray; NPHI, neutron porosity; RHOB, bulk density; ILD, induction deep; ILM, induction medium.

Table 2. *Parameters used in predicting P- and S-wave transit times at both wells*

Lithology	T^p (μs m^{-1})	T^s (μs m^{-1})	Density (kg m^{-1})	Aspect ratio
Sandstone	180	288	2650	0.12
Shale	344	672	2500	0.05
Brine	623		1050	

exercise and the S-wave log prediction. The curves in panels 1, 2 and 3 of the two figures represent shale volume (SVSH), porosity (SPOR) and water saturation (SW), respectively. The solid circles in panel 2 mark the porosities measured from core. Panel 4 compares the measured (DT) P-wave log with our prediction (PDT) using porosity (SPOR) and shale volume (SVSH). Panel 5 shows the S-wave log (PSDT) predicted from the P-wave (DT) and shale volume (SVSH) logs, together with the measured S-wave log.

No information about the measured S-wave logs, the geological setting, the petrophysical properties of the formations or the fluids was given at this stage. The test was, therefore, totally blind. The predictions, in general, are reasonably good. The normalized mean square errors (NMSE, Xu & White 1995a) are 0.058 for both the P-wave and S-wave predictions at Well A and 0.069 for the P-wave and 0.075 for the S-wave predictions at Well B. However there are places where the predictions depart significantly from the measurements.

Over depth intervals within the two reservoirs, 3530–3590 m in Well A and 2875–2920 m in Well B the predicted P- and S-wave logs overestimate transit times. As there is a very good correlation between the errors and the estimated shale volume, the most likely cause is error in the estimated shale volume. This was later confirmed by the geological descriptions which indicated the presence of abundant feldspars in the reservoir rocks.

On the whole the predicted S-wave logs tend to overestimate the measured S-wave transit times, indicating a general positive bias in the S-wave prediction. As a precaution against the possible presence of other minerals such as calcite, the sand grain matrix was not assumed to be pure quartz and the moduli assigned to it corresponded to a v_p/v_s value of 1.6. This value was later proved to be too high since the geological descriptions described the sandstones as very clean and consisting mainly of quartz.

In order to verify the above interpretations, we changed the v_p/v_s for the matrix to 1.5, a value for pure quartz, keeping other parameters

the same as before. We also used scheme 3 (predicting the S-wave log from the P-wave and porosity logs) instead of scheme 2 in the revised prediction. Figures 10 and 11 show the results. Panel 1 compares SVSH (solid line), the shale volume estimated from the γ-ray log, and VSH (dashed line), that predicted from the P-wave and porosity logs using the model. In general the two curves agree with each other except over particular depth intervals. The geological information which we have so far indicates VSH is a better estimate than SVSH at least over the reservoir depth intervals. Panel 4 compares the measured S-wave log DTSM (solid line) and the predictions PSDT (dashed line) from the P-wave log and porosity SPOR (this is equivalent to predicting from VSH and SPOR). The much improved S-wave prediction from this scheme over that using SVSH and the P-wave log (panel 5 of Figs 8 and 9) indicates that SVSH is the most error-prone of the logs. The NMSEs between the predicted and observed S-wave logs are reduced from 0.058 to 0.044 at Well A and from 0.075 to 0.051 at Well B.

Panel 5 in Figs 10 and 11 compares the measured S-wave log (DTSM, solid line) and that predicted using Castagna's 'mudrock line' (TS, dashed line). The NMSEs between DTSM and TS are 0.084 at Well A and 0.103 at Well B. The figures show that over shale depth intervals the 'mudrock line' simulates the measurements rather well, while over the sandstone intervals the predictions significantly overestimate measured transit times. This is understandable since the 'mudrock line' was, as its name implies, designed only for mudstones and shales. The comparison of panel 4 with panel 5 demonstrates how much the clay–sand mixture model can improve over a simple linear relation such as Castagna's 'mudrock line'.

Over the depth interval of 3450–3460 m in Well A there is no measured S-wave log. It is well known that tools with a monopole source cannot record useful S-wave logs through soft formations whose S-wave velocity is less than the P-wave velocity of the drilling mud (Castagna et al. 1993). There is much evidence suggesting that the monopole source is the

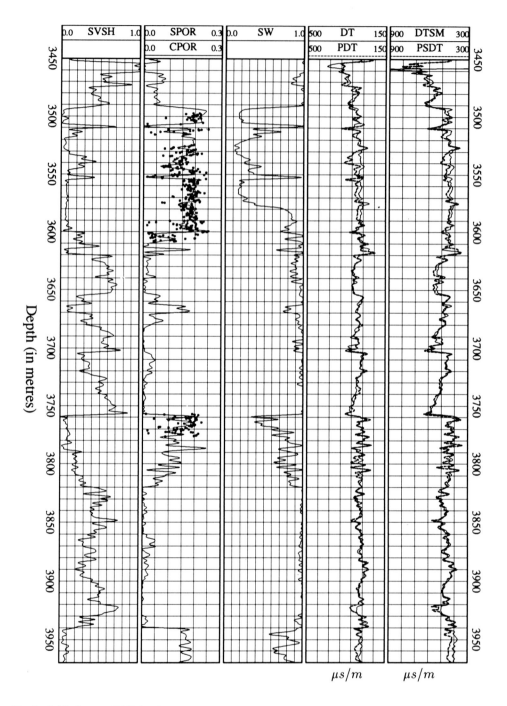

Fig. 8. A blind test at Well A. The P-wave sonic logs (PDT, dashed line in panel 4) was predicted from shale volume SVSH and porosity (SPOR) and the S-wave sonic log (PSDT, dashed line in panel 5) was predicted from the P-wave (DT) and shale volume (SVSH) logs. The solid circles (CPOR) in panel 2 represent porosities measured in the laboratory. Other notations: SW, water saturation; DTSM, measured S-wave sonic log.

main cause of the missing S-wave logs. Firstly the S-wave logs in these wells were recorded using a monopole source and, secondly, the S-wave transit time just below the missing depth interval is approximately $720\,\mu s\,m^{-1}$ which is close to the P-wave transit time for the drilling mud. In another case study similar S-wave behaviour was observed and an S-wave log recorded from a full-waveform tool with a monopole source was interpolated across soft

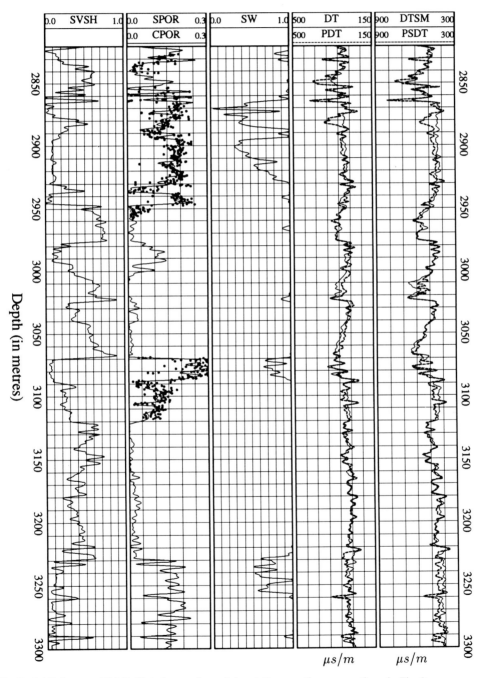

Fig. 9. A blind test at Well B. Notations and panel descriptions are the same as those in Fig. 8.

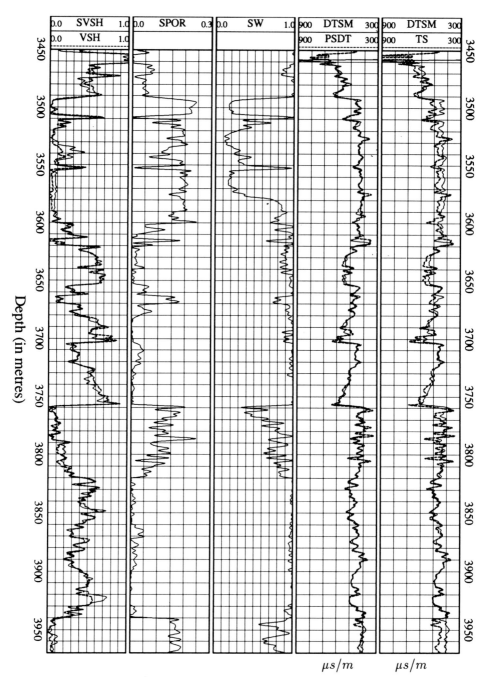

Fig. 10. Comparison of the measured S-wave transit times (DTSM) and those predicted from the clay–sand mixture model (PSDT, dashed line in panel 4) and Castagna's mudrock line(TS, dashed line in panel 5) at Well A. VSH in panel 1 (dashed line) represents the shale volume estimated using the model from P-wave sonic (DT) and porosity (SPOR) logs. The S-wave log (PSDT, dashed line in panel 4) was predicted from VSH and SPOR. Other notations: SVSH, shale volume estimated from γ-ray log; SW, water saturation; DTSM, measured S-wave sonic log.

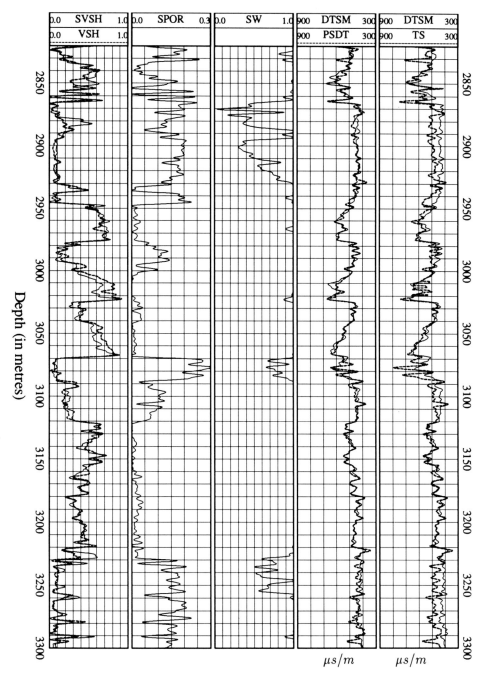

Fig. 11. Comparison of the measured S-wave transit times (DTSM) and those predicted from the clay–sand mixture model (PSDT, dashed line in panel 4) and Castagna's mudrock line (TS, dashed line in panel 5) at Well B. Notations and panel descriptions are the same as those in Fig. 10.

formation intervals using the clay–sand model on the basis of its accurate prediction of the valid segments of log.

The effect of fluid content

Generally sonic tools measure P- and S-wave velocities within a zone that is at least partially invaded. Castagna *et al.* (1995) recommend that sonic logs should be corrected for invasion and dispersion prior to seismic modelling. The effect of invasion is more marked over reservoir, especially gas reservoir, depth intervals than brine-saturated intervals because the bulk modulus of the mud filtrate is similar to that of brine but significantly different from that of gas or live oil. It is of great interest to investigate how invasion affects the P- and S-wave velocity measurements.

We demonstrate the effect of invasion by comparing our predictions with the measurements, as in Fig. 12. For this simulation, we assumed an oil with a P-wave transit time of $820 \, \mu s \, m^{-1}$ and a density of $700 \, kg \, m^{-3}$; other parameters are listed in Table 1. The dotted lines

in this figure correspond to curves calculated for values of normalized shale volume (as a fraction of the total volume of the grain matrix) ranging from 0.0 (clean sands) to 1.0 (pure shale) in steps of 0.2. 100% brine-saturation was assumed during the calculation. The dashed line represents the predicted $v_s - v_p$ trend for 100% oil-saturated clean sands. Log values at Well B were sorted according to shale volume VSH and the figure shows every fifth log value corresponding to a shale volume larger than 40% or less than 10%. Other log values were omitted to prevent the figure being over-crowded. Different symbols represent log values with different shale volume. The solid line corresponds to the 'mudrock line' (Castagna *et al.* 1985).

The figure shows that the majority of the log measurements with shale volume less than 10% (stars) lies between the clean sand line with 100% oil saturation (dashed line) and that with 100% brine-saturation. This clearly indicates the presence of hydrocarbon despite the effect of mud filtrate invasion and partial invasion of the reservoir rock within the investigation depth of sonic log tools.

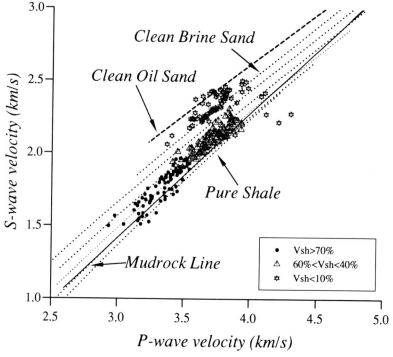

Fig. 12. Comparison of the measured $v_p - v_s$ relationships at Well B and those predicted from the clay–sand mixture model. The dashed line represents 100% oil saturated clean sandstones. The dotted lines correspond to curves calculated for values of the normalized shale volume (as a fraction of the total volume of the grain matrix) ranging from 0.0 (clean sands) to 1.0 (pure shale) in steps of 0.2. 100% brine-saturation was assumed during the calculation. The solid line represents Castagna's 'mudrock line'.

More quality control of sonic readings

In addition to mud filtrate invasion, the quality of sonic readings is also affected by factors such as cycle skipping, formation damage and washed-out hole, and unreliable P-wave and S-wave readings over soft-formation intervals. Castagna *et al.* (1995) address a number of

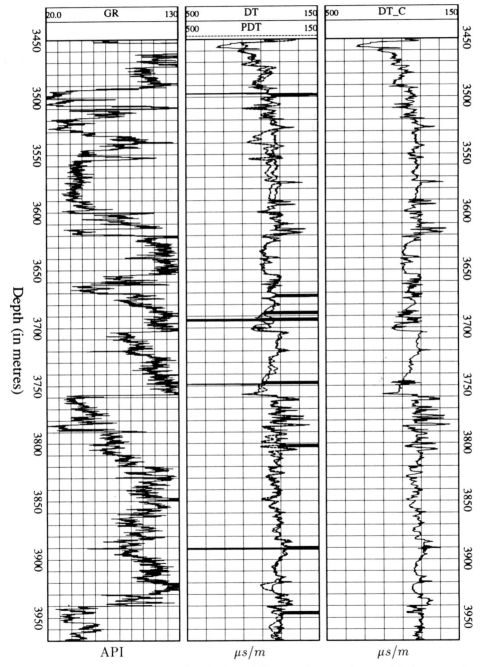

Fig. 13. Illustration of sonic log editing for seismic modelling. Panel 1, γ-ray log; panel 2, the measured P-wave sonic log without editing (DT) and the predicted P-wave sonic log (PDT); panel 3, the measured P-wave sonic log after editing.

important issues concerning the quality of sonic readings of gas sands. Of prime importance is the editing, correction and checking of sonic logs prior to seismic modelling.

The example of Fig. 13 demonstrates the application of petrophysical modelling to sonic log editing. The recorded P-wave sonic-log is very unreliable over some depth intervals (panel 2). One can edit these unreliable readings using a despiking procedure but, apart from the problem of distinguishing spikes that could possibly be valid readings (from thin coal seams, say), the question then arises of what values should replace the spikes. We prefer to approach the problem in a more physical way by:

(1) predicting the P-wave sonic log from the porosity and shale volume logs, tuning the parameters to make a robust best fit between the measurements and the prediction;
(2) detecting and then substituting bad sectors using the predictions guided by the mean error-of-fit.

Panel 2 compares the measured sonic log (DT, solid line) with our prediction (PDT, dashed line), and panel 3 shows the edited P-wave sonic log.

Conclusions

Castagna *et al.* (1995) addressed the most important issues in rock physics for DHI interpretation and modelling. Examples are shown in this paper to demonstrate how the clay–sand mixture model tackles these important issues.

(1) Velocity dispersion caused by fluid flow (both local and macro) mechanisms can be predicted by using low- and high-frequency versions of the model. Numerical results show that velocity dispersion increases with increasing porosity and clay content. The use of the high-frequency model in simulating laboratory measurements at high confining pressures and the low-frequency model in predicting sonic logs over well compacted depth intervals leads to very similar best fit aspect ratios (0.12 for sand-related pores, and 0.03 for clay-related pores). This implies that sonic logs can be regarded as low-frequency measurements and laboratory measurements as high-frequency measurements.

(2) Two blind tests of the model using two wells from the Norwegian North Sea showed that the model was capable of predicting S-wave logs from a standard suite of logs reasonably well without any prior information. The normalized mean square errors (NMSE) are 0.058 for the S-wave prediction at Well A and 0.075 at Well B.

(3) The predictions were improved once geological and petrophysical information was provided. The NMSEs were reduced to 0.044 at Well A and 0.051 at Well B. This implies that the predictions can benefit from a 'training' test using a well having a measured S-wave sonic log within the same geological regime. Then the parameters tuned to provide the best fit between the measured and predicted P-wave and S-wave logs at the test well can be applied to other wells in the region.

(4) The model was used to study the relationship between shear modulus μ_d and bulk modulus K_d. The numerical results suggest that $K_d = 0.9\mu_d$ for clean sands, as observed by Murphy *et al.* (1993), and that K_d is roughly equal to μ_d for reservoir quality sandstones with about 10% clay content. The K_d–μ_d relationship becomes increasingly non-linear with increasing clay content.

(5) The effect of fluid content on the P-wave and S-wave velocity relationship was also studied. Comparison between the log measurements and our predictions indicates that the reservoir rocks within the investigation depth of the sonic tools were only partially invaded. The presence of oil was clearly seen on the v_s–v_p crossplot.

(6) The application of the model to the quality control of sonic log readings was demonstrated by showing how bad sectors in the P-wave sonic at Well A could be replaced by predictions from the model. The examples of S-wave log prediction showed how the model could contribute to the integrated quality control of a suite of logs, in this case indicating intervals with unreliable estimates of shale volume, and they also gave an indication of how a predicted S-wave log could fill the gaps in a recorded S-wave log within soft formation where the S-wave velocity is less than P-wave velocity of the drilling mud.

Predictions of accuracy comparable to that from the model can be made for clean sandstones and pure shales via Castagna's sandstone and mudrock lines. The great advantage of the model is that it can accurately predict P- and S-wave velocities at any point between these extremes, from shaly sands to sandy shales, for different pore fluids, and at seismic and laboratory frequencies. Moreover it provides a physical basis for an integrated understanding of the effects of clay content, fluid fill and frequency.

The authors are indebted to the sponsors of the London University Research Programme in Seismic Lithology (1994–1997) and the Birkbeck College Research Programme in Exploration Seismology

(1991–94), Amoco (UK) Exploration Company, BP Exploration Operating Company Ltd, Elf UK plc, Enterprise Oil plc, Fina Exploration Ltd., Geco Geophysical Company Ltd, Mobil North Sea Ltd, Sun Oil Britain Ltd and Texaco Britain Ltd for their support of this research. We thank BP and Statoil for providing the well log data and for permission to publish these data.

References

BIOT, M. A. 1956a. Theory of propagation of elastic waves in a fluid saturated porous solid. I. Low-frequency range. *Journal of Acoustic Society of America*, **28**, 168–178.

——1956b. Theory of propagation of elastic waves in a fluid saturated porous solid. II. Higher-frequency range. *Journal of Acoustic Society of America*, **28**, 179–191.

BLANGY, J. P., STRANDENES, S., MOOS, D. & NUR, A. 1993. Ultrasonic velocities in sands – revisited. *Geophysics*, **58**, 344–356.

BRUNER, W. M. 1976. Comment on "Seismic Velocities in Dry & Saturated Cracked Solids" by R. J. O'Connell & B. Budianskey. *Journal of Geophysical Research*, **81**, 2573–2576.

CASTAGNA, J. P., BATZLE, M. L. & EASTWOOD, R. L. 1985. Relationships between compressional-wave and shear-wave velocities in clastic silicate rocks. *Geophysics*, **50**, 571–581.

——, HAN, D. H. & BATZLE, M. L. 1995. Issues in rock physics and implications for DHI interpretation. *The Leading Edge*, **14**, 883–885.

——, BATZLE, M. L., TUBMAN, K. M., GAISER, J. E. & BURNETT, M. D. 1993. Shear-wave velocity control. *In*: CASTAGNA, J. P. & BACKUS, M. M. (eds) *Offset-Dependent Reflectivity – Theory and Practice of AVO Analysis*. Society of Exploration Geophysics, Tulsa, OK, 115–134.

CHENG, C. H. & TOKSÖZ, M. N. 1979. Inversion of seismic velocities for pore aspect ratio spectrum of a rock. *Journal of Geophysical Research*, **84**, 7533–7543.

GASSMANN, F. 1951. Elasticity of porous media. *Vierteljahrschrift der Naturforschenden Gesellschaft in Zürich*, **96**, 1–21.

GREENBERG, M. L. & CASTAGNA, J. P. 1992. Shear-wave velocity estimation in porous rocks: theoretical formulation, preliminary verification and applications. *Geophysical Prospecting*, **40**, 195–209.

HAN, D, NUR, A. & MORGAN, D. 1986. Effect of porosity and clay content on wave velocity in sandstones. *Geophysics*, **51**, 2093–2107.

HUDSON, J. A. 1981. Wave speed and attenuation of elastic waves in material containing cracks. *Geophysical Journal of the Royal Astronomical Society*, **64**, 133–150.

KLIMENTOS, T. & McCANN, C. 1990. Relationships among compressional wave attenuation, porosity, clay content, and permeability in sandstones. *Geophysics*, **55**, 998–1014.

KUSTER, G. T. & TOKÖSZ, M. N. 1974. Velocity and attenuation of seismic waves in two phase media: Part 1: Theoretical formulation. *Geophysics*, **39**, 587–606.

MARION, D., NUR, A., YIN, H. & HAN, D. 1992. Compressional velocity and porosity in sand-clay mixtures. *Geophysics*, **57**, 554–563.

MAVKO, G. & JIZBA, D. 1991. Estimating grain-scale fluid effects on velocity dispersion in rocks. *Geophysics*, **56** 1940–1949.

—— & ——1994. Relation between seismic P- and S-wave velocity dispersion in saturated rocks. *Geophysics*, **59**, 87–92.

MUKERJI, T. & MAVKO, G. 1994. Pore fluid effects on seismic velocity in anisotropic rocks. *Geophysics*, **59**, 233–244.

MURPHY, W., REISCHER, A. & HSU, K. 1993. Modulus decomposition of compressional and shear velocity in sand bodies. *Geophysics*, **58**, 227–239.

WANG, Z. & NUR, A. 1992. Elastic wave velocities in porous media: a theoretical recipe. *In*: WANG, Z. & NUR, A. (eds) *Seismic and Acoustic Velocities in Reservoir Rocks, Theoretical and Model Studies*. S.E.G., Tulsa, 1–35.

VERNIK, L. 1994. Predicting lithology and transport properties from acoustic velocities based on petrophysical classification of siliciclastics. *Geophysics*, **59**, 727–735.

XU, S. & WHITE, R. E. 1995a. A new velocity model for clay–sand mixtures. *Geophysical Prospecting*, **43**, 91–118.

—— & ——1995b. Poro-elasticity of clastic rocks: a unified model. *In: The Transactions of the 36th Annual Logging Symposium, June 26–29 1995, Paris, France*, Paper V.

—— & ——1996. A physical model for shear-wave velocity prediction. *Geophysical Prospecting*, **44**, 687–717.

Measurements of the relationship between sonic wave velocities and tensile strength in anisotropic rock

T. APUANI[1], M. S. KING[2], C. BUTENUTH[3], M. H. DE FREITAS[3]

[1] *Università degli Studi, Dipartimento di Scienze della terra, Via Mangiagalli 34, 20133 Milano, Italy*
[2] *Department of Earth Resources Engineering and*
[3] *Department of Geology, Imperial College of Science, Technology & Medicine, London, SW7 2AZ, UK*

Abstract: This work investigates relationships between the mechanical behaviour under tension of a building stone widely used in Italy and the compressional and shear-wave velocities within the stone. The stone investigated is the Antigorio orthogneiss, which is characterized by an orthotropic petrographical anisotropy.

This rock was sampled in three orthogonal directions two of which lie in the plane of schistosity, the first parallel and the second perpendicular to the elongated structures in that plane, with the third sampling direction oriented normal to the plane of schistosity. In these directions rock cylinders were cored co-axially to provide hollow cylinders of 14 cm outer diameter together with their inner solid cores of approximatly 5 cm diameter. These inner cores were used for ultrasonic wave velocity measurements (V_p and polarized V_s) oriented at 90° to the core axis, and conducted at intervals around their circumference so as to obtain wave velocities through the gneiss in different orientations. The hollow cylinders were cut into hoops and tested by the hoop tension test to provide measurements of tensile strength in the same directions as those in which wave velocities had been measured. Thus non-destructive velocity tests and destructive tensile tests were conducted in the same orientations.The paper describes the results so obtained using dry samples.

One of the main requirements of the quarry industry and of rock working technology is a method for predicting the mechanical behaviour of pre-cut rock blocks and slabs in different directions. It is desirable to have a reliable non-destructive method to recognize any unexpected plane of weakness and to provide some confidence in predicting the future performance of cut blocks when in use. Planes of weakness can be associated with a petrological inhomogeneity such as a change in mineralogy and texture, and with cracks.

Compressional and shear wave velocities propagating in rock are a function of all these parameters in addition to being affected by pressure, temperature, moisture content and the nature of the pore fluid (Lama & Vutukuri 1978). For these reasons wave velocities in an anisotropic material are directionally dependent. Geophysical measurements are also related to the elastic properties of a material and their relationship, for homogeneous and isotropic material, is known (Telford 1963). Geophysical measurements provide a promising non-destructive method for assisting in predicting the mechanical behaviour of rock materials and it was with this in mind that the work described here was completed.

As the initiation of failure and propagation of cracks occurs in zones where the applied system of forces results in extension (Atkinson 1987), the authors have chosen to study the mechanical behaviour of the rock under tension. It is the relationship between the mechanical behaviour, under tension, and the variation of compressional and shear wave velocities in different directions in an orthogneiss widely used for building, that is the aim of this study.

The mechanical behaviour of the gneiss in extension was measured using the Hoop Tension Test (Xu et al. 1988) and analysed using the method described by Buthenuth et al. (1993). The determination of both compressional and shear wave velocities in different orientations, utilizing an ultrasonic wave propagation, used the experimental system similar to that described by King (1983) with pulses of shear waves polarised in one direction in addition to compressional waves. All the tests have been conducted on samples dried over phosphorus pentaoxide (P_2O_5).

Rock description and sampling strategy

The investigated rock belongs to the Antigorio geological and structural Unit, a Lower Pennine nappe of the western Alps. The samples studied came from the Moro Serizzo quarry, located about 10 km north of Domodossola – in the Ossola region, northern Italy (Fig. 1).

From Lovell, M. A. & Harvey, P. K. (eds), 1997, *Developments in Petrophysics*, Geological Society Special Publication No. 122, pp. 107–119.

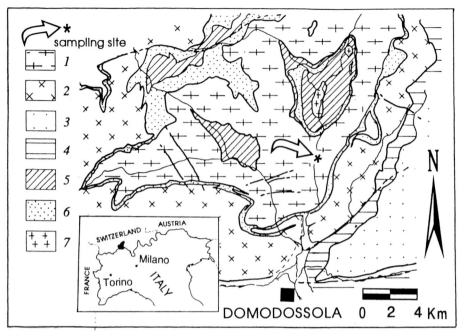

Fig. 1. Location and geological map of sampling site. 1, Antigorio Nappe; 2, Monte Leone Nappe; 3, Pioda di Crana Zone; 4, Orselina–Moncucco–Isorno Zone; 5, Lebendun Series, Baceno and Varzo Schist; 6, Mesozoic cover; 7, Verampio orthogneiss.

The rock studied is a medium grain grano-dioritic orthogneiss in the amphibolite facies of metamorphism. It possesses a strong penetrative schistosity, augen texture, and a lineation resulting from flattened and elongated minerals which give this rock its orthotropic petrographical anisotropy (Bigioggero 1977). Figure 2 illustrates this basic fabric and also shows three photomicrographs taken on the three orthogonal planes: (a) is parallel to the schistosity plane (xy plane); (b) is perpendicular to the schistosity plane and parallel to the maximum elongation of crystals (xz plane); (c) is normal to the schistosity and to the maximum elongation of the crystals (yz plane).

Blocks are quarried by drilling and light blasting, their extraction from the mass being related to the rock fabric and utilizing the long way (xz plane), the quartening way (yz plane), and the cleaving plane (yx plane), and thus to the textural characteristics described in Fig. 2. Nine rock cylinders, which were cored from adjacent *in situ* blocks, were collected from a total rock volume of about 2 m³. The cores were drilled orthogonal to each direction corresponding with the edges of the quarried blocks. Two of the reference coordinate directions, used to distinguish the samples, lie in the schistosity plane: the first (X) is parallel to the elongated

structures in the plane and the second (Y) is perpendicular to it; the third direction (Z) is normal to the schistosity plane (Fig. 2). Co-axial coring along these directions X, Y and Z, provided hollow cylinders (outer diameter *c.* 14 cm, inner diameter *c.* 6 cm, length *c.* 30 cm) together with their inner solid cores (diameter *c.* 5 cm). From each hollow cylinder two equal sets of four hoop-shape samples were cut with the following approximate thicknesses: 2, 2.7, 3.7, and 4.8 cm. The relative orientation between the hoop-shaped samples and the corresponding inner core is known. The sampling strategy is sketched in Fig. 3a and b.

All cut specimens were dried over P_2O_5 prior to testing until no further change in their weight was observed. The dry bulk density calculated from the specimen dimensions and masses lies in the range 2720 ± 10 kg m⁻³.

Ultrasonic wave velocity measurements

Apparatus and procedure

Nine cylindrical specimens were tested: three cored in each of the three orthogonal directions - in the Z-direction (specimens FA, FB and FC); in X-direction (specimens OA, OB and OC) and

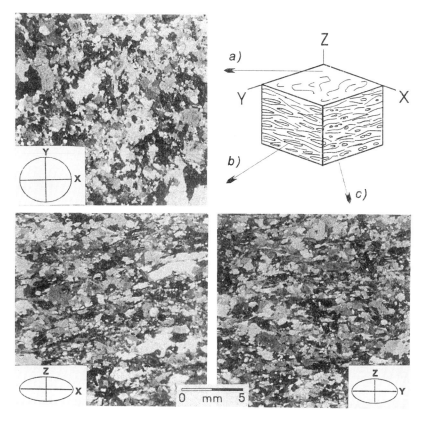

Fig. 2. Sketch of the texture of an Antigorio gneiss block with reference coordinate system shown (**a**) Photomicrograph of the schistosity plane (*xy*); (**b**) plane perpendicular to the schistosity and parallel to the elongation of crystals (*xz*); (**c**) plane perpendicular to the schistosity and to the elongation of crystals (*yz*). Strain ellipses are qualitative.

in the *Y*-direction (specimens LA, LB and LC), Fig. 3a. The top and bottom faces of each specimen were ground flat and parallel to within 0.07 mm across their diameter. The diameter for all specimens lies in the range of 49–51 mm, whilst their length lies in the range 262–300 mm. Locations, approximately 60 mm apart, were marked along each specimen, Fig. 3b. Measurements of P- and S-wave velocities, polarised either parallel or perpendicular to the specimen axis, and propagating across the diameter of the core were made at each marked plane, with propagation orthogonal to the core axis (station I II, III and IV) (Fig. 4a), the measurements being made through 360° on the plane at intervals of 22.5° (Fig. 4b).

Transmitter (T) and receiver (R) transducer holders were coupled to the side of the rock cylinder by two aluminium supports (a and b) Fig. 4b; each with its internal radius conforming to that of the rock and with a shear-wave couplant (SWC couplant, Panametrics Inc) to provide a good acoustic coupling between rock specimen and the transducers. The specimen was subjected to an axial stress of 0.5 MPa in order to stabilize the system mechanically and to provide a standard reference state of stress. Preliminary tests indicated that an increase in axial stress from zero to 6 MPa resulted in an increase of approximately 20% in the P-wave velocity in the Z-direction and an increase of less than 5% in the *X* and *Y* directions. The corresponding increases in S-wave velocity were considerably less being approximately 30% of the P-wave velocity increases. The transmitter and receiver transducers (CNS Electronics) are designed to propagate ultrasonic pulses of either compressional or polarized shear waves through the rock specimen. A Pundit (CNS Electronics) pulsing unit and switch box was used to excite pulses of either P (500–900 kHz) or S (300–700 kHz) waves.

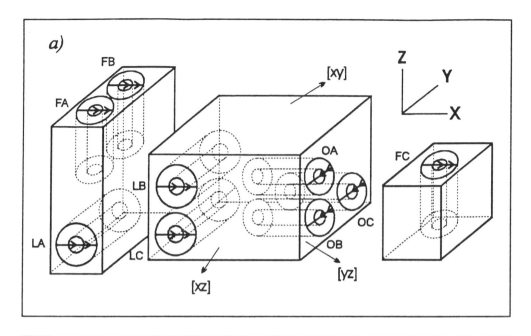

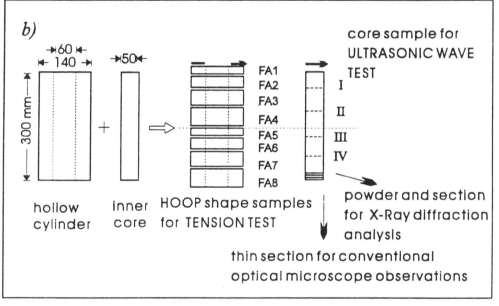

Fig. 3. Sketch of sampling strategy: (**a**) relative orientation of rock cylinders with respect to the reference coordinate system; (**b**) example of hoop sample and core relationship.

On traversing the rock specimen the pulses of mechanical energy were detected by the receiver transducer and, after amplification in a Tektronix AM502 broadband preamplifier, were fed to the Tektronix 7D20 digitizer of a Tektronix 7603 oscilloscope. The received waveform was displayed on the oscilloscope screen and recorded as digital data files on floppy disk by an Opus PCIII computer. The observed first arrival, first trough and first peak arrival times for both P and S waves were recorded. Typical P and S waveforms are shown in Fig. 5. The calculation of wave velocities follows the time-of-flight method procedure described by King (1983).

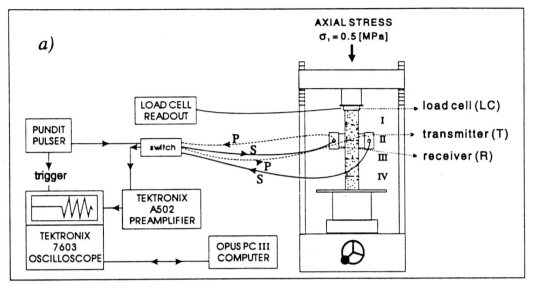

Fig. 4. (a) Block diagram of the ultrasonic wave test system; (b) plan view of transmitter (T), receiver (R), coupling pieces (a) and (b), and diametral wave travel paths at 22.5° used for measurements made around each core.

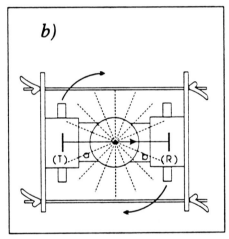

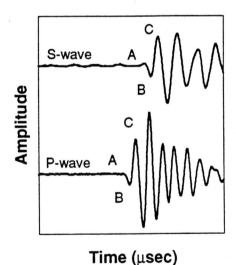

Time (μsec)

Fig. 5. Typical compressional (P) and shear (S) waveforms. A = first arrival; B = first trough; C = first positive peak.

Results

Figure 6 illustrates the radial velocities measured at the four stations along the length of one core drilled in the *Y*-direction (specimen LC): they were confirmed by measurements made on two other cores drilled in the same direction. These results demonstrated that the core was homogeneous and that the three cores in this direction had comparable geophysical properties. Similar results were obtained from the three cores oriented with their long axis in the *X*-direction, and again, in the *Z*-direction.

Table 1 lists the average values measured in a particular direction, obtained from the different stations along the length of the cores, where their long axis was in the *X*, *Y* and *Z* directions.

From Table 1 it can be seen that both V_{px} and V_{pz}, and V_{sxz} and V_{szy}, have values which are equal within the experimental errors, suggesting the orthogneiss behaves closely to that of a transversely isotropic material: it is intended to study this aspect in due course. The standard

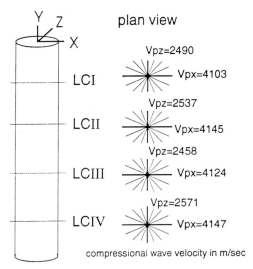

plan view

Vpz=2490

LCI   Vpx=4103

Vpz=2537

LCII   Vpx=4145

Vpz=2458

LCIII   Vpx=4124

Vpz=2571

LCIV   Vpx=4147

compressional wave velocity in m/sec

Fig. 6. Comparison between radial velocities measured at four stations along the length of one core (specimen LC).

Table 1. *Average values of compressional and polarized shear wave velocities propagating in the main directions:* X, Y, Z

Compressional and polarized shear wave velocities [m s^{-1}]
$V_px = 4209 \pm 199 \ (\pm 4.7\%)$
$V_pz = 2506 \pm 100 \ (\pm 4.0\%)$
$V_py = 3984 \pm 87 \ \ (\pm 2.2\%)$
V_{sxy} & $V_{syx} = 2634 \pm 61 \ (\pm 2.3\%)$
V_{sxz} & $V_{szx} = 2088 \pm 35 \ (\pm 1.7\%)$
V_{szy} & $V_{syz} = 2045 \pm 61 \ (\pm 2.2\%)$

deviation based on 12 values for each case was always less than 5% for the P-wave velocities and less than 3% for the S-wave velocities. Thus there is firm geophysical evidence to conclude that geophysically the samples cut from the three columns drilled in the X-direction can be compared between themselves, and likewise for the samples cut from the columns drilled in the

Y and Z directions. By analogy this implies that the tensile strength obtained from hoop-samples tested from a particular column (Fig. 3) can also be compared.

The variation of radial velocity with orientation around the axis of each core could be described by the general equation $V = a\cos^b(2\alpha + d) + c$ where the maximum and minimum values reflect the fabric being traversed by the propagating waves (Fig. 7). In most cases $d = 0$: the mirror plane passes through the X, Y and Z directions and $V_{max} \equiv V_x$, $V_{min} \equiv V_z$. Some deviation from this behaviour was observed for samples measured in the plane of schistosity (XY), where it was particularly difficult to gauge by eye the precise orientation of the X and Y directions and Fig. 7d. illustrates such an example

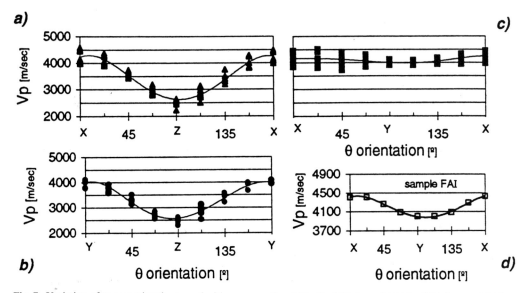

Fig. 7. Variation of compressional wave velocities propagating (**a**) in the XZ-plane, (**b**) in the YZ plane and (**c**) in the XY plane. All results obtained from core LA, LB, LC; OA, OB, OC and FA, FB, FC respectively are plotted. (**d**) Example of lack of coincidence between the orientation of maximum and minimum values and that of V_x and

where $V_{max} \equiv V(x + 10°)$ and $V_{min} \equiv V(y + 10°)$: however this presents no problems in interpreting the work because any deviation from the real X, Y or Z textural characteristics and the marker (which is also the reference line for the hoop tension test) was known and any anomaly in further results could be corrected.

The geophysical degree of anisotropy defined as $100 \times (V_{max} - V_{min})/V_{max}$ by Crampin (1989) on the three orthogonal planes is: $\Delta V_p[XZ] = 40.5\%$; $\Delta V_p[YZ] = 37.1\%$; $\Delta V_p[XY] = 5.4\%$. The variation observed with compressional wave measurements was also seen with shear wave velocity measurements. The direction of

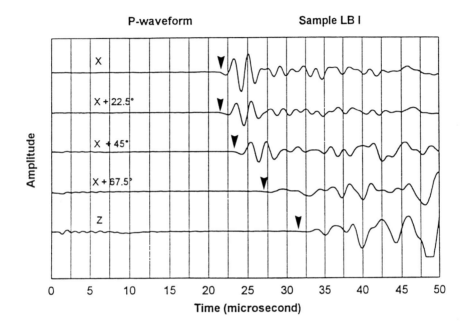

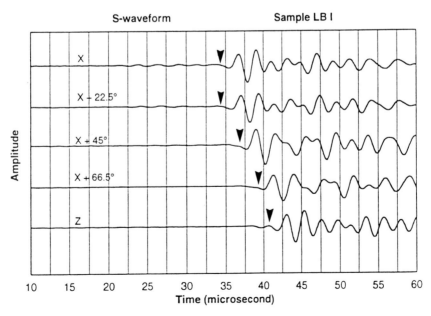

Fig. 8. Example of ultrasonic compressional (**a**) and shear (**b**) waveforms in the XZ plane (sample LBI).

maximum and minimum was the same even though the velocities differed. Figure 8 illustrates the nature of the wave trains measured radially for both P- and S-waves and is typical of the results obtained from other stations and other samples.

Hoop tension test

Apparatus and procedure

To reveal the behaviour in tension of the Antigorio gneiss the hoop tension test, introduced by Xu *et al.* (1988), has been used. Compared with the conventional direct tensile test, the hoop test procedures offer an easier method for testing anisotropic rocks in specific directions.

Figure 9 illustrates the apparatus. The loading device consists of two semi-cylindrical platens with an internal loading jack. The platens are inserted into the inner hole of the ring sample and separated along a loading line by a jack. Hydraulic pressure and jack piston position are

monitored, and recorded, with a time interval of no more than 1.2 s, by a GDS digital controller software until the rock ring sample fails. All the tests were carried out at a constant loading rate of 3.3 MPa/min in an environmental chamber containing P_2O_5. The loading directions chosen are shown in Figure 9b.

The tensile and shear stress distribution in a unidirectionally loaded hoop specimen, calculated from 2D finite-element modelling, is reported by John *et al.* (1991), Fig. 10. These authors also give an analytical solution of stress distribution based on continuum elastic mechanics supported by photoelastic studies. Maximum tensile stress is expected opposite the line of platen separation and at right angles to the direction of loading, and in the simplest case tensile failure occurs in the plane of platen opening. The movement of the platens parallel to the load line during a test is recorded by two linear variable displacement transponders mounted on the top of the platens and connected to an *XY* plotter: their main use in these tests was to register the sequence of fracturing.

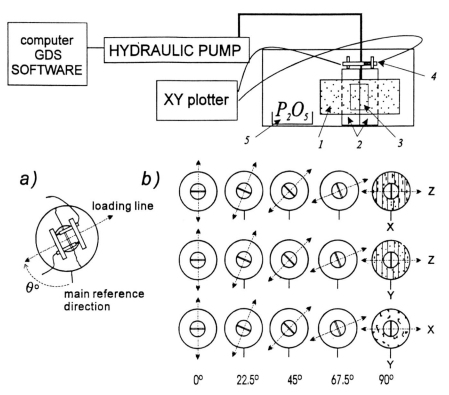

Fig. 9. Hoop test apparatus showing general experimental arrangement: 1, sample; 2, platens; 3, loading jack; 4, linear variable displacement transponders; 5, drying agent. (**a**) Plan view of sample set up. (**b**) Testing strategy (four samples in each of the directions 0° to 90° were tested).

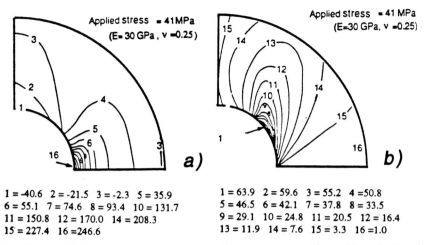

1 = -40.6 2 = -21.5 3 = -2.3 5 = 35.9
6 = 55.1 7 = 74.6 8 = 93.4 10 = 131.7
11 = 150.8 12 = 170.0 14 = 208.3
15 = 227.4 16 = 246.6

1 = 63.9 2 = 59.6 3 = 55.2 4 = 50.8
5 = 46.5 6 = 42.1 7 = 37.8 8 = 33.5
9 = 29.1 10 = 24.8 11 = 20.5 12 = 16.4
13 = 11.9 14 = 7.6 15 = 3.3 16 = 1.0

Fig. 10. Distribution of (**a**) principal stress in a hoop sample (tension expressed by positive values), and (**b**) shear stress distribution, from finite element analysis; values shown for contours 1 to 16 are in MPa (from John *et al.* 1991).

Pump pressure can be converted into the force applied to the piston within the loading jack and the volume change of the pump reservoir, into piston displacement. Maximum force, maximum displacement and total time up to failure were recorded and plotted as in Fig. 11. The ultimate force at failure (F) was plotted versus the failure area (A) for suites of specimens of differing area, and the tensile strength in any loading direction defined by the ratio $\sigma_t = \Delta F/\Delta A$ (Butenuth *et al.* 1993) as illustrated in the results described below.

Results

A total number of sixty samples were tested, all dried over P_2O_5. Their failure could be grouped into five categories according to the number and orientation of failure surfaces generated; Fig. 12

illustrates these classes. To calculate the tensile strength (σ_t) of a hoop it is necessary to plot the force at failure against the area at the time this force is obtained (Butenuth *et al.* 1993).The rock being strongly anisotropic, exhibited different

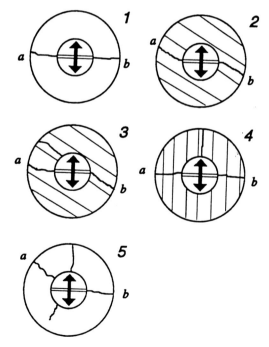

Fig. 12. Schematic presentation of different failure patterns for hoop samples tested by the Hoop Tension Test. 1, 2, 3, 4, 5 are the defined classes, *a* and *b* the surface areas used to calculate tensile strength.

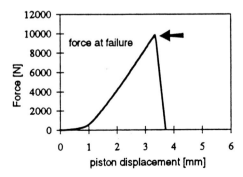

Fig. 11. Example of force and piston displacement recorded during a Hoop Tension Test.

patterns of failure in different directions and it was not always clear how some of these patterns should be analysed in terms of the area of failure at Fmax. Inspecting the samples it was decided to plot the failure area (designated by *a* and *b* in Fig. 12) according to the category of failure pattern. The quality of the data available are summarized in Table 2 together with values of

Table 2. *Summary of Hoop Tension Test results.*

Sample	Axis	Load line	*Pattern of failures no. 1	2	3	4	5	$\Delta F/\Delta A$ [MPa]	†
FA 5-8	Z	Y	✓					2.83	A
FB 5-8	Z	Y+22.5° X	✓					2.72	A
FB 1-4	Z	Y+45° X	✓					2.75	A
FC 1-4	Z	Y+67.5° X	✓					3.29	A
FA 1-4	Z	X	✓					2.56	A
OA 1-3	X	Y				✓		} 2.57	C
OA 4	X	Y					✓		B
OB 1-4	X	Y+22.5° Z			✓			1.48	B
OC 1	X	Y+45° Z				✓		} 1.31	A
OC 2-4	X	Y+45° Z		✓					A
OC 5	X	Y+67.5° Z				✓		} 1.23	A
OC 6-8	X	Y+67.5° Z		✓					A
OA 5-8	X	Z	✓					1.19	A
LA 1-2-4	Y	X				✓		} 2.19	A
LA 3	Y	X					✓		C
LB 5-8	Y	X+22.5° Z			✓			1.43	B
LB 1-2-4	Y	X+45° Z				✓		} 1.18	B
LB 3	Y	X+45° Z		✓					A
LC 1-4	Y	X+67.5° Z		✓				} 1.26	A
LC 3	Y	X+67.5° Z				✓			A
LA 5-8	Y	Z	✓					1.14	A

*According to Fig. 12.
†Data reliability: A, no alternative; B, one alternative; C, more than one.

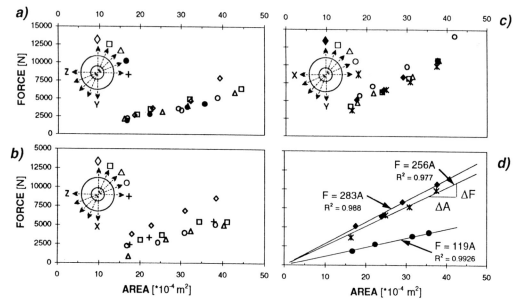

Fig. 13. Force at failure versus failure area loading at 22.5 rotations (**a**) in the *YZ* plane, (**b**) in *XZ* plane and (**c**) in the *YX* plane. (**d**) Example of tensile strength calculations.

tensile strength calculated. Figure 13 illustrates the force versus area diagrams obtained for samples loaded in the chosen directions from Y to Z (a), from X to Z (b) and from Y to X (c), and also illustrates three examples of the data used for calculating σ_t (d).

Comparison of results

To compare the ultrasonic wave velocity values in a known direction with the values of tensile strength for extension in that direction it is convenient to divide the results into two classes:

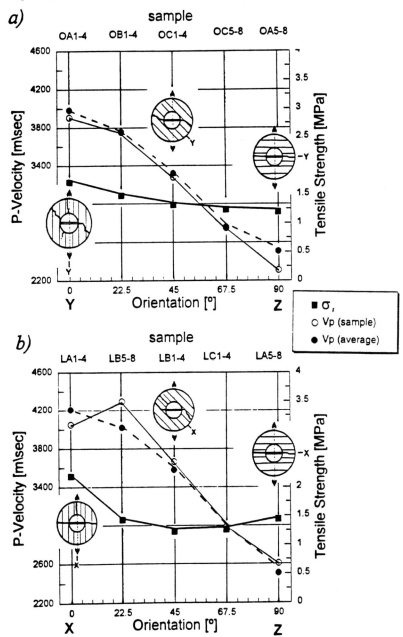

Fig. 14. Comparison between P-wave velocity (V_p) and tensile strength (σ_t). (**a**) Loading line and wave travel path at intervals of 22.5 from Y to Z and (**b**) from X to Z. Black squares are σ_t; black circles are V_p as average of all the measurements in a particular direction; empty circles are V_p calculated from the corresponding set of samples tested in extension.

measurements on the planes orthogonal to the schistosity plane (i.e. loading directions and wave travel paths from Y to Z and from X to Z) and those on the schistosity plane (i.e. loading directions and wave travel paths from Y to X).

Measurements in planes at 90° to schistosity

Figure 14a illustrates the results obtained in the YZ orientations, θ being the angle between the loading line, or wave travel path, and the reference directions. It can be observed that there is a corresponding decrease in tensile strength with the decrease in P-wave velocities. This is consistent with the relationship that both these measurements can be expected to have with the fabric of the rock (Fig. 2). The P-wave velocity decreases as the angle between its direction of measurement and the plane of schistosity increases, being 90° at Z. Likewise, the tensile strength is at its maximum in the direction 90° to the plane of schistosity (i.e. at Y) and reduces to its minimum in the plane of schistosity, at Z.

Figure 14b illustrates the results obtained in the XZ orientations and suggests that caution must be exercised when relating geophysical measurements in anisotropic material to other mechanical properties. The highest P-wave velocities are measured in the same orientation as that for the highest tensile strength but the same is not true for the lowest tensile strength. There is not a simple relationship between P-wave velocities and tensile strength, the minimum of both having an angular difference with respect to the orientation of fabric.

Measurements in the schistosity plane

The results obtained in the YX orientations, which are summarized in Fig. 15, reveal that tensile strength shadows P-wave velocity. However, even though these measurements were made in the plane of schistosity they prompt a note of caution. The average of fourteen measurements of P-wave velocities shows a rise in velocity as the orientation of measurements passes from direction Y to direction X. However, the tensile strength derived from $\Delta F / \Delta A$ obtained by testing four samples in each of the directions in which P-wave velocity was measured, (as illustrated in Fig. 13), reveals a rather different trend. Investigation of the raw data confirmed that the P-wave velocity measurements of those samples from which the measurements of strength were eventually obtained did have a consistent relationship with strength. The

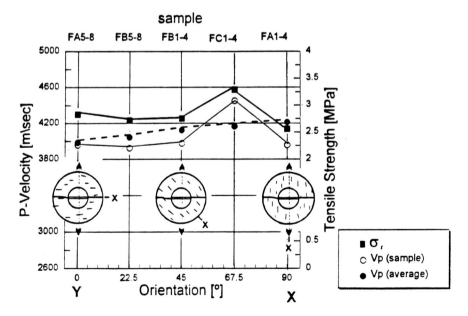

Fig. 15. Comparison between P-wave velocity (V_p) and tensile strength (σ_t). Loading line and wave travel path at intervals of 22.5° from Y to Z. Black squares are σ_t; black circles are V_p as average of all the measurements in a particular direction; empty circles are V_p calculated from the corresponding set of samples tested in extension.

lesson is clear; care has to be taken when working in the plane of anisotropy.

Conclusion

Although transversal anisotropic material lends itself well to characterization by geophysical measurements great care has to be taken when translating these measurements to a prediction of strength. The relationships revealed here appear to be simple, but there are basic questions which still cannot be answered; for example, would the relationships be the same if the fabric remained unchanged but the mineralogy changed; would the relationships remain unchanged if the mineralogy and fabric remained unchanged but the grain size changed? And what happens to the relationships when pore fluids are added, and changed? Such subjects can in part be studied with the aid of numerical simulations and the experimental data presented here provides a constraint for such models of material having the same characteristics as the Antigorio orthogneiss. Further work has to be completed before research can provide the quarrying industry with the 'simple' test it requires for assessing the strength of anisotropic rock in a non-destructive way.

The most straight forward relationships between P-wave velocities and tensile strength for the gneiss tested here were those obtained in the plane at 90° to the schistosity and maximum elongation of crystals.

The least straight forward relationship lay in the plane at 90° to the schistosity and parallel to the maximum elongation of crystals: the *XZ* plane.

Care must be taken even when working in the plane of transverse anisotropy when that plane itself contains an orientation to its fabric which can impart subtle differences to strength in the plane.

This research was made possible by a scholarship from University of Studies of Ferrara (Engineering Geology Department). The authors are grateful to P. S. Colback for his advice and encouragement, to the colleagues of the Engineering Geology Group for their helpful discussions and to the meticulous work of the technicians of Imperial College.

References

ATKINSON, B. K. 1987. *Fracture Mechanics of Rock*. Academic Press, London.

BIGIOGGERO, B., BORIANI, A. & GIOBBI, E. 1977. Microstructure and mineralogy of an orthogneiss (Antigorio gneiss-Lepontine Alps). *Rendiconti Società Italiana di Mineralogia e Petrologia*, **33**, 99–108.

BUTENUTH, C., DE FREITAS, M. H., AL-SAMAHIJI, D., PARK, H-D., COSGROVE, J. W. & SCHETELIG, K. 1993. Observation on measurement of tensile strength using the hoop test. *International Journal of Rock Mechanics Mining Sciences & Geomechanics Abstracts*, **30**, 157–162.

CRAMPIN, S. 1989. Suggestion for a consistent terminology for seismic anisotropy. *Geophysical Prospecting*, **37**, 753–770.

JOHN, S. J., AL-SAMAHIJI, D., DE FREITAS, M. H., COSGROVE, J. W., CLARKE, B. A., LOE, N. & TANG, H. 1991. Stress analysis of a unidirectionally loaded hoop specimen. *Proceeding 7th International Congress on Rock Mechanics. Aachen, Germany*, 513–518

KING, M. S. 1983. Static and dynamic properties of roks from the Canadian shield. *International Journal of the Rock Mechanics Mining Sciences & Geomechanics Abstracts*, **20**, 237–241.

LAMA, R. D. & VUTUKURI, V. S. 1978. *Handbook on Mechanical Properties of Rocks*, II. Transtech Publications.

TELFORD, W. M. 1963. *Applied Geophysics*. Cambridge University Press.

XU, S., DE FREITAS, M. H. & CLARKE B. 1988. The measurement of tensile strength of rock. *Proceeding of the International Society of Rock Mechanics. Symposium, Rock Mechanics and Power Plants, Madrid*, 125–132.

Prediction of petrophysical properties from seismic quality factor measurements

C. McCANN, J. SOTHCOTT & S. B. ASSEFA

Postgraduate Research Institute for Sedimentology, The University of Reading,
Whiteknights, Reading, Berkshire RG6 6AB, UK

Abstract: Over the past eight years the Geophysics Group of the Postgraduate Research Institute for Sedimentology has developed equipment for accurately measuring the seismic properties of centimetre-size samples of sedimentary rocks at ultrasonic frequencies (0.5–1 MHz) and under simulated *in situ* conditions. Relationships between the ultrasonic properties and the mineralogical/petrophysical properties of a wide range of rock types have been investigated. The results indicate that the attenuation of seismic waves at high frequencies is dominated by the 'imperfections' in the rock, defined in a broad way. For instance, at low pressures the attenuation depends on the volume concentration of microcracks; at high confining pressures the attenuation depends on the distribution of porosity between the macro inter-particle porosity and the micro intra-clay or intra-particle porosity. For some rocks, analysis of these relationships leads to the ability to predict petrophysical properties, including the permeability, from the seismic data. Techniques for up-scaling the relationships to the wavelengths of field seismic surveys are discussed.

Significant improvements in the accuracy and stability of the seismic imaging of subsurface sedimentary sequences in the past decade have resulted from new acquisition techniques such as 3D surface seismic, vertical seismic profiling with multi-element geophone arrays, amplitude-versus-offset, multi-geophone full waveform sonic logging and improved processing capability arising from increased computer power and dynamic range (Yalch 1988). The improved imaging techniques are leading to the increased frequency bandwidth necessary for the routine measurement of the energy loss of seismic waves. This can be expressed as the attenuation per unit distance (dB m^{-1}), the attenuation per wavelength (dB λ^{-1}) or as the inverse fractional energy loss per cycle, the quality factor Q. Q is widely used to express the ability of a rock to propagate seismic energy. A rock which is poor propagator of seismic energy would have $Q < 25$; a rock which is a good propagator would have $Q > 100$. Other measures of seismic loss which are routinely quoted are Q^{-1} and $(1000/Q)$ which are related in an obvious way to Q.

There are increasing numbers of papers reporting significant variations of attenuation with lithology and with subsurface conditions. Rapoport *et al.* (1994) and Ryjkov & Rapoport (1994) demonstrated high values of attenuation in productive hydrocarbon reservoirs at a frequency of about 50 Hz from surface seismic and VSP data respectively. In the unproductive zones the quality factors were consistently greater than 100 whereas in the productive zones the quality factors dropped to values between 35 and 60. Hauge (1981) demonstrated that seismic attenuation increased with sand/clay ratio in shales at a frequency of about 50 Hz from VSP data; Hauge's results are in good agreement with those of McDonal *et al.* (1958) from the Pierre Shale, Fig. 2. Marks *et al.* (1992) measured attenuation using VSP data at 50 Hz and correlated the observed loss per wavelength with lithology. Varela, Rosa & Ulrych (1993) showed how the subsurface image obtained from a seismic survey was improved by incorporating a quality factor correction. Klimentos (1995) established a relationship between compressional wave quality factor and the type of fluid (brine, oil, gas) in the pores of a sandstone reservoir from full-waveform sonic log data. Link & MacDonald (1993) measured attenuation at depths between 150 m and 215 m using high-frequency cross-hole seismic data and demonstrated high quality factors in shales and low quality factors in sands. Goldberg *et al.* (1985) measured attenuation from sonic log data at about 10 kHz on ocean floor sediments and demonstrated increasing attenuation with increasing compaction and the development of diagenetic minerals. Tang & Strrack (1995) used waveform inversion of borehole compressional wave logs to obtain stable values of compressional wave attenuation from borehole compressional wave logs. Their data showed that the quality factors of clean sandstones are high compared to those of shaley sandstones at sonic log frequencies.

The significant improvement in the stability and accuracy of seismic images of reservoirs

From Lovell, M. A. & Harvey, P. K. (eds), 1997, *Developments in Petrophysics*, Geological Society Special Publication No. 122, pp. 121–130.

arising from new acquisition and processing techniques has led to increasing interest in the inversion of seismic data into petrophysical properties. Equation 1 from Han *et al.* (1986) demonstrates the significant relationship between seismic compressional-wave velocity and porosity, and a weaker relationship with clay content.

$$V_p = 5.59 - 6.93\phi - 2.18V_{cl} \qquad (1)$$

where: V_p is the compressional wave velocity in km s^{-1}, ϕ is the porosity in %, V_{cl} is the volume percentage of clay minerals.

Given the strong relationship between velocity–porosity–clay content, why is velocity so poorly related to permeability, when a reasonable correlation often exists between porosity and permeability? Permeability increase corresponds to porosity increase (velocity decrease) or to clay decrease (velocity increase). Those changes of porosity and clay content which could increase the permeability would cause exactly opposite velocity effects. The importance of a geophysical diagnostic assessment of permeability, and its lack of relationship with seismic velocity, encourages the investigation of possible relationships between quality factor and permeability.

In order to try to understand the variation of attenuation with petrophysical properties in rocks, it has to be measured in controlled experiments in which accurate seismic data are complemented by detailed mineralogical and petrophysical analyses of the rocks. This can only be done in a systematic way by measurements on core. Ultrasonic measurements at 0.5 MHz are not fully satisfactory because of the two order-of-magnitude to five order-of-magnitude frequency difference from sonic logs and seismic surveys. These frequency differences inevitably mean that the relaxation times which control the attenuation mechanisms are very different. Nevertheless ultrasonic measurements of loss in conjunction with careful sedimentological, petrophysical and mineralogical analyses point out the types of loss which are important and give a guide towards an understanding of the low frequency mechanisms. The lack of good sedimentological and mineralogical control in many of the older sets of measurements meant that many of the associated attenuation measurements were of limited value.

The Geophysics Group of the Postgraduate Research Institute for Sedimentology, Reading University has devoted considerable effort to the development of techniques for measuring compressional-wave and shear-wave velocity and attenuation in reservoir rocks under realistic reservoir physical conditions. The seismic data have been complemented by full analyses of the mineralogical and petrophysical properties of the rocks. The results of the programme of research have been published in theses and in the open literature. This paper summarises and draws conclusions from some of these data about the basic principles of seismic attenuation in reservoir rocks.

Measurements of ultrasonic attenuation in reservoir rocks

Our principal measurement technique to date has been the reflection system originally devised by Winkler & Plona (1982). This method uses a cylindrical sample 5 cm in diameter and about 2 cm long with ends accurately surface-ground flat and parallel (Fig. 1). The sample is sandwiched between polymer buffer rods inside

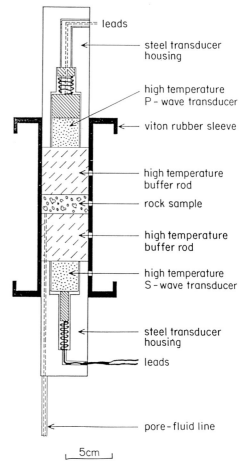

Fig. 1. Arrangement of the sample and buffer rods in the ultrasonic reflection system.

a hydraulic pressure cell. A transducer in transmit/receive mode generates a pulse at the top of the buffer rod. The travel times and amplitudes of the reflections from the top and base of the sample, after correction for the diffraction effects of the transducers and the impedance contrast with the buffer rods, are compared to give the velocity and attenuation of the sample. The system is equipped with a compressional-wave transducer transmitting into one buffer rod and a shear-wave transducer into the other. In this way both compressional and shear wave measurements can be made on the sample without removing it from the pressure cell. The system operates over a frequency range from 0.4 MHz to 1 MHz, with confining pressure up to 70 MPa. Any type of pore fluid compatible with nitrile rubber and stainless steel can be used in the system and the pore fluid pressure can be varied independently of the confining pressure. A number of modifications to the system have been developed and implemented. For example, a dual shear-wave transducer with orthogonal-polarized elements has been incorporated which can be rotated on the buffer rod surface whilst the sample is under pressure, to enable the velocity and attenuation of 'split' shear waves to be measured. Data acquired using this system have been reported by Peacock et al. (1994).

The principles underlying this system are simple, but the method is capable of giving very accurate measurements as demonstrated by McCann & Sothcott (1990). The system relies on there being welded contact between the polymer buffer rods and the flat faces of the specimen. This having been achieved, the system is 'self calibrating' in the sense that the reflected pulses from the top and from the base of the sample are treated identically. The interface between the transducer and the polymer buffer rod, the transducer itself and the electronics affect both pulses equally. The accuracy of routine measurements achieved with the equipment is ±0.3% in velocity and ±0.1 dB cm^{-1} in attenuation.

Attenuation versus effective pressure

Seismic attenuation varies significantly with the effective pressure, that is 'confining pressure minus pore fluid pressure'. For example, Toksoz et al. (1978) measured quality factors in brine saturated Berea sandstone. The compressional wave quality factor increased from about 20 at 5 MPa to nearly 80 at 60 MPa, whilst the shear wave quality factor changed from about 10 to 35 over the same pressure interval. Jones (1995) confirms that many sandstones show similar behaviour with the quality factors increasing from low values at room pressure to a relatively constant 'terminal' value at effective pressures greater than about 40 MPa. It has been qualitatively surmized since the work of Birch & Bancroft (1938a, b) that the increase in quality factor with pressure is due to the closure of micro-cracks. Peacock et al. (1994) demonstrated quantitatively for Carrara Marble that the compressional- and shear-wave quality factors were related to the number density of the micro-cracks as measured on photomicrographs of polished thin sections taken from acoustic samples.

'Terminal' attenuation at high pressure: sandstones and shales

At pressures where the cracks are closed the loss mechanism at ultrasonic frequencies is related to the distribution of porosity within the rock. Many reservoir rocks contain fine-grained, micro-porosity minerals in the pores between the main particles making up the framework of the rock. During burial, diagenetic processes can cause the dissolution of macro-particles and the subsequent re-precipitation of micro-porosity, fine-grained minerals. A typical process is the dissolution of feldspar into kaolinite clay, described by Giles & de Boer (1990). Klimentos & McCann (1990) give scanning electron micrographs of a clean sandstone (their fig. 5) and of a quartz sandstone of which the pores have been filled with a fine-grained, micro-porosity kaolinite clay by diagenetic processes (their fig. 4).

Figure 2 demonstrates the variation of compressional-wave quality factor with the change in the volume percentage of pore filling, fine-grained minerals. The data are from Best et al. (1994), Hauge (1981) and McDonal et al. (1958). The Best et al. data are laboratory ultrasonic measurements on sandstones under 60 MPa effective pressure. The fine-grained porefill in these sandstones includes clays, micritic calcite and sparry calcite cements. The data show that the compressional wave quality factor is high in clean sandstones. It decreases as the volume of micro-porosity porefill increases and then increases again as the porefill becomes the dominant mineral assemblage forming the rock, that is the rock changes from a shaly sandstone to a shale. It should be noted that, although the data are still sparse, the quality factors appear to reach a minimum at the 50%

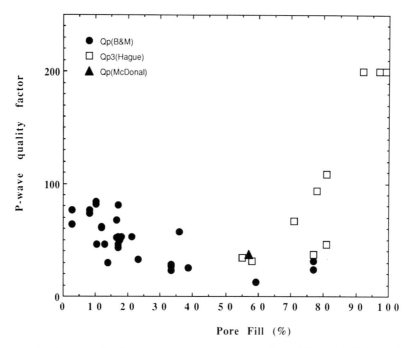

Fig. 2. Compressional wave quality factor versus volume percentage of pore filling minerals for sandstones. Data from Best *et al.* (1994) (filled circles), Hauge (1981) (open squares) and McDonal *et al.* (1958) (filled triangle).

porefill value. There are very few measurements of the acoustic properties of sandy shales at ultrasonic frequencies in the laboratory as Ëthe samples are very weak and tend to disintegrate very easily. Figure 2 therefore also includes field measurements of compressional-wave quality factors from Hauge (1981) and from McDonal *et al.* (1958). The shales investigated by Hauge were under an average effective stress of 16 Mpa ± 2 MPa. The Pierre Shale investigated by McDonal *et al.* was under an average effective stress of 1.5 MPa. The measurement frequency in both cases is between 50 Hz and 100 Hz.

The measurement frequencies of the laboratory and the field data plotted in Fig. 2 differ by four orders of magnitude. There is a significant difference in the effective stress on the rocks in the three data sets. Nevertheless, the three data sets consistently indicate an increase of the compressional-wave quality factor as the rock changes from a shaly sandstone to a pure shale, or conversely a decrease in quality factor as the sand content of the shale is increased.

The decrease in compressional-wave quality factor of sandstones with increasing clay content was first demonstrated by Klimentos & McCann (1990) from ultrasonic measurements on rocks

at a confining pressure of 40 MPa. Their quality factor–clay relationship has been confirmed by Tutuncu *et al.* (1994) for tight gas sandstones measured at ultrasonic frequencies in the laboratory. Tang & Strrack (1995) have demonstrated a significant difference between the quality factors of clean and shaley sandstones from sonic log data.

'Terminal' attenuation at high pressure: limestones

The decrease of quality factor due to the variation of the scale of porosity in reservoir rocks is demonstrated in the following previously unpublished data for Oolitic limestones from Assefa (1994). Figure 3 is a scanning electron microscope photograph of a calcite cemented oolitic limestone showing the microporosity of the ooiods (the intra-granular porosity) compared to the macro-porosity of the sparry calcite cement (the inter-granular porosity). Figure 4 shows compressional-wave quality factor plotted against the ratio of intergranular porosity to intra-granular porosity. The dependence of the quality factor on the combination of these two types of porosity is

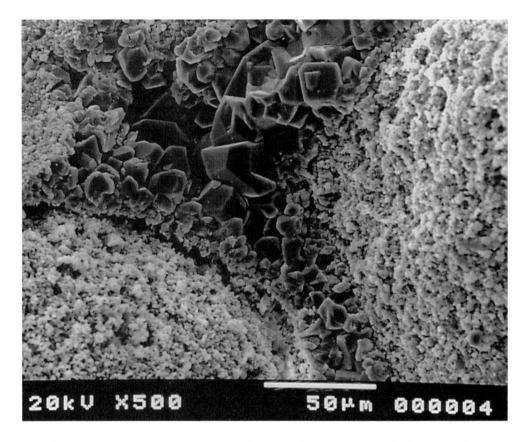

Fig. 3. Scanning electron microscope photograph of an oolitic limestone. Scale bar is 50 μm. Reproduced with permission from Assefa (1994).

very clear, attaining a minimum for rocks with equal percentages of intra-granular and inter-granular porosity.

Discussion of attenuation in sedimentary rocks

Tittman (1977) and Tittman et al. (1985) showed that the attenuation of seismic waves in vacuum-dry rocks is negligible. It is the viscous interaction of the pore fluids with the solid framework which causes attenuation. At ultrasonic frequencies 'perfect' fluid-saturated rocks are excellent propagators of seismic energy. Despite the potential for seismic loss due to the viscous interaction of the pore fluid and the solid (the Biot 1956 'global flow' loss), this only becomes significant in very high porosity, very high-permeability rocks (Klimentos & McCann 1990). In most rocks, attenuation starts to become significant due to the presence of discontinuities or disruptions in

the matrix, whether in the form of cracks or micro-porosity. These cause an increase in the attenuation and changes in the petrophysical properties. The disruptive components can take several different forms. At low pressures the third component is the distribution of micro-porosity cracks (Peacock et al. 1994). At high pressures, where the fractures are closed, micro-porosity minerals filling the pores of sandstones (Best et al. 1994) form the third component. Similarly Assefa (1994) shows that in limestones at high pressures, macro-porous calcite filling the interooid pores of micro-porous ooiods form the third component of the rock which cause high seismic attenuation. The third component may alternatively be in the fluid phase, with gas partially replacing the pore liquid (Frisillo & Stewart 1980).

To summarize, the chaotic elements of rocks cause seismic attenuation. Qualitatively, it is apparent that the same variability in the structures, compositions and porosity distributions of sedimentary rocks which cause them to

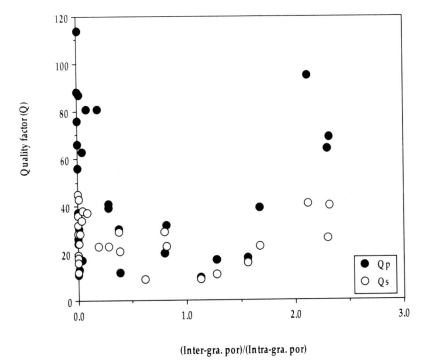

Fig. 4. Compressional-wave quality factor and shear wave quality factor versus the ratio of *inter*-granular porosity to *intra*-granular porosity for oolitic limestones. Reproduced with permission from Assefa (1994).

have significant seismic attenuation may also cause variations in their petrophysical properties. For some rocks, the common origin of these variations enables predictive relationships to be established.

Velocity–quality factor–permeability plot for shaly sandstones

Figure 5 is a cross-plot (after McCann 1994) of compressional-wave quality factor versus compressional-wave velocity for shales and shaly sandstones based on data from Best, McCann & Sothcott (1994), Hauge (1981) and Klimentos & McCann (1990). The rocks with acoustic data plotted in the shaded area, have an average permeability greater than 100 mD. The rocks plotted in the unshaded area have an average permeability of less than 10 mD. The velocity–quality factor cross plot enables the permeability of a rock to be estimated (crudely) from its seismic parameters. The basis of the crossplot and an interpretation of the relationships is given below.

Shales have low permeabilities because their pores are of small diameter. Their significant porosities lead to low velocities. They exhibit

the variable quality factors determined by variable sand content as shown in Fig. 2. They plot in the low permeability region on the left-hand side of Fig. 5. Low-porosity sandstones exhibit high velocities and high quality factors. Both result from the small amount of pore fluid available to interact with the solid framework. These sandstones are of low-permeability because there are few pores to transmit the fluid. These sandstones plot in the low-permeability region on the right hand side of Fig. 5. Medium-porosity, porefill-rich sandstones plot in the medium velocity region of the graph; these rocks also plot in the low quality factor region of the graph, as previously demonstrated in Fig. 2. These rocks have low permeability due to the high percentage of pore-filling clays and other fine-grained minerals. Clean, medium-porosity and high-porosity sandstones plot in the medium-velocity, high quality factor region of the graph. These are the only sandstones with high permeabilities.

Figure 5 appears to predict, at least in a crude way, the range of the permeability of a shaly sandstone from the values of its compressional wave velocity and compressional wave quality factor at ultrasonic frequency.

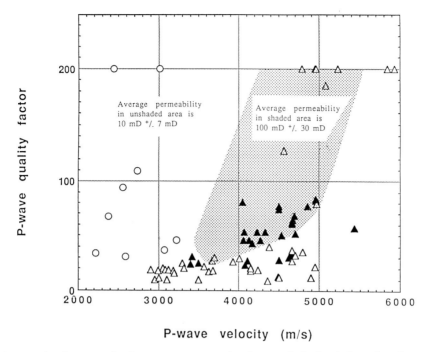

Fig. 5. Compressional wave quality factor versus compressional wave velocity for shales and sandstones. Figure reproduced from McCann (1984) by kind permission of the European Association of Geoscientists and Engineers. Data from Best *et al.* (1994) (solid triangles), Hauge (1991) (open circles) and Klimentos & McCann (1994) (open triangles). The stippled area is occupied by velocity and quality factor data for sandstones with an average permeability greater than 100 mD. The blank area is occupied by data for sandstones and shales with average permeabilities less than 10 mD.

Velocity–quality factor–permeability plot for oolitic limestones

Figure 6 shows a similar plot of compressional wave quality factor versus compressional wave velocity for oolitic limestones (the Great Oolite). The permeabilities of these rocks are very low. The filled squares indicate the acoustic data for the four rocks with permeabilities greater than 1 mD, open squares are rocks with permeabilities less than 1 mD. In contrast to the sandstones, the 'high'-permeability rocks have a wide range of velocities and low quality factors. The low-permeability rocks exhibit a wide range of velocities and quality factors, overlapping with the 'high' permeability rock data. The overlap of the data for the different permeability populations prohibits the possibility of using Fig. 6 to predict permeability from a velocity–quality factor plot. It is interesting to note that the average quality factors of the two permeability populations are significantly different. The 'high'-permeability limestones have an average quality factor of 15 ± 3 (standard error) whilst the low-permeability rocks have

an average quality factor of 35 ± 10. The only significant statement which could be made to reservoir engineers on the basis of this dataset is that a 'high'-permeability limestone will not exhibit a high quality factor. Measurement of high quality factor in a limestone sample therefore eliminates the possibility of it having significant permeability. The data set has to be extended to include more 'high'-permeability limestones to confirm this statement. The physical mechanism leading to low quality factors in limestones is the fluid flow between the inter-particle porosity and the intra-particle porosity as shown in Fig. 4.

Velocity–quality factor–permeability cross plots: discussion

The conclusion to be drawn from Figs 5 and 6 is that the permeability of a sandstone can be crudely predicted from acoustic data whilst the permeability of a limestone cannot. This result arises from the different mechanisms of acoustic attenuation in the two rocks; in the sandstone

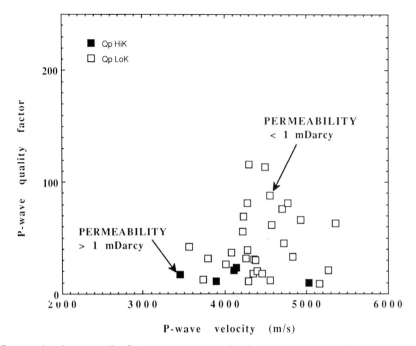

Fig. 6. Compressional-wave quality factor versus compressional-wave velocity for oolitic limestones (the Great Oolite). The filled squares indicate rocks with permeabilities greater than 1 mD. The unfilled squares indicate rocks with permeabilities less than 1 mD.

the pore-filling clays cause both high attenuation and low permeability whereas in the limestones the intra-particle porosity/inter-particle porosity ratio is related to the attenuation but not in a unique way to the permeability. These two data sets clearly demonstrate the importance of careful geological analysis of the samples and an understanding of the mechanisms of seismic loss to complement the ultrasonic measurements. A cross plot of quality factor versus velocity for the combined sandstone and limestone datasets would effectively disguise the significant sandstone relationship. The ultrasonic or seismic data have to be complemented by careful studies of the mineralogy and petrology of the rock suite. Vernik & Nur (1992) demonstrated this point very elegantly in their study of the relationships of compressional wave velocity to lithology and porosity of sandstones.

The success of the quality factor–velocity cross plot for sandstones demonstrates that it is worth investigating similar crossplots for other sedimentary rocks which are important as potential reservoirs. Do these mechanisms (or their equivalents) and the potential petrophysical relationships, apply at the lower frequencies of seismic and sonic surveys? The ultrasonic data show that attenuation is due to the presence of

discontinuities or disruptions in the rock matrix, whether in the form of cracks or micro-porosity. It is the variation of the scale of these disruptions which determine the variation of the attenuation with frequency and also the variation of the petrophysical properties with scale in the reservoir. The VSP, surface seismic and sonic log data of Rykov & Rapoport (1994), Hauge (1981), McDonal et al. (1958), Marks et al. (1992) and Klimentos (1995) show that quality factors, measured at frequencies in the range 50–10 kHz, vary significantly with lithology. The relationship between quality factor and clay content for sandstones observed by Best, McCann & Sothcott (1994) in the laboratory at 1 MHz has been qualitatively confirmed at 10 kHz sonic log frequencies by Tang & Strrack (1995) to the extent that they observed high quality factors in clean sandstones and lower quality factors in shaly sandstones. It is vital that further accurate field measurements of quality factor are undertaken to confirm and extend these data. Low-frequency measurements of quality factor on rocks in the laboratory, complemented by carefully measurements of the mineralogical and petrophysical properties of the samples are also necessary. The Geophysics Group at Reading has constructed a resonant bar equipment that will enable elastic and anelastic measurements to be made on rock samples at frequencies of 1–60 kHz, under simulated *in-situ* conditions of temperature and pressure. This equipment is now operational and undergoing calibration tests. It has been built as part of a research contract with considerable technical advice from past and present employees of Phillips Petroleum, who created a similar device in the early 1980s. We hope shortly to be able to report measurements on a range of rock types to complement our ultrasonic data.

Contribution No. 508 of the Postgraduate Research for Sedimentology, The University of Reading.

References

ASSEFA, S. B. 1994. *Seismic and petrophysical properties of carbonate reservoir rocks*. Ph.D. Thesis, University of Reading.

BEST, A. I., MCCANN, C. & SOTHCOTT, J. 1994. The relationships between the velocities, attenuations and petrophysical properties of reservoir sedimentary rocks. *Geophysical Prospecting*, **42**, 151–178.

BIOT, M. A. 1956. Theory of propagation of elastic waves in a fluid saturated porous solid: 2. Higher frequency range. *Journal of the Acoustical Society of America*, **28**, 179–191.

BIRCH, F. & BANCROFT, D. 1938a. The effect of pressure on the rigidity of rocks. I. *Journal of Geology*, **46**, 59–87.

—— & ——1938b. The effect of pressure on the rigidity of rocks. II. *Journal of Geology*, **46**, 113–141.

FRISILLO, A. L. & STEWART, T. J. 1980. Effect of partial gas/brine saturation on ultrasonic absorption in sandstone. *Journal of Geophysical Research*, **85**, 5209–5211.

GILES, M. R. & DE BOER, R. B. 1990. Origin and significance of redistributional secondary porosity. *Marine and Petroleum Geology*, 7, 378–397.

GOLDBERG, D., MOOS, D. & ANDERSON, R. N. 1985. Attenuation changes due to diagenesis in marine sediments. *SPWLA 26th Annual Logging Symposium*.

HAN, D., NUR, A. & MORGAN, F. D. 1986. Effects of porosity and clay content on wave velocities in sandstones. *Geophysics*, **51**, 2093–2107.

HAUGE, P. S. 1981. Measurements of attenuation from vertical seismic profiles. *Geophysics*, **46**, 1548–1558.

JONES, S. 1995. Velocity and quality factors of sedimentary rocks at low and high effective pressures. *Geophysical Journal International*, **123**, 774–780.

KLIMENTOS, T. 1995. Attenuation of P- and S-waves as a method of distinguishing gas and condensate from oil and water. *Geophysics*, **60**, 447–458.

—— & MCCANN, C. 1990. Relationships between compressional wave attenuation, porosity, clay content and permeability of sandstones. *Geophysics*, **55**, 998–1014.

LINK, C. A. & MACDONALD, J. A. 1993. Attenuation measurements using high frequency cross-hole seismic data. *Technical Abstracts of the 63rd Annual Meeting of the SEG, Washington DC*, 37–40.

MCCANN, C. 1994. Estimating the permeability of reservoir rocks from P-wave velocity and quality factor. *56th Meeting, European Association of Exploration Geophysicists, Extended Abstracts*, paper P172.

—— & SOTHCOTT, J. 1992. Laboratory measurements of the seismic properties of sedimentary rocks. *In*: HURST, A., WORTHINGTON, P. F. & GRIFFITHS, C. (eds) *Geological Applications of Wire-Line Logs 2*. Special Publications, Geological Society, London, **65**, 285–297.

MCDONAL, F. J., ANGONA, F. A., MILLS, R. L., SENGBUSH, R. L., VAN NOSTRAND, R. G. & WHITE, J. E. 1958. Attenuation of shear and compressional waves in Pierre Shale. *Geophysics*, **23**, 421–439.

MARKS, S. G., MCCANN, C., OLIVER, J. S. & SOTHCOTT, J. 1992. Compressional wave quality factors of reservoir sandstones at 50 Hz and 1 MHz (extended abstract). *Technical Abstracts of the 54th Annual Meeting of the European Association of Exploration Geophysicists, Paris*, 338–339.

PEACOCK, S., MCCANN, C., SOTHCOTT, J. & ASTIN, T. R. 1994, Experimental measurements of seismic attenuation in microfractured sedimentary rock. *Geophysics*, **59**, 1342–1351.

RAPOPORT, M. B., RAPOPORT, L. I., RYJKOV, V. I., PARNIKEL, V. E. & KATELY, V. A. 1994. Method AVD (Absorption and velocity dispersion) testing and using in oil deposit in Western Siberia. *56th Technical Meeting, European Association of Exploration Geophysicists, Extended Abstracts*, Paper BO56.

RYJKOV, V. I. & RAPOPORT, M. B. 1994. Study of seismic anelasticity from VSP. *56th Technical Meeting, European Association of Exploration Geophysicists, Extended Abstracts*, Paper P022.

TANG, X. & STRRACK, K.-M. 1995. Waveform inversion of seismic P-wave attenuation from borehole compressional wave logs. *57th Conference, European Association of Geoscientists and Engineers, Extended Abstracts*, Paper P102.

TITTMAN, B. R. 1977. Lunar rocks Q in 3000–5000 range achieved in the laboratory. *Philosophical Transactions of the Royal Society*, A285, 475–479.

——, BAULAU, J. R. & ABDEL-GAWAD, M. 1985. The role of viscous fluids in the attenuation and velocity of elastic waves in porous rocks. *In*: JOHNSON, D. L. & SEN, P. N. (eds) *Physics and Chemistry of Porous Media*. American Institute of Physics, New York, 000–000.

TOKSOZ, M. N., JOHNSTON, D. H & TIMUR, A. 1978. Attenuation of seismic waves in dry and saturated rocks: 1. Laboratory measurements. *Geophysics*, **44**, 681–690.

TUTUNCU, A. N., PODIO, A. L. & SHARMA, M. M. 1994. An experimental investigation of factors influencing compressional- and shear-wave velocities and attenuations in tight gas sandstones. *Geophysics*, **59**, 77–86.

VARELA, C. L., ROSA, A. L. R & ULRYCH, T. D. 1993. Modelling of attenuation and dispersion. *Geophysics*, **58**, 1167–1173.

VERNIK, L. & NUR, A. 1992. Petrophysical classification of siliciclastics for lithology and porosity prediction from seismic velocities. *Bulletin of the American Association of Petroleum Geologists*, **76**, 1295–1309.

WINKLER, K. & PLONA, T. J. 1982. Techniques for measuring ultrasonic velocity and attenuation spectra in rocks under pressure. *Journal of Geophysical Research*, **87**, 10776–10780.

YALCH, J. I. 1988. Evidence of improved acquisition and processing technology. *Geophysics: The Leading Edge*, **7**, 13–19.

Estimation of aspect-ratio changes with pressure from seismic velocities

Y. F. SUN & D. GOLDBERG

Borehole Research Group, Lamont-Doherty Earth Observatory of Columbia University, Palisades, NY 10964, USA

Abstract: Seismic velocities are modeled as a function of rock mineralogy, porosity, fracture density, aspect ratio and fluid saturation. We compare the model with seismic velocity measurements made by Nur and Simmons in 1969 and predict quantitative aspect ratio changes as a function of differential pressure. The velocity change with pressure is initially caused by the deformation and collapse of pre-existing cracks and pores followed by the initiation of new cracks. For low porosity Casco granite (0.25% pore and 0.45% crack), we estimate that crack aspect ratio increases from $5.9e^{-3}$ to 0.01, then decreases to $5.0e^{-5}$ for a pressure increase from 0 to 300 MPa. Most pre-existing cracks and pores close at differential pressures < 40 MPa, a lower pressure than predicted with standard approximations. These model results can be used to determine *in situ* porosity, fluid content, and pore pressure from seismic data by inversion and for reservoir monitoring from drilling and core data.

Although seismic methods have made significant contributions to the understanding of subsurface geology for many years, the correlation between seismic parameters and *in situ* petrophysical properties still remains poorly understood. Among the most common seismic parameters today, wave velocities in porous media and fractured media have been used extensively in both experimental and theoretical studies, but generally have not incorporated basic petrophysical variables into their formulations. Although various empirical relationships between wave velocities and porosity have been developed, theories capable of generalized parameter correlations have not yet emerged. The inversion of seismic data for additional petrophysical information such as mineralogy, rock composition, lithology; fluid content and saturation; wettability, fluid–rock coupling, and fracture density, aspect ratio, orientation, spacing, and connectivity is critical for complete reservoir analysis, modelling and monitoring.

The basic theory of seismic wave propagation is founded upon the equations of motion of classical continuum mechanics. The structural effects of individual defects, pores, or fractures and related dynamic instability have been studied only as boundary-value problems. Natural rocks, however, are mixtures of different minerals with complicated porous and/or fracture structures and it is the pores, cracks, fractures, fissures, joints, faults and other internal structures that are the vital elements for the storage and migration of subsurface fluids. Many theoretical studies of seismic properties of porous or fractured media, e.g., effective medium theories (e.g., Kuster & Toksöz 1974; O'Connell & Budiansky 1974; Bruner 1976; Hudson, 1980) and propagation theories (e.g., Kosten & Zwikker 1941; Frenkel 1944; Gassmann, 1951; Biot 1956; Dvorkin & Nur 1993), have been applied in experimental studies and in field examples with limited success. For example, Biot's theory (1956; including Gassmann's theory) can be used phenomenologically to predict porosity from seismic velocities better than the time-average equation (Wyllie *et al.* 1956), but physically cannot be used for rocks at low differential pressures (Gregory 1976; Murphy 1984; Sun 1994), limiting the ability to obtain high-resolution petrophysical and stratigraphic information from seismic or well-log data. To extend Biot's theory, Sun (1994) developed a topological characterization of structural media that provides a representation of the internal structure of a fractured porous medium at the microscopic scale and investigated the general mechanics and thermodynamics of fractured porous media.

In this paper, we use derived theoretical model from the general theory developed by Sun (1994) to interpret the seismic wave velocity versus differential pressure experiments performed by Nur & Simmons (1969). The results provide a quantitative understanding of the dynamic changes of aspect ratio with differential pressure.

From Lovell, M. A. & Harvey, P. K. (eds), 1997, *Developments in Petrophysics*, Geological Society Special Publication No. 122, pp. 131–139.

Theory

Observing the laws of general mechanics and thermodynamics, the model developed by Sun (1994) describes coupled effective macroscopic wave fields consisting of individual, microscopic fields that interact globally through volume averaging. From both these dynamic and constitutive equations, an extended Biot's theory for a two-phase fractured porous medium (neglecting the intrinsic viscoelastic effects of each individual phase) is derived:

$$
\begin{pmatrix} \rho^{11} & \rho^{12} \\ \rho^{21} & \rho^{22} \end{pmatrix} \frac{\partial^2}{\partial t^2} \begin{pmatrix} \mathbf{u}^1 \\ \mathbf{u}^2 \end{pmatrix}
$$

$$
= \begin{pmatrix} P & Q \\ Q & R \end{pmatrix} \nabla \nabla \cdot \begin{pmatrix} \mathbf{u}^1 \\ \mathbf{u}^2 \end{pmatrix}
$$

$$
- \begin{pmatrix} N & T \\ T & S \end{pmatrix} \nabla \times \nabla \times \begin{pmatrix} \mathbf{u}^1 \\ \mathbf{u}^2 \end{pmatrix}
$$

$$
- \begin{pmatrix} b^{11} & b^{12} \\ b^{21} & b^{22} \end{pmatrix} \frac{\partial}{\partial t} \begin{pmatrix} \mathbf{u}^1 \\ \mathbf{u}^2 \end{pmatrix}, \tag{1}
$$

where $\mathbf{u}^1$ and $\mathbf{u}^2$ are the solid and fluid displacements respectively. The parameters in Eq. (1) are all dependent upon fundamental geometrical parameters ϕ^{ab} and χ^{ab} ($a, b = 1, 2$). P, Q, R, N, S, and T are also functions of intrinsic moduli of the solid and fluid. Neglecting the global flow terms, i.e., $b^{ab} = 0$, $a, b = 1, 2$, the wave equation Eq. (1) admits four kinds of elastic waves: the fast and slow P waves and the fast and slow S waves. For the purpose in this report, we will address only the special solutions of Eq. (1). The parameters in Eq. (1) and $b^{ab}(a, b = 1, 2)$ are defined briefly as follows (see Sun 1994 for details).

For a two-phase isotropic fractured porous medium, the geometric parameter ϕ^{ab} can be specified as

$$
\begin{pmatrix} \phi^{11} & \phi^{12} \\ \phi^{21} & \phi^{22} \end{pmatrix}
$$

$$
= \begin{pmatrix} (1-\phi)(1-c_1) & \phi c_2 \\ (1-\phi)c_1 & \phi(1-c_2) \end{pmatrix}, \tag{2}
$$

where ϕ is porosity, i.e., the volume fraction occupied by fluid, c_1 is the content of solid

particles suspended in fluid, and c_2 is the content of fluid inclusions isolated in solid. χ^{ab} is symmetric, i.e., $\chi^{21} = \chi^{12}$, and χ^{12} is a real non-positive number. And

$$
\rho^{ab} = \rho^1 \chi^{1a} \chi^{1b} + \rho^2 \chi^{2a} \chi^{2b}, \qquad a, b = 1, 2,
$$

where ρ^1 and ρ^2 are defined as

$$
\rho^a = \phi^{a1} \rho_0^1 + \phi^{a2} \rho_0^2, \qquad a = 1, 2,
$$

and ρ_0^1 and ρ_0^2 are the intrinsic density of the solid and fluid respectively. The wave equation presented in the form of Eq. (1) involves many parameters which are the inherent properties of a fractured porous medium. As a forward problem, any given set of all these parameters specifies a concrete model. Nevertheless, it has to be simplified to be compared with experimental results. If it is assumed that all the pores and fractures in fractured porous medium are connected; that, more strongly, the space occupied by each phase is connected, i.e., $c_1 = c_2 = 0$, the tensor ϕ^{ab} is uniquely determined by porosity ϕ.

Using constraints on energy partition of relative motion and assuming that the flow in pores is only Poiseuille flow, tensor $b^{ab}(a, b = 1, 2)$ becomes (see Sun (1994) for details)

$$
\begin{pmatrix} b^{11} & b^{12} \\ b^{21} & b^{22} \end{pmatrix} = bF \begin{pmatrix} 1 & -1 \\ -1 & 1 \end{pmatrix},
$$

where b is a constant,

$$
F \equiv \chi^2 \frac{\chi^2 v^2 \alpha - \chi^1 v^1}{v^1 - v^2 \alpha},
$$

v^1 and v^2 are the particle velocities of solid and fluid respectively, $\chi^1 = \chi^{11} + \chi^{21}$, $\chi^2 = \chi^{12} + \chi^{22}$, and α is the dynamic tortuosity defined as

$$
\alpha = \frac{1}{|\cos\theta|},
$$

where θ is the angle between the velocity of the fluid and that of the solid. Therefore, $b^{ab}(a, b = 1, 2)$ are non-linear functions of particle velocities of both solid and fluid phase.

Since fluids do not possess intrinsic rigidity, it can be assumed that the intrinsic moduli of fluids are negligibly small. By further assuming that the shear moduli acquired by fluids through coupling with the solids are also negligibly small, Eq. (1) becomes

$$\begin{pmatrix} \rho^{11} & \rho^{12} \\ \rho^{21} & \rho^{22} \end{pmatrix} \frac{\partial^2}{\partial t^2} \begin{pmatrix} \mathbf{u}^1 \\ \mathbf{u}^2 \end{pmatrix}$$

$$= \begin{pmatrix} P & Q \\ Q & R \end{pmatrix} \nabla \nabla \cdot \begin{pmatrix} \mathbf{u}^1 \\ \mathbf{u}^2 \end{pmatrix}$$

$$- \begin{pmatrix} N & 0 \\ 0 & 0 \end{pmatrix} \nabla \times \nabla \times \begin{pmatrix} \mathbf{u}^1 \\ \mathbf{u}^2 \end{pmatrix}$$

$$- bF \begin{pmatrix} 1 & -1 \\ -1 & 1 \end{pmatrix} \frac{\partial}{\partial t} \begin{pmatrix} \mathbf{u}^1 \\ \mathbf{u}^2 \end{pmatrix}, \quad (3)$$

which is in the same form as Biot's equation of poroelasticity published by Biot in 1956. The notation used here in Eq. (3) is conformal to that used by Johnson & Plona (1982).

Equation (3) is different from the Biot's equation in that all the phenomenological parameters in Eq. (3) can now defined in terms of the fundamental geometrical parameters of the internal structures and the physical parameters of the constituents. Particularly, the shear modulus of the fractured porous medium is expressed explicitly as a function of not only the shear modulus of the solid but also the geometry of internal structures. Let

$$\psi^{ab} = \chi^{ca}\phi^{cb},$$

and

$$\beta = \psi^{11} - \frac{\psi^{12}\psi^{21}}{\psi^{22}},$$

then

$$N = \beta\mu_0^1,$$

$$R = (\psi^{21}\alpha_p + \psi^{22})K_0^{22},$$

$$Q = (\psi^{11}\alpha_p + \psi^{12})K_0^{22},$$

and

$$P = \beta K_0^{11} + \frac{\psi^{12}}{\psi^{22}}Q + \frac{4}{3}N,$$

where μ_0^1 is the intrinsic shear modulus of the solid, K_0^{11}, K_0^{22}, and α_p are functions of the intrinsic bulk moduli of the solid (K_0^1) and fluid (K_0^2). The procedure to derive these parameters is analogue to that given by Biot & Willis (1957).

Since Eq. (3) is in the same form as the Biot's equation, the solutions of the former is in the same form as those of the latter which have been attempted by many authors (e.g., Geertsma and Smit 1961; Stoll 1974; Johnson and Plona 1982). Neglecting the global flow, i.e., $b = 0$, Eq. (3) admits three kinds of waves, fast and slow compressional waves and shear wave. Let V_s denote the shear wave velocity, V_{p+} the velocity of the fast compressional wave, and V_{p-} the velocity of the slow compressional wave, then

$$V_s^2 = \frac{\rho_{22}}{\rho_{11}\rho_{22} - \rho_{12}^2} N,$$

$$V_{p\pm}^2 = \frac{\triangle \pm \sqrt{\triangle^2 - 4(PR - Q^2)(\rho_{11}\rho_{22} - \rho_{12}^2)}}{2(\rho_{11}\rho_{22} - \rho_{12}^2)},$$

where

$$\triangle = P\rho_{22} + R\rho_{11} - 2\rho_{12}Q.$$

Among the geometrical parameters mentioned above, the geometrical parameter β and the dynamic tortuosity α can be expressed explicitly in the wave velocity expressions

$$V_s^2 = \frac{\beta}{1 - \phi + (1 - \alpha^{-1})\chi\phi} \frac{\mu_0^1}{\rho_0^1}, \quad (4)$$

and

$$V_{p\pm}^2 = V_{p*}^2 \pm \left[V_{p*}^4 - \frac{\alpha^{-1}}{1 - \phi + (1 - \alpha^{-1})\chi\phi} \right.$$

$$\left. \cdot \left(\frac{P}{\rho_0^1} \frac{R/\phi}{\rho_0^2} - \frac{Q}{\rho_0^1} \frac{Q/\phi}{\rho_0^2} \right) \right]^{1/2}, \quad (5)$$

where

$$2V_{p*}^2 = \frac{1}{1 - \phi + (1 - \alpha^{-1})\chi\phi} \frac{P}{\rho_0^1}$$

$$+ \frac{(1 - \phi)\alpha^{-1} + (1 - \alpha^{-1})\chi\phi}{1 - \phi + (1 - \alpha^{-1})\chi\phi} \frac{R/\phi}{\rho_0^2}$$

$$+ 2 \frac{1 - \alpha^{-1}}{1 - \phi + (1 - \alpha^{-1})\chi\phi} \frac{P}{\rho_0^1},$$

where ρ_0^1 and ρ_0^2 are the intrinsic mass densities of the solid and fluid respectively, $\chi = \rho_0^2/\rho_0^1$, and ϕ is porosity. P, Q, and R are also explicit functions of the effective geometrical parameter β. As expected, both α and β depend upon aspect ratio and differential pressure. The pressure change affects the solid/solid and solid/fluid coupling. We use F_c to denote the coupling factor which is linearly proportional to differential pressure. Once the solid and fluid properties are known as well as porosity and differential pressure, the dependency of wave velocities on aspect ratio can be investi-

gated using Eqs (4) and (5) and estimation of aspect ratio change with pressure would be done if wave velocities are measured in experiments.

When porosity is very small, which is the case studied in this report, the slow waves can be neglected and the slow wave velocity is very small. For clarity, the velocities of S (shear) wave and fast P (compressional) wave in Eqs (4) and (5) can be rewritten equivalently as

$$V_s^2 = \frac{1}{2}\mathcal{F}\left(\phi, \mu_0^1, C_0^2, \frac{1}{a}, F_c, \chi\right)\frac{\mu_0^1}{\rho_0^1}, \qquad (6)$$

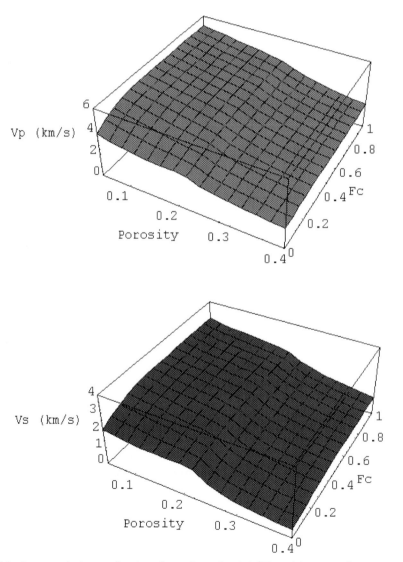

Fig. 1. Seismic wave velocity as a function of porosity and scaled differential pressure for water-saturated sands or sandstones, see text for details (after Sun 1994).

and

$$V_p^2 - \frac{4}{3}V_s^2 = \mathcal{G}_1\left(\phi, C_0^1, C_0^2, \frac{1}{a}, F_c, \chi\right)\frac{K_0^1}{\rho_0^1}$$

$$+\mathcal{G}_2\left(\phi, C_0^1, C_0^2, \frac{1}{a}, F_c, \chi\right)\frac{K_0^2}{\rho_0^2}, \quad (7)$$

where K_0^1 and μ_0^1 are the intrinsic bulk and shear moduli of the solid respectively, K_0^2 is the intrinsic bulk modulus of fluid, $C_0^1 = 1/K_0^1$ and $C_0^2 = 1/K_0^2$ are the compressibility of solid and fluid respectively, $\chi = \rho_0^2/\rho_0^1$, a is the average aspect ratio, and F_c is a coupling factor which depends upon wettability and pressure. All the unknown parameters such as wettability, solid connectivity, fluid connectivity, c_1 and c_2 can be lumped into two functional parameters which in turn are expressed in terms of porosity ϕ, χ, F_c, and aspect ratio a. The explicit expressions of non-linear functions $\mathcal{F}$, $\mathcal{G}_1$, and $\mathcal{G}_2$ can thus be

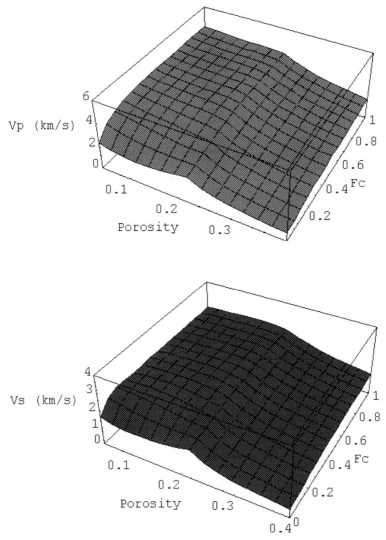

Fig. 2. Seismic wave velocity as a function of porosity and scaled differential pressure for gas-saturated sands or sandstones, see text for details (after Sun 1994).

easily obtained from their equivalent expressions in Eqs (4) and (5). The detailed algebraic derivations of these expressions are omitted in this report and the details of the procedure to solve the wave equations are referred to, e.g., Geertsma & Smit (1961) or Stoll (1974). The discussions on different behaviours for the low-frequency, immediate-frequency, and high-frequency ranges are referred to Johnson & Plona (1982).

Modelling and results

The expressions of seismic wave velocities in Eqs 4 and 5 (or 6 and 7) are used to derive continuous functions of wave velocities in terms of porosity, fluid content, mineralogy (elastic constants), aspect ratio, and coupling factor (wettability and differential pressure). In Figs 1 and 2 P and S wave velocities are shown as 3D functions of porosity and coupling factor in water-saturated and gas-saturated sandstones for porosities from 0% to 40%. The coupling factor is scaled from 0 to 1, linearly proportional to differential pressure. Unlike the Biot's theory or the time-average equation, these relationships between wave velocities and rock properties are valid for P and S waves in water-saturated and gas-saturated rocks at both high and low differential pressures. In Figs 1 and 2 we observe that porosity is the major controlling factor of the behavior of seismic wave velocities. The second major control on wave velocities is mineralogy (elastic constants of the solid constituents), which defines the lower and upper limits of wave velocity at a given porosity. The structural parameter F_c has an important effect on wave velocity, depending on the differential

pressure, porosity, and fluid content, and its effect is always greater for gas-saturated rock than for water-saturated rock. The difference due to fluid content is always greater at low differential pressure than at high differential pressure. The velocity relationships presented here have been used successfully in reservoir modeling under *in situ* conditions (Anderson *et al.* 1995) and compared well with experimental results in sandstones and limestones (Sun 1994). In model computation for sandstones, the mass density, bulk modulus, and shear modulus of the solid constituent are $2.6\,\mathrm{g\,cm^{-3}}$, $37.9\,\mathrm{GPa}$, and $30.0\,\mathrm{GPa}$, respectively. The mass density and bulk modulus of distilled water for the liquid phase are $1.0\,\mathrm{g\,cm^{-3}}$ and $2.24\,\mathrm{GPa}$. The mass density and bulk modulus of air are $0.0011\,\mathrm{g\,cm^3}$ and $0.127\,\mathrm{MPa}$.

In Fig. 3 P and S wave velocity slices at 4.6% porosity are shown versus the coupling factor (or scaled differential pressure) for both gas-saturated and water-saturated sandstone. Experimental results (dots) plotted are data from Murphy (1984) and agree well with the theoretical prediction (solid lines). At low differential pressure, V_s for water-saturated sandstone is notably higher than for gas-saturated sandstone, contrary to prediction from Biot's theory. This difference is particularly important in gas-saturated sandstone at transitional pressures ($0.1 < F_c < 0.4$) due to the closure of cracks and changes in pore structure. For higher porosity rock, this effect is not as important as in low-porosity microcracked rocks (Fig. 4).

In order to understand the dynamic effect of crack aspect ratio changes on velocity, we compare the model prediction with experimental data for low-porosity Casco granite (Nur & Simmons, 1969). We use the velocity

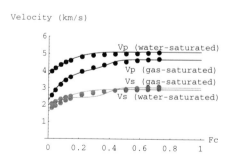

Fig. 3. Seismic wave velocity (solid lines) as a function of scaled differential pressure for sandstone of porosity 4.6% compared with experimental results (dots) performed by Murphy (1984, after Sun 1994).

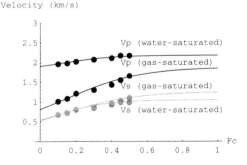

Fig. 4. Seismic wave velocity (solid lines) as a function of scaled differential pressure for sandstone of porosity 38.3% compared with experimental results (dots) performed by Domenico (1977) (after Sun 1994).

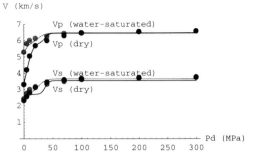

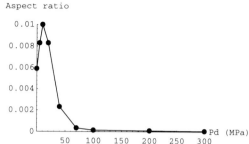

Fig. 5. Seismic wave velocity (solid lines) as a function of scaled differential pressure for Casco granite, compared with experimental results (dots) performed by Nur & Simmons (1969). The intrinsic density of Casco granite is 2.626 g cm^{-3}. The bulk and shear moduli of the solid grain are 66.0 GPa and 37.9 GPa, respectively. The density and bulk modulus of water as the saturating fluid are 1.0 g cm^{-3} and 2.25 GPa. The mass density and the bulk modulus of air are 0.0011 g cm^{-3} and 0.127 MPa. The initial crack density is 0.45% and pore porosity is 0.2%, resulting in total porosity 0.65%.

Fig. 6. Aspect-ratio variation as a function of differential pressure in Casco granite of porosity 0.65% (0.45% crack), derived from velocity measurements by Nur & Simmons (1969).

relationships in Eqs 4 and 5 by solving the coupled wave equation in Eq. 3 to calculate the expected P and S wave velocities for Casco granite. Since crack porosity is dominant in these rocks, wave velocities are very sensitive to geometrical factors, especially aspect ratio. Figure 5 shows the predicted P and S wave velocities (solid lines) and experimental results (dots), both in excellent agreement except for velocities over transitional pressures for the gas-saturated. Notice that the shear wave velocity for water-saturated granite, shown in Fig. 5 is always greater than that for dry granite at any given differential pressure. The same phenomena are also observed noticeably in Figs 3 and 4 when differential pressure is low. Biot's theory predicts that shear wave velocity for gas-saturated rock is always greater than that for the same rock but water-saturated, which is attributed only to the density change and is against experimental results as shown in Figs 3 and 5 (Murphy 1982; Nur & Simmons 1969). Gregory (1976) and Murphy (1982) have discussed that this contradiction between experiments and prediction may be due to the presence of micro-cracks. The extended Biot's theory as given by Sun (1994) takes account of the crack effects on wave velocities. When porosity is low and cracks are abundant, aspect ratio plays much more important role than fluid content, as well

as in the case of low differential pressure for rocks of low to intermediate porosity.

Figure 6 shows the aspect ratio change derived from this difference between the model and the data at each experimental pressure. The results suggest that initially the experimental samples were dominantly populated with cracks (0.45% crack and 0.2% pore porosity) that then began to collapse as differential pressure was increased from 0 to 5 MPa. The average aspect ratio of the rock correspondingly increased from $5.9e^{-3}$ to $8.3e^{-3}$ (i.e. smaller aspect ratio cracks closed). As differential pressures reached 10 MPa, cracks in the sample continued to collapse until almost all the original cracks with small aspect ratios were closed. Large aspect ratio pores and cracks became dominant with average aspect ratio as high as 0.01. When differential pressure was increased further to 20 MPa, large aspect ratio cracks continued to collapse and pores began to deform into cracks, reducing the average aspect ratio back to $8.3e^{-3}$. Then, as differential pressures reached 40 MPa, all the original cracks were fully closed and pores were deformed to become cracks with low aspect ratio of $2.3e^{-3}$. Above 40 MPa, newly-formed cracks deformed the rock and at 70 MPa, the average aspect ratio was about $3.3e^{-4}$. At differential pressure of 100 MPa, most new cracks were closed; an average aspect ratio of $5.0e^{-5}$ was reached at differential pressures of 300 MPa.

The dynamic process of rock deformation under changing differential pressure can thus be divided into four stages:

Stage I   the collapse of original cracks;
Stage II   the deformation of pores into cracks;
Stage III closure of original cracks;
Stage IV   formation of new cracks of small aspect ratio.

The differential pressure needed to close cracks of given average aspect ratio in a cracked solid is less than that estimated by Nur & Simmons (1969) using Walsh's approximation (1965), $P \approx Ea$, where E is Young's modulus of the solid, a is aspect ratio of a crack, and P is the needed pressure to close such a crack. Therefore, the internal deformation in an overpressured rock may be more rapid and severe than expected, a fact that may be useful both for drilling and reservoir analysis and for the study of earthquake mechanisms.

Conclusions

We model seismic wave velocities through a new formulation representing intrinsic rock properties such as mineralogy, porosity, connectivity, aspect ratio, coupling factor, and fluid content. Given porosity and fluid content information, we are able to estimate crack aspect ratio. Comparisons between experimental results of seismic wave velocities and the model allow prediction of the distribution of crack aspect ratio for Casco granite with differential pressure change. With further comparative work, this knowledge could be expanded for different rock types and crack distributions. Once aspect ratio, porosity and fluid content are determined, the *in situ* pore pressure can be evaluated in a particular reservoir or lithology. In conjunction with experimental or field information, our preliminary model results indicate that quantitative petrophysical properties may be inverted from seismic measurements and provide a significant advance in understanding for reservoir modelling, monitoring and drilling. The internal deformation in an overpressured rock may be more rapid and severe than expected by using conventional estimation, a fact that may be useful both for drilling and reservoir analysis and for the study of earthquake mechanisms.

We thank two anonymous reviewers for their critical and helpful comments. Major portions of this report constitute a part of Y.F.S.'s doctoral dissertation. Y.F.S. is grateful to Yu-Chiung Teng for discussion and support, and to E. A. Robinson for encouragement and help in completing his dissertation. Y.F.S. and D.G. were supported for this report under NSF contract JOI 66–84. Lamont-Doherty contribution number 5472.

References

ANDERSON, R. N., BOULANGER, A., HE, W., SUN, Y. F., XU, L., SIBLEY, D., AUSTIN, J. WOODHAMS, R., ANDRE, R. & RINEHART, K. 1995. Gulf of Mexico reservoir management, I., 4D seismic helps track drainage, pressure compartmentalization; II., Method described for using 4D seismic to track reservoir fluid movement. *Oil & Gas Journal*, (I) March 27, 55–58; (II) April 3, 70–74.

BIOT, M. A. 1956. Theory of propagation of elastic waves in a fluid-saturated porous SOLID, I. Low frequency range; II. High frequency range. *Journal of the Acoustical Society of America*, **28**, 168–191.

—— & WILLIS, D. G. 1957. The elastic coefficients of the theory of consolidation. *Journal of Applied Mechanics*, **24**, 594–601.

BRUNER, W. M. 1976. Comment on 'Seismic velocities in dry and saturated cracked solids' by Richard J. O'Connell and Bernard Budiansky. *Journal of Geophysical Research*, **81**, 2573–2576.

DOMENICO, S. N. 1977. Elastic properties of unconsolidated porous sand reservoirs. *Geophysics*, **42**, 1339–1368.

DVORKIN, J. & NUR, A. 1993. Dynamic poroelasticity: A unified model with the squirt mechanism and the Biot mechanisms. *Geophysics*, **58**, 524–533.

FRENKEL, J. 1944. On the theory of seismic and seismoelectric phenomena in a moist soil. *Journal of Physics (USSR)*, **8**, 230–241.

GEERTSMA, J & SMIT, D. C. 1961. Some aspects of elastic wave propagation in fluid-saturated porous solids. *Geophysics*, **26**, 169–181.

GASSMANN, F. 1951. Elastic waves through a packing of spheres. *Geophysics*, **15**, 673–685.

GREGORY, A. R. 1976. Fluid saturation effects on dynamic elastic properties of sedimentary rocks. *Geophysics*, **41**, 895–921.

HUDSON, J. A. 1980. Overall properties of a cracked solid. *Mathematical Proceedings of the Cambridge Philosophical Society*, **88**, 371–384.

JOHNSON, D. L. & PLONA, T. J. 1982. Acoustic slow waves and the consolidation transition, *Journal of the Acoustic Society of America*, **72**, 556–565.

KOSTEN, C. W. & ZWIKKER, C. 1941. Extended theory of the absorption of sound by compressible wall-coverings. *Physica*, **8**, 968–978.

KUSTER, G. T. & TOKSÖZ, N. M. 1974. Velocity and attenuation of seismic waves in two-phase media, I. Theoretical formulations. *Geophysics*, **39**, 607–618.

MURPHY, W. F. 1984. Acoustic measures of partial gas saturation in tight sandstones. *Journal of Geophysical Research*, **89**, 11549–11559.

NUR, A. & SIMMONS, G. 1969. The effect of saturation on velocity in low porosity rock. *Earth and Planetary Science Letters*, **7**, 183–193.

O'CONNELL, R. J. & BUDIANSKY, B. 1974. Seismic velocities in dry and saturated cracked solids. J. Elastic Moduli of dry and saturated cracked solids. *Journal of Geophysical Research*, **79**, 5412–5426.

STOLL, R. D. 1974. Acoustic waves in saturated sediment. *In:* HAMPTON, L. (eds) *Physics of Sound in Marine Sediments*. Plenum, New York 19–39.

SUN, Y. F. 1994. *On the foundations of the dynamical theory of fractured porous media and the gravity variations caused by dilatancies.* PhD dissertation, Columbia University.

WALSH, J. B. 1965. The effect of cracks on the compressibility of rocks. *Journal of Geophysical Research*, **70**, 381–389.

WYLLIE, M. R., GREGORY, A. R. & GARDNER, L. W. 1956. Elastic wave velocities in heterogeneous and porous media. *Geophysics*, **21**, 41–70.

Petrophysical estimation from downhole mineralogy logs

P. K. HARVEY[1], M. A. LOVELL[1], J. C. LOFTS[2],
P. A. PEZARD[3] & J. F. BRISTOW[4]

[1] *Borehole Research, Department of Geology, University of Leicester,
Leicester LE1 7RH, UK*
[2] *Schlumberger GeoQuest, Schlumberger Evaluation and Production Services (UK) Limited,
Loriston House, Wellington Road, Altens, Aberdeen AB1 4BH, UK*
[3] *Institut Méditerranéen de Technologie, Technopole de Chateau-Gombert,
13451 Marseille cedex 13, France*
[4] *Schlumberger GeoQuest, Schlumberger House, Buckingham Gate, Gatwick,
West Sussex RH6 0NZ, UK*

Abstract: A number of physical and chemical properties of rocks, many of which are important in petrophysics are simple, usually linear functions, of a rock's mineralogy. Examples include matrix density, porosity, magnetic susceptibility and cation exchange capacity. Mineralogy logs can be obtained directly from geochemical logs through an inversion process and, in turn, estimates of petrophysical properties such as those noted above can be obtained. The accuracy of such estimates is directly dependent on the quality of both the inversion and the geochemical measurements. Examples from producing oil fields and from deep-sea environments are used to show that accurate estimates of derived parameters can be obtained provided that appropriate inversion procedures are adopted, and that these estimates are generally better than those obtained by conventional log analysis. Logs of some parameters, such as cation exchange capacity, cannot be obtained except through the mineral inversion.

A number of physical and chemical properties of rocks, many of which are important in reservoir characterisation and evaluation, are simply related to the quantitative mineralogy of the rock. Potential applications include the estimation of matrix (grain) density, porosity, cation exchange capacity (Chapman *et al.* 1987; Herron 1987*b*; Herron & Grau 1987), thermal conductivity (Dove & Williams 1988), heat flow (Anderson & Dove 1987), magnetic susceptibility, fluid saturation (Hastings 1988), neutron capture cross-section (Herron 1987*b*), and, indirectly and probably in formation specific situations, permeability (Herron 1987*a*).

The relationship between mineralogy and the physical or chemical parameter is usually linear, and the parameter coefficients reasonably well known within certain bounds. For example, in the case of matrix density the parameter coefficients are the reciprocals of the densities of the mineral phases themselves so that for a simple dolomitic limestone containing a mixture of calcite and dolomite the (reciprocal of the) matrix density (ρ_m) is given as:

$$\rho_m^{-1} = \rho_{dol}^{-1} \cdot w_{dol} + \rho_{cal}^{-1} \cdot w_{cal}$$

where ρ_{dol} and ρ_{cal} are the densities of dolomite and calcite respectively, and w_{dol} and w_{cal} are the corresponding weight proportions. In addition is the closure constraint (that governs some of these properties) that:

$$w_{dol} + w_{cal} = 1.0.$$

While there are other approaches to estimating ρ_m direct calculation from the mineralogy is both simple and precisely constrained. In principle it is only necessary to know the mineral proportions and the density of the mineral phases. In practice there is the problem of how to estimate the mineral proportions, and then how to obtain, in this example, appropriate density values.

This contribution reviews some of the problems of obtaining accurate estimates of the proportions of actual minerals present in a formation, and some associated problems with the values of the parameter coefficients. These problems, and some possible solutions, are illustrated by recourse to a number of case histories.

Mineralogy logging

Mineralogy logs, which consist of a set of curves representing the variation in abundance of

From Lovell, M. A. & Harvey, P. K. (eds), 1997, *Developments in Petrophysics*, Geological Society Special Publication No. 122, pp. 141–157.

different minerals throughout the borehole section, can be obtained by the inversion, ideally, of a full suite of geochemical logs. In their absence other log combinations can sometimes be used to estimate the mineralogy of some very simple lithologies. The conversion of a rock analysis to a set of compatible minerals has been carried out for decades for one purpose or another, and a number of different approaches have been developed. The widest application has probably been in igneous petrology where norms, particularly CIPW norms (Cross *et al.* 1903; the acronym CIPW comes from the four authors, Cross, Iddings, Pirsson & Washington, of this classic paper), have been used for classificatory purposes. In the calculation of a norm the weight proportions of a set of theoretical minerals, which may or may not relate to what is actually in a rock, are obtained, usually by the application of a number of simple rules. In the estimation of petrophysical parameters from calculated mineralogy it is essential that this estimation is based not on normative mineral assemblages but on the actual minerals present in the rock (the 'mode'), so that there is compatibility with the parameter coefficients that are used.

The need to compute actual mineral assemblages ('chemical modes' of Wright & Doherty 1970) is more involved than the computation of simple norms. This problem has been discussed in the context of geochemical logging and downhole measurements by Harvey *et al.* (1990) and Harvey & Lovell (1991, 1992) and is summarized briefly in the following section.

Mineral inversion

Figure 1 shows a flow diagram of the inversion process, with the geochemical input to the left. For logging purposes geochemical data may be obtained (in Fig. 1) from Schlumberger's Geochemical Logging Tool (GLT) where the initial measured elemental yields are converted to more conventional weight percent oxides through appropriate geological assumptions and closure algorithms (Hertzog *et al.* 1989; Grau *et al.* 1989). In core studies geochemical log data may, of course, be obtained by conventional analytical methods. On the right in Fig. 1 are depicted the mineral modelling parameters; that is, the specification of the actual minerals present (that are to be entered into the inversion), the composition of each mineral, and specification of the chosen modelling strategy. The latter has been discussed by Harvey *et al.* (1990) and Lofts *et al.* (1994, 1995a) and will only be considered

here in discussion of the case histories. The relationship between the geochemistry of a rock or formation and the mineral parameters may be expressed as $\mathbf{Xp} = \mathbf{c}$ where $\mathbf{c}$ is the geochemical composition vector, p is the (unknown) vector of computed mineral abundances, and $\mathbf{X}$ is a matrix of mineral compositions for those minerals used in the inversion (the components matrix). Accepting $\mathbf{c}$ then the accuracy of any derived mineral abundances, and any subsequently derived parameters, depends on (a) knowing the actual minerals that are present, and (b) their compositions. Following the computation of mineral abundances for a given rock or formation some idea of the quality of the inversion can be gained from the standard error (s_e) measured between the original (input) chemistry, and the composition back-calculated from the derived (output) mineralogy (Harvey *et al.* 1990, p. 176).

The choice of minerals to be used in a given inversion usually requires the integration of core mineralogy data, such as X-ray diffraction or infra-red measurements. The latter can provide lists of possible mineral assemblages in a given formation, and even quantitative estimates of mineral abundance which can be useful in validating the computed models. With downhole geochemical logs which measure, for each depth interval, over a volume six or so orders of magnitude greater than the laboratory sample used for the mineral studies only in the most homogeneous of formations might a given laboratory derived mineralogy be expected to truly represent the actual mineral assemblage at a given depth. However, if (all) the possible or likely mineral assemblage are known each of these might be modelled in turn for a given chemistry.

The mineral inversion procedure is extremely sensitive (Lofts *et al.* 1995b) and evaluation of a very large number of mineral inversions over the past few years shows that a good fit between the input chemistry and derived mineralogy (as indicated by a low s_e, which here might be considered as less than 0.5%) virtually only occurs when the correct mineral assemblage, or any compositionally collinear equivalent assemblage, is being used in the model. The strength of this statement is, of course, diminished as the analytical precision on the geochemical data deteriorates as it can do sometimes with downhole measurements, particularly where hole conditions are poor. Where compositionally colinear equivalent assemblages are chosen (as better fits to the input chemistry) then what may or may not be seen as the correct mineral assemblage, will certainly be passed on to any

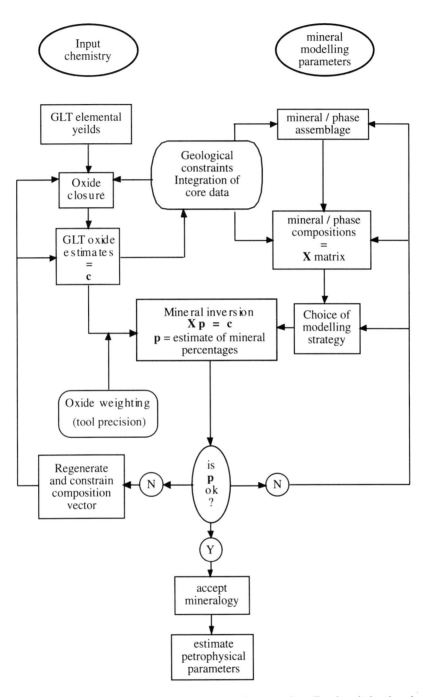

Fig. 1. Flow diagram showing the steps in the mineral inversion procedure. For downhole mineralogy logging the raw geochemistry (**c**) is provided by Schlumberger's Geochemical Logging Tool (right) and this in turn is inverted to give the 'chemical mode' (**p**). The inversion is constrained by the minerals in the model and their compositions (**X**, left).

Table 1. *Two possible solutions (A, B) for the mineralogy of the oxide analysis shown in the leftmost pair of columns*

Oxide	Analysis	Mineral	A	B	ρ
SiO$_2$	53.51	Quartz	16.12	6.83	2.65
Al$_2$O$_3$	16.28	Albite	3.99	3.74	2.61
TiO$_2$	1.66	Chlorite	4.90	1.04	2.80
Fe$_2$O$_3$	21.14	Calcite	0.93	0.45	2.71
MgO	2.34	Siderite	17.86	18.23	3.93
CaO	1.73	Illite	42.42	42.31	2.75
Na$_2$O	0.68	Kaolinite	12.24		2.62
K$_2$O	2.39	Montmorillonite		25.85	2.03
MnO	0.27	Rutile	1.27	1.28	4.23
		Density	2.937	2.779	

The fit between the original chemical analysis and the compositions derived from the computed mineralogies is excellent in both cases. In the last row is the computed density for each solution using 'book' figures (right column) for the densities of the individual minerals.

derived parameter as a real, and possible significant difference. This is demonstrated with a simple example, which is given in Table 1.

Table 1 shows two possible solutions for the mineralogy of a given analysis from a North Sea (Brent Group) sample. The latter has a relatively complex mineralogy, with variation possible particularly in the clay minerals present. The densities derived for the two solutions differ by $0.158\,\mathrm{g\,cm^{-3}}$.

Compositional colinearity may occur when three or more minerals included in a model lie on, or close to, the same compositional plane or vector (Harvey *et al.* 1991; Harvey & Lovell 1992). For example, in Fig. 2 five minerals (K-feldspar, muscovite, illite, quartz and kaolinite) are plotted within the compositional triangle [K$_2$O]–[Al$_2$O$_2$]–[SiO$_2$], where the [ ] denote molecular proportions. K-feldspar, illite and kaolinite, for example, lie essentially on a straight line. Any rock composition lying on that line could be solved in terms of {kaolinite–K-feldspar} or either of {kaolinite–illite} or {illite–K-feldspar}, depending on precisely where the rock lay. No unique solution can determine the rock composition in terms of all three minerals because of their precise compositional interrelationship; the latter may be more easily visualized as the potential mineralogical reaction:

K-feldspar + kaolinite = illite.

In Fig. 2 three other compositionally colinear situations, each now involving four minerals, will also be noted:

K-feldspar + kaolinite = muscovite + quartz

quartz + muscovite = K-feldspar + illite

quartz + muscovite + kaolinite = illite.

The example in Table 1 is more complex and involves a compositional equivalence of {quartz, chlorite, kaolinite} balancing {montmorillonite}.

The other important factor that can seriously affect the accuracy and sensitivity of a mineral inversion is the components matrix (**X**) used for the solution. With a few minerals, such as quartz and rutile, there is virtually no variation in composition and the theoretical compositions (SiO$_2$, TiO$_2$, respectively) can be used safely in **X**. For other minerals some strategy which allows variability in their composition will generally be needed. A single, fixed composition for, say, illite, is not generally sufficient.

As an example consider the assemblage quartz–kaolinite–K-feldspar–muscovite–dolomite. This was one of a number of synthetic mixtures used

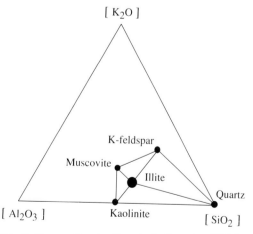

Fig. 2. Compositional colinearity in the system [SiO$_2$]–[Al$_2$O$_3$]–[K$_2$O]. Axes are in units of molecular proportions.

Table 2. *Eight possible mineral inversion solutions for a synthetic rock sample*

Phase	Density	1	2	3	4	5	6	7	8
Kaolinite	2.62	31.32	33.42	39.30	31.21	19.60	25.96	32.88	36.30
Quartz	2.65	12.58	10.49	7.43	12.58	16.81	14.23	11.70	9.94
K-feldspar	2.59	37.95	42.71	48.29	37.95	18.10	25.82	33.21	36.95
Muscovite	2.84	12.67	7.98	0.00	12.67	41.98	30.35	17.91	13.22
Dolomite	2.87	5.59	5.40	4.98	5.59	3.45	3.63	4.30	3.59
Density		2.657	2.641	2.620	2.654	2.719	2.692	2.664	2.650

The mineral composition matrix, **X**, is the same in all cases and uses the compositions of the minerals actually present in the sample, except for muscovite, where a range of muscovite compositions have been inserted into **X**.

by one of us (J.C.L.) in an evaluation of mineral inversion algorithms (Lofts 1993). For the preparation of the synthetic rocks pure minerals were taken and carefully analysed by X-ray fluorescence spectrometry; the mineral powders were then mixed together in model proportions. In this way the composition of the 'rock' is reasonably well constrained. In Table 2 eight different solutions are shown for this sample; in all cases the mineral compositions used for modelling quartz, kaolinite, K-feldspar and dolomite were the actual measured mineral analyses. For muscovite Solution 1 uses the ideal or theoretical composition of muscovite $(K_2Al_4(Al_2Si_6)O_{20}$-$(OH)_4)$, while Solutions 2 to 8 use natural muscovite compositions, with increasing variation from the 'ideal' composition and into more phengitic compositions. The muscovite used for Solution 5 is the 'pure' mineral muscovite used for the synthetic mixes, and the computed mineral assemblage for that solution is very close, as might be expected, to the original target mineral assemblage. The compositions of the muscovites used in this exercise are given in Table 3.

Using typical or 'book' densities for the minerals in Table 2 (column 2) the corresponding rock densities calculated from the computed mineralogy are given on the bottom row. There is a relatively small range in density, Solution 5

(the 'correct' solution) being the highest (2.719), and Solution 3 the lowest (2.620), a range of $0.099 \, \mathrm{g \, cm}^{-3}$. The calculation of density is here a simple linear relationship, so, for the matrix in a rock:

$$\rho_{ma}^{-1} = \sum \rho_i^{-1} w_i$$

where ρ_{ma} is the rock matrix density, ρ_i the density of the i-th mineral, and w_i the weight fraction of the i-th mineral in the inversion solution.

Another important property which could be derived if ρ_{ma} is known is porosity. Porosity (ϕ) is easily computed from the well known relationship:

$$\phi = (\rho_{ma} - \rho_b)/(\rho_{ma} - \rho_{fl})$$

where ρ_b is the bulk density, and ρ_{fl} the fluid density. The terms in this relationship, being based on differences, are very sensitive to small changes, in this case, in the computed matrix density. Table 4 shows the variation in porosity that would be determined for the eight solutions given in Table 2. Solution 5 is assumed correct, and other solutions calculated relative to this. Differences of over 5% porosity (in the most extreme solution – 3), a significant figure, are seen to occur simply because of the choice of an incorrect mineral composition. Indeed, in the

Table 3. *Compositions of the muscovites used for the corresponding mineral inversions summarized in Table 2*

Solution	1	2	3	4	5	6	7	8
SiO_2	45.25	45.87	46.01	45.78	47.08	47.11	48.42	48.54
TiO_2	0.00	0.00	0.00	0.24	0.20	0.74	0.87	0.66
Al_2O_3	38.40	38.69	35.64	34.21	30.81	29.68	27.16	21.99
Fe_2O_3	0.00	0.00	0.01	3.02	3.08	4.17	7.47	9.37
MnO	0.00	0.00	0.00	0.12	0.00	0.01	0.00	0.00
MgO	0.00	0.10	0.00	0.69	1.71	1.77	0.00	3.09
CaO	0.00	0.00	1.12	0.00	0.35	0.19	0.00	0.35
Na_2O	0.00	0.64	1.89	1.01	0.75	0.34	0.35	0.75
K_2O	11.82	10.08	8.19	10.24	10.11	10.32	11.23	11.27

Muscovite 1 is the composition for theoretical muscovite.

Table 4. *Variation in computed porosity due to differences in density arising from the incorrect choice of muscovite composition*

Porosity	1	2	3	4	5	6	7	8
0	3.7	4.7	5.9	3.9	0.0	1.6	3.3	4.1
10	13.3	14.2	15.3	13.5	10.0	11.5	13.0	13.7
20	23.0	23.7	24.7	23.1	20.0	21.3	22.7	23.3

Solution 5 is assumed correct and other solutions should be compared against this.
A fluid density of $1.05\,\mathrm{g\,cm^{-3}}$ was assumed.

case of Solution 3 the composition of the muscovite was such that muscovite was not even recognised in the solution as an essential mineral.

In a real situation examination of error measures such as s_e would have shown that Solution 5 was the best fit, and that Solution 3 was in error and probably unacceptable. There remains, however, a problem of just how to get the right mineral composition for minerals like muscovite and many of the clays and feldspars where the composition can be quite variable. For some clay minerals, illites and montmorillonites particularly, there is still insufficient information available about their natural compositional variation. One approach is to determine the composition of key minerals, by electron microprobe or other appropriate technique, so that there is some idea of the variation to be expected in a given field or formation. This variation can then be built into the inversion procedure. As an example, in modelling micaceous sands micas from the Brent Group, Thistle Field (northern North Sea) a sample of 45 phengitic micas were analysed to provide compositional distributions for modelling. The latter are summarized in Table 5.

This summary can be treated as the parameters of a probability distribution from which an acceptable mica composition can be taken during modelling. This approach has been

discussed by Lofts *et al.* (1994), and offers one approach to the way in which viable mineral compositions can be made available during mineral inversion.

Another factor to consider is the value of the parameter coefficients that are to be used. In the examples above these have been simple 'book' values of density. For muscovite, for example, a value of $2.84\,\mathrm{g\,cm^{-3}}$ (Table 2) was used. This is some sort of average value, and in view of the sensitivity of matrix density values as used above to calculate porosities it must be questioned as to whether there are any alternative methods to finding more appropriate mineral density values. For any mineral the density can be expressed as the as the ratio of the mass of atoms in the mineral unit cell to the volume of the unit cell; that is:

$$\rho_i = M_i Z m_H / V_i$$

where: M_i is the molecular weight of the i-th mineral, Z is the number of formula units in the unit cell, m_H is the mass of a hydrogen atom (1.66×10^{-24} grams) and V_i is the volume (usually expressed in nm^3) of the unit cell of the same mineral. Specific mineral densities can be calculated in this way directly from the mineral composition and Fig. 3 shows the distribution of these calculated densities for the 45 phengitic micas described above. The

Table 5. *Means, standard deviations and linear correlation matrix for 45 phengitic muscovite micas from the Thistle Field, Brent Group northern North Sea*

	SiO_2	Al_2O_3	Fe_2O_3	MgO	CaO	Na_2O	K_2O
Mean	46.74	29.8	3.88	1.76	0.05	0.42	10.19
Std. Dev.	1.95	3.03	1.77	0.75	0.17	0.68	1.14
SiO_2	1						
Al_2O_2	−0.298	1					
Fe_2O_3	−0.121	−0.774	1				
MgO	0.424	−0.645	0.281	1			
CaO	−0.167	0.134	0.011	−0.267	1		
Na_2O	0.015	0.414	0.402	−0.328	−0.022	1	
K_2O	0.218	0.196	0.483	0.424	−0.307	−0.4611	1

NB: iron is quoted as Fe_2O_3, here, to be consistent with modeling.

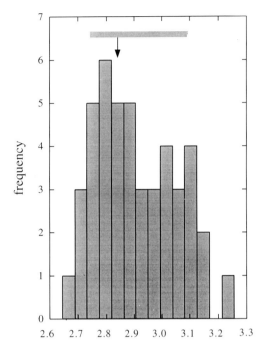

Fig. 3. Distribution of computed densities for the 45 phengitic muscovite micas from the Thistle Field, Brent Group (northern North Sea). See text for explanation.

horizontal bar across the top represents the range of muscovite densities quoted condensed from a number of classic mineralogy texts, and the down-arrow the value of $2.84\,\mathrm{g\,cm^{-3}}$ which was used above. The range of computed values is slightly greater than the 'textbook' range which is to be expected when the range of compositions are considered, together with some analytical error. The point is that if the chemistry and mineralogy can be matched satisfactorily in the inversion then a computed density is likely to be closer to the correct value than an 'average book' value.

Examples of petrophysical estimation from mineralogy logs

In the following part of this contribution a number of case histories are presented in which some petrophysical parameter is estimated from mineralogy logs. The process by which the latter were obtained (modelling strategy and parameters) is described in the light of the problems and aspects of mineral modelling discussed above.

Case history 1: bulk density determination in a sandstone sequence

This study comes from work on wells in the Thistle Field of the northern North Sea Brent where amongst the objectives were to test the use of geochemical log data for the estimation of petrophysical parameters, in particular, matrix density and porosity. In this section the estimation of matrix density is examined. Thistle Field Hole 211/18a-a33, penetrates the Brent Group sequence from the lower half of the Ness Formation, down through the Etive Formation to the upper half of the Rannoch Formation. The Rannoch Formation is interpreted as a prograding delta front (Johnson & Stewart 1985; Budding & Inglin 1981) comprising a very fine highly micaceous sandstone with up to 25% mica (Morton & Humphreys 1983); the sequence coarsens upwards from a shaly base, and locally contains diagenetic calcite doggers. The overlying Etive Formation consists of well sorted fine- to medium-grained subarkosic cross-bedded sandstones, locally very coarse and becoming more massive in the upper part. The Ness Formation is interpreted as a deltaic environment and is characterized by very fine- to coarse-grained sub-arenite to sub-arkose sandstones and siltstones with thin subordinate coals and organic shales.

Throughout the section quartz, K-feldspar (both orthoclase and microcline), albite, phengitic muscovite, biotite, authigenic kaolinite and locally siderite are the main minerals present. Minor phases include organic material, pyrite, illite–smectite, chlorite, zircon, rutile and garnets. In this first example the mineralogy was modeled from the geochemistry of core samples; the latter was obtained by X-ray fluorescence analysis. The oxides SiO_2, TiO_2, Al_2O_3, Fe_2O_3, MgO, CaO, Na_2O and K_2O, together with S a total of nine components, were used for the chemical input. For the minerals fixed compositions were used (i.e., hole specific compositions). For quartz, pyrite and rutile ideal (theoretical) compositions were used in the components matrix while for the remaining minerals (K-feldspar, albite, kaolinite, muscovite, biotite and siderite) 30 or so grains of each mineral were analysed by electron microprobe, and the mean composition taken to represent the 'Brent' mineralogy. From petrographic and X-ray diffraction data on the core samples a table of possible mineral parageneses was constructed and the mineral modelling proceeded by evaluation, for each sample modelled, of each of the possible mineral parageneses. The best fit assemblage was accepted in each case.

Table 6. *Commonly quoted mineral densities*

Mineral	Edmunson & Raymer (1979)	Ellis (1987)	Carmichael (1984)	This study
Quartz	2.65	2.65	2.648	2.65
K-feldspar	2.55	2.57	2.570	2.57
Albite	2.62	2.61	2.620	2.62
Muscovite (d)	2.83	2.84	2.831	2.83
Biotite (d)	3.12	3.22	2.900	3.12
Kaolinite (d)	2.44	2.62	2.594	2.61
Pyrite	5.00	5.01	5.011	5.01
Rutile	4.25	4.23	4.245	4.25
Siderite	3.94	3.96	3.944	3.94

For hydrous minerals that contain water as part of their atomic structure the density quoted is for a dry analysis, i.e., an analysis with only H_2O^+ present, denoted (d).

The inversion procedure was carried out using an Euclidian distance model (a modification of constituent analysis; Fuh 1973). The Euclidian distance model in essence treats the input in terms of the normalized Euclidian distances between mineral phases (end-members) present in the components matrix and therefore takes advantage of the multi-dimensional geometry inherent in the components matrix. The specifics of the algorithm are detailed by Harvey & Lovell (1992) and Lofts (1993). A number of strategies were applied to the basic algorithm in the examples shown here to overcome some problems such as compositional collinearity (Harvey & Lovell 1992), and varying mineralogy and mineral compositions (Lofts 1993).

With the mineral modelling complete matrix density was calculated using 'standard' values for the individual mineral densities. The values used are summarized in Table 6. Estimates of model-derived matrix density are compared to core-measured estimates of matrix density in Figs 4a and 5. A good correlation is observed over the section between both density estimates, showing that an accurate matrix density can be estimated from mineralogy (in this example to approximately $\pm 0.06 \, \mathrm{g\,cm}^{-3}$). The slight systematic error noticeable in Fig. 4a is due to a rounding of the laboratory-derived estimates. The largest discrepancies of $0.06 \, \mathrm{g\,cm}^{-3}$ occur for the highest matrix densities. This may be due to the inherent sample volume size differences that may affect the heterogeneity of each sample. That a good correlation exists with core data is to be expected because of the use of well constrained laboratory elemental data and

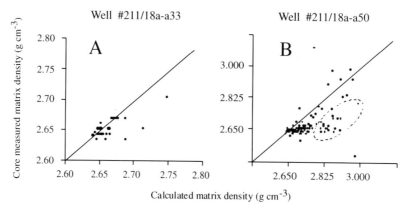

Fig. 4. Comparison of core measured estimates of matrix density to model derived mineralogy estimates, from Thistle Field well 211/18a-a33 (Plot A, left), and from well 211/18a-a50 (plot B, right). Mineralogy and derived matrix density computed from the chemistry of core samples (**a**), and from GLT derived elemental data (**b**). Circled area in plot B indicates the discrepant samples below 9340 ft in Fig. 5; see text for full explanation.

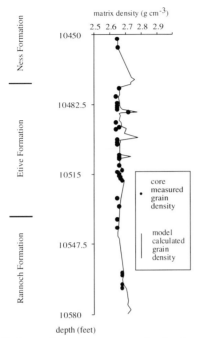

Fig. 5. Comparison of core measured estimates of matrix density to modelled derived mineralogy estimates, from well 211/18a-a33, Thistle Field. Depth in feet.

mineralogical data, but is also a confirmation of the level of information needed to perform accurate mineral inversions.

Another brief example of matrix density estimation is shown in Figs 4b and 6 this time using geochemical logs acquired directly from Schlumberger's Geochemical Logging Tool, the data being derived from hole 211/18-a50 of the Thistle Field. The lithologies are broadly similar to those described above, as is the basic modelling strategy. One difference, however, is that data on only six oxides and one element are available from the GLT (SiO_2, TiO_2, Al_2O_3, Fe_2O_3, CaO, K_2O, S) and these are insufficient to solve for the nine-mineral model used above. A reduction in minerals to seven was achieved by combining (K-feldspar + albite) to give 'total feldspar', and (muscovite + biotite) to give 'total mica'.

In Fig. 4b, except in the circled area, the correlation between estimates is again very good. From Fig. 6 which shows the corresponding downhole log, it is clear that the good correlation exists down to 9340 ft with only small systematic discrepancies of <0.02 g cm^{-3} between 9200–9340 ft. The two larger discrepancies in calculated estimates, at 9280 ft and

9308 ft, are due to localized siderite rich intervals and probably represent a sampling discrepancy. Below 9340 ft a larger overall discrepancy occurs in estimates, up to 0.2 g cm^{-3} (circled in Fig. 4a). These coincide with the presence of silty-sands and muds at the base of the Rannoch through to the Dunlin Formation which lies beneath the Brent Group. It is still not clear why this discrepancy occurs below 9340 ft but with the change in overall lithology there is also a deterioration in hole conditions (see below).

One interesting observation from Fig. 4b is the spread in density values (viz. 2.58–3.0 g cm^{-3}). Conventional estimates of porosity, calculated from one assumed value of matrix density (usually 2.65 g cm^{-3}) would clearly produce porosity values in error for the majority of samples in this section. This follows comments made above and is an observation noted in the Brent Group by Moss (1992).

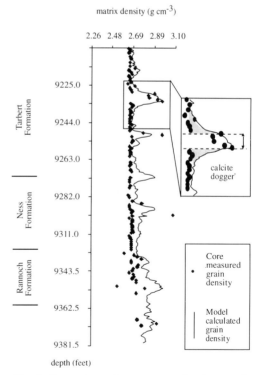

Fig. 6. Comparison of core measured estimates of matrix density and modelled derived mineralogy estimates, from well 211/18-a50, Thistle Field. Matrix density estimated from modelled mineralogy acquired using GLT derived elemental data. Depths are in feet.

Case history 2: porosity determination in a sandstone sequence

Continuing with Thistle Field wells described above the following section outlines the estimation of porosity in the 211/18-a50 hole, the mineralogy of which was calculated directly from geochemical log curves. These porosity estimates were calculated using matrix density estimated from modelled mineralogy above, the log derived bulk density estimate, ρ_b, and a fluid density, ρ_{fl}, of $1.1\,\mathrm{g\,cm^{-3}}$. These estimates of porosity are compared (Fig. 7a) to core-derived estimates of porosity (helium porosity). In general, there is a good correlation of both porosities, although some core estimates are under-estimated (circled in Fig. 7a). These represent the calculated porosity values below depth 9340 ft and are the effect of the poor matrix density estimates seen in Fig. 6.

Figure 7b shows a comparison of core (helium) porosity with a porosity estimate derived from the density tool. The correlation with core is poor in comparison to Fig. 7a and this is emphasiszd in the downhole logs shown in Fig. 8. Density tool estimates below 9340 ft appear widely in error and this may point to the density tool measurement being adversely affected through this interval. A corresponding increase in total thermal neutron absorption capture cross section (CSIG) log at this point also indicates possible environmental effects such as tool stand-off or mudcake buildup (Lofts 1993). Independent model derived estimates of matrix density produce, in this example, a more accurate porosity estimate than when a single matrix is assumed. This is

particularly encouraging as the mineralogy is derived directly from the GLT tool.

As an indirect application of mineral modelling, the calculation of porosity and matrix density as shown in the Thistle Field appears very encouraging. This is especially so for the upper part of the studied section of well 211/18a-50 (viz. 9216–9340 ft) where estimates of both parameters are in very good agreement with core and in the case of porosity, appear superior to those values estimated using a single matrix. The discrepancies below 9340 ft may be due to (a) core sampling inconsistencies (in the measurement of helium porosity in silty sands), (b) sample volume differences between core and log data, (c) errors in the bulk density log estimates at this interval, or (d) incorrect mineral estimates from the modelled mineralogy.

Most samples over thick and consistent lithological units (most samples above 9340 ft) have accurate estimates of matrix density. Discrepancies occur, however, at the boundary between two contrasting lithologies. This is attributed firstly to the averaging in the geochemical log elemental processing and secondly, the intrinsic vertical resolution of the tool measurement which is of the order of 0.5–1 m (Hertzog *et al.* 1989). In effect, the elemental response of a thin-bed horizon, during processing, will be averaged over six 15 cm sample intervals that lie directly above and below the sample. The result is a 'smearing' of the elemental data over those sample intervals. This, in turn, is carried through to the modelled estimates of mineralogy.

The averaging of mineral estimates (called here the bed boundary effect) has a clear effect

Well # 211/18a-a50

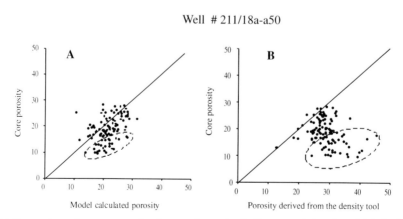

Fig. 7. (a) Comparison of core measured estimates of porosity (helium porosity) and model derived estimates, as calculated from GLT elemental log data well 211/18-a50, Thistle Field. (b) Comparison of core helium porosity with porosity derived from the density tool.

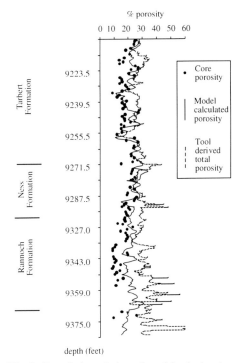

% porosity

depth (feet)

Fig. 8. Downhole variation of model calculated porosity derived from the density tool and core helium porosity. A good correlation between model derived porosity and core porosity is seen down to 9340 ft. The discrepancy below 9300 ft is probably due to the matrix density differences, affected by the silt-siderite lithology of this interval. These data are circled in Fig. 7a and b.

on estimates of matrix density in more localised horizons through the 211/18-a50 section. For example, the increase in calculated matrix density up to 2.9 g cm^{-3}, at the 'calcite dogger horizon', is verified by core measurements. Core measurements on the boundary indicate a sharp drop, to more typical sandstone densities (viz. 2.65 g cm^{-3}) as illustrated in Fig. 6 (insert). The model calculated matrix density values for sandstone samples surrounding the 'dogger' on the boundary, however, show a smooth drop in calculated matrix density, over 1–2 m before a more typical sandstone density is recorded. This smooth drop in density values is a good example of the bed boundary effect indirectly produced from GLT-derived chemistry, that has been passed through into the mineralogy. This, in turn, affects the calculation of matrix density and will therefore ultimately affect porosity. The effect of a 0.2 g cm^{-3} error (increase) in estimation of matrix density at 9244 ft, for example, is to double the porosity estimate.

Enhanced estimates of both grain density and porosity appear possible, over good intervals, using GLT-derived data. This will hold true as long as the mineralogy model is correct and the interval is reasonably homogeneous. It is not unreasonable to expect that estimates may, in fact, be superior to the core measured values by the nature of the sample volume measured. Great care must be taken in interpreting these derived parameters over bed boundaries and in thin-bedded, heterogeneous units as these bed boundary effects may seriously affect estimates.

Case history 3: bulk density determination in oceanic basement

During Leg 118 of the Ocean Drilling Program 500 m of oceanic Layer 3 gabbros were drilled at Site 735 on the edge of the Atlantis II Fracture Zone of the Southwest Indian ridge. This was the first significant recovery of seismic Layer 3 rocks from the ocean floor, and Hole 735B has been of considerable scientific importance in constraining ideas about models of the oceanic crust. With nearly complete core recovery and an extensive logging program 735B offers an excellent opportunity to test mineral models and derived parameters.

The lithology throughout the hole is gabbro that shows significant variation in its texture, from isotropic to highly deformed planer fabrics, and in its mineralogy, particularly with respect to the abundance of iron and titanium oxides. From examination of the core the 500 m sequence was broken into six lithological units, I–VI, which are briefly distinguishable as follows (Dick *et al.* 1991).

Unit I.   Massive foliated gabbronorite with thin olivine gabbro layers in the upper two-thirds and dominantly olivine gabbro in the lower third. The latter is regarded as transitional to Unit II.

Unit II.   Dominantly massive olivine gabbro with minor intrusive microgabbros and oxide–olivine rich bands.

Unit III.   Disseminated oxide-olivine gabbro transitional from Unit II, and comprising roughly equal proportions of olivine gabbro and disseminated oxide–olivine gabbro.

Unit IV.   Massive oxide-rich olivine gabbro, distinguished from Unit III by the absence of disseminated oxide–olivine gabbro, and the exceptionally high concentration of oxide. Total iron (as FeO) can exceed 30 wt%, and TiO_2, 9 wt%.

Unit V. Relatively uniform massive olivine gabbro with a scarcity of iron and titanium oxides. Separated from Unit IV by a distinct tectonic break.

Unit VI. Compound olivine and oxide–olivine gabbro with some minor intrusive rocks including microgabbros and troctolites.

The mineralogy throughout virtually the whole of the section comprises greater or lesser proportions of calcic plagioclase, diopsidic clinopyroxene, orthopyroxene, olivine and oxide (Hébert *et al.* 1991; Ozawa *et al.* 1991; Natland *et al.* 1991). All the main mineral phases show a surprisingly wide range of composition, the plagioclase ranging between about An_{30} and An_{70}, olivine between about Fo_{30} to Fo_{85} and other ferromagnesians phases (clino- and orthopyroxenes) showing a similar range of FeO–MgO variation. The oxide minerals, termed here ilmenite and magnetite, are actually ilmenite–hematite and magnetite–ulvaspinel solid solutions (Natland *et al.* 1991), which in these gabbros co-exist as oxide pairs, though through recrystallization have achieved compositions close to the ilmenite and magnetite end-members. The variation in oxide composition is demonstrated in Fig. 9.

The geochemical logs obtained on Leg 118 have been described and developed into a chemostratigraphy by Pelling *et al.* (1991). Good quality logs were obtained for Si, Al, Ca, Ti, Fe and S. Data were obtained also for

potassium but the K_2O concentration is low throughout, and potassium is not a primary element in any of the key mineral phases. In addition an estimate of magnesium, an important and significant element in basic igneous rocks, was obtained from comparison of calculated and measured photoelectron factors. Hence, for mineral modelling GLT data are available on seven elements which are important constituents of the minerals present in the gabbros, with K_2O in addition. Ignoring the K_2O data this implies a maximum of seven minerals which can be expected from the inversion process.

For purposes of mineral modelling from these GLT logs all the main minerals can be considered to come from essentially binary systems. Plagioclase, for example, in this case was considered as a single phase with an allowable composition between An_{25} and An_{75}. With the ferromagnesian minerals the different mineral groups can be considered here to be binary systems varying between Mg and Fe end-members, though published mineral data show that the full range need not be considered (e.g. olivine range: about Fo_{30} to Fo_{85} – see above). The chemical relationships between the Mg–Fe silicate phases is summarized in Fig. 10 where part of the tetrahedral [CaO]–[SiO_2]–[MgO]–[FeO] system is shown. The clinopyroxene, orthopyroxene and olivine mineral series are depicted as three oblique lines, each with the Mg end-member to the left, and Fe end-member to the right. For a given MgO/FeO ratio the three minerals can be modelled simultaneously; an example is shown in Fig. 10 as three squares at the apices of a compositional triangle which itself has a fixed MgO/FeO ratio. The constant MgO/FeO ratio used for modelling the 735B gabbros is an over-simplification of the measured mineral data but was considered sufficiently close to be acceptable. For the oxides a pseudo-binary system, as shown in Fig. 10 could be used, but would remove one degree of freedom from the 'free' iron concentration which was in turn being simultaneously distributed between the ferro-magnesian silicates. As a result the oxides were modelled separately as ilmenite and magnetite, the actual compositions being used in the inversion corresponding to the averages of each cluster in Fig. 10. Sulphur occurs only in pyrite in this sequence of rocks, so pyrite was included as an essential mineral in the modelling to mop up the sulphur. In this way it was possible to model the mineralogy of the gabbro sequence using the mineral assemblage: [plagioclase, orthopyroxene, clinopyroxene, olivine, ilmenite, magnetite, pyrite], as a set of

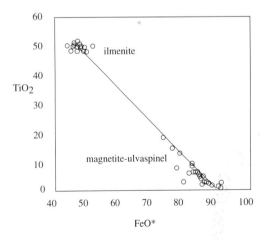

Fig. 9. Composition of oxide minerals from the gabbros of Hole 735B (Leg 118 of the Ocean Drilling Program). The line between the groups of compositions joins the theoretical compositions of ilmenite and magnetite.

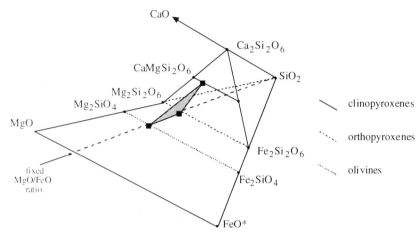

Fig. 10. Part of the [CaO]–[SiO₂]–[MgO]–[FeO] tetrahedral [CaO]–[SiO₂]–[MgO]–[FeO] system showing the compositional relationships between olivines, orthopyroxenes and clinopyroxenes, with Mg end-members to the left, and Fe end-members to the right. The grey triangle is in a plane of constant MgO/FeO ratio, and for a given model olivine, orthopyroxene and clinopyroxene compositions, shown as squares at the apices of the triangle, could be used to represent the compositions of coexisting ferromagnesium phases. The axes are not to scale.

seven minerals, and a fully determined system of equations. At each depth interval the plagioclase composition was iterated in steps of 5% Anorthite, and the MgO/FeO ratio in steps of 0.05, take into account the compositional variation on the ferromagnesian silicates, in the manner described by Harvey *et al.* (1990).

The estimates of bulk density were derived directly from the computed mineralogy using both combinations of 'book' values of density for the parameter coefficients, and densities calculated from the compositions of the minerals (described above) which occurred in the 'best fit' models at each depth interval. The latter were distinctly closer to bulk densities measured on mini-cores in the laboratory. A summary of the bulk density results for Hole 735B are given in Fig. 11. On the left is shown the density derived from the neutron porosity (NPHI), using $\rho_m = 2.9$, and on the right, the bulk density derived from the geochemical logs. The sets of closed circles are laboratory based measurements of bulk density made on mini-cores. There is reasonable agreement with both sets of data, though the GLT-derived curve is overall the better match particularly where high oxide concentrations occur (note lithological units IV and VI in this regard). The importance of the oxides in the modelling process is, of course, because they have significantly greater densities than the associated silicates, and much more care must be taken with the choice of mineral compositions as a whole because of the impor-

tance of partitioning, particularly iron, correctly between the oxides and silicates. While this is a relatively unusual application the same comments can apply to sediments rich in heavy minerals.

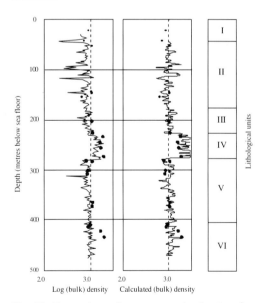

Fig. 11. Comparison of core measured estimates of matrix density, modelled derived mineralogy estimates and the log density from Hole 735B (Leg 118 of the Ocean Drilling Program) which was drilled into Layer Three Gabbros on the South-West India Ridge. Closed circles are core bulk density measurements.

Case history 4: magnetic susceptibility determination in oceanic basement

One of the objectives of obtaining a precise and accurate mineral inversion from the Hole 735B geochemical log data was to use the derived mineralogy to create a magnetic susceptibility log. Such a log was run in 735B during ODP Leg 118, and provided the opportunity to make a direct comparison with the geochemical computed log. Full details of the magnetic logging are provided by Pariso *et al.* (1991) and J. E. Pariso is thanked for providing the magnetometer logging data used here.

The computed magnetic susceptibility is essentially a function only of the oxide minerals and their concentrations; none of the other minerals present contribute significantly to this magnetic property. As described above in the mineral modelling of the 735B gabbros the oxide minerals, ilmenite and magnetite, were solved as separate phases. The ilmenite abundance is

governed essentially by the concentration of TiO_2, while the amount of magnetite is dependent upon the partitioning if iron between magnetite, pyrite, the ferromagnesian silicate phases as well as ilmenite.

The results of the magnetic susceptibility determination are summarized in Fig. 12. Figure 12a shows the geochemical logs for TiO_2 and FeO* on which the oxide estimates are largely dependent, Fig. 12b the computed mineralogy logs for ilmenite and magnetite, and Fig. 12c the magnetic susceptibility estimates derived from the oxide logs. Also plotted in Fig. 12c is the susceptibility curve measured by the magnetometer tool.

There is excellent correlation, as should be expected from the petrological data, between FeO and TiO_2 (Fig. 12a), with Unit IV showing particularly high values for both elements. The oxide–olivine rich bands in Unit II also show up clearly, with two more dominant bands just above and below the 100 mbsf (metres below sea

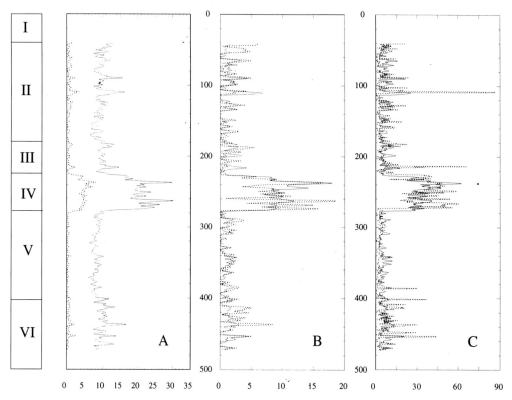

Fig. 12. Variation downhole in ODP Hole 735B from top (left) to bottom (right) at 500 m below sea floor (mbsf). Depth scale in mbsf. (**a**) Iron (solid line) and titanium (dotted line) geochemical logs; scale in weight percent oxide. (**b**) Computed oxide mineralogy, magnetite (solid line), ilmenite (dotted line); scale in weight percent abundance of the minerals. (**c**) Derived magnetic susceptibility (solid line), magnetic susceptibility log (dotted line); scale in 10^{-3} SI units. See text for full explanation.

floor) level. The relatively uniform Unit V is seen to be characterized by fairly constant FeO and TiO$_2$ levels and in the case of TiO$_2$, comparatively low concentrations. The derived oxide (ilmenite–magnetite) mineralogy (Fig. 12b) is in very good agreement with available petrographic data, and reflects the same features seen in the oxide curves. The final and important comparison is seen in Fig. 12c where the computed magnetic susceptibility, k_c, is plotted together with the measured magnetic susceptibility, k_m.

There is excellent agreement between the two curves, but a number of points of interest emerge. Over most of the hole k_m peaks occur at a slightly shallower depth (about 3 m) than k_c. This is clearly seen by an offset of the curves immediately above and below Unit IV, and at one of the dominant oxide–olivine bands just below the 100 mbsf point within Unit II. This is almost certainly an error in depth-shifting. In addition, in several places where corresponding peaks on the two curves can be matched k_m is almost always higher than k_c, and sometimes by as much a five times (see peaks at 215 and 403 mbsf). Part of this problem results from the much greater degree of smoothing that is generally applied to geochemical logs (in contrast to the same phenomenon caused by 'bed boundary' effects as encountered in the Brent sandstones above). A second possibility concerns the values of the parameter coefficients that are employed. It is far more difficult to obtain a set of reliable working coefficients for magnetic susceptibility than it is, for example, for density, where there is a choice of tabled, theoretical, measured or calculated values. However, as discussed earlier, even with density the actual parameter coefficients to be used in a given model may not be easy to obtain.

Summary and conclusions

(1) Several important petrophysical parameters can be calculated directly from the quantitative mineralogy of a formation. Examples presented here include: matrix density, and indirectly, porosity, which together have been considered by a number of authors (see Introduction), and magnetic susceptibility (this work). Other possible physical parameters include Cation Exchange Capacity (Herron & Grau 1987), thermal conductivity (Dove & Williams 1988), heat flow (Anderson & Dove 1987), fluid saturation (Hastings 1988), neutron capture cross-section (Herron 1987b), and possibly even permeability (Herron 1987a). With

good data these estimates can be a significant improvement on those obtained through conventional log analysis

(2) Good estimates of at least some of these parameters can be obtained from appropriate logging tool (e.g. Schlumberger's Geochemical Logging Tool (GLT) or Geochemical Reservoir Analyzer (GRA)) data or from chemically determined core samples, provided that appropriate mineral inversion methods are employed. Such methods must solve for the actual mineral assemblage present in the formation, and not some possible or theoretical set of minerals. The compositions of the minerals used in any solution must be compatible with the known geology, and the final fit to the chemistry to show a minimal difference.

(3) Given that the mineral abundance estimates are acceptable the accuracy of the derived petrophysical parameter estimates depends on how well the parameter coefficients are known. In the case of the magnetic susceptibility case discussed above there is considerable uncertainty about the parameter coefficients, and there is the likelihood that they vary quite a lot throughout the hole studied. Despite this good qualitative agreement was achieved between the computed and measured susceptibilities.

(4) The determination of porosity from density data is particularly sensitive to the accurate determination of matrix density. For density estimation single 'book' estimates of minerals with wide possible compositional ranges should be avoided where possible and be replaced by densities calculated from the compositions of the minerals.

References

ANDERSON, R. & DOVE, R. 1988. The determination of heat flow in a wellbore in the South Eugene Island area of offshore Louisiana: implications for fluid migration and hydrocarbon location in the subsurface. *Transactions of the Spectroscopy and Geochemistry Symposium*, Schlumberger-Doll Research, Ridgefield, CT. Paper K.

BUDDING, M. C. & INGLIN, H. F. 1981. A reservoir geological model of the Brent Sands in the southern Cormorant. *In*: ILLING, L. V. & HOBSON G. D. (eds) *Petroleum Geology of the continental shelf of North-West Europe.* Heyden, London, pp. 326–334.

CARMICHAEL, R. S. (ed.) 1984. *Handbook of Physical Properties of Rocks.* 3, CRC Press, Inc.

CHAPMAN, S., COLSON, J. L., FLAUM, C., HERTZOG, R. C., PIRIE, G., SCOTT, H., EVERETT, B., HERRON, M. M., SCHWEITZER, J. S., LA VIGNE, J., QUIREIN, J. & WENDLANDT, R. 1987. The emergence of Geochemical Well Logging. *The Technical Review*, **35**, 27–35.

CROSS, W., IDDINGS, J. P., PIRSSON, L. V. & WASHINGTON, H. S. 1903. *Quantitative classification of igneous rocks*. University of Chicago, Chicago.

DICK, H. J. B., MEYER, P. S., BLOOMER, S., KIRBY, S., STAKES, D. & MAWER, C. 1991. Lithostratigraphic evolution of an in-situ section of oceanic layer 3. *Proceedings of the Ocean Drilling Program, Scientific Results*, **118**, 439–515.

DOVE, R. E. & WILLIAMS, C. F. 1988. Thermal conductivity estimated from elemental concentration logs. *Transactions of the Spectroscopy and Geochemistry Symposium*, Schlumberger-Doll Research, Ridgefield, CT. Paper J.

EDMUNDSON, H. & RAYMER, L. L. 1979. Radioactive logging parameters for common minerals. *Log Analyst*, **20**, 38–47.

ELLIS, D. V. 1987. *Well logging for Earth Scientists*. Elsevier, New York.

GRAU, J. A., SCHWEITZER, J. S., ELLIS, D. V. & HERTZOG, R. C. 1989. A geological model for gamma-ray spectroscopy logging measurements. *Nuclear Geophysics*, **3**, 351–359.

FUH, T. M. 1973. The principal of constituent analysis, with special reference to the calculation of weight percentages of minerals in metamorphic rocks. *Canadian Journal of Earth Science*, **10**, 657–669.

HARVEY, P. K. & LOVELL, M. A. 1991 Nuclear Logging and Mineral Inversion in Sedimentary Sequences. *1991 IEEE Nuclear Science Symposium and Medical Imaging Conference*, 1103–1105.

—— & ——1992. Downhole mineralogy logs: mineral inversion methods and the problem of compositional colinearity. *In*: HURST, A., GRIFFITHS, C. M. & WORTHINGTON, P. F. (eds) *Geological Applications of Wireline Logs II*. Geological Society, London, Special Publications, **66**, pp. 361–368.

——, BRISTOW, J. F. & LOVELL, M. A. 1990. Mineral transforms and downhole geochemical measurements. *Scientific Drilling*, **1**, 163–176.

——, LOVELL, M. A., & BRISTOW, J. F. 1991. The Interpretation of Geochemical Logs from the Oceanic Basement: Mineral Modelling in Ocean Drilling Program (ODP) Hole 735B. *Nuclear Geophysics*, **5**, 267–277.

HASTINGS, A. F. 1988. Using the derived elemental concentrations to improve the accuracy of fluid saturations determined from well logs. *Transactions of the Spectroscopy and Geochemistry Symposium*, Schlumberger-Doll Research, Ridgefield, CT. Paper T.

HÉBERT, R., CONSTANTIN, M. & ROBINSON, P. T. 1991. Primary mineralogy of Leg 118 gabbroic rocks and their place in the spectrum of oceanic mafic igneous rocks. *Proceedings of the Ocean Drilling Program, Scientific Results*, **118**, 3–20.

HERRON, M. M. 1987a. Estimating the intrinsic permeability of clastic sediments from geochemical data. *SPWLA 28th Annual Logging Symposium*, paper HH, 1–23

——1987b. Future applications of elemental concentrations from geophysical logging. *Nuclear Geophysics* **1**, 197–211.

—— & GRAU, J. A. 1987. Clay and framework mineralogy, cation exchange capacity, matrix density and porosity from well logging in Kern County, California. *American Association of Petroleum Geologists Bulletin*, **71**, 567–575.

HERTZOG, R., COLSON, L., SEEMAN, B., O'BRIAN, M., SCOTT, H., MCKEON, D. GRAU, J. A., ELLIS, D., SCHEWITZER, J. & HERRON, M. M. 1989. Geochemical logging with spectrometry tools. *SPE Formation Evaluation*, **4**, 53–162.

JOHNSON, R. J. & STUART, D. J. 1985. Role of clastic sedimentology in the exploration and production of oil and gas in the North Sea. *In*: BRENCHLEY, P. J. & WILLIAMS, B. P. J., (eds) *Sedimentology: Recent Developments and Applied Aspects*. Geological Society, London, Special Publications, **18**, 249–310.

LOFTS, J. C. 1993. *Integrated Geochemical and Geophysical Studies of Sedimentary Reservoir Rocks*. PhD Thesis, University of Leicester.

——, HARVEY, P. K. & LOVELL, M. A. 1994. A Stochastic Approach to Mineral Modelling of Log Derived Elemental Data. *Nuclear Geophysics*, **8**, 135–148.

——, —— & ——1995a. Reservoir characterisation from downhole mineralogy. *Marine and Petroleum Geology*, **12**, 233–246.

——, —— & ——1995b. The Characterisation of Reservoir Rocks Using Nuclear Logging Tools: Evaluation of Mineral Transform Techniques in the Laboratory and Log Environments. *The Log Analyst*, **36**(2), 16–28.

MORTON, A. C. & HUMPHRIES, B. 1983. Petrology of the Middle Jurassic sandstones from the Murchison Field, North Sea. *Journal of Petroleum Geology*, **5**, 245–260.

MOSS, B. 1992. The petrophysical characteristics of the Brent sandstones. Geology of the Brent Group. *In*: MORTON, A. C., HASZELDINE, R. S., GILES, M. R. & BROWN, S. (eds) *Geology of the Brent Group*. Geological Society, London, Special Publications, **61**, 471–496.

NATLAND, J. H., MEYER, P. S., DICK, H. J. B. & BLOOMER, S. H. 1991. Magmatic oxides and sulphides in gabbroic rocks from Hole 735B and the later development of the liquid line of descent. *Proceedings of the Ocean Drilling Program, Scientific Results*, **118**, 75–107.

OZAWA, K., MEYER, P. S. & BLOOMER, S. H. 1991. Mineralogy and textures of iron-titanium oxide gabbros and associated olivine gabbros from Hole 735B. *Proceedings of the Ocean Drilling Program, Scientific Results*, **118**, 41–73.

PARISO, J. E., SCOTT, J. H., KIKAWA, E. & JOHNSON, H. P. 1991. A magnetic logging study of Hole 735B gabbros at the Southwest Indian Ridge. *Proceedings of the Ocean Drilling Program, Scientific Results*, **118**, 309–321.

PELLING, R., HARVEY, P. K., LOVELL, M. A. &
GOLDBERG, D. 1991. Statistical analysis of
geochemical logging tool data from Hole 735B,
Atlantis Fracture Zone, southwest Indian Ocean.
*Proceedings of the Ocean Drilling Program,
Scientific Results*, **118**, 271–284.

WRIGHT, T. L. & DOHERTY, P. C. 1970. A linear
programming and least squares computer method
for solving petrologic mixing problems. *Geological Society of America Bulletin*, **81**, 1995–2008.

Petrophysical estimation of permeability as a function of scale

PAUL F. WORTHINGTON

Gaffney, Cline & Associates, Bentley Hall, Blacknest, Alton, Hants GU34 4PU, UK

Abstract: The prediction of reservoir intergranular permeability from core and log measurements of other physical properties is demonstrably refined by taking a more scientific account of the scale of the predictors. Further, the expression of an empirical algorithm that relates permeability to a measured predictor can vary significantly from the core through the log to the well-test scales, with a reduction in data scatter as scale increases. In the most common case of porosity–permeability relationships, the coefficients and exponents that define the petrophysical algorithms are themselves functions of scale. The axial scale dependence of these predictive algorithms is quantified through filtering of regularly sampled core data for the specific case of primary flow and for four different distributions of characterized strata. The algorithms initially diverge as the scale of the processed data increases beyond the spatial resolution of the input core measurements, and they reconverge as the running means become more generally representative. The alternative method of predicting intergranular permeability from measurements of apparent formation factor also shows a scale dependence, which here is quantified through porosity-governed averaging of irregularly sampled core data. The algorithm so generated shows encouraging agreement with that established from a correlation of permeability inferred from drawdown tests with field-measured apparent formation factor, normalized to a reference water resistivity. The procedures used to investigate the scale dependence of petrophysical relationships are applicable to all primary clastic reservoirs that require to be characterized at the flow unit scale. They offer a *modus operandi* for establishing a fit-for-purpose, scale-reconciled relationship between permeability and a selected petrophysical predictor for a given reservoir situation. This approach should therefore lead to more reliable estimates of field permeability for input to reservoir simulation.

One of our overriding goals in petrophysics is to develop a quality assurance scheme for the core-controlled log analysis of reservoir rocks. At present we benefit from rigorous calibration checks on the laboratory and downhole measurements that form the basis of our subject, but the interpretation of these measurements, which can follow a variety of deterministic and statistical methodologies, has comparatively little quality control.

The principal difficulty in developing a quality assurance scheme for petrophysical interpretation is the lack of control points against which we might evaluate the reliability of the interpretation procedure at progressive stages. In the strict sense, the existence of such control points would pre-suppose that we already know the answer before the interpretation exercise commences. Clearly, this analogy to the calibration of measurements is not practicable in the realm of interpretation. An alternative approach, which is feasible from a petrophysical stand-point, would be to estimate reservoir parameters using two independent measurement and/or interpretation procedures and to adopt the degree of convergence of the two predictive methods as an indicator of quality. A specific objective of this kind is to attain agreement between permeability inferred from drawdown or build-up tests, on the one hand, and permeability interpreted from core-calibrated log responses, on the other.

The evaluation of the permeability of heterogeneous clastic rocks from core or downhole measurements of other physical properties remains one of the most important objectives in reservoir geoscience (Ahmed *et al.* 1991). The usual practice of industry is to relate porosity and/or some other indicator(s) to permeability through a regression of core data (Jensen & Lake 1985; Wendt *et al.* 1986; Nelson 1994), and then to apply the resulting relationship(s) to well-log data at a much larger scale. The 'permeability logs' so derived are subsequently depth-averaged so that zonal means might be compared with the interpretations of drill-stem tests. These comparisons often expose large discrepancies between predicted and measured field permeabilities. If we are to advance our quality assurance objectives in the area of petrophysical interpretation, these disparities will need to be reduced.

The purpose of this paper is to take some initial steps towards this goal by investigating how core petrophysical data might be most effectively applied to the petrophysical prediction of permeability from wireline-log data and how the log analyses might then be scaled up for

From Lovell, M. A. & Harvey, P. K. (eds), 1997, *Developments in Petrophysics*, Geological Society Special Publication No. 122, pp. 159–168.

comparison with well-test interpretations. The treatment does not set out to be exhaustive, for that would require a treatise based on understanding that does not yet exist, but rather to illustrate some simple yet practical applications that might reduce uncertainty and add value in reservoir description and its subsequent synthesis for simulation. In so doing, the subject matter is confined to the most commonly used field predictors of permeability in petroleum and groundwater petrophysics.

Essentially, we shall be concerned with intergranular flow in the near-wellbore zone of clastic reservoirs, for which core and log data are presumed to be available. For simplicity, it will be assumed that the wells are vertical and that any discernible layering is horizontal. It will also be presumed that all data have been environmentally corrected. Further, we suppose that there has been pre-identification of any facies-governed reservoir zones that might require mutually distinct predictive relationships (e.g. Craig 1991; Chork *et al.* 1994; Lin & Salisch, 1994) and that we are operating within one of those zones.

Basic algorithms for permeability estimation

The evaluation of the permeability of subsurface reservoirs is a primary ongoing objective in the petroleum and water industries. The prohibitively high cost of extensive programmes of reliable direct measurements, either on core samples in the laboratory or through targeted well tests, has led to permeability being estimated through downhole sensing in a way that is calibrated by a limited suite of direct laboratory or field measurements. The most common permeability indicators are based on petrophysical correlations (Fig. 1).

The origins of the petrophysical estimation of permeability are synthesized in the classic paper

of Archie (1942), which contained a specific form of the following expression that is generally known as Archie's first law:

$$\phi = (a/F)^{1/m} \qquad (1)$$

where ϕ is total porosity, F is the formation resistivity factor (the ratio of formation resistivity to formation-water resistivity, this ratio being taken as independent of reservoir salinity), and a and m are empirical constants appropriate to the formation, reservoir or reservoir zone under consideration. Equation (1) shows a linear negative gradient when plotted bilogarithmically. It was established using core data and was introduced in the days when there were no wireline porosity tools. At that time, the conventional approach to determining formation porosity was to calculate F from a resistivity log in a water zone and a measurement or interpretation of formation-water resistivity. Porosity was then calculated using equation (1), either in a reservoir-specific form or by using the recommended defaults of $a = 1$ and $m = 2$. In hydrocarbon reservoirs, the calculated porosity was assumed to apply also to the (uncored) hydrocarbon leg, where porosity could not be determined from resistivity logs in this way.

Archie (1942) also presented an analogous relationship between intergranular permeability K and formation factor:

$$K = (b/F)^{1/c} \qquad (2)$$

where b and c are empirical constants. Like equation (1), equation (2) shows a linear negative gradient when plotted bilogarithmically (Fig. 2), it was established using core data and could be applied only to resistivity logs run in water zones. As before, in the petroleum industry the calculated permeabilities from the water zone were extrapolated to the (uncored) hydrocarbon leg.

The advent of porosity tools in the 1950s allowed porosity to be determined directly from wireline logs in the hydrocarbon leg. Therefore equation (1) no longer had to be used as an extrapolation facility and its role in contemporary hydrocarbon petrophysics has evolved differently (Worthington 1986). For the same reason, permeability could now be estimated from other predictors measured in the hydrocarbon leg itself, and the most obvious predictor was, of course, porosity from wireline logs. A number of predictive equations emerged, one of which could be dervived from equations (1) and (2):

$$K = d\phi^e \qquad (3)$$

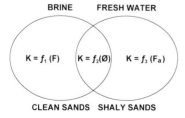

Fig. 1. Nature of single-predictor algorithms for estimation of permeability K. (F = formation factor; F_a = apparent formation factor; ϕ = porosity; f = generic function.)

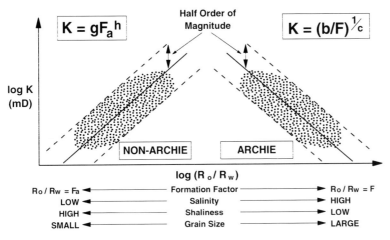

Fig. 2. Schematic relationships between formation factor F, apparent formation factor F_a and permeability K.

where the empirical constants d and e are such that:

$$d = (b/a)^{1/c} \qquad (4)$$

and

$$e = m/c. \qquad (5)$$

Equation (3), established using core data, shows a linear positive gradient when plotted bilogarithmically (Fig. 3). It allows estimates of permeability to be made from wireline porosities obtained in the hydrocarbon zone. Note that the conventional practice is to characterize equation (3) using helium porosities and air permeabilities and this implies that the resulting algorithm is set within the total porosity system. It also suggests unrealistically that the wireline predictand is air permeability. Improved interpretative algorithms would be secured through correlating total porosity or some effective porosity, such as that derived from nuclear

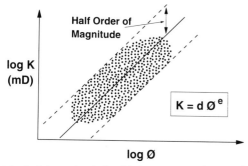

Fig. 3. Schematic relationship between porosity ϕ and permeability K.

magnetic resonance, with liquid permeability. Despite these concerns, equation (3) and similar forms have been widely used as permeability indicators in petroleum reservoirs, for which they are no longer tied back to the Archie equations (Nelson 1994; Worthington 1995).

The water industry has also drawn upon Archie's laws, but in a different way. First, the conditions governing equation (1) are rarely satisfied in groundwater exploration and development. These conditions include a saline interstitial water with equivalent NaCl concentration in excess of 20 000 ppm. At the much lower salinities targeted by hydrogeologists in the supply sector, equation (1) can break down catastrophically (Barker & Worthington 1972). Therefore, equation (1) is not widely used in evaluating the porosity of freshwater aquifers, even though its early application to water zones might initially have made it seem attractive.

Second, when applied to groundwater problems, equation (2) is also affected by departures from the Archie conditions governing formation water salinity. However, in this case, equation (2) progressively breaks down with decreasing salinity and then reconstitutes itself with positive linear gradient on a bilogarithmic plot as salinity is reduced further into the freshwater range (Fig. 2). The reconstituted expression then takes the form:

$$K = gF_a^{\,h} \qquad (6)$$

where g and h are empirical constants established from core analysis or directly from field data. The parameter F_a is termed an apparent formation factor. Although it remains the ratio of formation resistivity to formation-water resistivity, F_a is no longer presumed to be independent of water salinity, because the Archie conditions are not satisfied. A physical explanation of this trend reversal has been given by Worthington (1983) and it has been illustrated by Biella *et al.* (1983).

Equation (6) is the preferred permeability indicator in the water industry, and there have been many examples of its application (e.g. Alger 1966; Croft 1971), although the documented case histories do not always take proper account of the salinity dependence of F_a. Permeabilities should be measured using simulated formation water. Expressions of the form of equation (3) are less prominent in hydrogeology, because the standard logging suite for water wells does not include oilfield porosity tools, largely on grounds of cost.

Equations (1)–(6) have their origins substantially in core analysis. It is common practice, especially in the petroleum industry, to establish the empirical constants at the core scale and then to apply these at the larger scale afforded by well logs. In some cases, this scale transgression does not matter: in other cases, it does (Worthington 1994).

In this paper I shall examine the application of the core-derived equations (3) and (6) to permeability estimation at the wireline log scale (and beyond) and suggest how modified practices might lead to improved permeability predictors without increasing costs. Therefore the discussion will be set within the context of standard industry procedures.

Porosity as a permeability predictor

This investigation has drawn upon a database of measurements on some 400 core plugs from shallow wells that penetrate the Permo-Triassic Sherwood Sandstone of the United Kingdom. The database comprises porosities determined using a saturation method and permeabilities measured using simulated formation waters. At the outset, a five-layer sandstone model was conceived with the layers being separated by impermeable shales (Fig. 4). Each sandstone layer has a specified thickness of 2.28 m and it is divided into 15 sublayers of thickness 0.152 m. Each sublayer has a specified bulk density and porosity, and these densities and porosities are distributed cyclically within each layer. Each sublayer has been sampled by one horizontal core plug. Therefore, in this model, each core plug represents a sublayer of specified thickness and porosity. This information has been used to simulate an environmentally corrected density

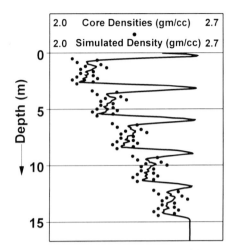

Fig. 4. Sandstone model with core porosities assigned cyclically to each of the five reservoir layers, which are separated by intraformational shales.

log response throughout the model system (Fig. 4). The density log describes an average of the core densities. With such a cyclic model, the average core porosity across a sufficiently large interval will be very similar to the average density log response for the number of digital levels sampled.

Permeabilities have been assigned to the hypothetical core plugs by seeking matching porosities within the Sherwood Sandstone core analysis database and selecting the corresponding measured permeability. For each sublayer porosity, the database was scanned, the first matching porosity to within the generally accepted measurement tolerance of ±0.5 porosity units was identified, the corresponding permeability was imported to the model, and the dataset in question was deleted from the source database prior to rescanning. The process was repeated until all the model sublayers had been assigned a permeability. In this way, it has been possible to impart field credibility to the model without losing the control on its form.

Figure 5a shows a bilogarithmic crossplot of porosity ϕ v. permeability K for the selected subset of the Sherwood Sandstone database that corresponds to the model succession. Regression of Y ($\log K$) on X ($\log \phi$) has furnished the expression:

$$\log K = 5.66 + 5.13 \log \phi. \qquad (7)$$

This type of regression has traditionally been favoured because the error in core-measured porosity, typically about 2.5% for a porosity of 0.20, is considerably lower than that for

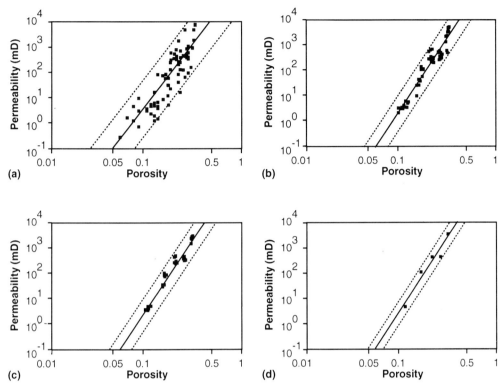

Fig. 5. Crossplots of porosity v. permeability with regression lines for the Sherwood Sandstone based on: (**a**) measured data from single core plugs; (**b**) five-point running means of core data; (**c**) nine-point running means of core data; (**d**) averages of core data for each of the five model zones.

core-derived permeability, typically 15–30% (Thomas & Pugh 1989). There is an uncertainty of about one order of magnitude associated with permeabilities predicted using equation (7). Although typical of many core-derived relationships that have not been sorted according to facies, the inherent uncertainty compares unfavourably with the realistic target uncertainty for predictive permeability algorithms of about half an order of magnitude, as illustrated in Figs 2 & 3.

Figure 5b illustrates a similar crossplot, but with the data of Fig. 5a having been subjected to an unweighted five-point running mean within each of the five model layers (Fig. 4). Porosities and (horizontal) permeabilities have been averaged arithmetically. This means that each data point no longer represents a single sublayer of thickness 0.152 m but rather five sublayers of overall thickness 0.76 m, a vertical distance that corresponds broadly to the spatial resolution of the density log. The regression line has the equation:

$$\log K = 6.22 + 5.60 \log \phi. \qquad (8)$$

The uncertainty in predicted permeability is now approaching the target of one half an order of magnitude. Equation (8) might therefore be suitable for application to porosities interpreted from density log response.

Figure 5c shows a bilogarithmic crossplot of porosity v. permeability in which the data of Fig. 5a have been subjected to an unweighted nine-point running mean within each of the five model layers (Fig. 4). Again, the averages are arithmetic. Each data point represents an interval of thickness 1.37 m, which corresponds broadly to the spatial resolution of the induction log. The regression line has the equation:

$$\log K = 6.26 + 5.64 \log \phi. \qquad (9)$$

The uncertainty associated with permeability prediction using equation (9) is less than one half an order of magnitude.

Figure 5d illustrates a bilogarithmic crossplot of porosity v. permeability in which the data of Fig. 5a have been arithmetically averaged with equal weightings over each of the five model layers. Each data point represents a layer of

164 P. F. WORTHINGTON

thickness 2.28 m. The regression line is described by the expression:

$$\log K = 6.35 + 5.70 \log \phi. \qquad (10)$$

Because of the cyclic nature of the model, these average porosities are virtually the same as those that would be obtained by averaging the log porosities over the same intervals.

Figure 6a shows the constant terms and coefficients from equations (7)–(10) plotted as a function of scale. Additional data have been incorporated for the three-point and twelve-point running means. For this particular combination of model succession and core-derived petrophysical relationship there is an immediate progression to higher constant terms and coefficients as one moves away from the core-plug scale. The regression algorithm seems to stabilise as scale increases and the effects of the cyclic succession are damped.

Figures 6b and 6c show similar plots for different distributions of the sublayers within each of the five model sand layers of Fig. 4 but with the same petrophysical relationship

between core-derived porosity and permeability. Figure 6b relates to a five-layer model succession with the sublayers re-sorted within each layer in order of decreasing porosity with depth, to reflect coarsening-upwards sequences. The same constants and coefficients would be obtained if the layers were re-sorted in order of increasing porosity with depth, to reflect fining-upwards sequences. In this case there is no immediately sharp increase in the constant terms and coefficients as one moves away from the core-plug scale. This is to be expected because adjacent sublayers now have similar characteristics and there should therefore be no immediate marked adjustment in the distribution of the plotted points.

Figure 6c relates to a model succession that also comprises these five layers but with the sublayers distributed at random within each layer. Randomization was effected by assigning a generated random number to each sublayer in the model represented by Fig. 6b. These numbers were then ordered monotonically within each layer so that the sublayers to which they corresponded then became randomly

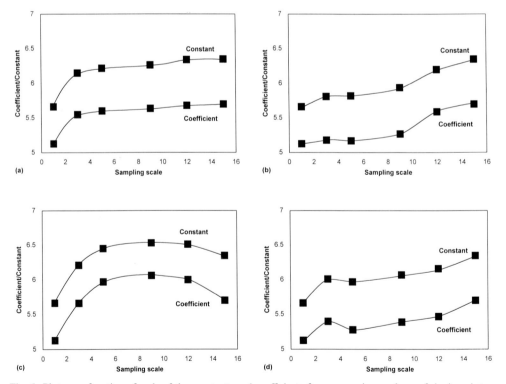

Fig. 6. Plots as a function of scale of the constants and coefficients from regression analyses of single-point core data, three-, five-, nine- and twelve-point running means, and zonal averages, for (**a**) the five-layer model of Fig. 4; (**b**) the five-layer model with each group of sublayers sorted in order of decreasing porosity; (**c**), (**d**) the five-layer model with the sublayer porosities distributed at random.

distributed. In this particular case the constant terms and coefficients initially increase, though not as sharply as for the cyclic model, before passing through maxima and then decreasing towards the values for the interval scale.

Given the constraints imparted by the model specification and the method of data sampling, Fig. 6c should not be presumed generally descriptive of the variation in constant terms and coefficients for random distributions of the constituent sublayers. Indeed, Fig. 6d shows more subdued variations for another random distribution of sublayers within each of the five model layers. The pronounced differences between Figs. 6c and 6d confirm a wide variety of scale dependences in the core-to-log range.

The constants and coefficients that relate to the smallest (core plug) and largest (interval) scales within Fig. 6 must remain unchanged regardless of how the sublayers are distributed, because they are based on the same data sets. At intermediate scales the constants and coefficients diverge and they reconverge as the running means become more representative. Therefore the application of a core-derived predictive relationship to log data should entail filtering of the core data at a scale that reflects the spatial resolution of the logging tool, otherwise an inappropriate algorithm might be used. This form of core data processing has been used to good effect in field investigations (e.g. Dawe & Murdock 1990; Hook *et al.* 1994).

Apparent formation factor as a permeability predictor

If the Archie conditions are satisfied, equation (2) serves as a predictive algorithm for permeability. However, equation (2) is substantially a consequence of the relationships between formation factor and porosity (equation 1), on the one hand, and between porosity and permeability (equation 3), on the other. Further, equation (2) cannot be applied in hydrocarbon zones without recourse to porosity measurement. Equation (2) is therefore seen as subordinate to equation (3).

If the Archie conditions are not satisfied, equation (6) serves as a predictive algorithm for permeability. The application of equation (6) is mostly to be found in the area of groundwater hydrology, where fresh waters do not allow the reservoir rocks to satisfy the Archie criteria. It can also be applied in the water zones of freshwater petroleum reservoirs, such as those of India, Indonesia, Malaysia and Argentina, to name but a few. However, equation (6) can only be applied directly in cases where there are no pronounced variations in the salinity, and therefore in the resistivity, of the interstitial electrolyte. This condition will usually be satisfied during programmes of laboratory measurement. It can be violated at the field application stage. For this reason, the field application of equation (6) must use a modified form that allows normalization of the quantity F_a to a value F_{an} that relates to a preset groundwater conductivity.

Figure 7a illustrates the positive trend on a crossplot of permeability v. apparent formation factor for irregularly-spaced core samples from three wells within a zone of Sherwood Sandstone aquifer where intergranular flow is believed to predominate. Well logs of closely spaced boreholes have indicated that marker beds within the sandstone can be traced horizontally over a distance of at least 20 m. The laboratory-measured apparent formation factors relate to an electrolyte resistivity of 20 Ωm. The uncertainty in predicted permeability exceeds an order of magnitude.

Figure 7b shows these same data, but with the data points averaged within the following specified porosity intervals: <0.140, 0.140–0.159, 0.160–0.179, 0.180–0.199, 0.200–0.219, ≥0.220. This method of data averaging has been used for many years. For example, it was applied to reduce data scatter in the classic paper of Carothers (1968). In the present case, apparent formation factors were averaged arithmetically. Permeabilities were averaged geometrically on the basis that the values within a given porosity window are random samples of the logarithmic normal distribution of permeability that represents the formation as a whole. The incremental averaging approach is more appropriate here, because its application to irregularly spaced core data effectively decouples the data set from a layer-cake description. The regression lines of Figs 7a and 7b are very different in character.

Figure 7c illustrates a crossplot of normalized, field-measured apparent formation factor v. hydraulic conductivity inferred from pumping tests for seven wells within this same part of the Sherwood Sandstone. The quantity F_a was known to be related to electrolyte resistivity R_w over the range of prevailing salinities through a generic expression of the form:

$$F_a = (A + BR_w)^{-1} \qquad (11)$$

where A (a function of porosity) and B (a function of pore surface area) are sample-specific petrophysical constants. Field values of F_a were normalized to F_{an} at the field-wide average R_w of 20 Ωm using the following expression, derived by eliminating B through the twofold application of equation (11):

$$F_{an} = [A + \{(R_w/20)((1/F_a) - A)\}]^{-1} \quad (12)$$

where $A = 0.0735$, derived from the empirical relationship $A = \phi^{1.61}$ at the field-average porosity of 0.20.

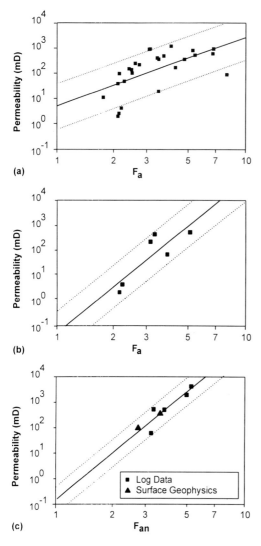

(a)

(b)

(c)

■ Log Data
▲ Surface Geophysics

Fig. 7. Crossplots of apparent formation factor v. permeability with regression lines for an intergranular flow unit of the Sherwood Sandstone based on: (**a**) measured data from single core plugs from three wells; (**b**) core data averaged within 2 p.u. increments of porosity; (**c**) field data based on drawdown tests and logging/geophysical surveys, with apparent formation factors F_{an} normalized to $R_w = 20$ Ωm.

The values of F_{an} show a well-defined positive trend with permeability (Fig. 7c). In five cases, the values of F_{an} represent normalized averages of longitudinal resistivity from well logs (Keller 1964). In two cases, the values of F_{an} have been determined from the interpretations of electrical soundings around the borehole sites. The conformance of both types of data to the same linear trend suggests that the established method of averaging resistivity log data is broadly compatible with the physical averaging process that is inherent in the larger-scale surface electrical data.

The regression lines of Figs 7b and 7c have virtually the same gradient and their vertical displacement is less than one half an order of magnitude. This comparison suggests a degree of equivalence between the two data sets that form the basis of these two linear fits. This, in turn, lends support to the incremental porosity method of averaging core permeabilities where sampling is incomplete and irregular. The field data distributions are significantly higher than those of the averaged core data, partly because the former represent intervals of higher intergranular permeability that are too friable to be sampled. Similar field relationships to that of Fig. 7c have been reported using as a predictor apparent formation factor from well logs (Kwader 1985) and from geoelectrical soundings (Kelly 1977).

Conclusions

This paper has dealt exclusively with single-predictor algorithms for the estimation of intergranular permeability in clastic reservoirs. Porosity is the principal permeability predictor, because its application is independent of the nature of the interstitial fluids and of the composition of the reservoir rock, once the porosity measurement itself has been made. On the other hand, the application of apparent formation factor as a permeability predictor is strongly reliant on the presence of a fresh interstitial electrolyte and/or a high degree of shaliness. Its use is therefore restricted. The Archie intrinsic formation factor F has been used as a predictor of permeability in clean brine-saturated reservoir rocks, but the feasibility of this approach is largely a consequence of the relationship between F and porosity, and, unlike the porosity algorithm, methods based solely on F are applicable only in a water zone. Therefore this approach has not been considered further here.

Predictive relationships are scale-dependent. If a given linear regression procedure is applied to bilogarithmic data pairs comprising one of our two predictors with intergranular permeability as predictand, the resulting constant terms and coefficients will change with increasing scale. Data scatter reduces as scale increases. The reduction in scatter is evident regardless of whether the increase in scale has been achieved through spatial averaging or through larger sensing volumes.

The effective scale of regularly spaced core data sampled at an interval that is below logging tool resolution can be increased along the borehole axis through the application of running means that are chosen to match the spatial resolutions of those logs that are to be used as field predictors of permeability. In this case, horizontal plug permeabilities are averaged arithmetically to produce a horizontal permeability at a larger scale. In so doing, it is presumed that each core plug represents a sublayer of thickness equal to the sampling interval and with a breadth that is greater than the diameter of investigation of the wireline permeability indicator that is to be used. The running means can be weighted to track the response functions of the tools in question. This latter refinement would marginally improve predictive performance.

The effective scale of irregularly spaced core data can be increased by the technique of averaging core petrophysical data over preset porosity intervals. The averages are arithmetic with the exception of permeability, which has a geometric mean. In the particular case of apparent formation factor as a predictor of permeability, the resulting data distribution has shown encouraging similarity to that established using field data alone.

The overriding message is to use a predictive relationship that is fit for purpose. Thus, for example, predictive relationships that are founded on unprocessed core plug data should strictly be applied only at the core plug scale. If a predictive relationship established using conventionally sampled core data is to be applied to porosity logs so that the latter might be used as permeability predictors, a 2 ft (0.61 m) running mean should be applied to the core data before the predictive relationship is established, so that the different spatial resolutions of core and logs might be rendered more compatible. Similarly, the application of such core-derived predictive relationships to the induction log response for purposes of permeability prediction might require that a 4 ft (1.22 m) running mean be initially applied to the core data. In the particular case of manual interpretation, where log responses are visually estimated over vari-

able intervals, any core-derived predictive relationship should be based on plug data averaged over a vertical distance that is typical of the thicknesses of the selected intervals.

As far as possible, predictive methods for permeability should be validated over key intervals. It is recognized that the database needed to do this will not always be available. Therefore it is especially important that permeability indicators continue to be seen as providing estimates rather than determinations of intergranular permeability. Our understanding of the permeability problem in petrophysics is not as advanced as that of the porosity problem, and this will remain the case for the foreseeable future.

Areas of potential benefit include expanding the petrophysical algorithms to include additional predictors such as shale parameters (Yao & Holditch 1993), water saturation (Saner et al. 1993), flushed-zone resistivity (Ball et al. 1994), and free-fluid porosity (Lomax & Howard 1994), although the same need for scale compatibility would apply. A successful application of these procedures would enhance the identification of hydraulic flow units (Amaefule et al. 1993: Johnson 1994). These developments should be guided by the realization that the most effective predictors are practically useful rather than theoretically exact, and therefore empiricism will continue to have a strong role to play in the estimation of permeability using petrophysical data. All of this points to an increased use of multiple predictors in a way that is not constrained by algorithmic or model preconceptions. For example, fuzzy regression is not subject to the exclusive crispness of membership or non-membership of classes of data that are candidates for correlation, because it allows a grading of those memberships through simple mathematical expression (e.g. Pu & Fang 1993). Again, the multi-correlative capability of artificial neural networks, with their ability to learn and self-adjust as they seek to discover complex relationships between input variables, also offers a vehicle for potentially enhanced permeability prediction from petrophysical data (e.g. Mohaghegh et al. 1994). Here, too, the optimum applications will be those that take proper account of the scales of measurement.

References

AHMED, U., CRARY, S. F. & COATES, G. R. 1991. Permeability estimation: the various sources and their interrelationships. *Journal of Petroleum Technology*, **43**, 578–587.

ALGER, R. P. 1966. Interpretation of electric logs in fresh water wells in unconsolidated formations. *Transactions SPWLA 7th Annual Logging Symposium*, CC1–25, Society of Professional Well Log Analysts, Houston, Texas.

AMAEFULE, J. O., ALTUNBAY, M., TIAB, D., KERSEY, D. G. & KEELAN, D. K. 1993. Enhanced reservoir description: using core and log data to identify hydraulic (flow) units and predict permeability in uncored wells. SPE Paper 26436, Society of Petroleum Engineers, Richardson, Texas.

ARCHIE, G. E. 1942. The electrical resistivity log as an aid in determining some reservoir characteristics. *Transactions American Institute of Mining and Metallurgical Engineers*, **146**, 54–62.

BALL, L. D., CORBETT, P. W. M., JENSEN, J. L. & LEWIS, J. J. M. 1994. The role of geology in the behaviour and choice of permeability predictors. SPE Paper 28447, Society of Petroleum Engineers, Richardson, Texas.

BARKER, R. D. & WORTHINGTON, P. F. 1972. Methods for the calculation of true formation factors in the Bunter Sandstone of northwest England. *Engineering Geology*, **6**, 213–228.

BIELLA, G., LOZEJ, A. & TABACCO, I. 1983. Experimental study of some hydrogeophysical properties of unconsolidated porous media. *Ground Water*, **21**, 741–751.

CAROTHERS, J. E. 1968. A statistical study of the formation factor relation. *The Log Analyst*, **9**(5), 13–20.

CHORK, C. Y., JIAN, F. X. & TAGGART, I. J. 1994. Porosity and permeability estimation based on segmented well log data. *Journal of Petroleum Science and Engineering*, **11**, 227–239.

CRAIG, D. E. 1991. The derivation of permeability-porosity transforms for the H. O. Mahoney Lease, Wasson Field, Yoakum County, Texas. *In*: LAKE, L. W., CARROLL, H. B. JR. & WESSON, T. C. (eds) *Reservoir Characterization II*. Academic Press, San Diego, California, 289–312.

CROFT, M. G. 1971. *A method of calculating permeability from electric logs*. US Geological Survey Professional Papers, 750-B, B265–269.

DAWE, B. A. & MURDOCK, D. M. 1990. Laminated sands: an assessment of log interpretation accuracy by an oil-base mud coring program. SPE Paper 20542, Society of Petroleum Engineers, Richardson, Texas.

HOOK, J. R., NIETO, J. A., KALKOMEY, C. T. & ELLIS, D. 1994. Facies and permeability prediction from wireline logs and core – a North Sea study. *Transactions SPWLA 35th Annual Logging Symposium*, AAA1–22, Society of Professional Well Log Analysts, Houston, Texas.

JENSEN, J. L. & LAKE, L. W. 1985. Optimization of regression-based porosity-permeability predictions. *Transactions CWLS 10th Formation Evaluation Symposium*, R1–22, Canadian Well Logging Society, Calgary, Alberta.

JOHNSON, W. W. 1994. Permeability determination from well logs and core data. SPE Paper 27647, Society of Petroleum Engineers, Richardson, Texas.

KELLER, G. V. 1964. Compilation of electrical properties from electrical well logs. *Colorado School of Mines Quarterly*, **59**(4), 91–110.

KELLY, W. E. 1977. Geoelectric sounding for estimating aquifer hydraulic conductivity. Ground Water, **15**, 420–425.

KWADER, T. 1985. Estimating aquifer permeability from formation resistivity factors. *Ground Water*, **23**, 762–766.

LIN, J. L. & SALISCH, H. A. 1994. Determination from well logs of porosity and permeability in a heterogeneous reservoir. SPE Paper 28792, Society of Petroleum Engineers, Richardson, Texas.

LOMAX, J. & HOWARD, H. 1994. New logging tool identifies permeability in shaley sands. *Oil & Gas Journal*, **92**(51), 104–108.

MOHAGHEGH, S., AREFI, R., AMERI, S. & ROSE, D. 1994. Design and development of an artificial neural network for estimation of formation permeability. SPE Paper 28237, Society of Petroleum Engineers, Richardson, Texas.

NELSON, P. H. 1994. Permeability-porosity relationships in sedimentary rocks. *The Log Analyst*, **35**(3), 38–62.

PU, Z. W. & FANG, J. H. 1993. Permeability estimation from log-derived porosity: part 1: via fuzzy linear regression. *GSA 42nd Annual Southeast Section Meeting, Tallahassee, Florida, Program Abstracts* **25**(4), 63, Geological Society of America, Boulder, Colorado.

SANER, S., KISSAMI, M. & AL-NUFAILI, S. 1993. Estimation of permeability from well logs using resistivity and saturation data. SPE Paper 26277, Society of Petroleum Engineers, Richardson, Texas.

THOMAS, D. C. & PUGH, V. J. 1989. A statistical analysis of the accuracy and reproducibility of standard core analysis. *The Log Analyst*, **30**(2), 71–77.

WENDT, W. A., SAKURAI, S. & NELSON, P. H. 1986. Permeability prediction from well logs using multiple regression. *In*: LAKE, L. W. & CARROLL, H. B. JR. (eds) *Reservoir Characterization*. Academic Press, Orlando, Florida, 181–221.

WORTHINGTON, P. F. 1983. The relationship of electrical resistivity to intergranular permeability in reservoir rocks. *Transactions SPWLA 24th Annual Logging Symposium* 2, RR1-16, Society of Professional Well Log Analysts, Houston, Texas.

——1986. The relationship of aquifer petrophysics to hydrocarbon evaluation. *Quarterly Journal of Engineering Geology*, **19**, 97–107.

——1994. Effective integration of core and log data. *Marine and Petroleum Geology*, **11**, 457–466.

——1995. Estimation of intergranular permeability from well logs. *Dialog*, **3**(3), 7–10.

YAO, C. Y. & HOLDITCH, S. A. 1993. Estimating permeability profiles using core and log data. SPE Paper 26921, Society of Petroleum Engineers, Richardson, Texas.

Prediction of petrophysical parameter logs using a multilayer backpropagation neural network

C. A. GONÇALVES, P. K. HARVEY & M. A. LOVELL

Borehole Research, Department of Geology, University of Leicester, University Road, Leicester LE1 7RH, UK

Abstract: Quantitative petrophysical characterization is one of the principal tasks of a reservoir analyst and is generally affected by the methods used. For example, different theoretical and empirical formulas are in most cases restricted to the specific areas where they were developed. Prediction of continuous petrophysical parameters is often time consuming and complicated because of geological variability such as facies changes due to sedimentary and structural changes.

In this work we propose a neural network approach which is used to predict quantitative petrophysical parameters from wireline logs of cored intervals. We then apply the knowledge learned during training to uncored intervals or other holes. Data from the Ocean Drilling Program and from two South American oilfield holes are used to test this technique.

The results show a good match between the neural network-derived petrophysical parameter logs and the actual core measurements. Problematic petrophysical measurements can be identified by a mismatch between the responses.

Artificial neural networks (ANN) are relatively new numerical tools in geoscience and have proved very useful in applications where conventional computing methods are inadequate (Haykin 1994). They have been used in a broad range of non-linear modelling and classification problems in different areas such as biological science and engineering. In petrophysical analysis, when core recovery is extensive, petrophysical parameters can be obtained for different sections of a hole through direct core measurement. However, when core recovery is poor or even when there is no core recovery at all, a method of deriving petrophysical parameters must be developed.

There are many techniques available for petrophysical parameter prediction. Algorithms using simply the log responses, complex lithology model analysis and geochemically derived mineral abundance are among them. However, all generally require the formation to be zoned into intervals of different characteristics in order for the methods to achieve a satisfactory result (Jenner & Baldwin 1994). Another important aspect is the local dependence of these methods. They are in general strongly related to the geological environment for which they were developed and are less likely to be successful when applied elsewhere.

As in the classical procedure for most neural networks, petrophysical parameters are estimated here by training the system with measured core data from recovered sections. During training, the objective is to reduce the difference between the core measurements and the neural network results. This is performed by adjusting the connection weights in the neural network according to the learning rate (Hetch-Nielsen 1990). The final weights obtained from the training stage are then applied to a different dataset and petrophysical parameter logs are generate. In the case of an offset hole, the neural network is trained with core measurements from one hole and then the weights may be applied in a nearby hole to predict petrophysical parameter logs.

The results are presented here as a comparison between the neural network response and the core measurements. Geological variations need to be trained by the neural network in order to obtain a reasonable accuracy in the results.

Artificial neural network (ANN)

Neural networks are considered computational systems, either hardware or software, which mimic the computational abilities of biological systems by using large numbers of simple interconnected artificial neurons (Anderson & Rosenfeld 1988). Aleksander & Morton (1989) go into more detail stating that 'Neural computing is the study of cellular networks that have a natural property for storing experimental knowledge. Such systems bear a resemblance to the brain in the sense that knowledge is acquired through training rather than programming and is retained due to changes in neural functions. The knowledge takes the form of stable states or

From Lovell, M. A. & Harvey, P. K. (eds), 1997, *Developments in Petrophysics*, Geological Society Special Publication No. 122, pp. 169–180.

cycles of states in the operation of the net'. They continue: '...A central property of such nets is to recall these states or cycles in response to the presentation of clues'.

The major difference between conventional computing and neural networks is that while conventional computing relies on the instructions we provide to the machines, which then execute a series of operations, neural computing relies upon training where the machine learns from experiences. In addition, in conventional computing the task is explicitly represented, whereas the representation is implicit within the links of the neural network.

An artificial neural network is an information-processing system that has certain performance characteristics in common with biological neural networks. A neural network is characterized by its architecture, or the way in which the neurons are interconnected, by the method of determining the weights on the connections and by its activation function (Fausett 1994). These defined characteristics, which distinguish neural networks from other approaches of information processing, are now considered in turn.

The concept of a neural network consists generally of a large number of processing elements called neurons, cells or nodes. In an artificial neural network each neuron is connected through connection links, each one with an associated weight. Each neuron has an internal state, called its activation function, which is a function of the input it has received. In a feedforward neural network, a neuron sends its activation as a signal to several other neurons. Even though neurons can only send one signal each time, this signal can be broad-cast to several other neurons (Fausett 1994). Consider neuron Y in the simple example in Fig. 1 which receives input from neurons X_1, X_2 and X_3. The output signals of these neurons are x_1, x_2 and x_3, respectively. The weights on the connections from X_1, X_2 and X_3, to neuron Y are w_1, w_2 and w_3. The neural network input y_in, to neuron Y is the sum of the weighted signals from neurons X_1, X_2 and X_3,

$$y\_in = w_1 x_1 + w_2 x_2 + w_3 x_3.$$

The activation y_in of neuron Y is then transformed by a non-linear activation function acting on the sum of the inputs, $y = f(y\_in)$. An example is the sigmoid function,

$$y\_out = 1/[1 + \exp(-y\_in)].$$

Backpropagation neural networks

Single-layer neural networks were first used in the mid-1950s (Rosenblat 1958). As shown in Fig. 1, the earliest neural networks had the input signals directly broadcast to the output units, with the signals being multiplied by the weights present in these connections and the activation function. However, the limitations of these systems as a 'hard delimeter' were a significant factor in their decline and lack of interest towards 1970 (Minsky & Papert 1969). By the mid-1980s, an increase was seen in the interest for a new system of neural networks, which include one or more hidden layers. In this new approach, called back-error propagation neural network (Rumelhart *et al.* 1986), the result given by the neural network at any stage during the training process is compared with an expected known result. The differences between them are used to calculate weight updates. The process is repeated until a good fit is achieved with the expected result.

As in the case of most neural networks, the aim of the backpropagation neural network is to train the net to achieve a balance between the ability to respond correctly to the input patterns that are used for training (memorisation) and the ability to give reasonable responses to an input that is similar, but not identical, to that used in training (Fausett 1994). This is known as generalization.

The training of a backpropagation neural network involves three stages: the feedforward of the input training data, the computation and backpropagation of the associated error, and the adjustment of the weights. This can be quite slow. After training, the application of the

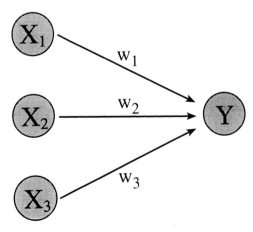

Fig. 1. Artificial neural network without a hidden layer (after Anguita 1993).

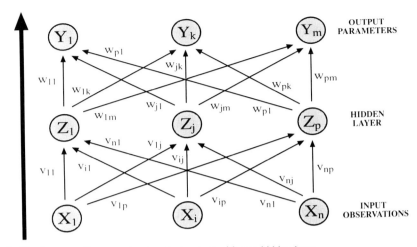

Fig 2. Typical backpropagation neural network structure with one hidden layer.

neural network involves only the feedforward process and therefore the output is obtained very rapidly.

A typical backpropagation neural network with one layer of hidden nodes is shown in Fig. 2. The output Y the hidden layer nodes Z and the input nodes X are presented as well as the associated weights v and w. Only the feedforward process propagation direction is shown. In the backpropagation stage the errors between the output nodes and the expected known values are propagated backwards. During the first stage, each input node X_i receives an input signal and broadcasts it to each of the hidden layer nodes $Z_1, \ldots, Z_p$. Each node then computes its activation function and sends its signal z_j to each output node. Each output node Y_k computes its activation y_k to

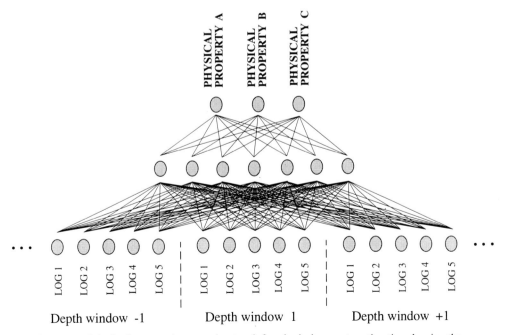

Fig. 3. Structure of the backpropagation neural network for physical property estimation showing the 'windowing' technique.

form the response of the neural network to the previous input pattern (Haykin, 1994).

During the training process, each output node compares its computed activation y_k with the expected target value t_k in order to determine the associated error. Based on this error a factor δ_k ($k = 1, \ldots, m$) is computed and used to distribute the error at the output nodes back to all nodes in the previous layer. In a similar manner, the factor δ_j ($j = 1, \ldots, p$) is computed for each hidden node (Battiti 1992).

Numerous variations of the backpropagation technique have been developed to improve the speed of the training process. In this work, we use an implementation of the backpropagation algorithm, called 'matrix back propagation' (Anguita 1993). The purpose of this version is to implement a faster backpropagation algorithm compared with the original one developed by Rumelhart *et al.* (1986). The backpropagation algorithm is taken from the classical method developed by Vogl *et al.* (1991) which uses a gradient descent method.

Neural network implementation

In a petrophysical parameters prediction approach, the structure of the backpropagation neural network has to be changed from the one showed before in Fig. 2.

The number of neurons in the input layer is given by the number of log curves and the output layer is defined by the petrophysical parameters to be trained. Considering that log data are an average over their vertical resolution and core measurements are taken from single points, the use of a 'windowing' technique (Jenner & Baldwin 1994) compensates for any minor depth matching error between log and core data and also allows resolution matching (Fig. 3). The number of depth windows in the input layer depends on the vertical resolution of the logs used as input data. In this work, the number of depth windows is obtained dividing the average vertical resolution of the log curves by the interval in which the log observations are made (0.1524 m).

An important aspect is the reliability of the results produced by the neural networks. Comparison of generated output with core measurements provide information on how good the neural network has been trained. However, this does not provide any information about the quality and accuracy of the neural network predictions in holes with no core recovery. When applied to offset holes, the results from neural networks are compared with other log derived

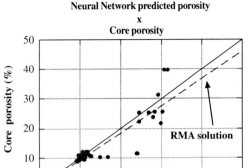

Fig. 4. Comparison of neural network predicted porosity with core porosity in ODP Hole 807C.

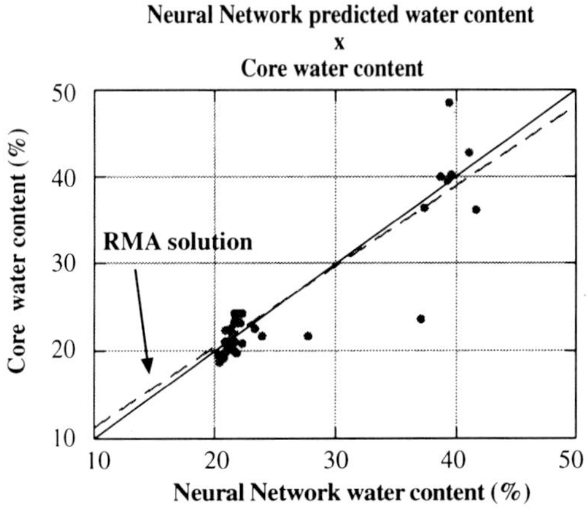

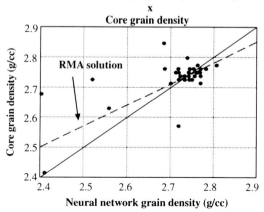

Fig. 5. Comparison of neural network predicted water content with core water content in ODP Hole 807C.

Fig. 6. Comparison of neural network predicted grain density with core grain density in ODP Hole 807C.

techniques to predict petrophysical parameters. For intervals where neural network prediction agrees well with log derived petrophysical parameters, the neural network has learned the training set in a way to correctly identify the variations in the log data.

Because the neural network is trained to predict petrophysical parameters based on log responses, the log curves used as input data in the training process must be sensitive to changes in the petrophysical characteristics of the formations. Changes in lithofacies, resulting in changes of petrophysical characteristics not reflected in the log responses will not be correctly predicted by the neural network. This is a problem when this technique is applied to predict petrophysical parameters in offset holes, where changes in facies of the formations can occur, and are difficult to train from a different hole.

Data normalization

The aim of normalization is to cause each log curve to have a common dynamic range. This

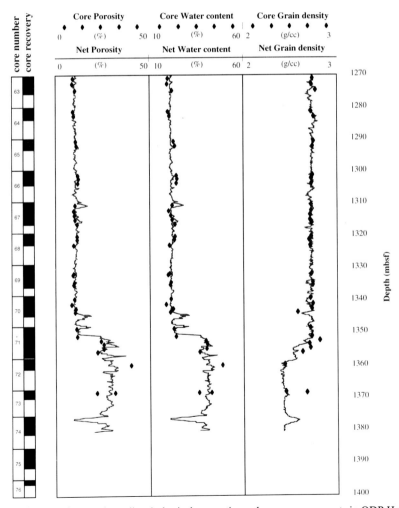

Fig. 7. Log curves for neural network predicted physical properties and core measurements in ODP Hole 807C.

removes the problem which arises when different units of measurements are used to express physical properties (i.e. resistivity and porosity). It also accommodates log curves which use the same scale of measurement but have different magnitudes. In a large number of experiments which were carried out on both normalized and unnormalized data for which there was *a priori* knowledge, it was found that normalized data tended to give improved results.

The method of normalizing log curves used in this work is that of reducing to a standard form with zero mean and unit standard deviation as follows:

$$z_{ij} = (x_{ij} - x_i)/s_i,$$

where z_{ij} is the normalized value for the ith log curve at the jth depth measurement, x_{ij} is the ith log curve at the jth depth measurement, x_i is the overall mean of all measurements of the ith log curve and s_i is the standard deviation associated with the ith log curve.

Field examples

ODP Hole 807C

ODP Hole 807C is located on the northern rim of the Ontong Java Plateau in 2805 m of water depth. For the purpose of this study the interval between 1270–1400 mbsf was selected in this hole. The reason is that it comprises in a 130 m interval three of the four main lithofacies observed in the whole section of the hole. The sedimentary sequence observed includes a carbonate sequence, a claystone and siltstone and a volcanic basement. Due to reduced core recovery (65%) a complete physical property characterization was not possible for the whole interval. Measurements of water content, grain density and porosity were taken at different depths depending on core recovery and core conditions. Below 1380 mbsf, due to the poor quality of the cores, no physical property measurements were made (Kroenke *et al.* 1991).

The structure of the neural network in this case is given by 4 neurons at the input layer, which represent the four log curves (ILM, SFLU, DT and RHOB – medium induction, shallow induction, slowness and density respectively) used as input data. The output layer has three neurons representing the three different physical properties to be estimated, which are porosity, water content and matrix density. As the vertical resolution of the log curves used

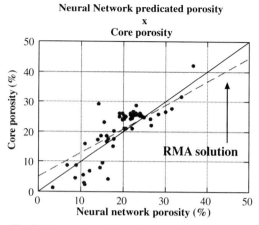

Fig. 8. Comparison of neural network predicted porosity with core porosity in ODP Hole 878A.

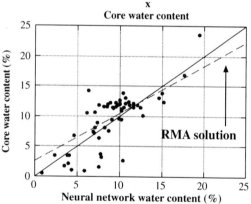

Fig. 9. Comparison of neural network predicted water content with core water content in ODP Hole 878A.

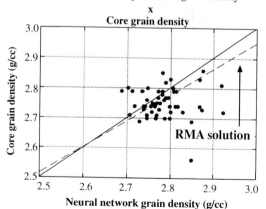

Fig. 10. Comparison of neural network predicted grain density with core grain density in ODP Hole 807C.

Hole 878A

Core and Neural Network log curves

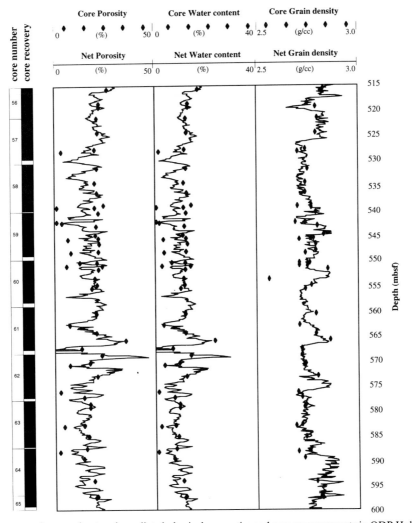

Fig. 11. Log curves for neural network predicted physical properties and core measurements in ODP Hole 878A.

ranges from 0.4 m (DT and RHOB) to 1.5 m (ILM), we selected five depth windows to be used in the training process.

After training, the neural network was applied to the whole interval. A comparison of the predicted physical properties to the actual core measurements is shown in Figs 4 to 6. The figures show the 1:1 ratio line (solid) and the reduced major axis (RMA) regression line (dashed). The latter is obtained by the fit which intersect the mean points of each variable and which has a

slope given by ($\lambda = S_y/S_x$ (S_y and S_x are the standard deviations). A quantitative measure of goodness of the fit between the two solutions can be obtained by computing the standard normal deviate (Z) and verifying the hypothesis that $\lambda_1 = \lambda_2$. Therefore, $Z = (\lambda_1 - \lambda_2)/(S_x^2 + S_y^2)^{1/2}$. For a 95% of confidence limit, Z should be less than 1.96. In general, a good agreement is observed for low porosity, low water content and high grain density. The values for Z are 0.007, 0.007 and 2.42 for water content, porosity and grain density respectively. It shows that the hypothesis of $\lambda_1 = \lambda_2$ is not only verified for grain density. Predicted physical property logs are then generated for the whole interval (Fig. 7). It is possible to observe good agreement between core and neural network data for the top of the interval. Below 1351 mbsf, the neural network predicted curves are still able to follow the main trend present in core data, although it does not match all core measurements. One of the reasons for that is the reduced amount of core data used to train the neural network for the lithofacies variation between 1351–1379 mbsf. The upper part of the interval (carbonate sequence) was trained with a greater number of core measurements.

ODP Hole 878A

ODP Hole 878A is located in a water depth of 1323 m, on the northeastern part of MIT Guyot (West Pacific). The interval between 515 and 600 mbsf was selected to be used in this work. It consists of a breccia predominantly siliciclastic at the top and calcareous at the bottom of the interval (Premoli Silva *et al.* 1993). The

overall variation in the polymictic breccia is as follows: volcanic clasts are dominant in the upper half and decrease in abundance towards the base. Some alternation occurs between volcanic-rich and volcanic poor horizons throughout the interval, but carbonate is dominant in the matrix and clasts below 577 mbsf. Premoli Silva *et al.* (1993) shows that the analysis of bulk carbonate content demonstrates the increase of carbonate in the matrix from the top to the bottom.

The structure of the neural network used in ODP Hole 878A was the same as used in the previous hole. Four neurons are present at the input layer, which represent the four log curves used (DT, RHOB, NPHI and SFLU – slowness, density, porosity and shallow resistivity respectively) and three neurons in the output layer representing the three different physical properties to be estimated. The vertical resolution of the log curves ranges from 0.4 m (RHOB) to 0.75 m (DT and SFLU). Therefore, in this case three depth windows were used for each input/output pair during the training process.

Despite the complex changes in lithofacies (mainly given by Si and Ca content and grain size variation within the polymitic breccia), the good core recovery and density of measurements

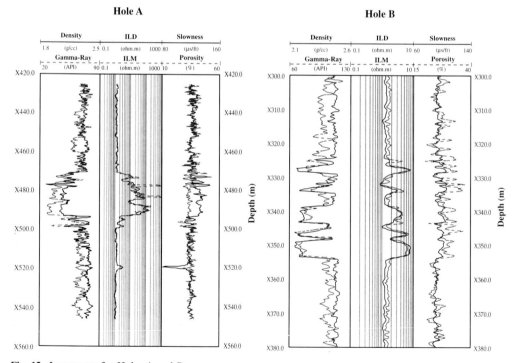

Fig. 12. Log curves for Holes A and B.

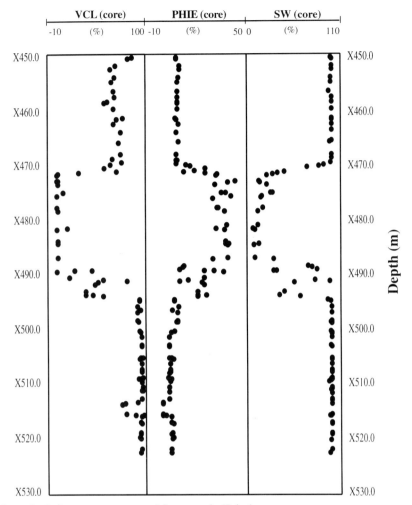

Fig. 13. Petrophysical parameters measured from core in Hole A.

along the whole interval shows a reasonable correlation between predicted physical properties and core measurements (Figs 8 to 10). Again, the RMA solutions were obtained and compared with the 1:1 ratio lines. The standard normal deviate ($Z = 3.91$) shows that the hypothesis of $\lambda_1 = \lambda_2$ is not only valid for grain density. The lithofacies variations within the breccia affect the log curves and consequently the physical property estimation. The neural network predicted physical property logs for the whole interval are shown in Fig. 11. The main variation in the physical property characteristics are represented. As expected because of their close relation, porosity and water content show nearly the same variations along

the interval, following in general the variations observed in core measurements and lithofacies changes.

Holes A and B

Two oilfield holes from SE Brazil are used to test the application of the backpropagation neural network in predicting petrophysical parameters in an offset hole. Hole A (Fig. 12) shows a continuous reservoir (between X470.0 and X495.0 m) with characteristic high resistivity ($c.\,100\,\Omega\text{m}$) while Hole B is characterized by a sequence of interbeded reservoirs. Core measurements from Hole A for clay volume (VCL),

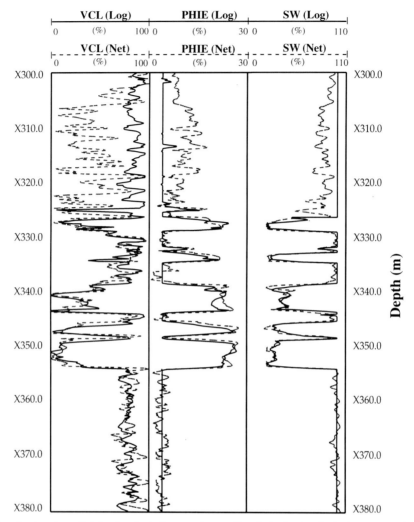

Fig. 14. Neural network predicted petrophysical parameters and log derived petrophysical parameters in Hole B.

effective porosity (PHIE) and water saturation (SW), and are shown in Fig. 13.

The log curves and core data in Hole A were used for training the neural network to predict petrophysical parameters. After training, the weights learned by the neural network for that hole were applied to predict the same petrophysical parameters in Hole B where no core measurements were available. The results for the neural network predicted petrophysical parameters log curves are observed in Fig. 14. In general, a good agreement between predicted and log derived VCL, PHIE and SW is observed. In the upper section of the interval in Hole B, however, there is a clear mismatch between both results. The mismatch is observed

for all three parameters. It suggests that changes in facies across the holes affected the log data in Hole B which have been not used to train the neural network. This identifies a problem faced with ODP data. In that case, a insufficient amount of trained data caused the mismatch between core and predicted physical properties for the lower section of Hole 807C. Here, the core data in Hole A do not represent the petrophysical changes across the holes reflected in the upper section of Hole B log responses. Another possibility is an interpolation problem, where there are insufficient data in the training set to describe the variability. Finally, different borehole conditions could also affect the log responses and be responsible for the mismatch.

Figure 15 shows the cross-plot of neural network and log derived petrophysical parameters within the reservoir zone (between X325.0 and X354 m) in Hole B. Despite changes in the reservoir characteristics, we still observe a good agreement between both results. The RMA solutions are presented and the $\lambda_1 = \lambda_2$ hypothesis is not valid only for PHIE ($Z = 4.53$).

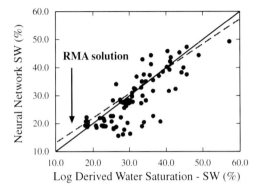

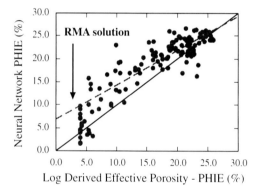

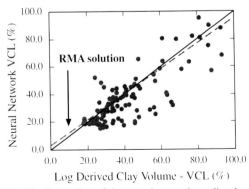

Fig. 15. Cross-plots of the neural network predicted petrophysical parameters and log derived petrophysical parameters for the reservoir zone in Hole B.

Conclusions

The backpropagation neural network is shown to give good results for petrophysical parameters estimation. Neurocomputing can also be considered a powerful tool for petrophysical evaluation in an offset hole as long as the lithofacies and petrophysical characteristic changes can be trained. The results presented here showed that petrophysical logs can be obtained for uncored intervals with a reasonable degree of success.

An important factor for its success was the use of a 'windowing' technique, which allowed a more realistic training set for the neural network. The results for all examples have shown good matches between actual core measurements and the neural network predicted physical properties. The physical property logs generated by the neural network match the main lithofacies variation present. Again, ODP Hole 878A shows some degree of variation in the generated physical property logs due mostly to the complex changes in lithofacies.

The neural network presented here also holds a great promise in helping well-to-well prediction of petrophysical parameters. The results obtained with the examples show good correlation with other log derived values even in the presence of lateral changes of facies which generally request some sort of skill in providing good results.

References

ALEKSANDER, I. & MORTON, H. 1989. *An introduction to neural computing*. Chapman & Hall, London.
ANDERSON, J. A. & ROSENFELD, E. 1988. *Neurocomputing, foundation of research*. MIT Press, Cambridge, MA.
ANGUITA, D. 1993. *MBP – Matrix back propagation, v. 1.1 – An efficient implementation of the BP algorithm*. DIBE, University of Genova.
BATTITI, R. 1992. *First and Second-order methods for learning: between steepest descent and Newton's method*. Neural Computation, **4**, 141–146.
FAUSETT, L. 1994. *Fundamentals of neural networks: architectures, algorithms and applications*. Prentice Hall, New York.
HAYKIN, S. 1994. *Neural networks: a comprehensive foundation*. MacMillan Press, New York.
HETCH-NIELSEN, R. 1990. *Neurocomputing*. Addison-Wesley, Reading, MA.
JENNER, R. & BALDWIN, J. R. 1994. Application of an artificial neural network an improved permeability profile. *SPWLA 16th European Formation Evaluation Symposium*, Aberdeen, Scotland, paper M.
KROENKE, L. W., BERGER, W. H., JANECEK, T. R. ET AL. 1991. *Proceedings ODP, Initial Reports*, **130**. Ocean Drilling Program, College Station, TX.

MINSKY, M. L. & PAPERT, S. A. 1969. *Perceptrons.* MIT Press, Cambridge, MA.

PREMOLI SILVA, I., HAGGERTY, J., RACK, F. *ET AL.* 1993. *Proceedings ODP, Initial Reports,* **144**. Ocean Drilling Program, College Station, TX.

ROSENBLATT, F. 1958. The Perception: A probabilistic model for information storage and organisation in the brain. *Psychological Review,* **65**, 386–408

RUMELHART, D. E., McCLELLAND, J. & THE PDP RESEARCH GROUP 1986. *Parallel distributed processing,* **1**. MIT Press, Cambridge, MA.

VOGL, T. P., MANGIS, J. K., RIGLER, A. K., ZINK, Z. T. & ALKON, D. L. 1991. Accelerating the convergence of the back-propagation method. *Biological Cybernetics,* **59**, 257–263.

The partitioning of petrophysical data: a review

BRIAN P. MOSS

Moss Petrophysics Limited, 1 Swaynes Lane, Merrow, Guildford, Surrey GU1 2XX, UK

.

Abstract: Petrophysical analyses are undertaken within a framework of ordered geological systems. The recognition of the natural geological ordering through the grouping and partitioning of petrophysical data is known to be a fundamental requirement of successful petrophysical interpretations.

The mathematical basis and examples of the use of a variety of techniques of value in partitioning data are reviewed. Beginning with the simple use of curve discrimination on conventional crossplots, the discussion proceeds by covering increasingly complex but relatively well-known methods including discriminant function analysis, cluster analysis and principal component analysis, and goes on to include multi-dimensional scaling, 'pigeon-holing', fuzzy clustering, characteristic analysis, projection pursuit and neural networks. The discussion highlights some of the benefits, and some of the pitfalls, of each technique through the included examples.

The discussion encourages the use of exploratory data analysis techniques in the processing and interpretation of petrophysical data and provides a reference framework from which can follow further evaluation of the value of the techniques in circumstances particular to the interested reader.

The analysis of petrophysical data seeks to describe the formation under study by quantifying certain of its properties such as porosity, pemeability, fluid saturation and mineralogy. The natural ordering of geological systems arising from the controlling influences of sedimentary environment, other rock-forming processes and rock mechanics imparts an ordering to the petrophysical properties of interest. Recognition of the inherent ordering, usually manifest as structure in the petrophysical data available for analysis, assists both in the identification of the petrophysical interpretation model to be applied, and in the characterization of the variation present in the formation to facilitate the quantification of geologically distinctive genetic units in terms of their petrophysical properties.

A definition of petrophysical rock types might be 'classes of rock characterized by differences in physical properties' and these differences most probably have resulted from the sum of all the previously listed effects. The properties of interest could be the basic formation properties that we seek to measure in petrophysical analyses, viz. density, resistivity, hydrogen, index, acoustic travel time, nuclear magnetic resonance etc., or they might be the formation parameters of porosity, permeability, capillarity, saturation etc. that we infer from out measurements of the basic properties. The fundamental controlling phenomena for nearly all of these properties are the quantity, shape, size and connectivity of the pore system (i.e. the pore geometry of the rock). Together with the mineralogy and texture of the rock fabric, both of which, jointly and

severally, control the pore geometry, these are the fundamental subjects of importance in petrophysical investigations.

However, it is necessary to remember that in our sampling of formations, be they sedimentary, igneous or metamorphic in origin, we are observing the sum of the various processes that affected the rocks from their initial creation through their particular burial history to their present-day depth and state of diagenetic and/or mechanical alteration. It is this combined effect of burial history, diagenetic influence and mineralogy that gives rise to the observation that, commonly, subdivisions of formations based upon solely lithostratigraphic criteria frequently fail to capture the petrophysical variability of the rock adequately.

In order to capture this combined influence we may approach the differentiation or partitioning of formations by computing proportions of minerals that they contain or by determining patterns in the data that form our 'view' of the formation. This review concentrates on techniques that address the latter.

All the techniques discussed here fall under the group title of 'multivariate analysis'. A definition of multivariate analysis is offered by Dillon & Goldstein (1984) as 'the application of methods that deal with reasonably large numbers of measurements (i.e. variables) made on each object in one or more samples simultaneously'. Whilst admitting that their definition is loose they stress that multivariate analysis deals with the simultaneous relationships among variables (their italics). As opposed to univariate (one variable) or bivariate (two variables)

From Lovell, M. A. & Harvey, P. K. (eds), 1997, *Developments in Petrophysics*, Geological Society Special Publication No. 122, pp. 181–252.

analyses where the mean and variance of a single variable or the pairwise relationships between two variables are studied respectively, multivariate methods of analysis are directed at the covariances and correlations between three or more variables.

Deductive v. inductive

Analysis methods that seek to differentiate the data by the computation of a set of component proportions whose identification is linked with the log data by some set of response equations are 'deductive' methods. Many commercial packages exist that perform analyses based upon multiple least squares analysis. The number of components included in the model and the number of data curves (variables) available dictates the solution procedure; measures to detect gross errors and mismatch are usually included in the techniques although mathematical consistency is not a guarantee of geological accuracy. By pre-specifying the model, the analyst always runs the risk that new or different circumstances from those that are expected will go unexplained or even undetected. Typically, the components defined in these methods seek to capture the mineralogy of the section; theoretically, there is no reason why differentiation into more subtle variability reflecting pore geometry cannot be performed, provided the data can discriminate such effects. Our typical data are highly correlated and most of the tools are designed to respond chiefly to the fluids rather than to the solid material of the formation. These two features of the data lead to the result that resolution into different minerals is always difficult and sometimes simply impossible; commonly, it is only possible to define 'pseudo-minerals' such as 'shale' or 'heavy' minerals, that are sometimes difficult to relate quantitatively with the actual mineralogy of a formation. Herein lies the great difficulty with using these approaches: how do we know *a priori* what mineral components and/or formation attributes are discernible in the data? Core data are always required to provide the framework mineralogy to be expected, but even then, deciding which components are actually distinguishable in the data is not easy.

The methods covered in this paper can be described as 'inductive' methods in that they 'allow the data to speak for themselves' (Doveton 1994, chapter 4). Classes or transformations are suggested by, or derived from, the data rather than from any physical model. The overriding purpose of all these methods is to isolate distinctive patterns and to derive classifications or new variables that will characterize and differentiate those patterns. Once the patterns or groupings are identified, then physical interpretation of their significance, if any, proceeds. Again, core data are often a prerequisite to correct interpretation.

The techniques can be subdivided into 'supervised' and 'unsupervised' methods. In the former the approach is to impose different categories or groups onto the data such that the variability of the data are captured by membership of the different groups; subsequently, the group-definition function, however derived, is used to process new data and determine their classification by group membership. Therefore, the 'supervised' methods may well be prone to the same difficulties in application as the deductive methods outlined above; the major difference between these paradigms, however, is one of expectation on the part of the analyst since, commonly, no physical model is invoked to extract groupings of data from the supervised inductive approaches, whereas physical component models are usually proposed in the deductive methods. In 'unsupervised' inductive methods the data are categorized solely on the basis of their inherent underlying structure without the imposition of any *a priori* subdivisions.

However derived, once the analyst has achieved a differentiation, he or she is then faced with the task of interpretation of the revealed patterns or structure in terms of geological significance. Whilst the log responses are reflections of geological variation expressed as variation in the physical properties of the rocks, it is wrong to suggest that these analysis techniques are capable of performing a lithofacies analysis *per se*. This is partly because of the trivial point that lithofacies analysis by definition requires the observation of actual rock samples and log data are remote samples of the formation, but also because the petrophysical variation is influenced by more factors than the formation lithology. The most that can be achieved from log (and core analysis) data would be an 'electro-facies' analysis (Serra & Abbott 1980), i.e. the grouping of depth levels on the basis of similarity of log responses or of laboratory measured properties. The derived electro-facies subdivisions possibly can be associated with lithological variation, and this association is often made easier if core descriptive data are available, but log and core analysis (i.e. experimental) data cannot infer lithofacies detail directly. Furthermore, as petrophysical variation commonly can be controlled by diagenetic fabric, rock strength, pore fluid type and other phenomena as well as by lithological composition and texture, the imposition of classes that are purely lithologically distinctive onto the log data tends to result in blurred subdivisions when expressed in terms of log responses.

Classification v. prediction

In many cases once the classification of data has been accomplished the analysis is not then considered complete. The classes are interpreted in the light either of geological general principles or of other available data (e.g. core descriptions) and then the interpreted data model is applied as a predictive tool to achieve some desired goal.

Generally, it is hoped that insight into the controls of formation variables that are not directly measured by the log data, such as permeability or pore connectivity as applied in resistivity modelling, may be obtained. The detailed subdivision of data can then be applied in the refined prediction of such properties from the basic log data.

Figure 1 represents a classification of the multivariate techniques covered here. Non-statisticians would tend to view the major subdivision of the use of the conventional methods as classification of data and/or prediction of variables. Statisticians may be more comfortable with labelling the major groups as interdependence and dependence methods, each of which class of technique is involved principally in classification of data (typically by data reduction) and in prediction, respectively.

Within these principal headings the methods differ in their detailed handling of the data and the choice of method to use depends largely upon the nature of the data and the number of variables under consideration.

Neural networks are an example of a technique that can equally well be applied to both principal objectives of analysis, classification and prediction, but commonly not from the same trained net.

Some considerations concerning petrophysical data

Pre-processing

It is axiomatic in the application of multivariate analysis methods (indeed, also in the application of any integrated, multiple log and core interpretation exercise) that the data being processed are accurately depth-matched. This is frequently a non-trivial task that has so far evaded satisfactory automation. Whilst good computer methods do exist (Zangwill 1982; Kerzner 1986) and can be used to give a useful first pass to large volumes of data, detailed verification of depth equivalence is usually accomplished by hand.

The effects of the environment in which the measurements are made are various, and sometimes lead to severe reduction in the signal-to-noise ratio. Tool vendors offer approximate correction algorithms by which it may be possible to overcome environmental effects and increase the signal-to-noise ratio. The propensity of a formation to become rugose or to cave in is commonly controlled by its lithological and/or textural make-up, and therefore is a diagnostic indicator of some value in itself, but unfortunately increased washout or rugosity may severely impair the quantitative value of the data in the affected intervals.

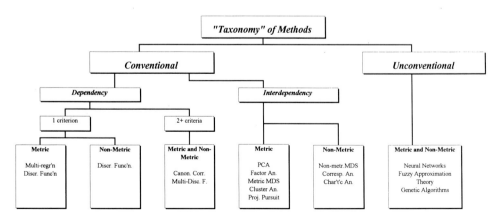

Fig. 1. A 'taxonomy' of methods. Redrawn after Dillon & Golstein (1994).

Methods have been devised (e.g. Kerzner 1986) by which to block log-derived data to transform the responses from a set of smoothed functions into stepped functions, with the steps occurring at bed boundaries. The idea is to create a formation model that more closely emulates the observation that geological phenomena commonly comprise discrete laminae, each of which possess distinct characteristics. Whilst micro-resistivity and acoustic-imaging tools can resolve features at the centimetre scale, in the case of most log measurements the minimum lamina thickness resolvable is at the decimetre (bedform) scale. The 'blocking' process leads to a reduction in data volume, since an identified 'bed' in logging terms now is characterized by a single measurement for each logging tool rather than by recordings throughout the bed comprising a set of measurements on some arbitrary depth increment reference. Care must be exercised in the application of the technique(s) to avoid loss of detail from finer resolution devices when selecting bed boundaries. The choice of values to be assigned to each lamina or bed may be difficult, because it must be based on the interpretation of smoothly varying, poorly resolved logging curves. Dipping beds pose particular problems when accessed by data representative of substantially different depths of investigation.

Curve-blocking operations have the benefit of reducing edge effects in laminated rocks exhibiting strongly contrasting laminae and are particularly useful in this situation, but this author has experienced less useful results in massive carbonates logged by poorly resolved older logging data, where bed definition was inadequate to derive consistent characterization. Increasingly, it is becoming accepted practice that the 'Earth model' defined by curve-blocking algorithms is used to recreate observed tool responses from a forward modelling procedure involving detailed knowledge of the tool response functions; *a priori* partitioning of petrophysical data may be of value in determining the Earth model that is appropriate for the data, or else the Earth model may be of value in removing edge effect scatter from the data distributions prior to applying partitioning methods of investigation.

Continuous v. discrete data

Many multivariate methods effectively assume that the measured variables are continuous, and, furthermore, that they conform to a multivariate normal distribution. Discrete, as opposed to continuous data, are analysed with their own set of techniques. Petrophysical data can be discrete or continuous. (Note that although porosity and permeability (etc.) data from core plugs may be considered 'discrete' data on the grounds that the plugs are physically separate samples and perhaps also sampled at irregular intervals, nevertheless these data are sampling the continuously varying formation properties of porosity, permeability or whatever and are therefore continuous variables. The same can be said of blocked data, see above.)

Continuous data are those data which comprise measurements that possess equality of length of steps between classes. We think in terms of two types of continuous data. Firstly, the interval scale, in which ratios of any two intervals are independent of the units of measurement and of the zero point since both are arbitrary. Interval variables allow us not only to rank order the items that are measured, but also to quantify and compare the sizes of differences between them. For example, temperature, as measured in degrees Fahrenheit or Celsius, constitutes an interval scale. We can say that a temperature of 40° is higher than a temperature of 30°, and that an increase from 20° to 40° is twice as much as an increase from 30° to 40°. Secondly, the highest order of measurement scale is the ratio scale, in which a true zero exists making the ratio of any two values constant, independent of the units system (e.g. the ratio of a measurement of any object in centimetres to the same dimension in inches is always constant at 2.54, whereas the ratio of two measures of an actual temperature, one in centigrade and the other in Fahrenheit, is not constant). Commonly, statistical data analysis procedures do not distinguish between the interval and ratio properties of the measurement scales.

On the other hand, discrete data are those which do not form part of a continuous numerical measurement scale. Discrete data commonly comprise counts of one thing or another, such as the number of vugs per core sample or the number of clay laminae of a given size per unit length of core. Also, grouped data can be created from the application of a set of categories to continuous data, e.g. 'high', 'moderate' or 'low' permeability. Discrete data are termed nominal if the classes are designated by names only, or ordinal if their values can be ranked in some numerical order. The possible categories of a discrete variable should be mutually exclusive and exhaustive. If a sample can be in one of only two possible states then it is termed a binary or dichotomous variable; if a depth sample in a formation can be classified by lithology (sandstone, limestone, dolomite), by grain size (coarse, medium, fine, very fine, clay) and by presence/absence of fractures then 'lithology' in this example is a nominal variable since no numerical rank can be assigned to the different lithotypes sensibly, 'grain size' is ordinal and 'presence of fracturing' is a binary variable. With such a data set the frequency with which observations fall within each combination of categories may be of interest; with two variables to consider, a two-way contingency table of frequencies is produced whilst data sets with more than two variables produce a multi-variate contingency table. Multivariate data of this sort are variously called categorical data, frequency data, count data, or attribute data. Such data tables are often transformed so that the elements of the table are regarded as conditional probabilities before analysis is undertaken.

In an example where the number of occurrences of open and closed fractures of each of two orientations are counted per unit length of core, a table may be produced where the rows correspond to samples (intervals of core) and the columns correspond to the different types and orientations of fractures counted. In this case the total number of fractures is the sum of all elements of the table, the sums of each row are the total number of fractures of all types per sample and the sums of the columns are the total number of each type/orientation of fracture across all samples. When there are n samples and m different combinations of open and closed fractures of two different orientations ($m = 4$ in this example) the various elements of the table can be described as shown in Table 1.

The row totals divided by the grand total give the marginal probabilities (P_i) that specific samples contain some sort of fracture, whereas the column totals divided by the grand total give the marginal probabilities (P_j) of the occurrence of specific types/orientations of fractures irrespective of the samples in which they might be found.

In addition to the marginal probabilities shown on the table we can compute: the ratios of the frequency recorded at each element of the table (x_{ij}) to the grand total number of fractures to find the joint probabilities of the occurrence of specific types of fractures in specific samples:

$$P_{ij} = \frac{x_{ij}}{\sum\limits_{i=1}^{n}\sum\limits_{j=1}^{m} x_{ij}} \qquad (1)$$

(for $i = 1 \ldots n$ samples and $j = 1 \ldots m$ combinations); then, if the joint probabilities are divided by their corresponding marginal probabilities, the following conditional probabilities result:

$$P_{(i/j)} = \frac{P_{ij}}{P_{.j}} \qquad (2)$$

where we are given that we shall find a fracture of a certain type and orientation and we want the probability that it will come from a given depth interval, and

$$P_{(j/i)} = \frac{P_{ij}}{P_{i.}} \qquad (3)$$

where we are given the depth interval and wish to know the probability that its fractures will be of a particular type and orientation.

As will be noted later, continuous data may be similarly represented for the purposes of specific analyses, but in these cases the ratio of the element value to its row and/or column total becomes a measure of the relative significance of the observation rather than of the probability of the occurrence of the observation.

Variance–covariance matrix (also known as a covariance matrix or a dispersion matrix)

The variance–covariance matrix of a set of individual data values is constructed from the basic descriptive components of the data variation within a single log and between pairs of logs. An individual measurement consists of the value for a particular log (or core) response at a particular depth. Log and core data are for the most part interchangeable in the analyses and the distinction is not usually of relevance to the discussion of the analysis method. Typically, petrophysical data comprise a number of tool responses ('logs' or core measurements) each of which has values at a number of depth intervals; commonly, there are many more depth intervals than there are types of log response. Each log or core measurement set can be considered as a vector of individual readings; the entire set of tool and core responses forms the data matrix **X**. (In this paper components of equations identified in the text in boldface will refer to either whole matrices or to vectors.)

Let $\mathbf{X} = (x_{ij})$ be an $n \times p$ data matrix comprising p log curves (variables) each with measurements at n depth levels. Each element (x_{ij}) of this array corresponds to a measurement of log $j(j = 1, \ldots, p)$ at a particular depth

Table 1. *Two-way contingency table, layout of joint probabilities*

Column row	Fracture styles...				Row totals	Marginal probabilities (P_i)
	$j=1$	2	3	$(m=)4$		
Core samples $i=1$	x_{11}	x_{12}	x_{13}	x_{1m}	$\sum_{j=1}^{m} x_{1j}$	$\dfrac{\text{Total row 1}}{\text{Grand total}}$
2	x_{21}	x_{22}	x_{23}	x_{2m}	$\sum_{j=1}^{m} x_{2j}$	$\dfrac{\text{Total row 2}}{\text{Grand total}}$
3	x_{31}	x_{32}	x_{33}	x_{34}	$\sum_{j=1}^{m} x_{3j}$	$\dfrac{\text{Total row 3}}{\text{Grand total}}$
$(n=)4$	x_{41}	x_{42}	x_{43}	x_{44}	$\sum_{j=1}^{m} x_{4j}$	$\dfrac{\text{Total row 4}}{\text{Grand total}}$
Column totals	$\sum_{i=1}^{n} x_{i1}$	$\sum_{i=1}^{n} x_{i2}$	$\sum_{i=1}^{n} x_{i3}$	$\sum_{i=1}^{n} x_{i4}$	Grand total $=\sum_{i=1}^{n}\sum_{j=1}^{m} x_{ij}$	
Marginal probabilities (P_j)	$\dfrac{\text{Total column 1}}{\text{Grand total}}$	$\dfrac{\text{Total column 2}}{\text{Grand total}}$	$\dfrac{\text{Total column 3}}{\text{Grand total}}$	$\dfrac{\text{Total column 4}}{\text{Grand total}}$		

$i(i = 1, \ldots, n)$. Associated with each variable x_j is its data mean $\bar{x}_j$ and a data variance s_j^2, where

$$\bar{x}_j = \frac{1}{n} \sum_{i=1}^{n} x_{ij} \quad \text{and} \quad s_j^2 = \frac{1}{(n-1)} \sum_{i=1}^{n} (x_{ij} - \bar{x}_j)^2.$$

$$(4a,b)$$

The latter parameter is the square of the data standard deviation s_j, a measure of the level-to-level variation of x_j (the denominator of this expression is changed to $(n-1)$ in appreciation of the fact that we are almost always dealing with a sample from an unknown population and this denominator renders the sample variance an unbiased estimate of the unknown population variance). The extent to which two variables x_j and x_k vary together is termed the covariance, s_{jk}, and is calculated from

$$s_{jk} = \frac{1}{(n-1)} \sum_{i=1}^{n} (x_{ij} - \bar{x}_j)(x_{ik} - \bar{x}_k). \quad (5)$$

Notice that $s_{jk} = s_{kj}$ and $s_{jj} = s_j^2$, this means that the matrix is symmetric and has diagonal elements equal to the variances of the constituent variables. These variances and covariances can be written in a compact form as a $p \times p$ dispersion matrix $\mathbf{S} = (s_{jk})$, otherwise known as a variance–covariance matrix or simply a covariance matrix. Whilst the off-diagonal elements of $\mathbf{S}$ may be of interest in their own right, they are frequently hard to interpret because the constituent variables are scaled in different units; the covariance matrix is generally only calculated as a stepping-stone to finding the correlation matrix, whose significance is discussed next.

Standardization

Figure 2 is a log plot of the wireline log and conventional core data from 3/3–3, a UK offshore well from the North Sea, in which the main reservoir beds are within the Brent Group of Middle Jurassic age. Also shown in Fig. 2 are the lithofacies and grain size curves that were established through examination of the cores. These are the data that form the main test case used to illustrate the computation techniques discussed in the paper.

In general, variables enumerated in different scales cannot be directly compared in a quantitative manner whilst retaining their original units. How would you measure the 'difference' between an acoustic travel time of $90\,\mu s\,ft^{-1}$ and a bulk density of $2.19\,g\,cc^{-1}$?

The most common method used to overcome the problem of different measurement units is to transform the data into standardized form by the following

$$z_{ij} = \frac{(x_{ij} - \bar{x}_j)}{s_j} \quad (6)$$

which is to say that we subtract the mean of the data set and divide the result by the standard deviation of the data set. Note that the transformed variable z_{ij} has zero mean and a standard deviation of unity. In fact, the values of the z_{ij} are scaled in standard deviations and are therefore dimensionless. If we now calculate a dispersion matrix of the standardized variables we find that we are actually calculating the $p \times p$ correlation matrix ($\mathbf{R} = (r_{jk})$) of the original, raw variables; this follows from recognition that the correlation coefficient between two variables is equal to their covariance (s_{jk}) divided by the product of their standard deviations ($s_j s_k$):

$$r_{ij} = \frac{s_{jk}}{s_j s_k} \quad (7)$$

which, if evaluated in terms of the standardized variables z_j and z_k defined above, is equivalent to the covariance of those standardized variables; the diagonal elements of the correlation matrix $\mathbf{R}$ are all equal to 1 (each variable exactly correlates with itself!) whilst the off-diagonal elements are the correlation coefficients between each pair of original variables, and the matrix is thus symmetric.

A different standardization approach is given by Walden (1992), in describing the processes involved in projection pursuit. He performs a 'centring and sphering' transformation of the data prior to analysis. This type of standardization changes the original data, which might be visualized as an ellipsoid with varying dispersion in orthogonal directions and varying correlation among the constituent variables, into a sphere that has equal dispersion in all directions and in which all variables are uncorrelated with each other.

It is important to realize that many multivariate techniques will give quite different results if one works with the raw variables rather than with the standardized variables.

It should also be remembered that a low correlation coefficient does not always imply independence between the two variables of interest, only that the degree of linear association between the two variables is small – there might well exist a strong non-linear association between them. In order to increase the likelihood that the data will behave in a multivariate normal fashion (see next section), therefore, it is often of value to apply other transformations to

B. MOSS

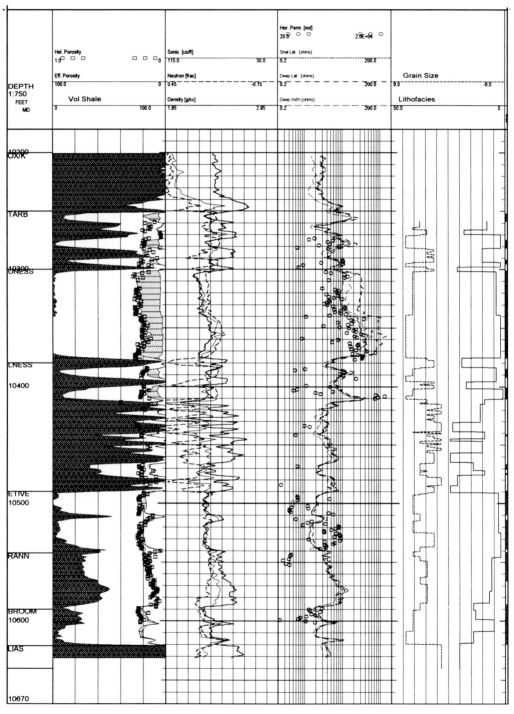

Fig. 2. 3/3-3 Brent Group wireline and core data set.

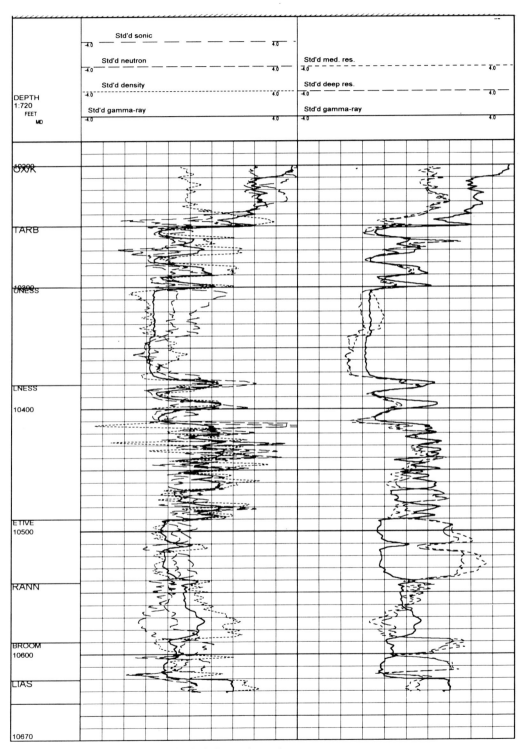

Fig. 3. 3/3-3 Brent Group standardized wireline and core data set.

the data prior to standardization so that each of the input variables is approximately normally distributed. (e.g. positively skewed data may be converted to logarithms, resistivity data may be converted to their inverse mth-root (where m is the Archie cementation exponent)) and so on. Some practitioners prefer to arrange things so that all data are positively correlated as well.

When comparing transformed, standardized data it becomes relatively easy to spot those variables or depth levels that are outliers from the main distribution, since these data fail to overlie when superimposed on depth plots of other standardized variables. If the outlying data comprise one or two variables over significant depth intervals then it may be supposed that these variables are responding differently to aspects of the formation and therefore contain different information about the formation, although possibly only as a result of borehole effects. If the outlying data are more sparsely distributed then the conclusion may be that the outlying data points are spurious, again possibly through borehole effects. A third possibility is that outlying data may reveal something of the underlying structure of the data and reflect real formation controls through lithological or petrophysical variation.

Figure 3 is an example plot of transformed, standardized porosity, gamma-ray and resistivity variables from 3/3-3. Notice in particular the tendency for greater separation between the curves in the various units as compared to the conventional display in Fig. 2. Also notice the strong anti-correlation between the resistivity curves (transformed into the inverse mth root of resistivity) and the gamma-ray data in the water leg ('oil-down-to' (ODT) is $c.\,10\,440$ ft). These are the data as 'seen' by the various analysis techniques. The correlation matrix corresponding to these data is shown in Table 2. Note from this table the strength of the density (RHOB) and gamma-ray (GR) correlations with the core data (CPHI, LOGCKX, LFAC, GRSZ) and the weak correlations between the neutron and density and neutron and core porosity (CPHI) data (see Table 3 for mnemonics). Figure 4 is a multiple crossplot display showing all the bivariate marginal distributions between the various standardized variables in the test data set. Whilst the correlations between the transformed resistivity data are very high (0.97) there are other indications that they may well possess independent data to some degree. Compare all the plot distributions in Fig. 4 with the corresponding correlation entries in Table 2. The two right hand columns of plots (and two lowest rows) correspond to the core description

lithofacies and grain size curves which are ordinal data as opposed to ratio scale data. The degree of overlap for all log responses amongst the grain size classes is notable.

Multivariate normal distribution

Recall that a normally distributed random variable has a 'bell-shaped' density function, defined by

$$\phi(x) = \frac{1}{\sqrt{2\pi}\sigma} \exp\left[-\frac{1}{2}\left(\frac{x-\mu}{\sigma}\right)^2\right]$$
$$-\infty < x < \infty \qquad (8)$$

where μ is the mean of x and σ is the standard deviation of x.

The joint density function of several independent normal variates is given by (from Morrison 1990),

$$\phi(x_2, \ldots, x_p) = \frac{1}{(2\pi)^{p/2}\sigma_1, \ldots, \sigma_p}$$

$$\times \exp\left[-\frac{1}{2}\sum_{i=1}^{p}\left(\frac{x_i-\mu_i}{\sigma_i}\right)^2\right] \qquad (9)$$

which, using matrix representations of the x_i, given by,

$$\mathbf{x}' = [x_1, \ldots, x_p], \quad \mu' = [\mu_1, \ldots, \mu_p],$$

$$\text{and} \quad \sum = \begin{bmatrix} \sigma_1^2 & \cdots & \sigma_{ij} \\ \vdots & \ddots & \vdots \\ \sigma_{ij} & \cdots & \sigma_p^2 \end{bmatrix} \qquad (10)$$

can be re-stated as

$$\phi(\mathbf{x}) = \frac{1}{(2\pi)^{p/2}|\Sigma|^{1/2}}$$

$$\times \exp[-\tfrac{1}{2}(\mathbf{x}-\mu)'\Sigma^{-1}(\mathbf{x}-\mu)]. \qquad (11)$$

Therefore, whereas the density function of a single random variable is completely characterized by the mean and standard deviation of that variable, the multivariate density function is characterized by the vector of means and the covariance matrix associated with the original data set.

The central limit theorem applies in the multivariate normal case such that large numbers of independent, *identically distributed* random vectors will tend to be jointly distributed in a multivariate normal distribution. The italics emphasize a minimum requirement for multivariate normality to hold (more is required of the data for rigorous multivariate normality

Table 2. *3/3-3 Brent Group wireline and core data, correlation and covariance matrices*

	GR	RHOB	DT	CNL	SQRLLS	SQRLLD	CPHI	LOGCKX	LFAC	GRSZ
Correlations										
GR	1	0.748	−0.086	0.632	0.257	0.413	−0.560	−0.770	0.600	−0.676
RHOB	0.748	1	−0.614	0.123	−0.081	0.079	−0.799	−0.766	0.332	−0.402
DT	−0.086	−0.614	1	0.557	0.399	0.320	0.628	0.272	0.270	−0.139
CNL	0.632	0.123	0.557	1	0.387	0.439	0.007	−0.354	0.633	−0.584
SQRLLS	0.257	−0.081	0.399	0.387	1	0.969	0.125	−0.213	0.171	−0.394
SQRLLD	0.413	0.079	0.320	0.439	0.969	1	−0.016	−0.356	0.257	−0.483
CPHI	−0.560	−0.799	0.628	0.007	0.125	−0.016	1	0.783	−0.138	0.235
LOGCKX	−0.770	−0.766	0.272	−0.354	−0.213	−0.356	0.783	1	−0.415	0.498
LFAC	0.600	0.332	0.270	0.633	0.171	0.257	−0.138	−0.415	1	−0.586
GRSZ	−0.676	−0.402	−0.139	−0.584	−0.394	−0.483	0.235	0.498	−0.586	1
Covariances										
GR	0.6306	0.5152	−0.0616	0.3738	0.2321	0.3657	−0.4208	−0.6126	0.3778	−0.4374
RHOB	0.5152	0.7531	−0.4803	0.0793	−0.0800	0.0760	−0.6559	−0.6656	0.2284	−0.2842
DT	−0.0616	−0.4803	0.8119	0.3735	0.4089	0.3218	0.5353	0.2455	0.1930	−0.1022
CNL	0.3738	0.0793	0.3735	0.5544	0.3275	0.3645	0.0046	−0.2643	0.3738	−0.3543
SQRLLS	0.2321	−0.0800	0.4089	0.3275	1.2917	1.2291	0.1346	−0.2419	0.1539	−0.3652
SQRLLD	0.3657	0.0760	0.3218	0.3645	1.2291	1.2449	−0.0170	−0.3977	0.2270	−0.4395
CPHI	−0.4208	−0.6559	0.5353	0.0046	0.1346	−0.0170	0.8947	0.7413	−0.1037	0.1811
LOGCKX	−0.6126	−0.6656	0.2455	−0.2643	−0.2419	−0.3977	0.7413	1.0025	−0.3294	0.4064
LFAC	0.3778	0.2284	0.1930	0.3738	0.1539	0.2270	−0.1037	−0.3294	0.6289	−0.3789
GRSZ	−0.4374	−0.2842	−0.1022	−0.3543	−0.3652	−0.4395	0.1811	0.4064	−0.3789	0.6638

Correlations and covariances are for casewise deletion of MD.
$N = 501$.
See Table 3 for mnemonics.

Table 3. *Wireline mnemonics*

Mnemonic	Log
GR	Natural gamma-ray (API)
RHOB	Bulk density (gm cc^{-1})
DT	Sonic travel time (μs ft^{-1})
CNL	Thermal neutron (frac. LPU)
SQRLLS	Inverse square root shallow laterolog
SQRLLD	Inverse square root deep laterolog
CPHI	Core porosity (frac.)
LOGCKX	Logarithm of ambient core horizontal air permeability
LFAC	Lithofacies coding
GRSZ	Grain-size coding

to hold) and most geoscientists are content to apply simple transforms to their data to obtain similar normal distributions for the variables of interest and then to assume multivariate normality for the joint distribution.

Measures of similarity/dissimilarity

Data with metric properties (ordinal, interval or ratio scales) can be analysed with distance-type similarity measures; non-metric data, on the other hand, are appropriately analysed with a matching-type measure.

In general terms the measure of distance between data points is the Minkowski metric and defined by

$$d_{ij} = \left\{ \sum_{k=1}^{p} |X_{ik} - X_{jk}|^r \right\}^{1/r} \quad (12)$$

which measures the distance between two objects (i.e. data points). The Euclidean distance is a special case of this metric where $r = 2$

$$d_{ij} = \left\{ \sum_{k=1}^{p} (X_{ik} - X_{jk})^2 \right\}^{1/2}. \quad (13)$$

Note that Euclidean distances are not scale invariant, therefore they should always be utilized on standardized data. Standardization has the benefit of retaining the relative distances or rankings of the original data points. Davis (1986) uses a standardized p-space Euclidean distance defined by

$$d_{ij} = \left\{ \frac{\sum_{k=1}^{p} (X_{ik} - X_{jk})^2}{p} \right\}^{1/2} \quad (14)$$

where p is the number of variables being considered.

If $r = 1$ in the Minkowski metric, then we have the city-block metric,

$$d_{ij} = \sum_{k=1}^{p} |X_{ik} - X_{ij}|. \quad (15)$$

The city-block metric makes no distinction between the case where two data points are each two units apart on two variables, or where one data point is one unit apart on one variable and three units apart on the second variable. Both cases will result in the data points possessing the same distance measure.

Another very popular distance measure is Mahalanobis' distance, D^2,

$$D^2 = (\mathbf{X}_i - \mathbf{X}_j)^T \mathbf{S}^{-1} (\mathbf{X}_i - \mathbf{X}_j) \quad (16)$$

where the i and j refer to data points and $\mathbf{S}$ is the pooled within-group covariance matrix. This distance measure will account for any correlation there might be between variables; it forms the exponent in the multivariate normal density function (see eq. (13)) and therefore is quite fundamental, and for this reason is frequently a measure of choice for analysing log (and core) data.

Matching-type measures might be referred to as association coefficients. They are generally applied to contingency tables derived from nominal scale data and usually take on values in the range 0 to 1. However, several different measures are available and each produces different results from the same input table. Dillon & Goldstein (1984) provide the example in Table 4.

If we count the presence or absence of six attributes on two samples we might create a table (Table 4), where a plus sign indicates that the attribute is present and a minus sign indicates absence. To assess similarity we could simply add all matches (counts in cells [a] and [c]) and divide by the total number of attributes to 3/6 or 0.5.

Alternatively, we could count just the possession of an attribute as significant, thus find the ratio of the number of attributes that are present for both samples to the number that are present for either sample. In this example, such a measure is $2/(2 + 1 + 2) = 0.4$

There are six commonly used coefficients of association, given in Table 5: (a, b, c, d are cells as in Table 4)

From our example the results of the equations that are the elements of Table 5 range from 0.33 to 0.8, depending on the criteria that are considered important.

(Distance measures may be transformed into association measures by $1/(1 + d_{ij})$, but the

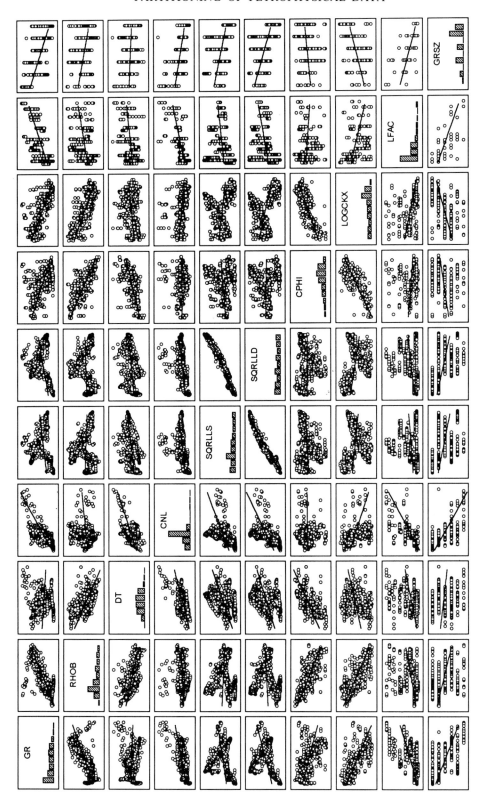

Fig. 4. Correlations amongst standardized variables, 3/3-3 Brent Group.

Table 4. *Hypothetical attribute counts*

Sample	Attribute					
	1	2	3	4	5	6
A	0	1	1	0	1	1
B	1	0	1	0	0	1

Which, upon counting the incidences of the attributes, becomes the two-way contingency table:

	Sample A		
	+	−	
Sample B			
+	[a] 2	[b] 1	3
−	[d] 2	[c] 1	3
	4	2	6

reverse transformation is not possible because of the necessity for distances to satisfy certain conditions, viz.

a function $d(x, y)$ of pairs of points of a set E is said to be a metric for E if it satisfies

$d(x, y) \geq 0.0; \quad d(x, y) = 0.0$ if $x = y$ (positivity)

$d(x, y) = d(y, x)$ (symmetry)

$d(x, z) + d(y, z) \geq d(x, y)$ (triangle inequality).

The correlation coefficient is another frequently used measure of association in many other procedures, but should be used with caution for clustering purposes. To see why, consider Table 6 (from Dillon & Goldstein 1984)

If two profiles are parallel but far apart, the correlation coefficient would be unity. However,

Table 5. *Measures of association*

(1)	$\dfrac{a+d}{a+b+c+d} = \dfrac{4}{6}$	$= \dfrac{2}{3}$
(2)	$\dfrac{a}{a+b+c} = \dfrac{2}{4}$	$= \dfrac{1}{2}$
(3)	$\dfrac{2a}{2a+b+c} = \dfrac{4}{(4+2)}$	$= \dfrac{2}{3}$
(4)	$\dfrac{2(a+d)}{2(a+d)+b+c} = \dfrac{8}{(8+2)}$	$= \dfrac{4}{5}$
(5)	$\dfrac{a}{a+2(b+c)} = \dfrac{2}{(2+4)}$	$= \dfrac{1}{3}$
(6)	$\dfrac{a}{a+b+c+d} = \dfrac{2}{6}$	$= \dfrac{1}{3}$

Table 6. *Hypothetical associations*

Sample	Variable			
	X_1	X_2	X_3	X_4
A	1	3	2	2
B	4	10	7	7
C	1	2	2	2

two profiles may be perfectly linearly related and not be parallel, yet still the correlation coefficient would be unity because of the linear relationship relating the scores. On Table 6 Samples A and B have a correlation coefficient of unity because sample B is simply $(3 \times A) + 1$. Samples A and C however, have a correlation coefficient of 0.82 and so would be considered more dissimilar despite the fact that the scores of A and C are identical except for a single variable, X_2.

Davis (1986) provides an example of just this situation, but his equivalent of samples A B and C above represent different measures on fossil samples. As noted before, samples A and B are highly correlated but samples A and C have the least distance between them; in Davis's example, samples A and B represent the same shape but A is smaller than B whereas A and C are different shapes but might be similar sizes.

Introduction to computation techniques

From a definition of multivariate analysis being the simultaneous evaluation of several observations on sets of samples it arises that in all methods we are manipulating an $n \times p$ data matrix comprising p observations on each of n samples. Much information concerning the structure of such a matrix is to be found from examining the eigenvalues and their associated eigenvectors extracted from the matrix.

The eigenvalues of a square matrix are defined as the roots of the determinantal equation (Dillon & Goldstein 1984):

$$h(\lambda) = |\mathbf{A} - \lambda \mathbf{I}| = 0. \tag{17}$$

We denote the p roots of $h(\lambda)$ by $\lambda_1, \lambda_2 \ldots \lambda_p$. Since for each $i = 1, 2, \ldots, p$ we have $|\mathbf{A} - \lambda_i\mathbf{I}| = 0$, it follows that $|\mathbf{A} - \lambda_i\mathbf{I}|$ is singular and hence that there exists a non-zero vector $\mathbf{X}$ satisfying

$$\mathbf{A} \cdot \mathbf{X} = \lambda_i \cdot \mathbf{X}. \tag{18}$$

Any vector $\mathbf{X}$ satisfying this equation is called an associated eigenvector of $\mathbf{A}$ for the eigenvalue λ_i.

In general, the eigenvalues may involve complex terms, but eigenvalues of square matrices are always real values. Rather than solve the roots of high-order polynomials, in practice the eigenvalues of a matrix are found from the following matrix manipulations:

Any symmetric matrix $\mathbf{A}$ can be written

$$\mathbf{A} = \mathbf{P}\Lambda\mathbf{P}^{-1} \qquad (19)$$

where Λ is a diagonal matrix of eigenvalues of $\mathbf{A}$ and $\mathbf{P}$ is an orthogonal matrix with columns equal to the standardized eigenvectors associated with the diagonal entries of Λ. This result is called the spectral decomposition theorem.

A unique solution in $\mathbf{P}$ is found by setting

$$\mathbf{P}\mathbf{P}^{T} = \mathbf{I}. \qquad (20)$$

In the event that the matrix for which the eigenvalues and eigenvectors are desired is not square but is $(n \times p)$, then $\mathbf{A}$ can be written

$$\mathbf{A} = \mathbf{P}\Delta\mathbf{Q}^{T} \qquad (21)$$

where $\mathbf{P}(n \times p)$ and $\mathbf{Q}(p \times r)$ are column orthonormal matrices and Δ is a diagonal matrix with positive elements which are the square roots of the non-zero eigenvalues of the square matrix $\mathbf{A}^{T}\mathbf{A}$. The r columns of $\mathbf{P}$ are the eigenvectors of $\mathbf{A}\mathbf{A}^{T}$, and the r rows of $\mathbf{Q}^{T}$ are the eigenvectors of $\mathbf{A}^{T}\mathbf{A}$. An orthonormal matrix is defined as a matrix $\mathbf{P}$ for which $\mathbf{P}^{T}\mathbf{P} = \mathbf{I}$, but $\mathbf{P}\mathbf{P}^{T} \neq \mathbf{I}$. That is, each column is orthogonal to every other column and each column displays unit sum-of-squares. This result is called singular value decomposition.

In many of the methods the raw data matrix is first transformed and standardized, and then may be converted into some kind of square matrix for ease of computation (e.g. the variance-covariance matrix is used in principal components analysis and the so-called R-mode and Q-mode matrices in factor analysis). Eigenvalues and eigenvectors are extracted from the square 'working matrix' that is created; several of the methods reviewed below differ only in the manner of derivation of this 'working matrix'.

Most of the analysis techniques attempt the further transformation of the (transformed and standardized) data into a set of new variables, generally fewer in number than the original variables, that reveal the intrinsic structure of the data. Techniques that are considered complete once the data reduction has occurred include principal components analysis, factor analysis, multi-dimensional scaling and canonical correlation. Other techniques go on to use the reduced data set to assign the original data values into some kind of discriminant grouping system and these include discriminant function analysis and cluster analysis. Fuzzy logic may be invoked to assist in the definition of data groupings; the 'pigeon-holing' method and neural networks are alternative methods for determining the significant patterns or structure within the data.

We begin the review of methods with a look at some basic plotting methods applicable to petrophysical data.

Crossplots

Crossplots of two variables are bivariate marginal distributions of the multivariate data and may not necessarily reveal the form of the multivariate distribution. If p variables are considered then the number of crossplots required to see all combinations is $p(p-1)/2\ldots$ if $p = 10$ then 45 crossplots are required! If the data require further subdivision on stratigraphic, lithological or other bases prior to analysis then the number of crossplots is multiplied by the number of such subdivisions. Viewing all requisite combinations is sometimes simply impractical.

Many specialized crossplots have been devised that concentrate upon detecting specific behaviour of log and core responses that may be used qualitatively or quantitatively in interpretation studies. An excellent review of these methods is given by Fertl (1981). Obviously, any arbitrary function of a particular data pair can be used to quantify limits to data clouds observed on the actual plots themselves. The analyst simply draws lines on the plots and expresses these limiting equations in terms of the data forming the crossplot. The original data are then tested against the limiting equations and assigned to different groups on the results of the tests. Depending upon the data used in the crossplots the groups thus defined can correspond to different fluids in the pores (resistivity–porosity or resistivity–spontaneous potential plots), to different matrix properties (multiple porosity tool crossplots or natural (or induced) gamma-ray–porosity plots) and so on. For many purposes these types of plots define the applicable petrophysical analysis techniques (Fertl 1981)

Although seldom used in petrophysical analyses, nevertheless of some practical value are plotting techniques such as 'spider's web', ladder plots, star diagrams, stick plots, line icons etc. In general, these types of plots are not available in mainstream log interpretation packages, which probably explains why they are not commonly used as display aids in petrophysics. Nevertheless they do form part of the toolset of most statistical evaluation packages. For some examples using the test data see Fig. 5a–c. Each individual icon plot in this figure, be it a star, line, bar chart or whatever, represents an individual 'case' in the study; in the nature of log and core data, a 'case' comprises all observations at a single depth level. Large amounts of data are summarized and patterns are revealed.

Hough transform

Hough (1962) introduced a very straightforward concept as a means to detect patterns of points in binary image data. Doveton (1994) discusses the application of the technique to 'Pickett' (resistivity v. porosity data) plots, in particular.

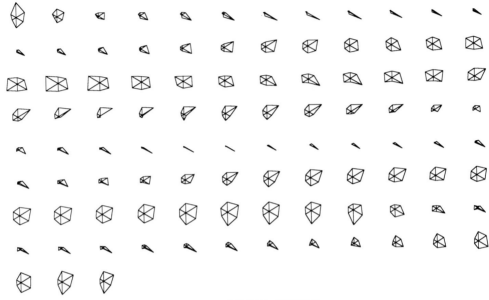

LEGEND (clockwise): GR, RHOB, DT, CNL, SQRLLS, SQRLLD,

(a)

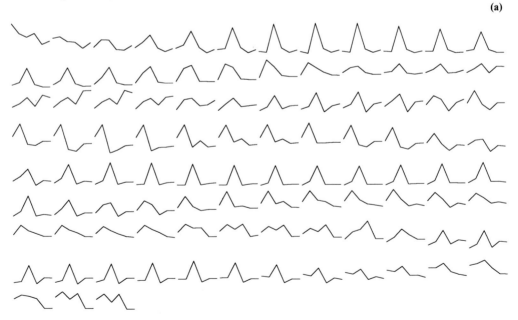

LEGEND (left to right): GR, RHOB, DT, CNL, SQRLLS, SQRLLD,

(b)

Fig. 5. 'Icon plots' of standardized variables, 3/3-3 Tarbert Formation.

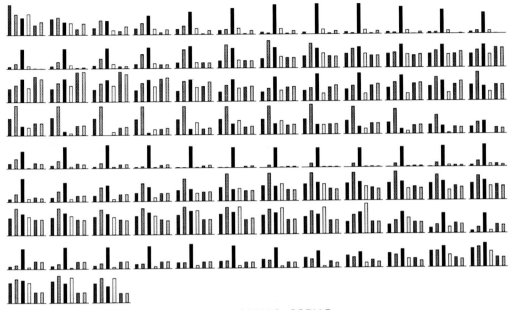

LEGEND (left to right): GR, RHOB, DT, CNL, SQRLLS, SQRLLD,

Fig. 5. (*continued*).

Essentially, plots of X v. Y data can be considered as representing points in 'image space'. Straight lines can be constructed that summarise the data cloud or sub-groups thereof, and can be used to characterize all or parts of the bivariate data cloud. The maximum number of straight lines that can be fitted through the data set is equal to the number of pairs of data points plotted on the graph. Computing the slopes and intercepts of all possible lines through the pairs of data points produces a bivariate table of data that, in turn, may be plotted and examined. This latter plot is considered to be a valid representation of the original data in 'parameter space' and is the outcome from the Hough transformation exercise. If n data points are plotted in image space then $(n^2 - n)/2$ points are created in parameter space. Systematic linear trends in the original data should be revealed as clusterings in parameter space.

Principal components analysis (PCA)

In this method we seek a few weighted linear combinations ('principal components') of the original observations that are each uncorrelated with the other such combinations and which together account for the total variation seen in the original observations. In this manner we describe a large proportion of the total variation ('account for' a large proportion of the total

variation) usually with a smaller number of variables than existed in the original data. The new variables are weighted linear combinations of the original observations. The ratio of the variance of each new variable to the total variance of all the original observations expresses the amount of that total variation in the original data that has been accounted for by each new variable. The 'first' principal component (PC_1) is that new variable (weighted linear combination of the observations) for which the ratio of its variance to the total original variance is the largest of all the new variables' ratios. The 'second' principal component (PC_2) is that new variable (weighted linear combination of observations) which is uncorrelated with PC_1 and has the next largest ratio. Because it is uncorrelated with PC_1, the second principal component accounts for the largest proportion of the variance remaining after the 'extraction' of the PC_1. Subsequent components are similarly defined and all components are uncorrelated with each other.

Although there are as many components calculated as there are original variables, in data that are highly correlated with each other (a typical feature of log data sets) a very high proportion of the total variation in the original data observations can usually be matched with only a few components (Moss & Seheult 1987). Commonly, the first and possibly the second

Table 7. *Eigenvalues and data loadings on principal components*

	Eigenval.	% Total variance	Cumul. eigenval.	Cumul. %		Factor 1	Factor 2	Factor 3	Factor 4	Factor 5	Factor 6	Factor 7	Factor 8
1	3.626828	45.33535	3.626828	45.33535	GR	**-0.897695**	0.188053	0.271914	0.208244	0.045738	0.136572	-0.143224	-0.014779
2	2.781538	34.76922	6.408366	80.10458	RHOB	**-0.876089**	-0.373002	0.018948	0.223210	0.106142	0.002698	0.178553	-0.000115
3	0.982834	12.28543	7.391201	92.39001	DT	0.343384	**0.816483**	0.357870	-0.203772	0.138650	0.140828	0.082403	7.61×10^{-5}
4	0.299107	3.738847	7.690309	96.12886	CNL	-0.416993	0.665366	0.573966	0.038044	-0.186493	-0.128223	0.035825	0.003811
5	0.136277	1.703467	7.826586	97.83233	SQRLLS	-0.262938	**0.825415**	-0.489885	0.022362	-0.033299	-0.013030	0.023971	-0.084916
6	0.092873	1.160923	7.919460	98.99325	SQRLLD	-0.422962	**0.778980**	-0.451613	0.042469	-0.019429	0.017530	-0.001975	0.088572
7	0.065241	0.815522	7.984702	99.80878	CPHI	**0.806083**	0.457418	0.083497	0.280736	0.191683	-0.130823	-0.036732	0.002646
8	0.015297	0.191219	8	100	LOGCKX	**0.929429**	0.018196	-0.010739	0.286062	-0.175175	0.142648	0.053342	0.001287
					Expl. var.	3.626828	2.781538	0.982834	0.299107	0.136277	0.092873	0.065241	0.015297
					Prp. totl.	0.453353	0.347692	0.122854	0.037388	0.017034	0.011609	0.008155	0.001912

* Emboldened loadings have absolute values >|0.700000|.
Extraction: Principal components.

principal components dominate and account for much larger proportions of the total variance than the other components combined. However, it is still instructive to study the weights comprising the less dominant components since they can reveal subtle structure in the original data set that is completely obscured by the intercorrelations amongst the original variables.

The principal components of a data set are the eigenvalues of the correlation matrix that describes the data observations, and the eigenvectors associated with the eigenvalues are the weights to be applied to each of the original variables.

The data observations may be regarded as n points in p-dimensional space. We seek new variables $t_1, \ldots, t_p$ which are linear combinations of the weighted, standardized original data observations. The weights comprise the elements of a $p \times p$ matrix $\mathbf{L} = (l_{jk})$ such that

$$t_1 = l_{11}u_1 + \cdots + l_{1p}u_p = \mathbf{l}_1^T \mathbf{u}$$

$$\vdots \qquad\qquad (22)$$

$$t_p = l_{p1}u_1 + \cdots + l_{pp}u_p = \mathbf{l}_p^T \mathbf{u}$$

are uncorrelated.

In matrix form we may write $\mathbf{t} = \mathbf{L}\mathbf{u}$ where $\mathbf{t}^T = (t_1, \ldots, t_p)$, $\mathbf{L} = (l_1, \ldots, l_p)$ and $\mathbf{u}^T = (u_1, \ldots, u_p)$. To satisfy the requirement that the $t_1, \ldots, t_p$ are uncorrelated the dispersion matrix (i.e. the variance-covariance matrix) of t should be diagonal. (This is because in a 'diagonal' matrix all off-diagonal elements are zero, and, since the off-diagonal elements of the dispersion matrix are the covariance terms, setting these to zero ensures that the $t_1, \ldots, t_p$ are uncorrelated.) This condition is achieved by evaluating

$$\mathbf{L}^T \mathbf{R} \mathbf{L} = \Lambda = \mathrm{diag}(\lambda_1, \ldots, \lambda_p)$$

$$\text{for some} \quad \lambda_j >= 0 \quad (j = 1, \ldots, p) \quad (23)$$

and a unique solution in $\mathbf{L}$ is found by setting $\mathbf{L}\mathbf{L}^T = \mathbf{I}$ (the $p \times p$ identity matrix).

The calculation of the principal components of a data set therefore involves finding the eigenvalues, $\lambda_1, \ldots, \lambda_p$, of the correlation matrix $\mathbf{R}$ of the data. The eigenvalues of $\mathbf{R}$ are none other than the variances of the $t_1, \ldots, t_p$ since they are the diagonal elements of the dispersion matrix of the $t_1, \ldots, t_p$). The eigenvectors associated with these eigenvalues are the weights that should be applied to the original observations in order to construct the new linear combinations that form the components.

Table 7 lists the eigenvalues that characterize a subset of eight variables from 3/3-3 Brent Group data. The eight variables used were gamma-ray, density, sonic, neutron, transformed resistivities, core porosity and the logarithm of core horizontal permeability. Note how rapidly the cumulative variance accounted for by the new components mounts up, so that after four new components some 96% of the original variance is 'matched' or 'explained'. The question of how many components to retain is subjective by nature. Plots such as shown in Fig. 6 (called a 'scree' plot) can be of value in reaching a decision; only those components plotting to the left of the break of slope on this plot are considered significant; eigenvalue number four might have significance but it would appear marginal. Actual values of eigenvalues are sometimes used: they can be thought of as being the ratio of the amount of variance explained by a component to the contribution from any single original variable; if the eigenvalue is greater than unity then it is considered to have contributed more to the explanation of the variance than is expected from a single original variable. On this measure only three components are significant.

It would be possible to replace our eight original variables with these three (or four) principal components and yet retain most of the variability in the original data. Interpretation of the meaning of the new components is far from straightforward, however. Crossplots of the component scores (i.e. the evaluation of the weighted sum of all variables at each depth level) against the original variables can be helpful. Commonly with wireline data, the first principal component, accounting for the major portion of the original variance, is a surrogate for formation porosity (Moss & Seheult 1987), which a moment's thought about the nature of log measurements suggests is to be expected. Examining the loadings, Table 7 (the magnitudes of the coefficients of the original variables in the weighted sum that builds the new components) can also suggest physical meanings for the components. Figure 7 shows a crossplot of components and their loadings with respect to the original variables. Component one is seen to carry the largest loadings from the density, gamma-ray, porosity and transformed permeability data; recall from Table 2 that density and gamma-ray are maximally correlated with porosity (and transformed permeability) and thus it is quite plausible for component one to be representing the porosity. A crossplot of core porosity against the first component (Fig. 8) confirms the strong trend, but also reveals some other subdivision, probably lithological, on the first component. Figure 9 illustrates component

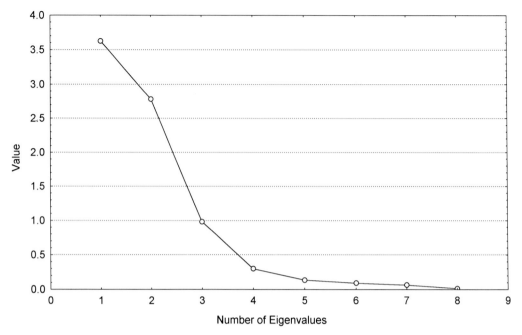

Fig. 6. 'Scree' plot of eigenvalues: eight variable case, 3/3-3 Brent Group.

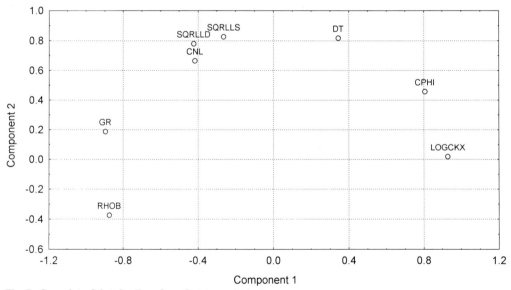

Fig. 7. Crossplot of data loadings from first two principal components.

scores plotted by depth and compared to the density data for reference.

It should be remembered that principal component analysis is limited by being a linear technique. As a result, the method may be unable to capture complex non-linear correlations and may conceivably over-estimate the reduced dimensionality as a result.

Factor analysis

This is a group of techniques for investigating interdependencies in the data. They all attempt to simplify relationships that exist amongst a set of observations by uncovering common dimensions or factors that link together the observations. The distinction between factor analysis

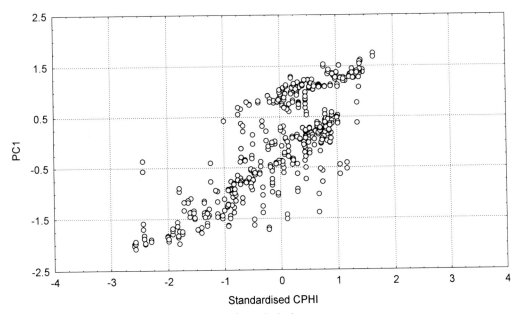

Fig. 8. Crossplot of standardized core porosity v. first principal component.

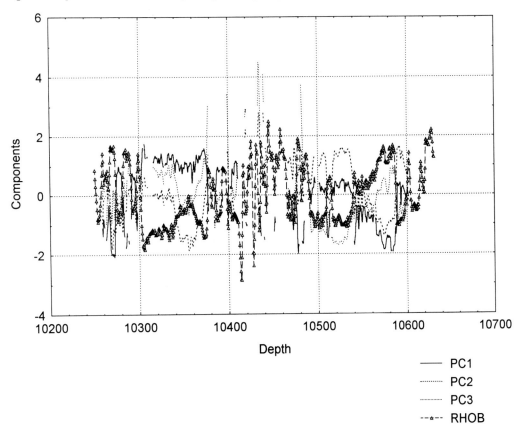

Fig. 9. Principal components 1 to 3 plotted against depth, 3/3-3 Brent Group.

(FA) and principal components analysis (PCA) is that the former is suited to determining qualitative and quantitative distinctions within the data and for testing *a priori* assumptions about the number of common factors within a given data set, whilst the latter technique is suited to determining a small set of linear combinations of the original variables that account for most of their observed variance. In large data sets these data reduction methods may provide very similar results but with smaller data sets the presence of an error term within the factor analytic model attains greater significance and the results may differ from PCA.

Factor analysis may be used in exploratory mode in which the analyst is attempting to find hidden structure within the data, or may be used in confirmatory mode in which some prior model of the common structure within a set of data is tested for the degree to which it represents the data.

The relationships within a set of p variables are considered to reflect the correlation of each variable with m underlying ('unobservable') factors that are mutually uncorrelated. Variance in the p variables is controlled by two components, namely, the variance in the m unobservable but common factors (correlated to some degree with all the variables), and by a contribution from a unique source affecting only the independent p variables. The model can be expressed as

$$X_j = \sum_{r=1}^{m} a_{jr} f_r + \varepsilon_j \qquad (24)$$

where f_r is the rth common factor, m is the specified number of factors and ϵ_j is the random variation unique to the p original variables X_j. The coefficient a_{jr} is the loading of the jth variate on the rth factor.

If we assume that the X_j are multivariate normal then the variance-covariance matrix (found in different ways in R-mode and Q-mode analyses, see below) comprises a $p \times p$ matrix with diagonal elements (variances of the p variables),

$$s_{jj}^2 = \sum_{r=1}^{m} a_{jr}^2 + \varepsilon_{jj} \qquad (25)$$

and off-diagonal elements (covariances),

$$\mathrm{cov}_{jk} = \sum_{r=1}^{m} a_{jr} a_{kr}. \qquad (26)$$

Factor methods either operate by extracting the eigenvalues and associated eigenvectors from a square matrix produced by multiplying a data matrix (or some transformation of a data matrix) by its transpose (Davis 1986), or by the rather more complex method of maximum likelihood estimation which will not be discussed further (the reader is referred to the literature, e.g. Dillon & Goldstein 1984). In matrix terms then, the problem may be expressed as

$$\mathbf{s}^2 = \mathbf{A}^{\mathrm{R}} \cdot (\mathbf{A}^{\mathrm{R}})^{\mathrm{T}} + \mathrm{var}\, \varepsilon_{jj} \qquad (27)$$

which says that the matrix of variances and covariances ($\mathbf{s}^2$) is the product of a $p \times m$ matrix (number of factors by the number of variables) of factor loadings ($\mathbf{A}^{\mathrm{R}}$) multiplied by its transpose and added to a $p \times p$ matrix of unique variances, which is diagonal.

Multiplying a $p \times m$ matrix by its transpose will produce a $p \times p$ matrix which has only m positive eigenvalues and associated eigenvectors. If $p = m$ then the unique variances matrix (var ε_{jj}) will vanish and the problem becomes equivalent to principal components analysis. If $m \leq p$ then both the matrix of parameters (loadings), $\mathbf{A}^{\mathrm{R}}$, and the matrix of unique variances, var ε_{jj}, must be estimated and the problem is indeterminate if the number of factors is not known beforehand. Since this is often the case in geoscientific problems then the use of factor analytic schemes in an exploratory sense is not quite so straightforward as is the use of principal components analysis.

Exploratory factor analysis proceeds through the principle of conditional linear independence. This asserts that once m common factors have been eliminated from the data, then the correlation between every pair of variables is zero. The method proceeds by finding the first common factor, eliminating it from the original variables and testing for non-zero partial correlation amongst any pair of variables after the elimination; if a non-zero correlation is found then a second factor is determined, both factors are eliminated from the variables and the new partial correlations determined. This continues until the partial correlations amongst the remnants are all zero.

The communality of a variable is that portion of its variance that is accounted for by the common factors. Consider our basic factor equation to be

$$X_i = c_i + e_i \qquad (28)$$

where c_i is $\sum_{j=1}^{m} a_{ij} f_j$ is that part of each variable common to the other $p - 1$ variables, and the term e_i is that part of each variable that is unique, then, because the common and unique parts of a variable are uncorrelated and the variance of the common factors is unity (they are standardized components) the variance of the original variable can be partitioned as follows

$$\text{var}(X_i) = \text{var}(c_i) + \text{var}(e_i). \qquad (29)$$

Communality is defined as $\text{var}(c_i)$ and is equivalent to

$$\text{var}(c_i) = \sum_{j=1}^{m} a_{ij}^2 \qquad (30)$$

which is the sum of the squared elements of the matrix of factor loadings. The uniqueness of a variable is the term $\text{var}(e_i)$, and measures the degree to which a variable is unexplained by the common factors. Thus the total contribution of a single factor to explaining the total variance of the combined set of variables is the eigenvalue of the factor f_i which is found from

$$V_j = \sum_{i=1}^{p} a_{ij}^2 \qquad (31)$$

and these are the squared factor loadings for the factors, summed across all variables. The total communality is found from

$$V = \sum_{j=1}^{m} V_j \quad \text{and} \quad V_c = \frac{V_j}{V} \qquad (32)$$

where V_c is the proportion of the variance amongst all the variables that is accounted for by the factor f_j as a fraction of the entire variance accounted for by all the common factors.

Therefore, factor loadings represent the correlation between a given factor and a variable and provide information as to which variable is involved with which factor and to what degree. In large-scale problems the pattern matrix of factor loadings tends to be complicated and interpretation of the factor loadings may not be straightforward. Interpretation is often assisted by isolating for each variable which factor has the highest loading and then determining the practical significance of that highest loading i.e. divide the highest loading value by the sum of the loadings for that variable across all factors and then apply some cut-off value of this ratio, say 0.3, for the highest loading to be significant). Having analysed the factor loadings in this manner the analyst attempts to assign meanings to the pattern of significant factor loadings.

Factor loadings may be rotated by any of a number of techniques, the goal of which is to simplify the structure by re-orientating the factors so that they are clearly marked by high loadings for some variables and low loadings for the rest. A common rotation strategy is the varimax normalized strategy and has been applied in the example discussed below. The technique is designed to orientate the factors so that the variance of the loadings on each factor is maximized. The normalization is performed by dividing each raw loading by the square roots of the respective communalities. This rotation is aimed at maximizing the variances of the normalized factor loadings across variables for each factor; this is equivalent to maximizing the variances in the columns of the matrix of normalized factor loadings.

A distinction can be drawn between R mode techniques that investigate the correlations and covariances between variables and Q mode techniques that look at similarities between pairs of observations. (Mathematically based treatises tend not to make this distinction, e.g. see Chatfield & Collins 1980.) The relevant square matrices are found as follows

$$\mathbf{R} = \mathbf{X}^T\mathbf{X} \qquad (33)$$

where the elements of $\mathbf{R}$ will consist of the sums of squares and the cross-products of the p variables (if $\mathbf{X}$ is the original data matrix, otherwise, if $\mathbf{X}$ is standardized, the elements of $\mathbf{R}$ will be correlations) and if $\mathbf{X}$ comprises n observation points, each with p variables, then $\mathbf{R}$ is $p \times p$ viz.:

$$r_{jk} = \sum_{j=1}^{n} x_{ij}x_{ik} \qquad (34)$$

and, similarly,

$$\mathbf{Q} = \mathbf{X}\mathbf{X}^T \qquad (35)$$

which has elements

$$q_{il} = \sum_{j=1}^{p} x_{ij}x_{lj}. \qquad (36)$$

Since in most data sets there are many more observations than there are variables, $\mathbf{Q}$ is often substantially larger than $\mathbf{R}$.

The Eckart–Young theorem (Eckart & Young 1936), a lucid account of which is in Davis (1986), provides the relationships between R-mode and Q-mode factor analyses, and shows that once the smaller R-mode-analysis has been performed, the larger Q-mode solutions can be obtained directly. Unfortunately, this is true only if the scaling procedures used in the creation of the R-mode and Q-mode matrices are the same, and this only occurs if particular measures are taken. In correspondence analysis (see later section) of contingency tables the transformation of the variables is the same for

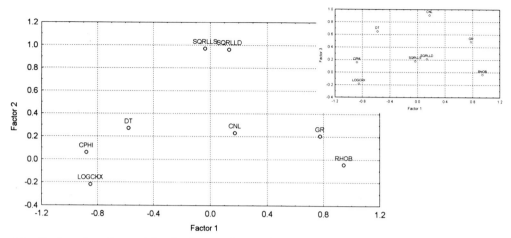

Fig. 10. Crossplots of rotated data loadings on the principal factors, three-factor extraction case.

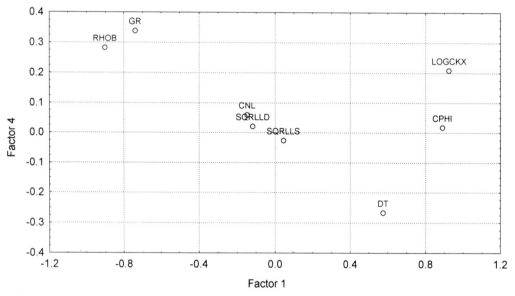

Fig. 11. Crossplot of data loadings on the first and fourth principal factors, four-factor extraction case.

both the columns and the rows, i.e. each element is divided by the product of the square roots of the row and column totals. Therefore, simultaneous R-mode and Q-mode factor analyses can be applied to these transformed data, following Eckart–Young. Other transformations are possible in order to perform simultaneous calculations, see Davis (1986).

As an example of a Confirmatory factor analysis see Tables 2, 5, 6 and 7 and Figs 10 to 12. In this example we compare the results of factor studies on the 3/3-3 data set in which firstly three factors (as suggested in the principal components analysis example) and then four

factors are extracted from the eight variable wireline and core data set. Finally, two more variables are added to the data set and their impact is assessed firstly through PCA then through Factor Analysis.

Figure 6 is a so-called 'scree' plot of the eigenvalues extracted from the matrix of standardized data that includes gamma-ray, density, sonic, neutron, deep and shallow transformed resistivities and the core porosity and permeability data. The plot strongly suggests that there are three components of significance that adequately summarize the original data and there is a possibility that the fourth component is also of

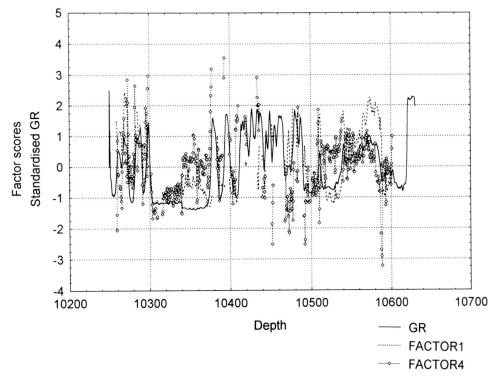

Fig. 12. Depth plot of first and fourth principal factors, four-factor extraction case.

significance. If a factor analysis is conducted with the aim of extracting just three factors the results can be examined to determine the extent to which the new factors have actually accommodated the original variance structure (the magnitude of the communalities) and the degree to which the residual 'unique' variation contains further structure that may suggest further significant factors can be extracted. A number of algorithms exist by which to extract factors, this example has used the so-called iterated communalities or MINRES solution (Statsoft 1994). The technique replaces the diagonal elements of the correlation matrix with the multiple R-square statistic (i.e. the multiple correlation coefficient (R^2) between each variable and all of the other variables); after the initial extraction of factors based on this modified correlation matrix, the factor loadings are substituted into the diagonal elements and new factors extracted. The new loadings are inserted in the diagonal and the process continues until successive calculations of loadings are within some narrow tolerance limit. The end result is the set of loadings on the specified number of factors that minimizes the residual sum of squares, i.e. that minimizes the unexplained or unique variance.

Table 8 presents the results of the three-factor case. From the communalities table note how the communalities increase as successive factors are brought in; by comparing the magnitudes of the 'communalities after three factors' with the multiple R^2 statistic for each variable, it can be seen that for almost all of the variables just three factors have accounted for as much of the variation in each variable as is otherwise found from the combination of all of the other original data variables. The only exceptions are for the core data (CPHI and LOGCKX). This result confirms the strong significance of the first three principal components, since they apparently account for as much of the total variability in the original data as could be found from multiple regression involving all the data.

The rotated loadings on these three factors should be compared with the loadings on the principal components (Table 7). The loadings represent the individual correlations between the original data variables and the new factors or components. The loadings for the three factors are larger than for the first three principal components and the proportion of total variance explained by the factors is higher, both of which are indications of the greater optimization in the three factor solution. However, the overall

Table 8. *Eigenvalues communalities, and data loadings on principal factors, three-factor solution*

Eigenvalues				Communalities					Factor loadings (varimax normalized)*		
Eigenval.	% Total variance	Cumul. eigenval.	Cumul. %		From 1 factor	From 2 factors	From 3 factors	Multiple R-square	Factor 1	Factor 2	Factor 3
1 3.487253	43.59066	3.487253	43.59066	GR	0.787650	0.808757	0.880410	0.884457	**0.775699**	0.206025	0.486058
2 2.699707	33.74634	6.186960	77.33701	RHOB	0.737948	0.891452	0.891459	0.885468	**0.942606**	-0.043024	-0.033170
3 0.915751	11.44689	7.102712	88.78390	DT	0.102712	0.709350	0.836384	0.822113	-0.579378	0.272501	0.653029
				CNL	0.180978	0.568235	0.917692	0.781188	0.173458	0.232301	**0.913039**
				SQRLLS	0.082549	0.784428	0.977494	0.968127	-0.036952	**0.971409**	0.180258
				SQRLLD	0.202359	0.823616	0.989997	0.970504	0.132616	**0.962934**	0.212524
				CPHI	0.590502	0.796421	0.804304	0.810395	**-0.880259**	0.060645	0.160527
				LOGCKX	0.802551	0.804699	0.804970	0.823555	**-0.851946**	-0.216654	-0.179493
				Expl. var.					3.375608	2.094015	1.633088
				Prp. totl.					0.421951	0.261751	0.204135

* Emboldened loadings have absolute values >|0.7000000|.
Extraction: Principal factors (MINRES).
See Table 3 for mnemonics.

Table 9. *Hierarchical factorization results, three-factor solution*

| | Secondary and primary (unique) factor loadings | | | |
	Second. 1	Primary 1	Primary 2	Primary 3
GR	0.422428	**0.761861**	0.058180	0.343722
RHOB	0.002393	**0.942769**	−0.040872	−0.031129
DT	0.487077	−0.595704	0.097443	0.484541
CNL	0.642781	0.152175	0.004538	0.693791
SQRLLS	0.655874	−0.058905	**0.736008**	−0.046306
SQRLLD	0.676854	0.110013	**0.720649**	−0.020673
CPHI	0.080597	−0.883157	0.029247	0.130331
LOGCKX	−0.264055	−0.843364	−0.125071	−0.091315

* Emboldened loadings have absolute values >|0.7000000|.
See Table 3 for mnemonics.

structure for the two solutions is closely similar as revealed by the patterns in the individual loadings' values. From Table 8 it can be seen that Factor 1 mostly strongly correlates with the density, gamma-ray and core data, Factor 2 is dominated by the resistivity data and Factor 3 is most correlated with the neutron and sonic data and less strongly with the gamma-ray data. None of the variables misses being adequately represented by at least one of the factors which, given the high communalities, is very much to be expected. From these loadings we may surmise some physical interpretations of the factors: Factor 1 correlates with porosity (with the strong loading of gamma-ray on Factor 1 reflecting the lower porosities in the shales), Factor 2 is basically the resistivity data, and Factor 3 is most likely a lithological control through the shale and mica variations throughout the section. Notice how in this clastic section the neutron 'porosity' tool does not feature in the loadings on Factor 1, the surrogate for formation porosity, but seems to have its dominant effect in the background lithological influence revealed through Factor 3. Figure 10 is a crossplot of these rotated factor loadings and should be compared with the equivalent plot for the unrotated PCA results (Fig. 7).

The process of rotation of the factors maximises the distinctiveness of each in terms of the other factors. The rotation produces factors that are clearly marked by high loadings from some variables and low loadings from others. It is possible to relax the rule that the rotated factors must be mutually orthogonal and thereby investigate whether the unexplained variance may contain further factors. Essentially, the rotated 'primary' factors are themselves factor-analysed with the result that the 'secondary' unique component of variance is now the

variance in the rotated factors, and the 'secondary' common component of variance represents any further factors that are hidden in the residual variance unexplained by the 'primary' factors. Table 9 contains the results of such a procedure carried out for the three factor solution. From inspection of Table 9 it can be suggested that, indeed, there does exist a further factor that can be extracted that would be significantly correlated with several of the variables, although not with the density and core porosity; from the suggested loadings the new factor might be representative of a fabric-related control on resistivity and permeability, perhaps involved with sorting or grain size as reflected in gamma-ray, neutron and sonic variation.

Table 10 presents the results from a repeat analysis in which four factors are specified to be extracted. Now the communalities mostly exceed the R^2 values indicating that the new solution is more effective at explaining the variation in the original data. However, notice from the loadings table for the four factor solution that the loadings on the fourth factor are not as suggested by the secondary loadings in Table 9 in that the resistivity data have only very low loadings on the new factor. The requirement to extract four factors has lead to a solution in the first three factors that is subtly different from the three-factor case and this has resulted in a somewhat different loading pattern on the 'extra' factor. The loadings on the fourth factor are still dominated by the gamma-ray, permeability and sonic, however, (Fig. 11) but now the density is relatively well correlated with the fourth factor, probably through the curve being so well correlated with permeability (see Fig. 4 and Table 2). From Fig. 12 it can be seen that Factor 4 shows some interesting responses in the lower part of the Upper Ness Formation

Table 10. *Eigenvalues communalities, and data loadings on principal factors, four-factor solution*

	Eigenvalues					Communalities					Factor loadings (varimax normalized)*			
	Eigenval.	% Total variance	Cumul. eigenval.	Cumul. %		From 1 factor	From 2 factors	From 3 factors	From 4 factors	Multiple R-square	Factor 1	Factor 2	Factor 3	Factor 4
1	3.555642	44.44552	3.555642	44.44552	GR	0.824426	0.848073	0.940063	0.989997	0.884457	**0.739370**	0.207973	0.534296	0.338532
2	2.715717	33.94646	6.271359	78.39198	RHOB	0.723876	0.874166	0.874181	0.893899	0.885468	**0.901676**	-0.030547	-0.019350	0.282082
3	0.905634	11.32042	7.176993	89.71241	DT	0.106561	0.764889	0.934467	0.965142	0.822113	-0.575278	0.245759	**0.708878**	-0.267003
4	0.243196	3.039960	7.420190	92.75237	CNL	0.168484	0.522432	0.794880	0.796145	0.781188	0.147213	0.246223	**0.842811**	0.059308
					SQRLLS	0.082403	0.787747	0.986767	0.987045	0.968127	-0.044399	**0.975376**	0.181720	-0.026313
					SQRLLD	0.199107	0.816373	0.980407	0.981784	0.970504	0.118038	**0.958532**	0.220553	0.020544
					CPHI	0.580038	0.784685	0.792663	0.816180	0.810395	**-0.889234**	0.059793	0.146808	0.017727
					LOGCKX	0.870743	0.872990	0.873560	0.989994	0.823555	**-0.924993**	-0.214033	-0.211979	0.208893
					Expl. var.						3.374560	2.084740	1.646840	0.314048
					Prp. totl.						0.421820	0.260592	0.205855	0.039256

* Emboldened loadings have absolute values $>|0.7000000|$.
Extraction: Principal factors (MINRES).
See Table 3 for mnemonics.

Table 11. *Eigenvalues communalities, and data loadings on principal factors, four-factor ten-variable solution*

	Eigenvalues					Communalities					Factor loadings (varimax normalized)*			
	Eigenval.	% Total variance	Cumul. eigenval.	Cumul. %		From 1 factor	From 2 factors	From 3 factors	From 4 factors	Multiple R-square	Factor 1	Factor 2	Factor 3	Factor 4
1	4.282842	42.82842	4.282842	42.82842	GR	0.343105	0.854495	0.892861	0.899734	0.888957	−0.585751	**0.715115**	0.195872	−0.082901
2	2.871186	28.71186	7.154028	71.54028	RHOB	0.708480	0.801024	0.803511	0.910576	0.890636	**−0.841712**	0.304211	−0.049864	−0.327207
3	1.125865	11.25865	8.279893	82.79893	DT	0.429667	0.601900	0.664575	0.969774	0.832189	0.655490	0.415009	0.250350	0.552448
4	0.252589	2.525894	8.532482	85.32483	CNL	0.000440	0.675762	0.730895	0.799593	0.789604	0.020980	**0.821780**	0.234803	0.262102
					SQRLLS	0.005368	0.032610	0.985131	0.989992	0.968468	0.073266	0.165052	**0.975972**	0.069718
					SQRLLD	0.005124	0.070402	0.980139	0.981426	0.970681	−0.071583	0.255495	**0.953801**	0.035879
					CPHI	0.882460	0.883622	0.887358	0.887469	0.814035	**0.939393**	−0.034081	0.061123	0.010522
					LOGCKX	0.688972	0.824769	0.867896	0.880854	0.824591	**0.830043**	−0.368507	−0.207670	−0.113832
					LFAC	0.016013	0.596106	0.599391	0.600149	0.558458	−0.126542	**0.761638**	0.057313	0.027534
					GRSZ	0.055214	0.505908	0.603198	0.612912	0.564016	0.234977	−0.671337	−0.311913	0.098556
					Expl. var.						3.134846	2.711757	2.168356	0.517522
					Prp. totl.						0.313484	0.271175	0.216835	0.051752

* Emboldened loadings have absolute values >|0.7000000|.
Extraction: Principal factors (MINRES).
See Table 3 for mnemonics.

(*c.* 10 340–10 380 ft) and in the lower Etive Formation (*c.* 10 520–10 540 ft).

If the problem is expanded slightly to include the ordinal variables denoting the lithofacies interpretation and the grain size value, both of which are deduced from the core descriptions, then the pattern of the major factors changes slightly again. Table 11 records the solution of the 10 variable problem solving for four factors. From the loading it can be observed that Factor 1 is principally as it was in the eight variable problem but with reversed sign, Factor 2 is now dominated by gamma-ray, neutron and the lithofacies and grain size data, with significant contributions from sonic, density and the resistivity data, Factor 3 is clearly the resistivity data and Factor 4 seems to be correlated mostly with the sonic, density and neutron data. Physically, Factor 2 on this analysis would seem to be indicative of sedimentary fabric-related controls on the data.

Correspondence analysis

This is both a generic term covering non-parametric methods of analysis applicable to contingency tables for which the assumption of multivariate normality is inappropriate, and a specific technique for performing a principal components analysis of two-way contingency tables.

The generic term arises from the French school, lead by Jean-Paul Benzécri, whose style of statistical analyses is given the French title of 'Analyses des données'. Most of the calculation procedures available for continuous data sets have directly comparable algorithms within the group of techniques known generically as 'correspondence analysis'. In particular, the French school refers to canonical analysis, multidimensional scaling, principal components, discriminant analysis and cluster analysis as applied to contingency tables rather than to continuous data. (Lebart *et al.* 1984).

When the term correspondence analysis (CA) is applied to a specific technique, consideration is often restricted to the non-parametric equivalent of principal components analysis applied to contingency tables. Correspondence analysis can be viewed as finding the best simultaneous representation of two data sets that comprise the rows and columns of a discrete data matrix. Whereas PCA is used for tables containing continuous variables, CA is best applied to contingency tables (cross-tabulations) and provides a satisfactory description of binary coding.

The construction of contingency tables is discussed in an earlier section; the important consideration is that the elements of the table are transformed into conditional probabilities whilst retaining the same relationships between the rows and columns of the transformed table as exist in the original data matrix. If the rows and columns of such a table are independent, then the expected value of the individual observations should be equal to the product of the marginal probabilities of their respective rows and columns. We can then compare the actual observations at each element of the table with these expectations in a process analogous to a χ^2 test. By comparing the results of this test for different observations we can construct tables of similarities between the observations expressed as the extent to which the observations deviate from their expected values based on their marginal distributions. Such a similarity measure is sometimes called the 'χ^2 distance' to show its provenance (Lebart *et al.* 1984). The similarity measure ('χ^2 distance') is a form of correlation coefficient and is given by

$$r_{jk} = \sum_{i=1}^{n} \left(\frac{P_{ij} - P_i P_j}{\sqrt{P_i P_j}} \right) \left(\frac{P_{ik} - P_i P_k}{\sqrt{P_i P_k}} \right) \quad (37)$$

where P_{ij} is the observed joint probability in row i and column j of the body of the contingency table, and $P_i P_j$ is the 'expected' probability computed as the product of the marginal probabilities. Each term in brackets of the above can be expressed as an observed value minus its expected value, all divided by the square root of the expected value (Davis 1986). The link with the familiar χ^2 statistic comes from considering

$$\left[\frac{O_{ij} - E_{ij}}{\sqrt{E_{ij}}} \right]^2 = \frac{(O_{ij} - E_{ij})^2}{E_{ij}} = \chi^2 \quad (38)$$

where O and E refer to observed and expected values respectively.

Computing the similarity measure r_{jk} between all pairs of columns j and k produces a square $n \times n$ matrix, where n is the number of observations. The resultant similarity matrix is symmetrical and has diagonal elements all equal to 1, with off-diagonal elements in the range 0–1. Extraction of the eigenvalues and associated eigenvectors of this square matrix provides us with the principal axes of correspondence analysis, which are directly analagous to the principal components similarly derived as the eigenvalues and eigenvectors of a square correlation matrix of continuous data. However, of some importance is the fact that the transformed contingency table upon which the correspondence analysis is performed is a closed system, since the sum of all the probabilities must equal unity. As a result, one of the eigenvalues must be zero and the ensuing characterization of the original data set will be expressed as, at most, $n - 1$ new variables (and hopefully considerably fewer than this).

Practical application of the technique can be developed using a slightly different formulation of the similarity measure which will provide the same eigenvectors but which also simplifies the matrix algebra (Davis 1986). Davis suggests the use of

$$r_{ik} = \sum_{i=1}^{n} \frac{P_{ij}P_{ik}}{P_i\sqrt{P_jP_k}} \qquad (39)$$

instead of equation (37) above. The joint probabilities (P_{ij}) are found from dividing the elements of the original $n \times m$ discrete data matrix $\mathbf{X}$ by the grand total of the matrix elements,

$$\mathbf{B} = \frac{1}{\sum\sum x_{ij}} \cdot \mathbf{X}. \qquad (40)$$

We then define an $m \times m$ square diagonal matrix $\mathbf{M}$ that contains the column totals of $\mathbf{B}$ along its diagonal, with off-diagonal elements set to zero, and an $n \times n$ square diagonal matrix $\mathbf{N}$ that contains the row totals of $\mathbf{B}$ along the diagonal and zeros elsewhere (i.e. these two matrices contain the row and column marginal probabilities of the elements of $\mathbf{X}$), and transform the $\mathbf{B}$ matrix with

$$\mathbf{W} = (\mathbf{N}^{-1/2})\mathbf{B}(\mathbf{M}^{-1/2}). \qquad (41)$$

Therefore $\mathbf{W}$ contains transformed elements w_{ij} corresponding to every x_{ij} of the original matrix $\mathbf{X}$ and we can compute the cross-products of the columns as $\mathbf{R} = \mathbf{W}^T\mathbf{W}$ and the cross-products of the rows as $\mathbf{Q} = \mathbf{W}\mathbf{W}^T$. The eigenvalues of $\mathbf{R}$ and $\mathbf{Q}$ will be the same but $\mathbf{Q}$ will have $(n - m)$ additional zero eigenvalues. The eigenvectors of $\mathbf{R}$ can be converted to factor loadings by multiplying each eigenvector by the square root of the associated eigenvalue (see earlier discussion of the principles of factor analysis). Multiplying the elements of $\mathbf{W}$ by the factor loadings produces the factor scores that may be cross-plotted and examined for patterns revealing hidden structure in the original data matrix.

Because of the transformation of our original count data to a closed system (i.e. to probabilities), correspondence analysis is particularly suitable for the analysis of percentage data in that it minimises the closure effect of the requirement that, e.g. lithological composition, percentages sum to 100%.

It is possible to apply the technique to suitably transformed continuous data. In essence the transformation of dividing each data element of continuous data by its row and column total does not produce probabilities as such but produces quantities that reflect the relative magnitude of the observations with respect to other observations. Thereafter, the analysis proceeds as above. Since the grand total of the variables comprises different measurements the units systems of those measurements will have profound effect on influencing the final result. Mean correction or standardization is required prior to the analysis to avoid undue weighting due to measurement magnitudes, so the transformation process is probably best considered as an arbitrary technique to close the data and reduce all measurement elements to the same scales (i.e. relative magnitudes).

Canonical correlation

This technique seeks to determine the linear association between a set of predictor variables and a set of criterion measures. Therefore, in this technique the variables are grouped and interest centres on the relationships between groups of variables and groups of criterion measures.

The process is considered to be a generalization of multiple regression; whereas instead of a single dependent variable (Y) being related to a set of independent variables (X) through a linear combination of the latter, in canonical correlation we find the linear combination of the set of dependent variables (criterion measures) that is maximally correlated with a linear combination of the set of independent variables (predictors). Note that both sets of variables are measurements on the same objects. A good petrophysical example of when to use the technique is in comparison of a set of core measurements with a set of wireline measurements in the same formation. The technique specifically examines combinations of variables and is particularly appropriate when the variables that comprise the set are highly correlated in the first place. In such conditions canonical analysis can reveal complex relationships that reflect the structure between the predictor and criterion variables. It is the only technique designed to compare two sets of variables.

The procedure effectively generates pairs of new variables (canonical variables or variates or roots), being the linear combinations of the set of Ys and of the set of Xs that are best correlated together (i.e. the pair of weighted linear combinations that exhibit the highest canonical correlation of any possible pair of combined variables forms the first canonical variate). This is somewhat similar to the idea of a principal component analysis, except that canonical correlation maximises correlations as opposed to maximising variances. There are q possible variates that can be extracted where q is the number of variables in the smaller set. These q variates are extracted so that they are independent of each other. The process is linear, so the new variables are found as the weighted combination of the original variables. Once found, weighted combinations can be calculated out to provide canonical scores which may be of interest in interpretation of results.

The sets of observations (all on the same objects) form a data matrix whose dimensions are $n \times (p + q)$ where p represents the number of Y variables and q is the number of X variables. The matrix of variances and covariances ($\mathbf{S}$) can be considered to comprise four parts:

$$\mathbf{S} = \begin{bmatrix} \mathbf{S}_{yy} & \mathbf{S}_{xy} \\ \hline \mathbf{S}_{yx} & \mathbf{S}_{xx} \end{bmatrix} \qquad (42)$$

and has dimensions of $(p + q) \times (p + q)$. $\mathbf{S}_{yy}$ is the $p \times p$ variance-covariance matrix of the set of Y variables; $\mathbf{S}_{xx}$ is the $q \times q$ variance-covariance matrix of the set of X variables; $\mathbf{S}_{xy}$ is the $p \times q$ matrix of covariances between the X and the Y variables and $\mathbf{S}_{yx}$ is its transpose. The resulting matrix $\mathbf{S}$ is just like any other variance-covariance matrix in that it is symmetrical about its diagonal, whose elements are variances, whilst the off-diagonal elements are covariances.

The objective is to find the linear combination of the Ys that maximally correlates with a linear combination of the Xs. We can state the desired combinations as being

$$X^* = \mathbf{a}^T x = a_1 x_1 + a_2 x_2 + \cdots + a_q x_q \quad (42a)$$

and

$$Y^* = \mathbf{b}^T y = b_1 y_1 + b_2 y_2 + \cdots + b_p y_p \quad (42b)$$

with our interest being in finding the vectors of unknown weights $\mathbf{a}$ and $\mathbf{b}$.

The correlation between these two, as a function of the vectors $\mathbf{a}$ and $\mathbf{b}$ is given by

$$r(\mathbf{a}, \mathbf{b}) = \frac{\mathbf{a}^T \mathbf{S}_{xy} \mathbf{b}}{\left\{ (\mathbf{a}^T \mathbf{S}_{xx} \mathbf{a})(\mathbf{b}^T \mathbf{S}_{yy} \mathbf{b}) \right\}^{1/2}}. \qquad (43)$$

In order to compare variables (observations) whose measurements are of different magnitudes it is necessary to standardize the variables. This step has the results that the X^* and Y^* variables have unit variance, that $\mathbf{a}^T \mathbf{S}_{xx} \mathbf{a} = \mathbf{b}^T \mathbf{S}_{yy} \mathbf{b} = 1$, and that $E(X^*)$ and $E(Y^*)$ are 0.0. It is also

true that after standardization our variance-covariance matrices, $\mathbf{S}_{xx}$, $\mathbf{S}_{yy}$, $\mathbf{S}_{xy}$, $\mathbf{S}_{yx}$, become the correlation matrices $\mathbf{R}_{xx}$ (correlations amongst the X variables), $\mathbf{R}_{yy}$ (correlations amongst the Y variables), $\mathbf{R}_{xy}$ and $\mathbf{R}_{yx}$ (correlations between the X and Y variables). Hence, the solution for the unknown weight vectors that maximise the correlation between the new, combined variables is found from the following system of equations.

First, define the pooled correlation matrix,

$$\Lambda = \mathbf{R}_{xx}^{-1} \mathbf{R}_{xy} \mathbf{R}_{yy}^{-1} \mathbf{R}_{yx}. \qquad (44)$$

It is the eigenvectors associated with the largest eigenvalue of this pooled correlation matrix that provide the elements of the vector of weights $\mathbf{a}$ of the X^* variable.

Similarly, define

$$K = \mathbf{R}_{yy}^{-1} \mathbf{R}_{yx} \mathbf{R}_{xx}^{-1} \mathbf{R}_{xy} \qquad (45)$$

whose largest eigenvalue is numerically the same as that for Λ but that has different associated eigenvectors which define the vector of weights $\mathbf{b}$ for the Y^* variable.

There are as many eigenvalues in each pooled correlation matrix as there are variables in the smaller variable set (X or Y) making up the matrix. The eigenvectors associated with each eigenvalue form different weight vectors that produce different linear combinations of the original variables. Thus we create q new combined variables, where q is the number of original variables in the smaller set under investigation. Now, the numerical values of the eigenvalues of the pooled correlation matrices are equal to the squared canonical correlation coefficients that arise between the new variables formed by applying the weights represented by the associated eigenvector elements. Thus the two sets of weights (eigenvector elements) associated with the largest eigenvalue of each pooled correlation matrix automatically give us the new combinations of the original variables that are maximally correlated.

It can be shown that

$$\mathbf{a} = \frac{\mathbf{R}_{xx}^{-1} \mathbf{R}_{xy} \mathbf{b}}{\sqrt{\lambda}} \qquad (46)$$

and

$$\mathbf{b} = \frac{\mathbf{R}_{yy}^{-1} \mathbf{R}_{yx} \mathbf{a}}{\sqrt{\lambda}} \qquad (47)$$

thus, only one pooled correlation matrix need be computed to derive both the requisite weight vectors. Since we are handling standardized data, the canonical variates found are dimensionless.

The canonical weights can be used, with caution, to assess the relative importance of each original variable from one set with regard to the other set in obtaining the maximum correlation between new combined variables. The caution arises because if only one or two of the original variables are highly correlated with the canonical variates, then the correlation between variates will reflect the specific relationships of the dominant variables and not the general relationships that include the less strongly correlated variables. The resultant proportion of variance explained by the variates will be relatively small in such instances, thus the canonical weights may not be an accurate reflection of the underlying structure.

Canonical loadings can be computed which represent the correlation between the original variables and the new canonical variates. Canonical loadings are found from pre-multiplying the vector of canonical weights by the appropriate matrix of within-set correlations,

$$\text{e.g.} \quad \mathbf{r}_{x^*x}^{(j)} = \mathbf{R}_{xx}\mathbf{a}^{(j)} \tag{48}$$

where $\mathbf{a}^{(j)}$ is the vector of canonical weights for the X-set variables on the jth canonical variate; similar expressions for the Y-set can be constructed.

As noted above, the fact that pairs of variates possess large canonical correlations is not always indicative of there being an interpretable underlying data structure. It is often instructive to calculate the proportion of variance in the original variables that is explained by its canonical variate. The proportion of variance in the original Y-set variables that is explained by a particular canonical variate (the jth variate) is given by

$$R_{(j)y}^2 = \frac{\mathbf{r}_{y^*y}^{(j)\mathrm{T}}\mathbf{r}_{y^*y}^{(j)}}{p} = \sum_{i=1}^{p}\frac{(r_{y^*y_i}^{(j)})^2}{p}. \tag{49}$$

That is to say that the proportion of explained variance is found from the sum of the squared canonical loadings divided by the number of variables comprising the set. Likewise, for the X-set of variables we can calculate a proportion of explained variance as

$$R_{(j)x}^2 = \frac{\mathbf{r}_{x^*x}^{(j)\mathrm{T}}\mathbf{r}_{x^*x}^{(j)}}{p} = \sum_{i=1}^{p}\frac{(r_{x^*x_i}^{(j)})^2}{p}. \tag{50}$$

Of potentially greater interest, bearing in mind that a principal objective of the analysis is to see how well one set of variables is predicted by another set, is to calculate the amount of variance in one variable set that is accounted for by the other set of variables. Simply taking the squared canonical correlation (i.e. the numerical value of the eigenvalue

extracted from the pooled correlation matrix, see above) is not sufficient since this refers to the variance shared by the canonical variates themselves rather than to the variance of the original variables. An index, referred to as the redundancy coefficient (Dillon & Goldstein 1984, p. 350, referencing work of Stewart & Love 1968), can be computed that is equivalent to the squared multiple correlation coefficient, R^2, used in regression. This notion of redundancy relates the proportion of the variance in one variable set that is accounted for by a canonical variate from the other variable set. The redundancy coefficient is found by pre-multiplying the proportion of explained variance attributed to the jth variate by the squared canonical correlation (i.e. the eigenvalue) expressing the relationship between the variates from the different sets. For the proportion of Y-set variance explained by the jth X-set variate:

$$R_{y|x}^2 = \lambda_j \sum_{i=1}^{p}\frac{(r_{y^*y_i}^{(j)})^2}{p} = \lambda_j R_{(j)y}^2. \tag{51}$$

Furthermore, it is often of value to interrogate separately the cross-loadings, which are the relationships between an original variable in one set with a canonical variate of another set. These quantities are found from the redundancy coefficient as

$$r_{x^*y_i}^{(j)} = \lambda_j \sum_{i=1}^{p}(r_{y^*y_i}^{(j)})^2. \tag{52}$$

The significance of canonical correlations can be tested when the sample is large. To test the null hypothesis that one set of variables is unrelated to the other set of variables Bartlett (1951) defined

$$\Lambda - \prod_{j=1}^{q}(1 - \hat{\lambda}_{(j)}) = \frac{|\mathbf{S}_{yy}^{-1}\mathbf{S}_{yx}\mathbf{S}_{xx}^{-1}\mathbf{S}_{xy}|}{|\mathbf{S}_{yy}|} \tag{53}$$

and a χ^2 distribution of Λ is approximated from

$$X^2 = -[(n-1) - \tfrac{1}{2}(q+p+1)]\ln\Lambda. \tag{54}$$

The hypothesis is rejected, i.e. there is a relationship, if $X^2 > \chi_a^2$ with qp degrees of freedom. If this is the case then the contribution of the first variate is removed from the Λ and the significance of the remaining variate pairs is assessed. In general, the test is conservative; unless the canonical correlations removed are close to one then subsequent correlations are likely to prove significant under this test. It is often better in practice to assign a threshold level in either the size of the canonical correlation coefficient, or in the amount of variance explained, or in the magnitude of the redundancy coefficient as a preferable guide to identifying useful correlations to be retained after an analysis.

Both canonical weights and canonical loadings can be used to assess the relationships between the original variables and the canonical variates. Unfortunately, canonical weights may be unstable in the case of multicollinearity amongst the original variables; furthermore, they do not necessarily accurately reflect the association between variables across sets. Canonical loadings may also suffer from instability, and they are prone to being interpreted as suggesting hypothesized linkages: i.e. the reasoning $Y_i \rightarrow Y^*$ and $X_k \rightarrow X^*$, therefore, since $X^* \rightarrow Y^*$ the inference is made that $Y_i \rightarrow X_k$ (Dillon & Goldstein 1984). In some cases such linkage may be meaningful, but in general this reasoning gives the false impression that the relationship(s) between two across-set variables is much stronger than is actually the case.

Stability in weights may be investigated either by splitting the sample set and comparing canonical correlations in the sub-samples or by using the weights from one sub-sample to compute variates in the other sub-sample and comparing the resultant correlations. Close agreement would indicate weight stability.

The following precautionary measures may lead to improved interpretation of canonical analysis results: establish that there is shared variance between the variable sets; identify any high bivariate correlations between variables prior to analysis; identify sign reversals and magnitude variations between canonical weights and loadings for a given pair of variates; perform cross-validation to investigate stability; use weights if one is interested in relative contributions of variables in the variates, and use loadings if the variates are to be interpreted as factors (i.e. when it is the relationships that are of interest).

As an example we have chosen to look at the canonical correlations between the set of wireline data and the set of conventional core data from the 3/3–3 data set. The data were therefore divided into a set of six wireline curves (gamma-ray, density, neutron, sonic and two transformed resistivity curves) and the four core data curves (porosity, permeability, lithofacies and grain size). All data were standardized before the analysis commenced. Table 12 presents the results in the form of canonical weights, canonical loadings, and correlations. The number of canonical variates formed equals the number of data variables in the smaller data set, in this case four. The canonical weights represent the coefficients of the original data variables in the linear combinations producing the new variables that lead to maximal correlation between the two data sets. To re-iterate the procedure, we have combined the six original log variables by means of weighted linear sums into four new variables, and simultaneously we have combined the four core data variables into four new variables, with the result that the two sets of new variables are maximally correlated with each other. For each set each new variate is numbered one to four and are extracted successively; the new variates are independent of (uncorrelated with) each other and account for unique contributions to the underlying variance structure in decreasing amounts. Significance tests of the variates are performed by stepwise removal of the pairs of variates and assessment of the significance of the amount of variance contained in the remaining variates. In our case all variates appear statistically significant, although note the earlier comments about the conservative nature of such testing.

Table 12 contains the canonical correlations, weights, scores and loadings which constitute the results of the analysis.

It is customary to report the canonical correlation between the first pair of variates as the overall correlation, and thus the efficacy of the characterization exercise. In our case we achieve nearly 90% correlation between the first pair of new variates. The graph of this correlation is shown in Fig. 13.

Examination of the canonical weights can provide insight into the interpretation of any physical meaning for the variates. Large weights imply large contribution from the particular variable. They can be viewed as partial correlations of the original variables with the variates. The weights can be used directly to compute the canonical scores which are the actual values of the summations of the weighted sums at each depth level. Bear in mind that the input data for this analysis were standardized. Plotting of the scores can be instructive, just as for the other data reduction examples discussed earlier. Figure 14 illustrates the first pair of variates plotted with gamma-ray for the example data and clearly shows that a strong anti-correlation exists between the gamma-ray and the variates.

Canonical loadings are potentially more useful in determining the physical basis, if any, for the particular choice of variates extracted. Directly analogous with factor loadings, canonical loadings represent the correlations of each variate with the variables in the analysis. Thus, from Table 12, the first variate in the wireline set

Table 12. *Canonical correlations results: weights, loadings, correlations; six log variables v. four core variables*

Correlations, left set (Logs)

	GR	RHOB	DT	CNL	SQRLLS	SQRLLD
GR	1					
RHOB	0.747603	1				
DT	-0.086064	-0.614278	1			
CNL	0.632220	0.122763	0.556762	1		
SQRLLS	0.257229	-0.081095	0.399294	0.386936	1	
SQRLLD	0.412704	0.078535	0.320132	0.438727	0.969234	1

Correlations, right set (Core data)

	CPHI	LOGCKX	LFAC	GRSZ
CPHI	1	0.782736	-0.138182	0.235037
LOGCKX	0.782736	1	-0.414827	0.498183
LFAC	-0.138182	-0.414827	1	-0.586371
GRSZ	0.235037	0.498183	-0.586371	1

Canonical correlations between PAIRS of variates

	1st variates	2nd variates	3rd variates	4th variates
Value	0.890987	0.754263	0.412203	0.190135

Correlations, left set with right set

	CPHI	LOGCKX	LFAC	GRSZ
GR	-0.560310	-0.770461	0.599992	-0.676107
RHOB	-0.799139	-0.766086	0.331906	-0.401932
DT	0.628073	0.272070	0.270056	-0.139190
CNL	0.006546	-0.354445	0.633009	-0.584062
SQRLLS	0.125200	-0.212560	0.170776	-0.394426
SQRLLD	-0.016119	-0.356027	0.256536	-0.483481

Canonical weights, right set

	Variate 1	Variate 2	Variate 3	Variate 4
CPHI	0.201083	-1.267472	0.710599	0.903167
LOGCKX	0.532046	0.620297	-0.906090	-1.535493
LFAC	-0.223340	-0.278424	-1.235090	0.071468
GRSZ	0.294708	0.208589	-0.629399	1.116866

Canonical weights, left set

	Variate 1	Variate 2	Variate 3	Variate 4
GR	-0.306421	-0.078954	0.309039	-1.422800
RHOB	-0.593799	0.022912	-0.525887	2.343051
DT	-0.084411	-0.799567	-0.615608	1.780517
CNL	-0.162049	-0.158117	-0.400219	-0.794304
SQRLLS	0.316421	-0.345995	1.195799	-0.304108
SQRLLD	-0.475933	0.197367	-0.148198	0.531476

Canonical loadings, right set

	Variate 1	Variate 2	Variate 3	Variate 4
CPHI	0.717665	-0.694442	0.024105	-0.046090
LOGCKX	0.928909	-0.152384	-0.151085	-0.301794
LFAC	-0.644642	-0.482909	-0.588349	-0.071266
GRSZ	0.737987	0.382968	-0.189559	0.522279

Canonical loadings, left set

	Variate 1	Variate 2	Variate 3	Variate 4
GR	-0.960561	-0.100521	-0.037728	-0.185424
RHOB	-0.853961	0.479190	-0.074438	0.154517
DT	0.190479	-0.969851	-0.111959	0.070159
CNL	-0.562039	-0.697678	-0.214466	-0.299358
SQRLLS	-0.271943	-0.557312	0.773632	0.058624
SQRLLD	-0.440461	-0.494107	0.724388	0.055059

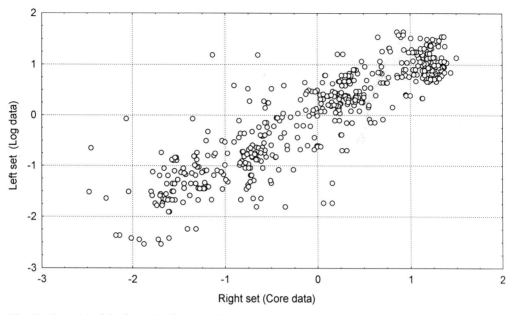

Fig. 13. Crossplot of the first pair of canonical variates.

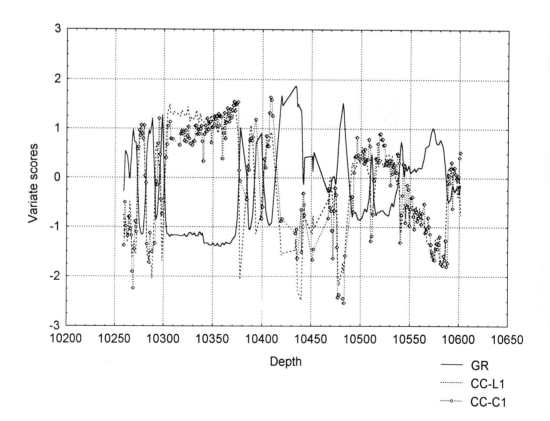

Fig. 14. Depth plot of the first pair of canonical variates.

Table 13. *Canonical correlations results: variance extracted and redundancy*

	Variance extracted (proportions), left set		Variance extracted (proportions), right set	
	Variance extracted	Redundancy	Variance extracted	Redundancy
Variate 1	0.378676	0.300616	0.584526	0.464031
Variate 2	0.370305	0.210672	0.221334	0.125920
Variate 3	0.198123	0.033664	0.101374	0.017225
Variate 4	0.026543	0.000960	0.092764	0.003354
Totals	0.973650	0.545911	1.000000	0.610530

(left set) is dominated by the gamma-ray and density data, and this is qualitatively apparent from Fig. 14 also. Whereas the weights are the individual contributions of the variables to the new variates, the loadings are the correlations between variables and variates, and, as such, offer insight to the physical meanings of the latter. In our example (Table 12) the first pair of variates is a combination of the effects of porosity, grain size and permeability, the second pair is dominated by neutron and sonic from the log data and porosity and lithofacies from the core data and it is tempting to suggest that this affirms the fabric-selective information content in the neutron and sonic data. The third pair of variates is dominated by the deep resistivity data and the lithofacies, again implying fabric-selective information in the resistivity data. The fourth pair of variates is only weakly correlated, but seems to contain information about permeability and grain size and is most strongly correlated with the neutron amongst the log data, again suggestive of fabric-related information in this log.

Table 13 provides further diagnostics from the analysis which are the proportions of variance, within each data set, that are attributable to the respective new variates. That is to say that these numbers are the amounts of within-set variance in each set that are 'explained' by the particular new variate. These values are computed as the average of the loadings of the original variables on each new variate. If these extracted variance quantities are then multiplied by the eigenvalues (squared canonical correlations) we arrive at the redundancy in each set of data, given the other data set. From Table 13 it would appear that just over 50% of the variability in the log data set is duplicated (redundant) when compared to the core data set. This suggests that the core data can form a standard against which to compare or calibrate the log data since such a high proportion of each data set's variance is

shared. Conversely, Table 13 also suggests that over 60% of the core data variability is already in the log data set, which might set a dangerous precedent for those concerned with cost reduction who might use such findings to reduce the amount of core being cut! Some of the non-redundant variability in the data sets is undoubtedly noise and environmental effects on the data, particularly the respective volumes of investigation of the various tools and plug samples.

Figure 15 presents a comparison between the first principal component (PC_1), the first extracted factor (Factor 1) and the first (log set) canonical variate (CC-L1). PC_1 and CC-L1 are almost identical; Factor 1, although it is sign reversed compared to the other two new variables, is nevertheless closely similar once this reversal is removed. Principal components analysis and canonical correlation are unsupervised techniques designed to reflect the underlying variance structure without prompting from the analyst. Thus the fact that they come up with very closely similar results is comforting. The factor analysis in our example constitutes a supervised technique whereby the analyst has pre-specified the number of factors to extract and so the result is therefore more optimised to suit the data, which almost certainly gives rise to the subtle differences in the comparison between Factor 1 and the other new variables in Fig. 15.

Multidimensional scaling (MDS)

The goal here is to detect meaningful underlying dimensions that allow the analyst to explain observed similarities or dissimilarities between the log signatures of the depth levels. MDS attempts to rearrange the objects of study (data points) in a space with a particular number of dimensions so as to reproduce the observed differences between the input data points and to

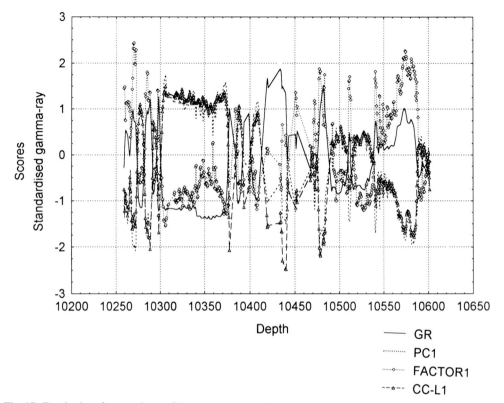

Fig. 15. Depth plot of comparisons of the major extracted data dimensions.

explain those differences in terms of the underlying dimensions.

The classic example used to explain the process is to take a table of distances between major towns and cities of the UK (or USA etc.) and perform an MDS in which two output dimensions are required. The analysis proceeds to position each town in terms of the same two dimensions such that all the input distances between pairs of towns are honoured, and the result is a two-dimensional map of the country where the output dimensions are interpreted as 'northings' and 'eastings'. Note that the output dimensions are entirely arbitrary and may require rotation before full correspondence with true northings and eastings is revealed. The analyst is always left to interpret the derived dimensions in terms of a physical model after an MDS exercise.

MDS rearranges the data points in the solution space defined by the requisite number of dimensions to maximise the degree to which the new configuration can match the similarities (= distances in the example) between the input data points. The principal measure of how well a given configuration matches the observed similarities is the stress measure:

$$St = \left[\frac{\sum_{i<j} (\hat{d}_{ij} - d_{ij})^2}{\sum_{i<j} d_{ij}^2} \right]^{1/2}. \tag{55}$$

This is a measure of the dispersion about a line of complete correspondence between original (d_{ij}) and reproduced $(\hat{d}_{ij})$ similarities (i.e. distances) (see Figs 17 to 19). The stress quantity is minimized by an iterative monotonic regression.

One of the more difficult aspects of this technique is the decision on how many dimensions to specify. In general, increasing the number of dimensions decreases the stress factor, i.e. improves the fit between the observed and reproduced similarities. Plotting the stress value for different numbers of dimensions specified in a so-called 'scree' plot (Fig. 16) reveals a trend from which the incremental improvement for additional dimensions can be deduced.

In terms of our multivariate data sets, MDS computes distance functions between the possible data pairs and then rearranges the distances in terms of the specified number of dimensions.

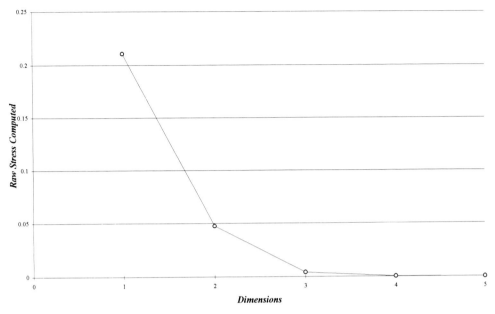

Fig. 16. 'Scree' plot of raw stress values after multidimensional scaling.

The commonest distance measures are either the Euclidean distance

$$D_{ij} = \sum_{k=1}^{m} (x_{ik} - x_{jk})^2 \qquad (56)$$

or the so-called Gower distance

$$G_{ij} = \frac{1}{m} \sum_{k=1}^{m} \left(1 - \frac{|x_{ik} - x_{jk}|}{Range_k} \right). \qquad (57)$$

This gives a number in the range 0.0–1.0, and therefore is not concerned about the type of data (interval, ratio, ordinal, or nominal) that comprise the original data. Indeed, mixtures of data of different rank can be assimilated into the same $n \times n$ association matrix, which makes the technique rather widely applicable.

Most geoscientific analyses using MDS have concentrated upon quantitative petrology and palaeontology, but Matyas (1995) has used the technique to perform rock type analyses from log data. He uses only two dimensions to characterize his data set which is drawn from log data that are zoned *a priori*, and so his analysis is aiming at confirming the dissimilarities between the zones suggested by the analyst, rather than aiming at exploring the possible dimensionality of the data set. Notwithstanding this difference in objective, Matyas finds that MDS is a useful technique for differentiating crystalline basement from overlying sediments when the long-established *M–N* cross-plot (Burke *et al.* 1969) fails to differentiate the beds.

The method proceeds by minimising the stress function through successive iterations. Particular minimization strategies need not concern us in this discussion. To illustrate the technique we again turn to the 3/3–3 data set. The basic intention of the process is to match an observed matrix of similarities between data items and, therefore, the procedure does not use the actual variables themselves. Instead, in the example we use the matrix of distances computed from a hierarchical cluster analysis (see later section) which represent the similarities between the individual logs and core measurements. Table 14 presents the matrix of similarities.

The method computes new distance measures for the number of dimensions specified by the user. This number will be smaller than the number of variables in the data set, and comparing the raw stress values from different dimensionalities can assist the decision concerning how many dimensions to retain. Figure 16 illustrates the scree plot arising from the example. Clearly, at most three new dimensions can accommodate the observed similarities, and perhaps only two need actually be retained. The calculations set up the required number of dimensions by projecting the data similarities onto the new dimensions in an iterative fashion until the distances computed in the new dimensionality are as close as possible to the original distances.

The estimated distances can be plotted against the input distances in an effort to determine the accuracy of the reconstruction. Figure 17 shows

Table 14. *Input Euclidean distance matrix for the multidimensional scaling exercise*

	GR	RHOB	DT	CNL	SQRLLS	SQRLLD	CPHI	LOGCKX	LFAC	GRSZ
GR	0	13.48107	28.19291	14.78939	27.87832	24.55559	35.05961	38.47701	15.87152	35.77614
RHOB	13.48107	0	35.55616	24.05478	33.53084	30.55874	38.73099	39.59197	21.61956	33.61423
DT	28.19291	35.55616	0	17.89607	25.59187	26.66282	18.13310	25.98841	23.21772	30.82794
CNL	14.78939	24.05478	17.89607	0	25.31815	23.74495	27.62084	33.03310	14.76170	33.95655
SQRLLS	27.87832	33.53084	25.59187	25.31815	0	6.414808	30.96176	37.26999	29.21952	37.31552
SQRLLD	24.55559	30.55874	26.66282	23.74495	6.414808	0	32.98875	39.03913	27.20584	38.27200
CPHI	35.05961	38.73099	18.13310	27.62084	30.96176	32.98875	0	14.40353	30.17222	25.49828
LOGCKX	38.47701	39.59197	25.98841	33.03310	37.26999	39.03913	14.40353	0	34.57786	21.75137
LFAC	15.87152	21.61956	23.21772	14.76170	29.21952	27.20584	30.17222	34.57786	0	34.91269
GRSZ	35.77614	33.61423	30.82794	33.95655	37.31552	38.27200	25.49828	21.75137	34.91269	0

Raw stress values

Dimensions	Raw stress computed
1	0.210512
2	0.047354
3	0.004146
4	0.000165
5	4.7×10^{-6}

See Table 3 for mnemonics.

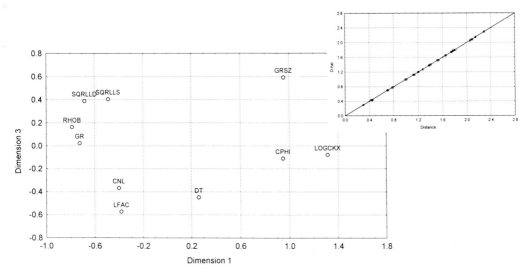

Fig. 17. Crossplots of data loadings and reconstructed distances, 3D MDS solution.

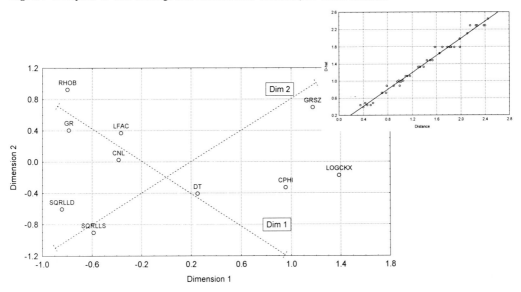

Fig. 18. Crossplots of data loadings and reconstructed distances, 2D MDS solution.

the crossplot of the first and third dimensions with the variables projected onto these axes; inset is the crossplot of estimated (*d-hat*) distances versus the input distances and we see a very tight distribution about the diagonal line of equality. Adding more dimensions would further tighten this line but the observed three-dimensional fit can only marginally be improved, hence the insignificant contributions to stress reduction from the higher dimensions as seen on the scree plot (Fig. 16). In Fig. 18 are similar plots for the two-dimensional case. Note again the tightness of the *d-hat* distribution about a fit line, but now

the line is no longer the diagonal. This implies that there is some residual misfit between the observed and predicted distance measures in the two-dimensional case. Finally, Fig. 19 is included to illustrate that the projection onto a single dimension fails to capture satisfactorily the original distances.

The meaning of the dimensions requires interpretation on the part of the analyst. The values of the co-ordinates plotted on Figs 17 and 18 are directly comparable with the loadings calculated from other techniques; however the directions of the dimensional axes are quite

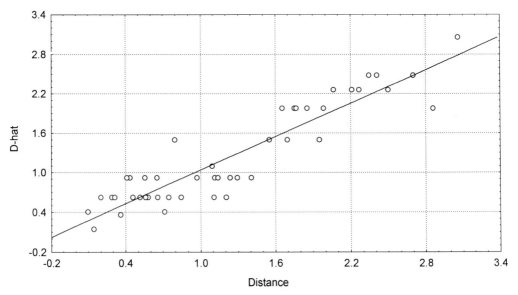

Fig. 19. Crossplot of reconstructed distances, 1D MDS solution.

arbitrary. Comparing the crossplots of principal component loadings or factor loadings with the distribution of the variables in the multidimensional analysis results suggests strongly that MDS has indeed captured the same principal dimensions as other methods. A slight rotation clockwise of the axes' orientations as suggested by the dashed lines in Fig. 18 (the new axes are not at right angles because of the plot's aspect ratio) reveals that dimension 1 is dominated by the porosity terms whilst the second dimension suggests that the resistivity data contain significant information about grain size and permeability. The orientation of the third factor is again suggestive of an underlying fabric-related control in the data since we see the grouping of the sonic, neutron and the lithofacies data.

Multidimensional scaling can operate on data of any rank. When performed on nominal or ordinal data the term non-metric multidimensional scaling is applied, whereas interval and ratio data constitute metric multidimensional scaling. Non-metric multidimensional scaling is sometimes called principal coordinates analysis (Chatfield & Collins 1980). Strictly speaking, however principal coordinates analysis is actually a Q-mode factor analysis operating on the individual data values across variables (in our case this would be the depth levels) (Davis 1986).

Projection pursuit

A very comprehensive reference for the application of this technique to reservoir characteriza-

tion is Walden (1992). Essentially, the process successively splits the data set under study into optimally dissimilar clusters. It is an interactive process, requiring the intervention of the analyst after each split who must select whether to continue with the next split or whether to stop the process. The outcome is a data set subdivided into discrete clusters whose number is determined by the analyst. Therefore the method can be classified as a supervised clustering method.

The process comprises a number of steps. The data are first standardized so that the mean of the transformed data is zero and the units are expressed as standard deviations; in addition to this centring exercise, the data are 'sphered' so that the covariance matrix becomes an identity matrix (i.e. the diagonal elements are unity and off-diagonal elements are zero). Walden (1992) contains a detailed description of the process used and further suggests that robustification of the standardization and a 5% trimming to remove outliers can lead to computational efficiency gains. Next, an initial projection axis is specified by the analyst and the data are projected onto this axis. In two dimensions (two variables) this is equivalent to projection onto a line on a crossplot of the two standardized variables. The initial axis can be at any orientation through the data cloud. The computation proceeds to optimize the orientation of the projection so that the smoothed histogram of the projected data is 'minimally Gaussian'. By this is meant that the form of the non-Gaussianity should be that of bimodality, or, at least, multi-modality. The data are then split into two groups that correspond to the bimodality of the smoothed histogram. Each group in turn is then available for further splitting, at the control of the analyst, until the data comprise a single logical grouping when further splitting does not occur (or until the analyst halts the process after a given number of groups has been found). At each stage the data in a sub-group under study are re-standardized (centred and sphered) within the context of the sub-group population being split (i.e. the mean and standard deviation of the sub-group of data are used).

Kernel density estimation is applied to the projected data to determine the smoothed distribution of the projection, and the index of non-Gaussianity that is used is 'Shannon's entropy'.

The probability density function of the projected data is found from the following, which is evaluated using a fast Fourier transform (Walden 1992),

$$f_N(z; a_s) = \frac{1}{Nh} \sum_{i=1}^{N} K\left(\frac{z - z_i(a_s)}{h}\right) \qquad (58)$$

where h is the window width or smoothing parameter and K is the Gaussian kernel function given by

$$K(y) = \left[\frac{1}{\sqrt{(2\pi)}}\right] e^{-y^2/2}. \qquad (59)$$

The entropy function is calculated, based on the kernel density estimate, from

$$e_N(a_s) = \int_{-\infty}^{\infty} f_N(z; a_s) \ln f_N(z; a_s)\, dz \qquad (60)$$

where an estimate $\hat{e}_N(a_s)$ of $e_N(a_s)$ can be found from Simpson's rule.

The initial projection vectors used to begin the splitting process could be chosen to be the eigenvectors of the transformed data matrix, although other initial guesses should always be investigated. Since the transformation process renders the entire data set into one having unit dispersion, all the eigenvalues are unity and the associated eigenvectors are unit vectors. Unlike the ranking imposed on the eigenvectors of the original correlation matrix in terms of their ability to account for proportionately different amounts of the original variance, there is no hierarchical structure to the unit eigenvectors produced from centred and sphered data. As a consequence, therefore, it is possible to choose an initial projection axis that leads to a local minimum in the optimization process, rather than to the global minimum, and for this reason more than a single initial guess should be used. Walden (1992)

suggests use of systematically or even randomly chosen vectors (standardized to unit length).

Because the method uses only linear projection axes it is not well-suited to non-convex clusters (e.g. curved and twisted sausage- or banana-shaped clusters, see Fig.25).

The analyst must decide at which point further splitting becomes of no interest. He may have *a priori* information that suggests how many clusters to expect, or, alternatively, the smaller sub-groups may be insignificantly different from Gaussian distributions for further splitting to be computed.

Discriminant functions

This method consists of finding a transform which gives the minimum ratio of the difference between a pair of group multivariate means to the multivariate variance within the two groups (Davis 1986). If we regard our two groups as clusters in multivariate space then the task is to find the one orientation along which the group means are furthest apart whilst the within-group variances are least enlarged (least 'inflated').

Computationally the same as linear regression, nevertheless discriminant function analysis (DFA) is philosophically different on three counts: regression assumes that the independent variable is fixed whilst the dependent variable (Y) is normally distributed, whereas DFA assumes normality in the independent variables and the dependent (i.e. grouping) variable is fixed (0 or 1); whereas regression predicts the mean of the dependent variable from the fixed values of the independent variable, DFA finds a linear combination of the independent variables that minimises misclassification of the individuals in groups; regression analysis invokes a formal model to derive statistical estimates with desirable properties whereas DFA provides a strategy for classifying individuals into groups with accuracy.

Two fundamental assumptions are required of the data in order to perform DFA:

(1) the p independent variables must have a multivariate normal distribution; and,
(2) the $p \times p$ variance-covariance matrix of the independent variables in each of the two groups must be the same.

One method that can be used to find the discriminant function is to regress the differences between the multivariate means of the two groups against the pooled variances and covariances of the original variables (Davis 1986). In this approach we seek to solve

$$\mathbf{s}_p^2 \cdot \lambda = \mathbf{D} \qquad (61)$$

where s_p^2 is an $m \times m$ matrix of the pooled variances and covariances of the m variables, λ is a vector of unknown coefficients to be found by the regression (N.B. by convention, λ is used in DFA, but it does *not* refer to an eigenvalue) and $\mathbf{D}$ is the column vector of m differences between the two groups.

The vector λ may be found from

$$\lambda = \mathbf{D} \cdot (s_p^2)^{-1}. \tag{62}$$

The vector $\mathbf{D}$ may be found from

$$D_j = \bar{A}_j - \bar{B}_j = \frac{\sum_{i=1}^{n_a} A_{ij}}{n_a} - \frac{\sum_{i=1}^{n_b} B_{ij}}{n_b} \tag{63}$$

where A_{ij} is the ith observation on variable j in group A; $\bar{A}_j$ is the mean of variable j in group A. Similarly for group B.

The matrix of pooled variances and covariances is found by computing a matrix of sums of squares and cross-products of all variables in group A and the same in group B. The appropriate matrix for group A is

$$\mathbf{SPA}_{jk} = \sum_{i=1}^{n_a} (A_{ij}A_{ik}) - \frac{\sum_{i=1}^{n_a} A_{ij} \sum_{i=1}^{n_a} A_{ik}}{n_a} \tag{64a}$$

and for group B is

$$\mathbf{SPB}_{jk} = \sum_{i=1}^{n_b} (B_{ij}B_{ik}) - \frac{\sum_{i=1}^{n_b} B_{ij} \sum_{i=1}^{n_b} B_{ik}}{n_b}. \tag{64b}$$

The matrix of pooled variance can now be found as

$$s_p^2 = \frac{(\mathbf{SPA} + \mathbf{SPB})}{(n_a + n_b - 2)}. \tag{65}$$

The elements of the vector λ are the weights to be applied to the original variables to transform them into a new variable (by convention denoted Y) in which the two groups are maximally separated whilst their within-group variance is least inflated. It is possible to test the significance of the separation in the group means expressed in terms of the new discriminant variable and the square of the Student's t test is used.

The assignment rule is applied as follows. For two groups whose samples are of the same size ($n_1 = n_2$) the point of separation is found from

$$Y_c = \frac{\lambda'\bar{\mathbf{x}}_1 + \lambda'\bar{\mathbf{x}}_2}{2}$$

$$= \tfrac{1}{2}(\bar{\mathbf{x}}_1 - \bar{\mathbf{x}}_2)'\mathbf{S}^{-1}(\bar{\mathbf{x}}_1 + \bar{\mathbf{x}}_2) \tag{66}$$

and the assignment rule states that an individual is in group (G_1) if

$$|\lambda'(\mathbf{x} - \bar{\mathbf{x}}_1)| \leq |\lambda'(\mathbf{x} - \bar{\mathbf{x}}_2)| \tag{67}$$

else the individual is in group (G_2).

If the samples of the two groups are of different sizes then a weighting must be applied to avoid serious misclassification. In this case the point of separation becomes

$$Y_c^* = \frac{n_2 \bar{Y}_1 + n_1 \bar{Y}_2}{n_1 + n_2}. \tag{68}$$

To become optimal the classification rule must take into account the prior probabilities of group membership and the cost of misclassification. If the prior probability of an individual data point being in group (G_i) is $P(G_i)$ ($i = 1, 2$), and the cost of misclassification into group j when the actual group membership is of group i is C_{ij}, then the assignment rule becomes

'An individual belongs in group (G1) if

$$\frac{|\lambda'(\mathbf{x} - \bar{\mathbf{x}}_1)|}{|\lambda'(\mathbf{x} - \bar{\mathbf{x}}_2)|} \geq \ln k \quad \text{where} \quad k = \frac{P(G_2)C_{12}}{P(G_1)C_{21}} \tag{69}$$

else the individual is in group (G_2)'.

The mean value of the discriminant function for each *a priori* grouping is commonly referred to as the group centroid. It is found from application of the weight vector to the 'mean score vector' which comprises the mean values for each variable in each group. Thus we compute:

$$\bar{Y}_i = \lambda'\bar{\mathbf{x}}_i \tag{70}$$

as the group centroid, where the subscript refers to the different groups. From this value for each group, in the two-group case we find Mahalanobis's generalized distance as:

$$\bar{Y}_1 - \bar{Y}_2 = \lambda'\bar{\mathbf{x}}_1 - \lambda'\bar{\mathbf{x}}_2$$
$$= (\bar{\mathbf{x}}_1 - \bar{\mathbf{x}}_2)'\mathbf{S}^{-1}\bar{\mathbf{x}}_1 - (\bar{\mathbf{x}}_1 - \bar{\mathbf{x}}_2)'\mathbf{S}^{-1}\bar{\mathbf{x}}_2$$
$$= (\bar{\mathbf{x}}_1 - \bar{\mathbf{x}}_2)'\mathbf{S}^{-1}(\bar{\mathbf{x}}_1 - \bar{\mathbf{x}}_2). \tag{71}$$

The squared Student's t statistic (T^2) is found from

$$T^2 = \frac{n_1 n_2}{n_1 + n_2}(\bar{\mathbf{x}}_1 - \bar{\mathbf{x}}_2)'\mathbf{S}^{-1}(\bar{\mathbf{x}}_1 - \bar{\mathbf{x}}_2). \tag{72}$$

This is distributed as an F distribution with p and $(n_1 + n_2 - p - 1)$ degrees of freedom. (p is the number of independent variables.)

Discriminant loadings may be computed that are less prone to instability in the face of highly intercorrelated variables amongst the set of independent variables than are the discriminant weights.

Rescale the discriminant weights by the square roots of the diagonal elements of $\mathbf{S}$, the total sample variance-covariance matrix:

$$\lambda_j^* = \mathbf{C}\lambda_j \tag{73}$$

and then pre-multiply the rescaled weights by the correlation matrix of the simple pair-wise product-moment correlations between the original variables:

$$\hat{l}_j = \mathbf{R}\lambda_j^*. \tag{74}$$

As an example of the application of DFA we shall attempt to determine the most effective discrimination for the grain size classes from amongst the set of log data. To begin with, the analyst should examine categorized plots (scatter crossplots, histograms and boxplots) that show the log responses divided across the target classes. Figures 20 and 21 illustrate these kinds of plot for the gamma-ray and the transformed shallow resistivity tool. From these can be gleaned an understanding of the degree of overlap in the log data between classes and whether any log curve is better than others at discrimination. In 3/3–3 there is no clear discrimination of grain-size class on the basis of a single curve (also, see Fig. 4).

The analysis was set up so that gamma-ray, density, sonic, transformed resistivities core porosity and core permeability were all recognised as independent variables in the investigation for the discrimination of grain-size class.

The method can be guided to proceed in a step-wise fashion, where at each step the program includes the variable that has the largest contribution to make out of all variables that are not already in the model. The contribution of a term is measured by the F test (i.e. a ratio of variances) and is a measure of the extent to which a term makes a unique contribution to group membership. In our example all variables were included and it was of interest to note the order of inclusion. Table 16 tracks the progress of the inclusions. Initially, with no variables in the model, the 'F-to-enter' column records the initial levels of contribution and the 'Tolerance' column contains all values 1. 'Tolerance' in this context is equal to

$$T = 1 - R^2 \tag{75}$$

where R^2 is the multiple correlation coefficient. 'tolerance', therefore, is a measure of the redundancy in a particular term, given the other terms in the included set. If this value is low then the term is correlated with the terms already in the set and the new term's contribution will be lessened. Wilks' lambda measures the ratio of within-group variance to total variance and partial Wilks' lambda assesses the individual contribution of each variable, rather like a partial correlation coefficient. A value of zero for either indicates perfect discriminatory ability.

After the first step, Table 16 shows that gamma-ray (GR) was included in the discriminant set. Now the tolerance values and the F-to-enter values have changed for the other variables to reflect the new arrangement. Note in particular that the tolerance statistic is reducing.

After Step 2 when the transformed shallow resistivity (SQRLLS) is included in the discriminant set, further adjustment of the tolerance values and the F-to-enter values has occurred. Now the tolerance of the transformed deep resistivity (SQRLLD) is very low, which is a direct reflection of its strong correlation with the shallow resistivity (see the correlation matrix of the original data, Table 2). The Wilks' statistics indicate that the gamma-ray is the dominant contributor at this stage.

And so the analysis proceeds. An interesting situation arises in Step 5 when the transformed deep resistivity is included in the discriminant set on the basis of its apparent contribution significance (F-to-enter value is highest of all variables left to enter), yet the tolerance value suggests that the curve is almost completely redundant. This suggests a weakness in the choice of criterion for a curve to enter the discrimination. However, the last part of Table 16 is a repeat analysis where the SQRLLD curve is omitted. Comparing this result with Step 8 of the first pass suggests that the impact of leaving out the curve is minimal (because it is so redundant its effect is not noticeable), the order of the other curves is not affected, nor is their contribution. The Wilks' statistics suggest that the contributions to the discrimination are fairly evenly spread across the variables and that the overall discriminatory power of the function is fairly good. Note how the partial lambdas increase as more terms are added in reflection of the 'broadening' of the function and the consequent reduction in an individual variable's contribution.

What we have undertaken here is a multi-group discriminant analysis, as opposed to a two-group discriminant analysis. We are now able to perform a canonical correlation study of the discriminant function and its ability to separate groups individually and derive canonical variates for particular groups. There will be one less variate than there are groups and each variate found will be independent of other variates. The variates are extracted in decreasing effectiveness of discrimination with the first variates possessing the greatest discriminatory powers. Figure 22 shows a crossplot of the first two canonical variates extracted from the second solution in our example (no SQRLLD),

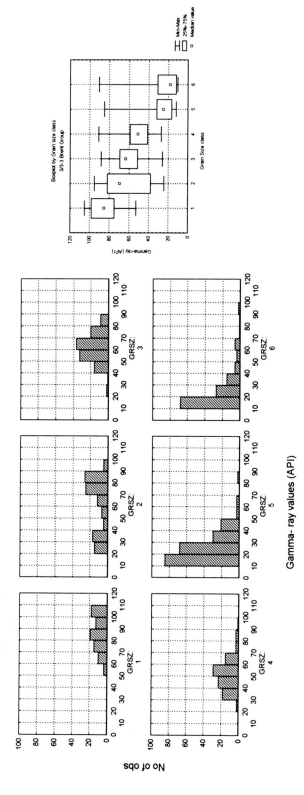

Fig. 20. Gamma-ray histograms by grain-size class, 3/3-3 Brent Group (see Table 15 for definitions).

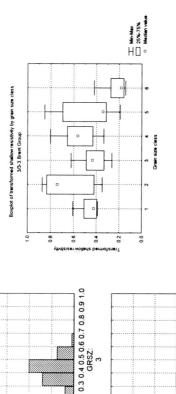

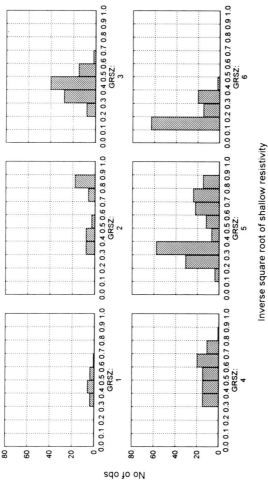

Inverse square root of shallow resistivity

Fig. 21. Shallow resitivity (transformed) histograms by grain-size class, 3/3-3 Brent Group.

Table 15. *Lithofacies and grain size classes, 3/3-3 data set*

Grain-size classes

Class no.	Class name
1	Clay
2	Silt
4	Fine sand
5	Medium sand
6	Coarse sand

Lithofacies scheme

Class no.	Major lithofacies
1	Intraformational conglomerate
2	Large-scale cross-bedded sandstone
6	Low-angled cross-bedded sandstone
8	Parallel-laminated sandstone
9	Ripple cross-laminated sandstone
12	Massive/deformed sandstone
14	Bioturbated sandstone
15	Rootletted sandstone
16	Heterolithic facies
18	Sandy claystone
20	Structureless claystone
23	Laminated claystone
24	Mottled claystone
25	Coal

Intervening values on the nominal scale of class numbers comprise combinations of these principal lithofacies.

with an inset showing the fourth and fifth variates crossplotted. The symbols represent the various grain-size classes (see Table 15), with G_1:1 being the clay size and G_6:6 being the coarse-sand size. The major graph shows how little overlap is achieved between the groups

in terms of the combined discriminatory powers of the first two variates, while the inset shows that the last two variates have much weaker discriminatory ability. The major graph also reveals some clustering in the variates that goes across grain-size classes, indicating that there are different controls on the log data than grain size.

The first discriminant function (variate 1), which has the greatest amount of discriminatory ability, is yet only capable of distinguishing between the following sub-groups of grain-size class: group 1 (with some overlap of group 2), groups 2 and 5 together, groups 3 and 4 together and group 6 separately. Categorized histograms of the function in the groups are shown in Fig. 23. It has to be said that this is not a spectacular success. The abilities of this group of log data in discriminating between the observed grain size classes are poor. Alternatives were tried firstly with fewer logs (GR, SQRLLS, CNL), but this had higher Wilks' lambda values and therefore was less discriminatory, and, secondly, with the addition of the core lithofacies curve which led to only a marginal improvement over the seven variable case. The problem may be illustrated by comparison of the boxplots in Fig. 24 with those of Figs 20 and 21. Figure 24 shows the mean/standard deviation displays, whereas Figs 20 and 21 show the interquartile ranges and the entire extent of the data distributions. Discriminant function analysis assumes multi-variate normality in the variables and is operating on standardized data, hence its 'view' of the data is with respect to data means and standard deviations. On this basis (Fig. 24) the groupings are reasonably distinct and so we see strongly significant apparent discriminating ability (even though the data in certain of the groups are by no means

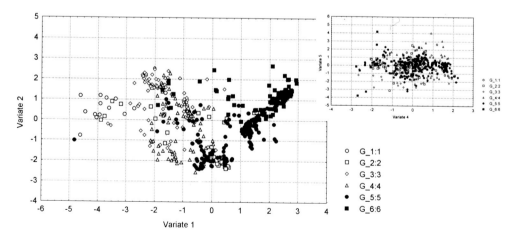

Fig. 22. Crossplots of data loadings on the canonical variates arising from the multi-group discriminant function analysis.

Table 16. *Summary output from multi-group discriminant function analysis*

Step 0, N of vars in model: 0;

Variables currently not in the model (333dat.sta)
Df for all F-tests: 5495

	Wilks' lambda	Partial lambda	F-to enter	p-Level	Toler.	1-Toler. (R-sqr.)
GR	0.395666	0.395666	151.2104	0	1	0
RHOB	0.537506	0.537506	85.18391	0	1	0
DT	0.684457	0.684457	45.64002	0	1	0
CNL	0.530460	0.530460	87.63019	0	1	0
SQRLLS	0.609254	0.609254	63.49370	0	1	0
SQRLLD	0.609306	0.609306	63.47967	0	1	0
CPHI	0.700400	0.700400	42.34767	0	0	1
LOGCKX	0.492348	0.492348	102.0770	0	1	0

Variables currently not in the model (333dat.sta)
Df for all F-tests: 5494

	Wilks' lambda	Partial lambda	F-to enter	p-Level	Toler.	1-Toler. (R-sqr.)
RHOB	0.290155	0.733333	35.92732	0	0.601788	0.398211
DT	0.269400	0.680878	46.30675	0	0.987401	0.012598
CNL	0.283809	0.717293	38.93997	0	0.811727	0.188272
SQRLLS	0.252767	0.638839	55.85554	0	0.979178	0.020821
SQRLLD	0.273799	0.691995	43.97548	0	0.942269	0.057730
CPHI	0.317436	0.802282	24.34869	0	0.785847	0.214152
LOGCKX	0.312081	0.788747	26.46196	0	0.651038	0.348961

Variables currently not in the model (333dat.sta)
Df for all F-tests: 5493

	Wilks' lambda	Partial lambda	F-to enter	p-Level	Toler.	1-Toler. (R-sqr.)
RHOB	0.205301	0.812213	22.79663	0	0.545566	0.454433
DT	0.195811	0.774669	28.68003	0	0.908160	0.091839
CNL	0.192067	0.759960	31.16072	0	0.782716	0.217283
SQRLLD	0.216471	0.856405	16.53236	0	0.044973	0.955026
CPHI	0.214393	0.848185	17.64819	0	0.736742	0.263257
LOGCKX	0.199052	0.787492	26.60752	0	0.649649	0.350350

Step 1, N of vars in model: 1; Grouping: GRSZ (6 grps)
Wilks' lambda: 0.39567 approx. F (5495) = 151.21 $p < 0.0000$

	Wilks' lambda	Partial lambda	F-remove (5495)	p-Level	Toler.	1-Toler. (R-sqr.)
GR	1	0.395666	151.2104	0	1	0

Step 2, N of vars in model: 2; Grouping: GRSZ (6 grps)
Wilks' Lambda: 0.25277 approx. F (10988) = 97.715 $p < 0.0000$

	Wilks' lambda	Partial lambda	F-remove (5495)	p-Level	Toler.	1-Toler. (R-sqr.)
GR	0.609254	0.414880	139.3410	0	0.979178	0.020821
SQRLLS	0.395666	0.638839	55.85553	0	0.979178	0.020821

Table 16. (*continued*)

Step 3, N of vars in model: 3; Grouping: GRSZ (6 grps)
Wilks' lambda: 0.19207 approx. F (151 361) = 74.228 $p < 0.0000$

	Wilks' lambda	Partial lambda	F-remove (5493)	p-Level	Toler.	1-Toler. (R-sqr.)
GR	0.359784	0.533842	86.09870	0	0.809682	0.190317
SQRLLS	0.283809	0.676749	47.09637	0	0.944183	0.055816
CNL	0.252767	0.759860	31.16073	0	0.782716	0.217283

Variables currently not in the model (333dat.sta)
Df for all F-tests: 5492

	Wilks' lambda	Partial lambda	F-to enter	p-Level	Toler.	1-Toler. (R-sqr.)
RHOB	0.180218	0.938306	6.469816	7.72×10^{-26}	0.372198	0.627801
DT	0.174366	0.907839	9.989184	4.06×10^{-9}	0.433053	0.566946
SQRLLD	0.167571	0.872459	14.38466	3.67×10^{-13}	0.043534	0.956465
CPHI	0.182695	0.951204	5.047737	0.000160	0.608327	0.391672
LOGCKX	0.156197	0.813240	22.59736	1.98×10^{20}	0.614889	0.385110

Step 4, N of vars in model: 4; Grouping: GRSZ (6 grps)
Wilks' lambda: 0.15620 approx. F (201 632) = 61.253 $p < .0000$

	Wilks' lambda	Partial lambda	F-remove (5492)	P-Level	Toler.	1-Toler. (R-sqr.)
GR	0.202499	0.771349	29.16865	5.99×10^{-26}	0.501930	0.498069
SQRLLS	0.227861	0.685492	45.14637	0	0.944180	0.055819
CNL	0.199052	0.784704	26.99746	3.74×10^{-24}	0.740837	0.259162
LOGCKX	0.192067	0.813240	22.59736	1.98×10^{-20}	0.614889	0.385110

Variables currently not in the model (333dat.sta)
Df for all F-tests: 5491

	Wilks' lambda	Partial lambda	F-to enter	p-Level	Toler.	1-Toler. (R-sqr.)
RHOB	0.148091	0.948103	5.375177	7.99×10^{-5}	0.303816	0.696183
DT	0.139987	0.896223	11.37092	2.13×10^{-10}	0.401964	0.598034
SQRLLD	0.136744	0.875455	13.97014	8.80×10^{-13}	0.042513	0.957486
CPHI	0.139690	0.894321	11.60388	1.30×10^{-10}	0.300280	0.699719

Step 5, N of vars in model: 5; Grouping: GRSZ (6 grps)
Wilks' lambda: 0.13674 approx. F (251 825) = 51.732 $p < 0.0000$

	Wilks' lambda	Partial lambda	F-remove (5491)	p-Level	Toler.	1-Toler. (R-sqr.)
GR	0.158243	0.864138	15.43916	4.09×10^{-14}	0.456449	0.543550
SQRLLS	0.168039	0.813763	22.47391	2.56×10^{-20}	0.043572	0.956427
CNL	0.172592	0.792295	25.74369	4.24×10^{-23}	0.725968	0.274031
LOGCKX	0.167571	0.816034	22.13809	4.98×10^{-20}	0.600463	0.399536
SQRLLD	0.156197	0.875455	13.97014	8.80×10^{-13}	0.042513	0.957486

Variables currently not in the model (333dat.sta)
Df for all F-tests: 5490

	Wilks' lambda	Partial lambda	F-to enter	p-Level	Toler.	1-Toler. (R-sqr.)
RHOB	0.129468	0.946792	5.507422	6.03×10^{-5}	0.303251	0.696748
DT	0.121751	0.890361	12.06769	4.87×10^{-11}	0.399098	0.600901
CPHI	0.122607	0.896623	11.29891	2.49×10^{-10}	0.300121	0.699878

Steps 6 and 7 omitted

Step 8, N of vars in model: 8; Grouping: GRSZ (6 grps)
Wilks' lambda: 0.10776 approx. F (402 129) = 35.523 $p < 0.0000$

All variables included

	Wilks' lambda	Partial lambda	F-remove (5488)	p-Level	Toler.	1-Toler. (R-sqr.)
GR	0.112443	0.958354	4.241231	0.000875	0.279857	0.720142
SQRLLS	0.131128	0.821793	21.16458	3.56×10^{-19}	0.042990	0.957009
CNL	0.122620	0.878816	13.45841	2.60×10^{-12}	0.362505	0.637494
LOGCKX	0.137213	0.785353	26.67530	7.30×10^{-24}	0.281449	0.718550
SQRLLD	0.124074	0.868515	14.77569	1.6×10^{-13}	0.042042	0.957957
DT	0.116769	0.922847	8.159620	2.06×10^{-7}	0.239841	0.760158
CPHI	0.116820	0.922446	8.205600	1.87×10^{-7}	0.249713	0.750286
RHOB	0.112377	0.958916	4.181568	0.000991	0.204325	0.795674

Final step of repeat analysis omitting the SQRLLD variable

Step 7, N of vars in model: 7; Grouping: GRSZ (6 grps)
Wilks' lambda: 0.12407 approx. F (352 059) = 37.794 $p < 0.0000$

	Wilks' lambda	Partial lambda	F-remove (5489)	p-Level	Toler.	1-Toler. (R-sqr.)
GR	0.133904	0.926592	7.748036	4.98×10^{-7}	0.298966	0.701033
SQRLLS	0.158369	0.783449	27.03247	3.63×10^{-24}	0.842218	0.157781
CNL	0.138862	0.893506	11.65636	1.17×10^{-10}	0.371967	0.628032
LOGCKX	0.157916	0.785696	26.67551	7.21×10^{-24}	0.283192	0.716807
CPHI	0.134311	0.923781	8.069232	2.50×10^{-7}	0.250697	0.749302
DT	0.133375	0.930267	7.330994	1.22×10^{-6}	0.241775	0.758224
RHOB	0.129397	0.958866	4.195406	0.000962	0.204329	0.795670

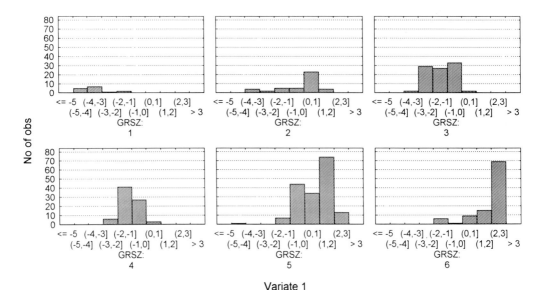

Fig. 23. Histograms of the distributions of variate 1 in different grain size class; qualitative indicator of the effectiveness of the discrimination.

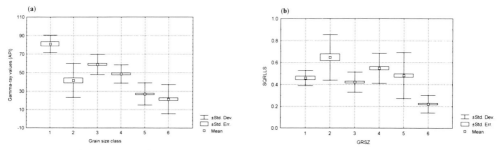

Fig. 24. Categorized boxplots of (**a**) gamma-ray and (**b**) transformed shallow resistivity data in grain-size classes.

normally distributed). However, when the discriminant function so derived is actually applied, the actual degree of overlap shown in Figs 20 and 21 causes blurring of the predicted classes.

Clearly, this analysis should only proceed if the data can be subdivided first into some sort of meaningful subgroup, such as major lithotypes like shales and sands, prior to the attempt at discriminating for grain size class. In this way some of the variability and overlap may be removed.

'Pigeon-holing'

Oliveira & Dufour (1989) and Rodrigues *et al.* (1988) introduce a method of interpretation that can be run in either predictive or classification mode. The approach is not to make any assumptions about the relationships that might exist between variables, but simply to construct

databases of responses and interpreted parameters and then to compare new data with the known response set in order to infer the same interpreted parameters from the new data. In predictive mode the database is compiled from log and core responses together with the values of porosity, permeability, saturation etc. that are available as a training set.

Whereas regression as a predictive tool is a function of the correlation coefficient between dependent and independent data, the pigeon-holing technique considers only the range of values (standard deviation) in the predicted variables for a given range in the independent data. This constitutes a conditional expectation. Widely scattered data responses will provide poor predictors in either technique. However, if two data groups are distributed as in Fig. 25, i.e. the data comprise two categorical variables exhibiting three levels of response, then regression prediction techniques, and other linear

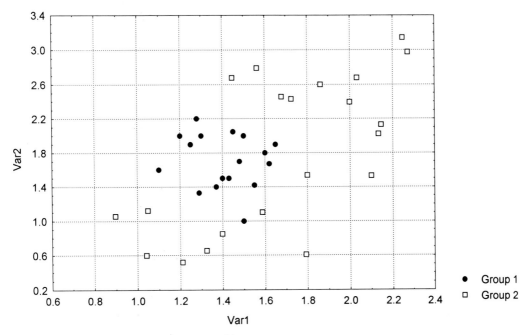

Fig. 25. Hypothetical non-linear clustering problem.

discriminant techniques, will find it all but impossible to resolve the correct groupings until a third variable is present. Without the correct groupings prediction may be ambiguous. By maintaining the co-ordinate information expressed in terms of the bivariate log and core responses, the pigeon-holing method is able to discriminate such a configuration with some high degree of accuracy, even without the third variable.

A database to be used for predictive analysis must contain only ratio scale data. However, one that is to be used for classification work may contain ordinal class data of any rank. Hence, ordinal class category data may be incorporated along with binary presence-absence criteria in the same database that holds the wireline data. The discriminating powers of the ordinal and binary data are automatically incorporated within the analysis of the higher rank data.

Essentially, the algorithm is presented with an example data set, either divided into categories of interest or containing the variable to be predicted, and it then proceeds to sub-divide the input variables into discrete ranges ('bins') and counts the number of cases within each bin. These counts are then combined with those of all other variables and a distribution of populated bins is established for the categories or target variable. New data are then presented to the same set of binned variables and the locations

(bin-counts) of the new data are compared with the training example. Membership of categories for the new data, or magnitude of the target variable from the new data, constitute the output from the model.

A potential drawback with the technique is the so-called 'curse of dimensionality' (Bishop, 1995). This refers to the fact that the number of bins in such a system will grow exponentially with the number of variables. Too many bins and the system is too sparsely populated for reliable prediction or classification. Judging on the density of bins (overall number of bins) is a trial and error process. However, when the system is set up satisfactorily then very impressive results, especially in permeability prediction from the combined responses of a set of wireline data, can be achieved.

The reader is referred to the literature for examples as the software algorithm is proprietary and unavailable to the author.

Characteristic analysis

This is a discrete multivariate procedure for combining and interpreting data (Chaves 1993). Its geoscientific applications have origins in the early 1970s with Botbol (1971) and later with McCammon *et al.* (1983) and Chaves (1988, 1989). In 1993, then current research was investigating generalizations of the method using fuzzy set theory and fuzzy logic to allow application to reservoir description problems.

The technique requires some kind of discrete sampling to be available, e.g. the spatial subdivision of maps into cells. Characteristic analysis attempts to establish the joint occurrences of various attributes or parameters that in some way are indicative of the desired phenomenon. Chaves (1993) indicates four steps that comprise the analysis methodology:

(1) sample the study area on a regular grid basis;
(2) express the 'favourabilities' of the variables in each cell in binary or ternary form;
(3) choose a model that indicates the cells that include the target;
(4) produce a combined favourability map of the area that should point out any new occurrence of the favourable criteria.

The favourability of individual variables is expressed in binary form (+1 for favourable, 0 for unfavourable or unevaluated) or in ternary form (+1 for favourable, 0 for unevaluated, −1 for unfavourable).

The aim is to characterize the model, e.g. the known occurrences of hydrocarbons in a target area – through relations observed among the chosen variables. Similar relations observed in new areas are hoped to portray similar levels of favourability with respect to the study objective. For each cell in the model the favourability of individual variables is assigned to create a single data matrix of n ternary (or binary) transformed variables in m cells of the grid. The favourability of a given cell becomes the weighted linear combination of the transformed variables comprising the data set within the cell, which can be written

$$f(x) = a_1 x_1 + a_2 x_2 + \cdots + a_n x_n \qquad (76)$$

with weights a_i $(i = 1, 2, \ldots, n)$ and n transformed variables $(n = 1, 2, \ldots, n)$.

If $\mathbf{X}$ is the $m \times n$ data matrix of cell values, then the coefficients a_i in the above are found from

$$(\mathbf{X}'\mathbf{X})a = \lambda a \qquad (77)$$

where λ is the largest eigenvalue of $\mathbf{X}'\mathbf{X}$. The a_i are the elements of the eigenvector associated with λ and are scaled so that $f(x)$ lies between −1 and +1 in ternary cases and between 0 and +1 in binary cases.

Using these equations the favourability of a cell outside the training set is evaluated.

The set of favourabilities may be considered as a fuzzy set

$$T = \{z, f_T(z)\}, \quad z \in Z \qquad (78)$$

where Z is the object space of all possible occurrences of the phenomenon (e.g. porosity greater than some threshold value) and $f_T(z)$ is the degree of pertinence (= possibility) of z in T. This approach allows the analysis to move away from Boolean algebra and to express the favourability of an individual variable using fuzzy sets rather than binary (or ternary) ratings. In this manner $f(x)$ favourability values can be derived from

$$f(x) \equiv \tau = V_1 \cup V_2 \cup \cdots \cup V_n \qquad (79)$$

where V_i are favourability maps of individual variables, defined as

$$V = \{x, fv(x)\}, \quad x \in X \qquad (80)$$

where $fv(x)$ represents the possibility that x belongs to X, the favourable area for the variable. Chaves (1993) contains examples of the technique.

Clustering

Cluster analysis is perhaps the longest established multivariate technique in use on petrophysical data. It is predicated on the belief that a given set of data can be clustered into groups that differ in meaningful ways from each other. Given the geological framework within which all petrophysical analyses are undertaken this belief is commonly true at any scale of investigation.

The cluster analysis procedure can be thought of as comprising a number of steps. First, an $n \times p$ matrix of data measurements (p variables at n depths) is transformed into an $n \times n$ matrix of similarities (or of distances) between pairs of data points across the data set. Then a clustering method is selected to define the groups of data – the essential aim being to minimize within-group variance relative to between-group variation. Finally, the uncovered groups are contrasted by comparison of means or other descriptive statistics of interest.

The definition of similarity or distance is of interest, and the choice of clustering method is often not straightforward. The literature on the subject is voluminous, covering many fields of research and study. Whilst there are dozens of similarity measures and many clustering criteria, only a couple of each will be mentioned here and are selected because they are known to have been used in a geoscientific context.

There is no universal agreement on what constitutes a cluster. Clusters may be thought of as continuous regions of data points within n or p-dimensional space separated by regions with few or no data points. However, the simple criterion of closeness, i.e. objects in a cluster should be closer to each other than to objects in other clusters, can be restrictive in that it really is capable of identifying spherical clusters only – elongate clusters may be real (e.g. clean sand points having a range of porosities form elongate clusters on density: neutron etc. cross-plots) and would be missed by such a criterion.

Cluster analysis seeks to reduce the data to a set of groups that are fewer in number than the original set of data points. Whereas PCA is a procedure with the same aim that operates mainly on the variables, cluster analysis is most commonly considered in terms of the samples (depth levels) containing the measurements, rather than the variables. Having said that, most clustering methods can be applied to variables quite effectively and apart from regrouping into linear composites as occurs within PCA.

In contrast to discriminant function analysis, cluster analysis need not require any a priori groups; the data are analysed to determine the extent to which differentiation is possible, whereas discriminant function analysis assumes that the data form well-defined groups and seeks to measure how those groups differ.

Having chosen an appropriate similarity measure, a clustering algorithm must be selected. There are a rather large number of these to choose from, and the reader is directed to the literature (e.g. Kaufman & Rousseeuw 1990). Davis (1986) suggests four main types of algorithm.

(1) Partitioning methods. In p-dimensional space find relatively densely populated regions separated by more sparsely populated regions, and construct mathematical separators in the latter.

(2) Arbitrary origin methods. Begin by computing an $n \times k$ similarity matrix between data points and k arbitrary points serving as group centroids. The observation closest (most similar) to a starting point is combined with it to form a cluster; observations are iteratively added to their nearest cluster and the centroid of the clusters continuously recomputed.

(3) Mutual similarity. Group together observations that have a common similarity to other observations.

(4) Hierarchical clustering. Begins by joining the most similar observations, then proceeds by joining the next most similar observations until all observations are assigned to one of two groups. A matrix of $n \times n$ similarities is computed, pairs with highest similarities are merged and their similarities with all other observations are averaged; the process continues until the original similarity matrix is reduced to 2×2. The levels of similarity at which observations are merged is used to construct a dendrogram. (see Fig. 26)

Hierarchical methods have always enjoyed wide usage in geoscientific applications through application on numerical taxonomy of fossils. Log responses have also been analysed by hierarchical clustering to determine 'rock types' (e.g. Serra & Abbott 1980).

By way of example of hierarchical cluster analysis, ten variables from the 3/3–3 data set (gamma-ray, density, sonic, neutron, transformed resistivities, core porosity, core permeability and the descriptive lithofacies and grain size curves) were subjected to analysis. The overall objective is to classify the variables into groups with similar response characteristics. Different criteria of partitioning were used, namely single linkage and weighted-pair averages. The former is simply defined as nearest-neighbour separation (i.e. the distance between two clusters is taken as the separation between their closest objects), whilst the latter measure is calculated as the average distance between all pairs of objects in two different clusters, weighted by the number of objects in each cluster. Thus the weighted-pair averages method attempts to account for uneven distribution of objects amongst clusters, which in our first application (clustering the variables) would be expected as certain of the variables are more closely correlated than are others and will naturally form larger groups of data.

Figures 26 and 27 present the hierarchical trees that illustrate the results from the single linkage and weighted-pair methods respectively. The process involves initially establishing groupings amongst the data, then progressively relaxing what are considered to be distinct differences between groups such that the groups begin to coalesce. The hierarchical tree pictures basically illustrate the relative similarities at which the groups coalesce. The plots are arranged with decreasing similarities ranging from left to right across the x-axes. Therefore, groups that are joined towards the left of the plots are fairly similar to each other, so that only a slight relaxation of criteria is needed before they coalesce, whereas groups that do not join until

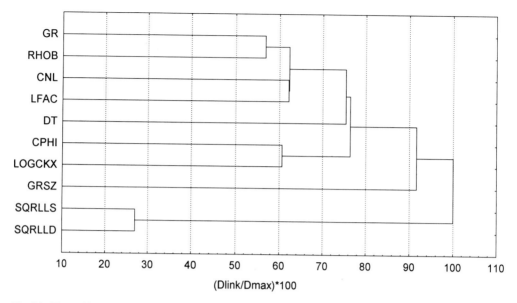

Fig. 26. Hierarchical tree diagram, single linkage clustering exercise on the variables.

the right-hand side of the plots are more dissimilar. As one progresses across the plots, larger and larger groups are being formed with greater amounts of within-group variance. It may be of interest to apply a threshold similarity (linkage distance) above which the clustering is not valid.

The single linkage results (Fig. 26) are similar to the groupings revealed on the factor loadings crossplot of Fig. 10. The resistivity data are closely similar to each other but are separate from the other data; the density, gamma-ray, lithofacies and neutron data are reasonably closely linked, as are the core porosity and permeability data, whereas the sonic and grain size curves are rather separate from the main groups.

Comparing the single linkage tree with that of Fig. 27 from the weighted-pair technique shows that the latter method moves the resistivity data slightly closer to the porosity/lithofacies group, but retains the overall grouping structure that

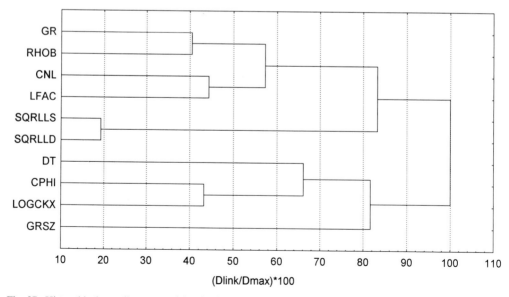

Fig. 27. Hierarchical tree diagram, weighted-pair average clustering exercise on the variables.

was found before. Other slight differences may be observed between the trees but these may not be significant (e.g. the sonic data seem to be clearly closer to the core porosity/permeability data in the weighted-pair method compared to the single linkage method). The main result appears to be that hierarchical data clustering of the variables captures the essence of the correlation matrix of the original data. Compare the relative similarities in either of the hierarchical trees with the original correlation matrix for the data (Table 2) and it can be seen that the same basic structure is identified.

An alternative analysis strategy is available through cluster analysis, one that constitutes a supervised approach in which the objects of study are the depth levels comprising the variables' responses rather than the variables themselves. The aim of this approach is to classify the depth levels on the basis of similarity of log responses, which probably constitutes the more common question to be addressed by cluster methods in geoscientific investigations. The method is an example of the second type of clustering algorithm listed above and is called K-means cluster analysis. The method requires the analyst to specify the number of groups that should be found from a data set and is illustrated by data from 3/3-3 . We restrict our attention to the Lower Ness Formation between 10 370 and 10 490 feet and set out to classify all depth levels into one of four groups. The method will form exactly four clusters from

amongst the depth levels and will ensure that the four clusters it finds are as distinct as possible given the input data. It may then be possible to assign physically meaningful labels to the groups so defined.

Figure 28 plots the results from the four group K-means analysis. Each depth level in the Lower Ness has been assigned to a group. From a comparison of the groupings with the log data and the core descriptive data that are available for this data set it is apparent that the following groups have been uncovered: group 1 corresponds to the intraformational conglomerates and large scale cross-bedded units, group 2 corresponds with other sandstones comprising mainly bioturbated and rootletted horizons, group 3 is coals (or conglomeratic beds containing coal clasts) and group 4 constitutes the shale beds. Some of the single depth classifications, e.g. at 10 376, 10 384, 10 400, 10 484 and 10 488 feet, are most likely to be the result of difficulties in bed boundary definition in the log data. Figure 29 presents another view of the groups, this time with the average value of each variable in each group plotted. Separation between the group mean values for a particular curve indicates that that curve is effective at identifying the particular groups. The converse is true if the group means are close together for a particular curve.

A repeat analysis was conducted in which six groups were specified, the aim being to attempt the subdivision of the shale intervals simultaneously with the other lithological groupings in

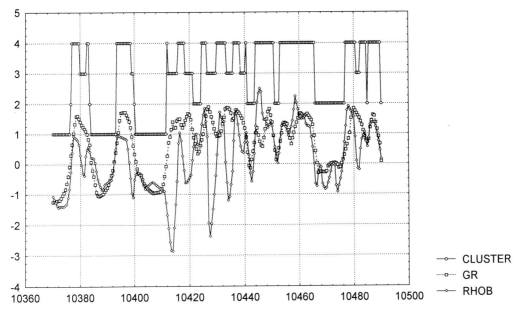

Fig. 28. K-means clustering of cases, four-cluster solution.

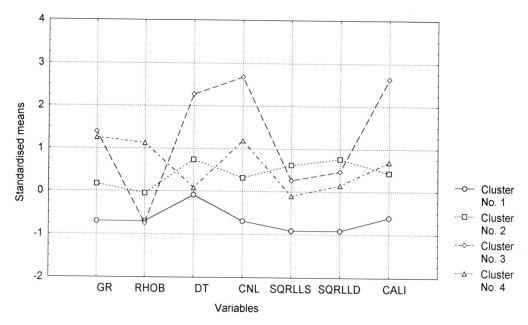

Fig. 29. Data means for clusters, K-means clustering of cases – four-cluster solution.

an attempt to provide a possible basis for correlation with other wells at some future point in time. The results are presented in Figs 30 and 31. Figure 30 shows the disposition of the groupings by depth, compared with the density and gamma-ray curves as before. From this plot it would seem that the same four groups are also identified as were found when only four groups were requested, only their numbering scheme is now somewhat different:

Group numbers:

four-cluster case	six-cluster case
1	6
2	3
3	4
4	5

For the extra groups, group 1 would appear to be the more strongly defined shale beds

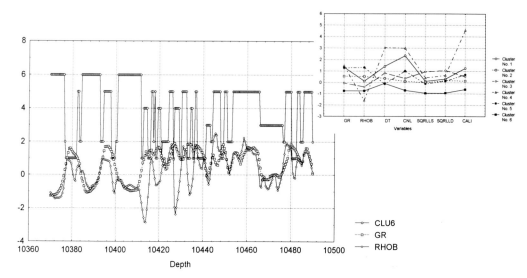

Fig. 30. Depth plot and data mean plot, K-means clustering – six-cluster solution.

(although an apparent coal at 10 485 is mis-classified if this is the case), whilst group 2 may be influenced by bed boundary problems in the log responses. This assessment of group 2 arises from the fact that the mean of each variable in this group is very close to the mean of the overall curve data (these are standardized variables so each has a mean of zero). This may well imply that group 2 merely reflects the poor resolution of bed boundaries, rather than a separate litho-logical grouping. Notably, all groups appear less well defined and more prone to mis-classification than was apparent in the four-cluster analysis.

The comparison may be further examined by the crossplot and histograms shown in Fig. 31. Here we compare the depth level grouping assignments under the six-cluster analysis case with the depth level grouping assignments under the four-cluster analysis case. It is apparent that groups 3, 4, 5 and 6 from the six-cluster analysis clearly map into the four-cluster analysis group-ings with no ambiguity whereas groups 1 and 2 from the six-cluster analysis are distributed across several of the four-cluster analysis groups.

The conclusion to be drawn would appear to be that the data do not support the subdivision of the Lower Ness Fm into more than four distinct lithological groupings; note that the core description lithofacies curve defines eight differ-ent lithofacies through the Lower Ness Fm. Apparently, these log data are incapable of resolving this level of detail. Clustering techni-ques tend to be sensitive to the presence of outliers in the data; therefore these should be removed prior to the analysis. Also, it should be remembered that cluster analysis, similarly with principal components analysis, frequently forms the first stage of statistical investigations in other disciplines. Consequently, applying the techni-que to petrophysical data should naturally lead into further investigation and interpretation of the data; casual acceptance of the clusters obtained is not recommended.

Kernel density estimation

The human eye is often very good at detecting clusters of points on cross-plots, at least in a qualitative sense. Kernel density estimation is a procedure whereby the areal density of data points on a cross-plot can be quantified and contoured, and groupings of points within clusters can be revealed from the patterns of contours that the method computes. The proce-dure estimates the probability density function that is centred on each data point and that takes account of data points within a 'window' of the cross-plot around the data point of interest. As with other, grid-based, contouring algorithms the choice of 'window' size is a critical parameter in controlling the appearance of the final contour 'map' of the cross-plot. If the window is too large then the data are smeared and small-mode clusters will be merged and smoothed into a relatively featureless surface; if the window is too tight then too much detail is preserved and the resultant surface will be quite noisy and contain excessive detail.

Mwenifumbo (1993) applied the method to the analysis of temperature and gamma-ray data; Brunsdon (1995) has illustrated the method on more conventional map data, where he calculates a 'risk surface' from probabilistic point-referenced data.

The method defines the estimate of the probability density function centred at any given data point from

$$f(x) = \frac{1}{nh} \sum_{i=1}^{n} k\left(\frac{x - x_i}{h}\right) \qquad (81)$$

where n is the number of observations in the sample. Thus n is critically controlled by the parameter h, a number that represents the variance of the function k: if h is large then k has a large variance or spread. If the value of h is too large then the estimated distribution becomes nondescript; in fact, if h exceeds the range of the x_i data, then the estimated surface function f will be equal to k and will not be any reflection of the actual data values. Conversely, if h is too small, for example having a value less than the average separation of the data points, then the estimated function f will appear to be a series of isolated spikes centred on the actual data points.

There should be some 'optimal' choice for h that is derived from consideration of the actual data being analysed. Brunsdon develops a method based on maximum likelihood estima-tion, in which the likelihood function is max-imized when the 'correct' value of h is selected. Maximum likelihood functions are expressed as $f(x; \mathbf{a})$ where $\mathbf{a}$ is a vector of parameters for some distribution f. The likelihood of a set of observations $\{x_i\}$ is the product of the prob-ability densities for each x_i

$$l(\{x_i\}; a) = \prod_{i=1}^{n} f(x_i; a). \qquad (82)$$

The estimate of $\mathbf{a}$ can be found by selecting a value such that the function is maximized. In our case, although the exact form of f is not known it may be thought of as being dependent

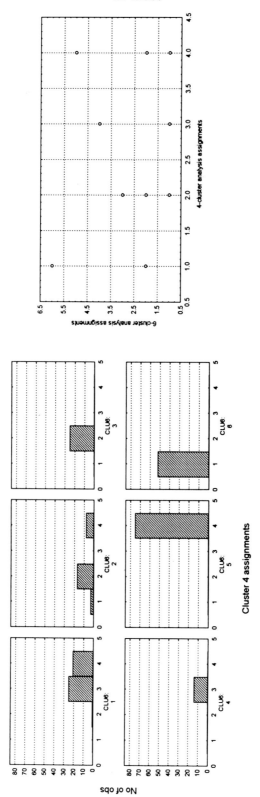

Fig. 31. Cluster assignments, four-cluster v. six-cluster solutions; clusters 1 and 2 in the six-cluster solution are ambiguous.

upon h and therefore it is possible to write a likelihood function $l(\{x_i\}; h)$ which could be maximized in terms of h. However, since the kernel density estimate function is an average of kernels around all points within h it can be thought of as comprising two terms, the kernel around x_i and the rest:

$$f(x_i) = \frac{1}{nk} k(0) + \frac{1}{nh} \sum_{j=1; j \neq i}^{n} k\left(\frac{x_i - x_j}{h}\right). \quad (83)$$

The first term is a function of zero because the numerator in the argument of k is $(x_i - x_i)$. Now, as the value of $f(x_i)$ tends to infinity as h tends to zero, then the maximum likelihood estimate will be found at $h = 0$! A more satisfactory result will be obtained if the first term in the foregoing equation is dropped. We then will be defining a new estimate $f^{(i)}(x)$ as

$$f^{(i)}(x) = \frac{1}{nh} \sum_{j=1; j \neq i}^{n} k\left(\frac{x - x_j}{h}\right) \quad (84)$$

which is an estimate of the probability density function based on all values in the sample except the ith, i.e. except on the actual data point being examined at any given calculation. Now if the likelihood function is modified to

$$l^*(\{x_i\}; h) = \prod_{i=1}^{n} f^{(i)}(x_i; h) \quad (85)$$

then this can be described as a likelihood estimate using the kernel technique to estimate the probability density at each x_i given all of the remaining observed values. Because the technique involves the estimation of the probability density function at each x_i with reference to the rest of the data set, the term cross-validation is applied (Brunsdon 1995).

With a single value for h applied across the entire data set there are likely to be areas where the data density is greater than in other areas and therefore where the value of h results in too much smoothing of the contours. The converse is also possible in more sparsely distributed parts of the data set. After Brunsdon (1995), the kernel estimator can be modified to the following

$$f^{(i)}(x) = \frac{1}{n} \sum_{j=1; j \neq i}^{n} \frac{1}{h_i} k\left(\frac{x - x_j}{h_i}\right) \quad (86)$$

where the h_i is estimated from a fixed-h solution at each x_i and is found from $l_i h$,

$$l_i = \left(\frac{f^*(x_i)}{h}\right)^{\alpha} \quad (87)$$

where h takes some fixed value and α is negative. Cross-validation may be useful in refining the value of α to be used.

In a slightly different approach, Mwenifumbo (1993) chooses to use a function called the 'Epanechnikov kernel' as the function k in the kernel density estimation equation. This is defined as

$$K_e(t) = \frac{3}{4\sqrt{5}} \left\{1 - \tfrac{1}{5} t^2\right\} \quad (88)$$

for $-\sqrt{5} \leq t \leq \sqrt{5}$, otherwise $K_e(t) = 0$, for the univariate case and

$$K_e(x, y) = \frac{1}{2\pi} \left\{1 - (x^2 + y^2)\right\} \quad (89)$$

for $(x^2 + y^2) \leq 1$ and $K_e(x, y) = 0$ otherwise, for the bivariate case.

Mwenifumbo (1993) comments on the importance of h in controlling the final kernel estimate and selects his optimum values by trial and error, rather than by a likelihood function.

The method is best suited to one- or two-dimensional data, since its value lies partly in the display of the kernel estimate function. Doveton (1986) gave the following transformation by which three variables can be reduced to two variables

$$X = \frac{(V_2 - V_1)}{\sqrt{2}} \quad (90a)$$

and

$$Y = \frac{(2V_3 - V_1 - V_2)}{\sqrt{6}} \quad (90b)$$

where V_1, V_2, and V_3 are the three variables. X and Y then lie in a plane containing the triangle whose vertices represent the maxima of the three variables.

Crisp v. fuzzy

In conventional logic all statements are either true or false; a thing is either A or not-A; the sky is blue or not-blue, the molecule is part of your finger or is not, the pore space is open or is not, the rock is sandstone or is not, and so on. This bivalent black and white view of the world has underpinned scientific thinking since the ancient Greeks. Symbolic logic and much of mathematics are firmly based on bivalency. A number is either even or is odd, $2 + 2 = 4$ is a completely true statement.

However, a moment's thought will reveal how difficult it is to apply such rigidity to real things.

The sky may be almost any shade of blue, or it may be shades of pink, red or yellow, or even white, depending upon atmospheric conditions and the time of day, but it is still sky. The pore space may be filled partially or completely with a variety of minerals; some 'pores' may be completely occluded whilst others in adjacent samples remain open and yet others may contain fibrous clay minerals and be only partially 'open'. Anyone who has attempted to classify rock samples will acknowledge immediately that the boundaries between classes are thoroughly blurred.

The fact is that 'everything is a matter of degree' (from an anonymous quote in Kosko 1993). The formal term for this state of affairs is multivalency, the somewhat less formal term is fuzziness. Fuzzy set theory was developed in the 1960s by L. Zadeh (1965) out of work earlier this century by Bertrand Russell, Lukasiewicz, Heisenberg and Black (Kosko 1993), who were all trying to capture the essence of 'vagueness' in mathematical terms. Zadeh (1965) first coined the term 'fuzzy' and applied it to sets whose elements belonged to them to different degrees. Whereas for a 'crisp' (bivalent) data set the elements of the set belong to it 100%, in fuzzy sets a membership function provides each element in a set with the degree of its membership of that set. The membership function represents the possibility of the element being within the particular set.

The concept of *possibility* in this sense is different from the *probability* that an element belongs to a set. This distinction can be illustrated by considering an example (adapted from Fang & Chen 1990) . Suppose that we have 100 core plug samples from the same formation and have studied their porosity values. Table 17 gives a hypothetical set of core plug results where the porosity values per plug are shown together with their associated probabilities (P_x) and possibilities (π_x) of occurrence (both scaled between 0 and 1).

The probability approach requires calculation of the frequencies of the porosity values from the plug samples; let us assume that 80 samples lie in the range 15–19.99% porosity, 10 samples contain 5–9.99% porosity and 10 samples have 10–14.99% porosity. On these data there is a 0.8 probability that the formation porosity lies in the range 15–19.99% (i.e. 80% of the samples have porosity in this range) and zero probability that porosity exceeds 19.99%. The possibility approach, however, requires no direct observations but can be inferred from regional knowledge of the formation. If we can say that the porosity of the formation always lies between 5 and 20%, then we can ascribe the possibility level of 1 to each of these porosity classes; then, although we may not have actually observed values outside this range in our sample, we can yet assign non-zero possibilities to other porosity classes since, in theory at least, there is a finite chance of them occurring. Thus regional knowledge might include evidence that porosities could exceed 20% although we have not actually encountered this in our sample, and so we could say that there is a finite (e.g. 80%) possibility of porosity falling in the range 20–22%, and so on.

By this means we can admit that our sampling may be incomplete and that, based upon general knowledge of the set of similar formations in the area, region, basin or whatever, there is a finite possibility that porosity outside our direct observations may be encountered. By so doing we are constructing a trapezoidal membership function for porosity in this formation which encapsulates the belief that although we are fairly sure of the range of likely porosities, there is uncertainty over the boundaries of the porosity distribution. This is illustrated in Fig. 32. An alternative, triangular, membership function is also shown which represents the case where we are certain of the limits of the data range and can also anticipate what the average value of the range will be: the peak of the triangle, whether it be central or asymmetric within the distribution, has a membership value of unity implying that the fuzzy number of interest is approximately that value (the so-called 'fuzzy C mean' discussed by Granath 1984). In principle there is no requirement for

Table 17. *Hypothetical core porosity sampling data*

	\multicolumn{7}{c}{$x = \{1, 2, \ldots, 7\}$ porosity classes}						
ϕ class	0–4.99	5–9.99	10–14.99	15–19.99	20–24.99	25–29.99	30–35
Counts	0	10	10	80	0	0	0
$\pi_x(u)$	0.25	1	1	1	0.8	0.5	0.25
$P_x(u)$	0	0.1	0.1	0.8	0	0	0

u = nominal porosity class; $\pi_x(u)$ is the possibility of occurrence of class (u); $P_x(u)$ is the probability of occurrence of class (u).

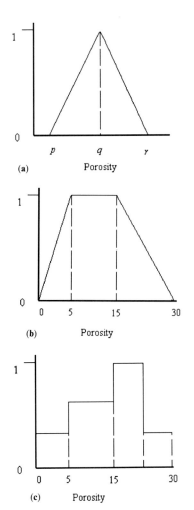

Fig. 32. Theoretical fuzzy membership distributions. (a) Triangular; (b) trapezoidal; (c) polygonal.

the membership function to be a simple geometric shape as shown, the function could consist of many data points giving a series of ranges and associated possibility values. Indeed, our hypothetical example would perhaps best be modelled by a more complex shape that captures the notion that, although our formation quite possibly has porosity in the range 5–19.99%, sample evidence suggests that there is greater likelihood (higher probability) that the porosity lies in the restricted range 15–19.99%. This function might have a shape such as shown in the lower part of Fig. 32.

Crisp numbers can be represented as a vertical line of value unity at the number itself, which says that the membership function of the number is 100% and zero elsewhere; crisp

intervals are similarly represented as rectangles and these reflect the fact that although a range is admitted in the parameter the boundaries of the range are known with certainty.

(It should be noted that a considerable number of scientists believes that fuzzy logic is an aberrant diversion from probability theory and is unscientific. Fuzzy set adherents, on the contrary, have cogent arguments demonstrating that probability is a special case of possibility.)

The application of fuzzy logic in practice requires the establishment of a series of rules relating fuzzy variables. Kosko (1992, 1993) provides details about the construction of rule sets and explores the link between fuzzy systems and neural net programs. Both process inexact information in inexact ways. Fuzzy systems estimate response functions with partial knowledge of system behaviour, perhaps encoded from heuristic input from experts. Neural networks recognize patterns without the need for explicit rules; indeed, neural nets may adaptively infer the sort of heuristic knowledge otherwise provided by experts.

Fuzzy systems directly encode structured knowledge in a numerical framework. Each entry describes a fuzzy associative memory (FAM) 'rule' or input-output transformation (Kosko 1992). FAMs are defined as clusters in the solution space, each one constituting an if-then rule. Clusters may overlap and may be almost any size, but the more efficiently that they cover the response curve the better the fuzzy system is at tackling the problem for which the response curve is a solution. The fuzzy approximation theorem says that a curve can always be covered by a finite number of clusters (Kosko 1992). Loosely defined conditions or rules give rise to large clusters; few of these are required to cover the curve but they turn the response line into a broad response area and so provide uncertain solutions. Tighter clusters provide less uncertain solutions but need more clusters.

Building a fuzzy system therefore involves mapping one fuzzy response variable onto another such that the variables overlap some desired response function that is expressed in terms of the input variables. The tighter the input fuzzy distributions or categories or ranges, then the more ranges or categories that are required to 'cover' (= predict) the response function. Wider fuzzy distributions arise when we are less certain and result in more poorly defined calculations. An important consideration is the fact that the system is 'model-free' in the sense that no specific response function or model is formulated in order to be able to map the outputs from a given set of inputs to the system. Rather it suffices to know which range of values of one variable corresponds to some range of values in a second variable. For example, we might partition the range of responses of two logging tools across a given formation on the basis of expected lithological and/or porosity groupings. Consider Fig. 33, in which ellipses are superimposed on some arbitrary function describing the joint behaviour of the two variables enabling overlapping data ranges to be partitioned within the data. From this arrangement we can construct simple rules of the type: *if* Var1 is *y*, *then* Var2 is, *x and* lithology is concrete or whatever. Alternatively, *if* Var1 is *y and* Var2 is *x* then lithology is concrete and porosity is quite high, would be another simple rule arising from typical crossplots in use in log interpretation.

Combining log and core data into families of bivariate responses leads to the reproduction of rule sets that describe the observed data in terms of expected lithology and porosity (or whatever). Such an approach is the basis of artificial intelligence classification methods investigated in the early 1980s, yet which were not taken up, largely because of the difficulty of establishing the rule sets from the observations. The rule sets being established were 'crisp' and therefore a data point could be in only one set or lithology group. However, with a fuzzy rule set data points belong to each cluster to some degree. Data points belong to most clusters to zero degree.

In fact, we are back to the problem of classification and partitioning of our observations. We derive our input data ranges from observations of the clustering amongst our data. We must decide and construct the ranges of input data that correspond with each other and with the lithological/porosity groupings of interest. The clusters overlap to a greater or lesser

degree. Kernel density functions can be used to estimate the fuzzy distribution or membership function of each cluster. Each data point can then be assigned a possibility in one or more clusters. The overall likelihood that a multivariate data set at a particular depth level belongs to one particular category (lithology, porosity, etc.) is calculated by adding the fuzzy distributions of the candidate categories to produce a combined fuzzy distribution. This combined distribution would then be amenable to kernel density estimation of the location of its modal value and thus the location of the appropriate category.

Zadeh (1965) extended the bivalent indicator function I_A of non-fuzzy subset A of X

$$I_A(x) = \begin{cases} 1 & \text{if} \quad x \in A \\ 0 & \text{if} \quad x \notin A \end{cases} \tag{91}$$

into a membership function, or multivalued (multivalent) indicator, $m_A: X \rightarrow [0.1]$

$$I_{A \cap B}(x) = \min(I_A(x), I_B(x))$$
$$I_{A \cup B}(x) = \max(I_A(x), I_B(x)) \tag{92}$$
$$I_{A^c}(x) = 1 - I_A(x)$$

where A^c denotes the complement of A, or the set not-A.

The membership value $m_A(x)$ measures the degree to which element x belongs to set A, otherwise known as the *elementhood* of x in A,

$$m_A(x) = Degree(x \in A). \tag{93}$$

Kosko (1992, 1993) develops the fuzzy entropy theorem as a quantification of the degree of fuzziness within a given fuzzy set. The formal expression of the theorem is the ratio:

$$E(A) = \frac{c(A \cap A^c)}{c(A \cup A^c)}. \tag{94}$$

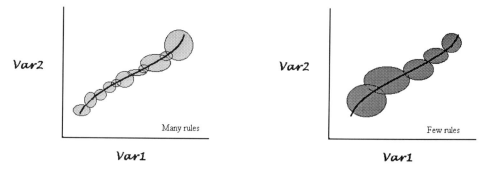

Fig. 33. Sketch to illustrate different ways to map fuzzy rule sets onto response functions.

This ratio of the intersection of sets to the union of those sets has the value zero in crisp sets that permit no overlap (= intersection). With fuzzy sets there is overlap and so the numerator is always greater than zero. In terms of log responses we would be interested in the proximity of our observed data values at each depth level to some expected value or end-point defining the category of interest.

The application of fuzzy systems thus requires the establishment of rules that cover the likely ranges of data pertaining to a particular investigation. By allowing the range limits to be fuzzy, i.e. allow the ranges to overlap, it is possible to avoid rigidity in the set of rules. This permits statements of the form: *if* density is *y and* resistivity is low, then formation is most likely to be 'group A', to coexist with a smaller (but non-zero) possibility of the formation being 'group B'. In general, fuzzy systems work with more than two input data items. Bringing other data into the model can result in a shift of the maximum possibility response away from 'group A' and into some other characteristic representation of the formation.

Most of the multi-variate analysis techniques discussed in this review are suitable for application to the task of establishing partitions (i.e. clusters) in the observations and hence to derive rule sets of input data ranges. Commonly, the technique of neurocomputing is used for fuzzy systems to establish patterns or clusters in the data as the basis for fuzzy rules.

Neural nets (neurocomputing)

The term neurocomputing refers to implementation of a group of algorithms with the common feature that they consist of many nodes (mathematical neurons) connected within a structured system forming the program. The general term for such a program is a 'neural net' or 'neural network'. In a neural network each node of the net is a calculation site, taking inputs, processing them and passing outputs somewhere else.

There is a quite fundamental difference between a neural network program and conventional algorithms such as constitute the techniques discussed previously. With the latter the programmer has a prescribed algorithm describing the transformation of the inputs to the desired outputs, and the program is an encoded version of the defined algorithm. In neural networks the approach is to build a completely general system of connected equations tailored only to the size of the problem to hand (numbers of inputs and outputs); once built, examples of

the desired outputs and a set of given inputs are entered into the system and the strengths of the many connections are altered by the program until the transformation from input to desired output is effected to some acceptable level of accuracy. The result is a 'trained' network that encapsulates the intrinsic mathematical model that maps the inputs to the outputs of the training example. Since no pre-specified algorithm is required, the system constitutes a 'model-free' solution to the transformation problem.

A very common structure for the system is a multiple-layered one comprising an input layer, one or more hidden layers, and an output layer (Fig. 34). One example of this type of net has the name multi-layer perceptron (MLP) and was first developed in the 1960s. Each node in a layer is connected to all the nodes in the next higher layer. Data are input directly to the input layer and these calculation nodes pass processed results to the first hidden layer. The first hidden layer passes its output to all nodes on the next higher layer, which may be another hidden layer or may be the output layer. The data that reach the output layer are the predictions of the net based upon the observed or input data. Each connection between nodes is termed a 'synapse' and comprises a transfer function that might be linear or non-linear, and that can vary in strength, or weight. A common function used for the connections is a sigmoidal function of the form,

$$x = \frac{1}{(1 + e^{-cx})} \qquad (95)$$

where c is the sum of the weighted inputs and x controls the slope of the function. (In some applications this is referred to as the 'temperature' since it controls the transfer from neuron to neuron, e.g. Jenner & Baldwin 1994.) Alternatively, a simple binary step function can be used where the threshold level of the step is adapted iteratively by the net minimization logic.

Choice of the number of nodes and layers is an important factor in the efficiency of the calculations. The number of nodes in the input layer is governed by the number of types of input data (e.g. density, neutron, facies class, depth etc. are types of data in this sense), whilst the output layer is dependent upon the number of parameters being estimated. The size and number of the hidden layers are often decided by trial and error. For training sets of between 100 and 1000 samples, networks with 1 or 2 hidden layers and comprising between 10 and 100 nodes seem to encompass the range used for one-dimensional log analysis.

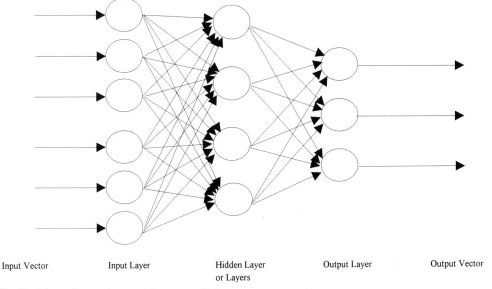

Input Vector   Input Layer   Hidden Layer   Output Layer   Output Vector
                               or Layers

Fig. 34. Schematic neural network layout, multi-layer perceptron architecture.

Hall & Scandella (1995) provide a recent summary of the issues and potential benefits from using neural networks in several different kinds of log evaluation problem. They bring a variety of neural network architectures to bear in different cases. Of the many architectures available, two quite distinct algorithms are reported to have been applied in petrophysical analyses more often than others, viz. back propagation (feed forward) neural network (BPNN) (Jenner & Baldwin 1994) and adaptive resonance theory (ART) (Wong *et al.* 1995). In each case problems for analysis may be handled in either supervised or unsupervised mode; the supervised mode requires input-output pairs to be present in the training data set, whilst the unsupervised mode requires only an input training data set. Other architectures of note that offer interesting potential are the generalized regression neural network outlined in Hall & Scandella (1995) and the radial basis theory described in Bishop (1995). The Chinese petroleum industry is active in research into the application of neural networks to petrophysical problems amongst others (e.g. Cheng-Dong *et al.* 1993).

The principle employed by BPNN methods is that successive calculations performed as the data traverse the net structure transform the weights of the node-node connections in such a way that the transformed output data match some desired output example data set to a chosen degree of exactitude. The mismatch between output data example and output node values is expressed as an objective function which is minimised during successive iterations. A commonly used error function is the root-mean-square-error (RMSE) term,

$$RMSE = \sqrt{\frac{\sum_{i=1}^{n_p} \sum_{j=1}^{n_o} (T_{ij} - Y_{ij})^2}{n_p \cdot n_o}} \quad (96)$$

where T_{ij} are the target (example) data, Y_{ij} are the network output estimates, and the summations are over n_p input-output pairings and n_o output nodes.

The program alters the weights of the node–node ('synaptic') connections based upon the effect that the altered structure has on the objective function. When some pre-defined level of match is achieved the process stops and the output node values are 'close' to the output data example set. Once the net has been trained to predict the target data set at the desired accuracy, it should be possible to use it for prediction of the same variables from new data. Provided the new data are within the ranges ('experience') of the data used to train the net initially it is possible to achieve highly accurate predictions. However, a feature of BPNN systems is that they are poorly tolerant of new data outside the range of the previous experiences and require re-training if a new data set contains unknown examples.

The choice of initial weights may influence whether the net converges onto a local minimum or the global minimum and other adjustable parameters, such as learning rates and a 'momentum' term, add complexity to the control of the net.

An alternative control scheme is used in the ART models (Fausett 1994; Carpenter & Grossberg 1994; Wong *et al.* 1995). Again, these algorithms are multi-layer, comprising an input layer, a 'prototype' layer and an output layer. The prototype layer contains no nodes at initialization. Each category is assigned to an output node and therefore the number of outputs equals the number of known categories in the problem and we start off with a supervised classification scheme. As data are presented to the input layer, a prototype node is constructed in the middle layer and assigned to the particular category representing the input signal. The number of prototype nodes created is at least equal to the number of output categories, but may exceed that if sub-categories are uncovered in the input data. Each sub-category will be assigned its own prototype node in the prototype layer. Unlike BPNN methods, ART architectures can allow more than one prototype node to represent a particular category. In this manner the category 'sandstone' may contain sub-groups such as coarse-grained and fine-grained which may be detectable by the network but 'hidden' from the analyst and so the network offers the potential to operate like an unsupervised conventional pattern recognition algorithm such as cluster analysis as the training phase proceeds.

With an ART network once a new category has been identified, all prototype nodes' activation and match functions are re-evaluated. These functions calculate the degree to which the weight vector is a fuzzy subset of the input vector (the activation function value, T), and the degree to which the input vector is a fuzzy subset of the weight vector (the match function value, M). The M value of the active node (the one with the highest T value) is compared to a parameter called the 'vigilance parameter', or ρ, and if $M > \rho$ then the node is considered too weak to encode the pattern of the category (the network is in a state of resonance) and a new node is created; or, if $M < \rho$, then the network is considered to be in a state of mismatch reset, which means that the prototype node is considered strong enough to encode the new category. When a node previously encoding for some other category becomes the winning node for a new category, the network is in the state of category mismatch, and the value of the vigilance parameter is deemed to be too small. The program automatically increases ρ in specified steps as the analysis proceeds: the higher the value of vigilance then the greater the number of prototypes that form.

Whatever the architecture of the network and the connection rules employed, a strong advantage in these techniques lies in the fact that they are 'model-free' algorithms that require no distributional assumptions to be met on the part of the data. This is in contrast with most other methods reviewed earlier that require multi-variate normal data in order to work most effectively.

As an example of the potential results from employing neural networks, consider Fig. 35. This displays the comparison between reconstructed and observed core data for 3/3-3 following some example analysis work carried out for the author by J. Hall of AGIP S.p.A. The overall comparison is very impressive, indeed the reconstruction is so close to the observed lithofacies curve (track 5) that the two curves seem to overlie throughout the section. Grain size and permeability are similarly very well characterized, as seen in Figs 36 and 37 which crossplot observed versus predicted data. Of particular interest is the comparison crossplots in Fig. 37 which shows the crossplot of the simple porosity-based prediction of permeability versus the core permeability data.

The neural algorithm used was of the generalised regression architecture and was presented with a subset of the section on which to train its responses. Results such as this require more study to understand the conditions necessary for such a degree of correspondence to be attained; however they serve to whet the appetite for the development of neural computing techniques as general petrophysical analysis tools. Of particular interest would be how robust such a result proves to be in step-out wells. The ideally trained network should not only have learnt a restricted representation of the training data that would be hard to pass on to neighbouring data sets, but rather should have divined some statistical model of the processes (the underlying controls on the data structure) that generate the training data.

Genetic algorithms

A genetic algorithm (GA) is a simple tactic for computer learning that is inspired by natural selection (Singleton 1994). A GA can evolve a variety of computer-based objects and solutions that are used in the optimization of some

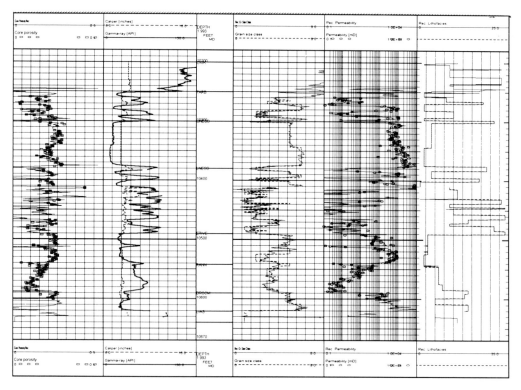

Fig. 35. Depth plot of neural network-derived reconstructions of porosity, permeability, grain-size class and lithofacies grouping.

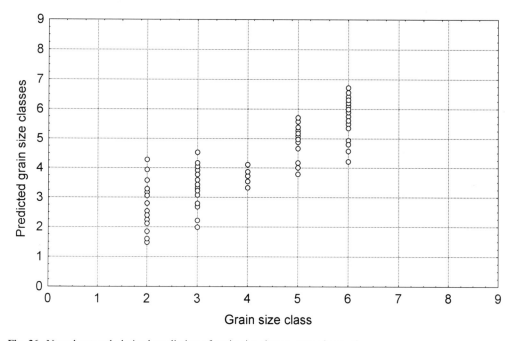

Fig. 36. Neural network-derived prediction of grain-size class v. core observations.

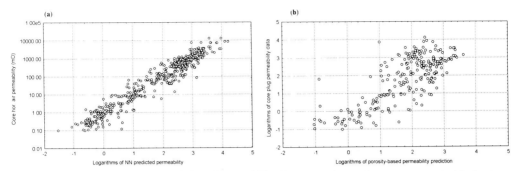

Fig. 37. Neural network-derived prediction of permeability (**a**) v. core observations as compared with simple porosity-based prediction (**b**) of permeability.

objective, for instance the minimization of a loss function. The GA can pick the 'best' individuals from this set by a process of repetitive trial and error and change them, producing a new set of objects (solutions) that are based upon the 'best' of the starting set. The intent is that some of the new generation of solutions will be an improvement on the older generation. The new, improved stock of objects becomes the feed for the next alteration or generation, and so on.

A GA consists of three operations: evaluation, selection and reproduction. Evaluation is the process of assigning some measure of how good a particular object (solution) is, compared to others of its generation. The resultant values of the 'fitness function' are ranked and the selection of the 'best' group of objects is based upon the magnitude of their fitness functions. Reproduction in this context refers to the *mutation* or the *crossover* of components making up the solution objects. Mutation involves changing the solution object in some small way by a random alteration process. For instance, a 1 might be replaced with a zero, a number might be changed or a function might be changed. Crossover is the combination of parts of the 'best' solution objects from two input objects, in order to make a new object.

Insofar as genetic algorithms are applied in computer programming, the requirement is to define operators that perform the evaluation, selection and reproduction tasks. An operator is a function that is followed by zero or more arguments, a so-called S-expression. Each argument is, in turn, an S-expression. Singleton (1994) uses examples from Lisp, an advanced programming language. In this, $2 + 2$ can be represented as (ADD 2 2) and $2 + 3 \times 5/2$ can be represented by (ADD 2 (MULTIPLY 3 (DIVIDE 5 2))); such functions are nodes whilst constants (functions with no arguments) are called terminals.

Functions may perform actions such as (TURN RIGHT, x degrees), and the system can be expanded to define strings of actions, conditional control statements (IF . . . THEN . . . ELSE), WHILE loops, FOR loops and memory management tasks of a full computer language. At this level, the set of defined functions can be directed at any task in the same manner as can an ordinary computer language. The difference between genetic algorithms and ordinary programs is that the S-expression functions are all defined to return the same argument type, usually a floating point number, and they all accept this type of input. This situation defines a state of closure. Therefore, any expression may be substituted for any other expression, and the system would still function syntactically. Whether the substitution gives a better answer is not central to the task, yet only those substitutions that do give rise to improvements are selected for the next generation of 'expression-swapping'. For example, the following two expressions (ADD 2 (MULTIPLY 3 (DIVIDE 5 2))) and (ADD (MULTIPLY 7 11) 23) could be crossed in many ways, one of which is to take out the 3 and insert the (MULTIPLY 7 11) expression to get the expression (ADD 2 (MULTIPLY (MULTIPLY 7 11)(DIVIDE 5 2))). With genetic programming, if such a swap leads to an improved solution it is retained, otherwise it is discarded.

Genetic induction is the name given to the process whereby symbolic regression is used to learn prediction and classification from real-world examples. Symbolic regression attempts to find an arithmetic formula matching some desired expression. It is accomplished by setting up a training database that includes examples of the things to be predicted, and fields for each input variable in the model. Evaluate a set of randomly chosen expressions and perform successive crossovers between expressions, all the time selecting only those new expressions that reduce some minimization function comparing the interim result and the desired prediction. Once the system is predicting the training set acceptably, new data can be introduced into the algorithm with the expectation that valid predictions will result.

Computation requirements are excessive, even for moderately sized problems. However, the system is ideally suited to massively parallel computers. As these computers become more commonplace, genetic algorithms may become the core application of the new machines. Research is being conducted into aspects of genetic algorithms such as defining the right Fitness Functions to achieve rapid convergence (= evolution) and finding the types of structures that respond best to mutation and crossover. Much work is needed on designing the structures of the algorithms i.e. their modularity and hierarchy.

Apart from the daunting size of typical petro-physical analyses problem sets, our typical tasks of prediction and classification seem well suited to being addressed by this class of algorithm. A recent application of genetic algorithms of petrophysical data is given in Fang et al. (1996).

Conclusions

A number of conclusions can be summarized.

Careful data preparation is essential to maximize the potential effectiveness of the conventional techniques, since nearly all of them require multi-variate normal data in order to operate at their best.

The findings of the conventional techniques are mostly pinned directly on the correlation matrix of associations between the input variables. Astute reading of this table by experienced practitioners may often reveal much about the structure in the data under investigation.

However, once reduced to the underlying controlling factors, that may or may not have ready interpretation in terms of formation parameters of interest to the analyst, the wireline data can be shown to contain information on a deeper level than the conventional parameters of porosity, shale and water saturation. Each of these parameters is going to be affected by lithology and pore fabric controls, and it would appear from some of the results of the example analyses carried out on 3/3-3 data that subtle structure, quite possibly pore fabric-related, can be extracted beyond the conventional parameters.

Data groupings, factors etc. that might be related to pore-fabric controls remain ambiguous, generally, unless detailed core descriptive data are to hand.

In a suite of eight conventional wireline tools and routine core data results, all the conventional data reduction methods were satisfied with extracting three dimensions from the input data correlation matrix. One method (factor analysis) could support four principal factors and another method (multidimensional scaling) seemed satisfied with two. Such parsimony is a result of the intercorrelations in log data and has implications for 'multiple mineral' deductive analysis algorithms.

Neural computing algorithms seem to offer an interesting and potentially fruitful development path towards providing evaluation routines capable of extracting more complete structure from our typically highly correlated data sets. The combination of neural networks and fuzzy associative rules offers particular promise.

The author gratefully acknowledges the assistance of J. Hall and the AGIP S.p.A. management in preparing and running the neural network example carried out on 3/3-3 data.

References

BARTLETT, M. S. 1951. The goodness of fit of a single hypothetical discriminant function in the case of several groups. *Annals of Eugenics*, **16**, 199–214

BISHOP, C. M. 1995. *Neural Networks for Pattern Recognition*. Clarendon Press, Oxford, UK.

BOTBOL, J. M. 1971. An Application of Characteristic Analysis to Mineral Exploration. *In*: *Decision Making in the Mineral Industry*. Canadian Institute of Mining and Metallurgy, Special Volumes, **12**, 92–99.

BRUNSDON, C. 1995. Estimating Probability Surfaces for Geographical Point Data: An Adaptive Kernel Algorithm. *Computers and Geosciences*, **21**, 877–894.

BURKE, J. A., SCHMIDT, A. W. & CAMPBELL, R. L. 1969. The Litho-Porosity Cross Plot. *The Log Analyst*, **10**(6), 25–41.

CARPENTER, G. A. & GROSSBERG, S. 1994. Fuzzy ARTMAP: A Synthesis of Neural Networks and Fuzzy Logic for Supervised Categorisation and Nonstationary Prediction. *In*: YAGER, R. A. & ZADEH, L. A. (eds) *Fuzzy Sets, Neural Networks and Soft Computing*. Van Nostrand Reinhold, New York, 126–165.

CHATFIELD, C. & COLLINS, A. J. 1980. *Introduction to Multivariate Analysis*. Chapman and Hall, London.

CHAVES, H. A. F. 1988. Characteristic Analysis Applied to Petroleum Assessment of Basins. *ILP Research Conference on Advanced Data Integration in Mineral and Energy Resources Studies*, Sotogrande, Spain, 1988.

——1989. Generalisation of Characteristic Analysis as a Tool for Petroleum Assessment of Basins. *ILP Research Conference on Advanced Data Integration in Mineral and Energy Resources Studies, Sotogrande, Spain 1988.* US Geological Survey Special Publication.

——1993. Characteristic Analysis as an Oil Exploration Tool. *In:* DAVIS, J. C. & HERZFELD, U. C. (eds) *Computers in Geology – 25 years of Progress.* International Association for Mathematical Geology, Studies in Mathematical Geology, **5.** Oxford University Press, New York, 99–112.

CHENG-DONG, Z., ZHEN-WU, J., XI-LING, W. & XING-SHUI, Z. 1993. Estimating Reservoir Parameters from Well Logs in Reservoir Studies Using a Large Scale Linear Neural Network: an Integrated System. *Transactions, 15th European Formation Evaluation Symposium, Stavanger 1993,* Paper AA.

DAVIS, J. C. 1986. *Statistics and Data Analysis in Geology.* 2nd Edition. Wiley, New York.

DILLON, W. R. & GOLDSTEIN M. 1984. *Multivariate Analysis: Methods and Applications.* Wiley, New York.

DOVETON, J. H. 1986. *Log Analysis of Subsurface Geology, Concepts and Computer Methods.* Wiley, New York.

——1994. *Geologic Log Analysis Using Computer Methods.* AAPG Computer Applications in Geology, **2.**

DUBOIS, D. & PRADE, H. 1988. *Possibility Theory.* Plenum Press, New York.

ECKART, C. & YOUNG, B. 1936. The approximation of one matrix by another of lower rank. *Psychometrika,* **1,** 211–218.

FANG, J. H. & CHEN, H. C. 1990. Uncertainties are Better Handled by Fuzzy Arithmetic. *AAPG Bulletin,* **74,** 1228–1233.

——, KARR, C. L. & STANLEY, D. A. 1996. Transformation of geochemical log data to mineralogy using genetic algorithms. *The Log Analyst,* **32,** 2, 26–31.

FAUSETT, L. 1994. *Fundamentals of Neural Networks.* Prentice Hall International Inc., Englewood Cliffs, New Jersey.

FERTL, W. H. 1981. Openhole Crossplot Concepts – A Powerful Technique in Well Log Analysis. *SPE of AIME, Journal of Petroleum Technology,* **33,** 535–549.

GNANADESIKAN, R. 1977. *Methods for Statistical Data Analysis of Multivariate Observations.* Wiley, New York

GRANATH, G. 1984. Application of Fuzzy Clustering and Fuzzy Classification to Evaluate the Provenance of Glacial Till. *Journal of Mathematical Geology,* **16,** 283–301.

HALL, J. & SCANDELLA, L. 1995. Estimation of Critical Formation Evaluation Parameters Using Techniques of Neurocomputing. *Transactions of Society of Professional Well Log Analysts 36th Annual Logging Symposium, Paris,* June 1995, paper PPP.

HOUGH, P. V. C. 1962. *A Method and Means for Recognising Complex Patterns.* US Patent 3069654.

JENNER, R. & BALDWIN, J. L. 1994. Application of an artificial neural network to obtain an improved permeability profile. *In: Transactions of the 16th European Formation Evaluation Symposium,* Aberdeen, 11–13 October 1994, Paper M.

KAUFMAN, L. & ROUSSEEUW, P. J. 1990. *Finding Groups in Data, An Introduction to Cluster Analysis.* Wiley, New York.

KERZNER, M. G. 1986. *Image Processing in Well Log Analysis.* International Human Resources Corporation, Boston, and D. Reidel Publishing Company, Boston.

KOSKO, B. 1992. *Neural Networks and Fuzzy Systems.* Prentice Hall Inc., Englewood Cliffs, New Jersey.

——1993. *Fuzzy Thinking.* Flamingo (HarperCollins, London).

LEBART, L., MORINEAU, A. & WARWICK, K. M. 1984. *Multivariate Descriptive Statistical Analysis.* Wiley, New York.

MATYAS, V. 1995. Application of Kruskal Multidimensional Scaling (MDS) to Rock Type Identification from Well Logs. *The Log Analyst,* **36**(1), 28–34.

MCCAMMON, R. B., BOTBOL, J. M., SINDING-LARSEN, R. & BOWEN, J. W. 1983. Characteristic – 1981: Final Program and a Possible Discovery. *Mathematical Geology,* **15,** 59–83.

MORRISON, D. F. 1990. *Multivariate Statistical Methods.* McGraw-Hill Publishing Company, New York.

MOSS, B. P. & SEHEULT, A. 1987. Does Principal Component Analysis have a Role to Play in Log Interpretation? *Transactions Society of Professional Well Log Analysts 20th Annual Logging Symposium, 28 June–1 July, 1987, London,* paper TT.

MWENIFUMBO, C. J. 1993. Kernel density estimation in the analysis and presentation of borehole geophysical data. *The Log Analyst,* **34,** 5, 34–45

OLIVEIRA, A. & DUFOUR, J. 1989. Estimating Petrophysical Parameters from a Probabalistic Database. *Transactions of the 12th International Formation Evaluation Symposium, Société pour L'Avancement d l'Interprétation des Diagraphies, Paris, October 1989.* Paper Z.

PALAZ, I. & SENGUPTA, S. K. (eds) 1992. *Automated Pattern Analysis in Petroleum Exploration.* Springer-Verlag, New York.

RODRIGUES, E., TETZLAFF, D. M., MEZZATESTA, A. & FROST, E. 1988. Estimating Rock Properties by Statistical Methods. *SPE International Meeting on Petroleum Engineering, Tianjin, People's Republic of China, November 1–4 1988.*

SERRA, O. & ABBOTT, H. 1980. The Contribution of Logging Data to Sedimentology and Stratigraphy. *SPE of AIME, Transactions 55th Annual Fall Technology Conference,* paper SPE 9270.

SINGLETON, A. 1994. Genetic Programming with C++. *BYTE, International Edition,* February, 1994, 171–176.

STATSOFT, INC. 1994. *STATISTICA[TM] Manuals and Program.*

STEWART, D. K., & LOVE, W. A. 1968. A general canonical correlation index. *Psychological Bulletin,* **70,** 160–163.

WALDEN, A. T. 1992. Clustering of Attributes by
Projection Pursuit for Reservoir Characterisation.
In: PALAZ, I. & SENGUPTA, S. K. (eds) *Automated
Pattern Analysis in Petroleum Exploration.*
Springer-Verlag, New York, 173–199.

WONG, P. M., GEDEON, T. D. & TAGGART, I. J. 1995.
The Use of Fuzzy ARTMAP for Lithofacies
Classification: A Comparison Study. *Transactions
of the Society of Professional Well Log Analysts
36th Annual Logging Symposium, Paris June,
1995*, Paper U.

YARUS, J. M. & CHAMBERS, R. L. (eds) 1994.
Stochastic Modelling and Geostatistics. AAPG
Computer Applications in Geology, **3**.

ZADEH, L. A. 1965. Fuzzy Sets. *Information and
Control*, **8**, 338–353.

ZANGWILL, J. 1982. Depth Matching – a computerised
approach. *Transactions, Society of Professional
Well Log Analysts 23rd Annual Logging Symposium, July 6–9, Corpus Christi, TX, 1982.*
Paper EE.

Electrical conductivity, spontaneous potential and ionic diffusion in porous media

A. REVIL[1], P. A. PEZARD[2] & M. DAROT[3]

[1] *Cornell University, Department of Geological Sciences, Snee Hall, Ithaca, New York,*
14853 USA (e-mail: andre@goethite.geo.cornell.edu)
[2] *Laboratoire de Pétrologie Magmatique, CNRS (URA 1277), Cerege, BP 80,*
13545 Aix-en-Provence, Cedex 04, France (e-mail: pezard@imtmer1.imt-mrs.fr)
[3] *EOPGS, Physique des Matériaux, CNRS (URA 1358), 5 rue René Descartes,*
F-67084 Strasbourg Cedex, France

Abstract: A consistent set of equations describing electrical conduction, spontaneous potential and ionic diffusion phenomena in porous media, with a particular emphasis given to surface electrical phenomena, is presented. The porous medium is assumed to consist of an insulating matrix and an inter-connected pore volume that is saturated with an electrolyte. When in contact with an electrolyte solution, mineral surfaces have an excess of charge that is balanced by mobile ions in an electrical diffuse layer (EDL). Electrical conduction and ionic diffusion in this diffuse layer can contribute substantially to the effective electrical conductivity of the porous medium, and can play an important role in the strength of several spontaneous electrical potentials (i.e., membrane potential, streaming potential, and thermo-electric potential) and ionic diffusion phenomena. Our surface electrical properties model is based on a thermodynamic description of surface chemical reactions and electrical diffuse layer processes. As an example, we consider an amphoteric mineral surface described by a three-site-type model. We derive the fractional occupancies of positive, negative and neutral sites, and the fractional ionic diffuse layer densities on the surface as a function of the salinity and the pH. The surface charge density at a given pH, is found to be dependent on the electrolyte concentration. Finally, parameters of interest in describing macroscopic effects (effective electrical conductivity, spontaneous potential coupling coefficients and effective diffusion coefficient), are related to the previously mentioned properties, via the mineral surface electrical potential (called the Stern potential).

The electrical conductivity and spontaneous potential of porous media play important roles in many scientific fields, such as surface chemistry (e.g., Fixman 1980; Hayes *et al.* 1988), geology, and geophysics including oil industry logging, earthquake prediction, geothermal systems and oceanic crust study (e.g., Doll 1949; Wyllie 1949; Mizutani & Ishido 1976; Corwin & Hoover 1979; Pezard 1990; Antraygues & Aubert 1993). This paper describes electrical conduction and spontaneous potential in saturated porous media, with a particular emphasis given to surface electrical properties. The effective conductivity of a porous medium saturated with a multi-ionic electrolyte is studied, and is found to be a function of certain micro-geometrical parameters, and specific ionic surface conductances which describe surface electrical conduction properties in the vicinity of mineral surfaces. Membrane potential is also investigated, together with ionic diffusion due to the link between these two properties. Finally, streaming and thermo-electric potentials are studied.

Surface electrical properties

Natural porous materials are formed by various minerals, the most common being silicates, carbonates and oxides. All these minerals develop an electrical diffuse layer when in contact with an electrolyte. For example, clay minerals have large surface areas, and electrical surface charge deficiencies at pH 7. This creates a strong negative electrical field perpendicular to the surface which attracts cations and repels anions in the vicinity of the pore-matrix interface. This zone near the clay surface, where counter-ions are present, constitutes the electrical diffuse layer (noted EDL hereafter), and we will call 'free electrolyte' the part of the pore fluid in the interconnected pore space which is not submitted to the electrostatic influence of the mineral surface charge. A sketch of the EDL is given in Fig. 1. The counter-ions of the diffuse layer are maintained at some distance from the mineral surface by the hydration water around each cation and the water

From Lovell, M. A. & Harvey, P. K. (eds), 1997, *Developments in Petrophysics*, Geological Society Special Publication No. 122, pp. 253–275.

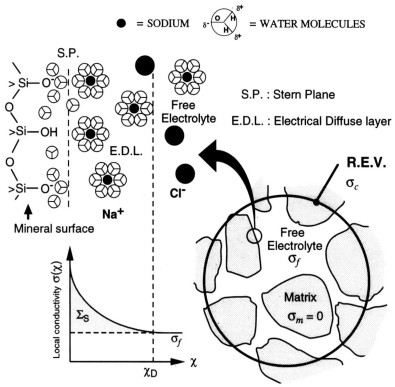

Fig. 1. Schematic representation of the electrical double layer (Stern layer and EDL) situated in the vicinity of the mineral surface. The parameter σ_c is the effective conductivity of a representative elementary volume (R.E.V.), whereas σ_f and σ_m are respectively the free electrolyte and matrix conductivities.

adsorbed on the surface which forms the 'Stern layer' (see Fig. 1). The surface between the Stern layer and the EDL is called the 'Stern plane' (Fig. 1). The electrical potential on the Stern plane is called the Stern potential, and will be noted by φ_d. The Stern layer and the diffuse layer define the so-called 'double layer' (e.g., Bockris 1973). This section deals with the electrical properties of the mineral surface and its associated EDL.

Internal surface densities

On a mineral surface, the fractional occupancy of surface sites is defined by $\Omega_i^0 \equiv n_i^0 / \sum_j n_j^0$, where n_i^0 is the number of sites of species (i) on the surface. By definition, we have: $\sum_i (\Omega_i^0) = 1$. We can introduce the total surface concentration of ionic sites Γ_S^0 (in m^{-2}) by:

$$\Gamma_S^0 \equiv \sum_i \Gamma_i^0 \equiv \sum_i n_i^0 / S, \qquad (1)$$

where Γ_i^0 represents the surface concentration of surface species (i) and S is the surface area of the

interface between the mineral and the free electrolyte. From crystallographic considerations together with various experimental results, the surface site density Γ_S^0 is close to 4–8 sites nm^{-2} for quartz (Iler 1979; Lyklema 1985), 10 sites nm^{-2} for calcite (e.g., Stipp & Hochella 1991; Van Cappellen *et al.* 1993, and references therein), 4–8 sites nm^{-2} for geothite and alumine (e.g., Zhang *et al.* 1994, and references therein). The relationship between Ω_i^0 and Γ_i^0 is: $\Omega_i^0 = \Gamma_i^0 / \Gamma_S^0$. The mineral surface charge density (in $C\,m^{-2}$) can be calculated from the ionic density of the mineral surface species by,

$$Q_S^0 = \sum_i (\pm 1) e Z_i^0 \Gamma_i^0$$
$$= e\Gamma_S^0 \left(\sum_i (\pm 1) Z_i^0 \Omega_i^0 \right), \qquad (2)$$

where $(\pm 1) e Z_i^0$ is the charge of the site of type (i) (Z_i^0 is the site valency taken positive or null, respectively for a charged or neutral surface site, e is the elementary charge).

Diffuse layer densities

We will consider that the thickness of the EDL is much smaller than (a) the local radius of curvature of the mineral surface, and (b) the throats size (which is called the 'flat and thin EDL assumption'). Poisson's equation in the diffuse layer expresses the electrostatic potential $\varphi(\chi)$ as a function of the local coordinate χ (χ is the local distance from the mineral surface, see Fig. 1),

$$\frac{\partial}{\partial \chi}\left[\varepsilon_S(\chi)\frac{\partial}{\partial \chi}\varphi(\chi)\right] = -\rho(\chi), \qquad (3)$$

where $\varepsilon_S(\chi)$ is the dielectric permittivity at the position χ in the diffuse layer (in $F\,m^{-1}$) and $\rho(\chi)$ is the excess charge density (in $C\,m^{-1}$) which is defined by $\rho(\chi) \equiv \sum_i (\pm 1)eZ_iC_i^S(\chi)$. In the present paper we suppose dielectric saturation and visco-electrical phenomena can be neglected in the diffuse layer (Pride & Morgan 1991). Consequently: $\varepsilon_S(\chi) \approx \varepsilon_f$ where ε_f is the dielectric constant of water. If the bulk electrolyte solution is in thermal equilibrium and its dilution large enough, then the concentrations in the EDL can be assumed to obey the Boltzmann's formula (Landau & Lifshitz 1980, p. 114),

$$C_i^S(\chi) = C_i^f \exp[(\pm 1)Z_i\tilde{\varphi}(\chi)], \qquad (4)$$

where T is the temperature in Kelvin, C_i are the ionic concentrations in the EDL (superscript ('S'), or in the free electrolyte (superscript 'f'), $\tilde{\varphi}(\chi) \equiv -e\varphi(\chi)/k_bT$ is the normalized dimensionless local potential in the diffuse layer at the position χ and in absence of any external electrical field, k_b is the Boltzmann's constant $(1.381 \times 10^{-23}\,J\,K^{-1})$. From Eq. (3) and Eq. (4) the well-known Poisson–Boltzmann equation is written as,

$$\frac{d^2\varphi}{d\chi^2} + \sum_i \frac{(\pm)eZ_iC_i^f}{\varepsilon_f}$$
$$\times \exp[-(\pm 1)Z_i\tilde{\varphi}] = 0, \qquad (5)$$

with appropriate boundaries conditions given by $\varphi(0) = \varphi_d$ (where φ_d is the Stern plane potential) and $\partial\varphi/\partial\chi)_{\chi_D} = 0$ (where χ_D is the size of the concentration disturbed layer). The resolution of Eq. (5) gives the electrostatic potential distribution in the EDL due to the mineral surface charge density Q_S^0. This distribution in the EDL has approximately an exponential form given by,

$$\varphi = \varphi_d \exp\left(-\frac{\chi}{\chi_d}\right). \qquad (6)$$

where χ_d is the Debye screening length defined by,

$$\chi_d \equiv \sqrt{\frac{\varepsilon_f k_b T}{2e^2 I_f}}, \qquad (7)$$

$$I_f \equiv \frac{1}{2}\sum_i Z_i^2 C_i^f. \qquad (8)$$

The parameter I_f is the ionic strength of the electrolyte solution ($I_f = C_f$ for a binary symmetric electrolyte like NaCl, where C_f is the salt concentration). Actually, Eq. (6) is the solution of Eq. (5) linearized by taking $|(\pm 1)eZ_i\varphi_d/k_bT| \ll 2$. The validity of this simple exponential distribution for φ (the 'Debye–Hückel approximation') and the relation between the mineral surface charge density Q_S^0 and the Stern plane potential φ_d are discussed in Appendix A of Pride (1994). The equivalent surface charge density of the EDL is given by the integration of the excess charge density,

$$Q_S = \int_0^{\chi_D} \rho(\chi)\,d\chi. \qquad (9)$$

Pride (1994) has demonstrated that,

$$\int_0^{\chi_D} \exp((\pm 1)Z_i\tilde{\varphi}(\chi))\,d\chi$$
$$\approx 2\chi_d \exp\left(\frac{(\pm 1)Z_i\tilde{\varphi}_d}{2}\right), \qquad (10)$$

where $\tilde{\varphi}_d \equiv -e\varphi_d/k_bT$ is the normalized Stern plane potential. Using Eqs (9), (10), and the definition of the excess charge density in the EDL, and the Boltzmann distribution [Eq. (4)] for the ionic concentrations in the EDL, we can find (Pride 1994; Revil & Glover 1997)

$$Q_S \approx 2\chi_d \sum_i (\pm 1)eZ_iC_i^f$$
$$\times \exp\left[-\frac{(\pm 1)eZ_i}{2k_bT}\varphi_d\right]. \qquad (11)$$

The excess conductivity in the vicinity of the pore–matrix interface is represented by the specific surface conductance, Σ_S, which is defined by,

$$\Sigma_S = \int_0^{\chi_D} (\sigma(\chi) - \sigma_f)\,d\chi, \qquad (12)$$

where $\sigma(\chi)$ is the spatially varying conductivity in the EDL which equals σ_f for $\chi \geq \chi_D$, where χ_D is the thickness of the EDL. The local conductivity $\sigma(\chi)$ is defined by $\sigma(\chi) = \sum_i (eZ_i\beta_i^S(\chi)C_i^S(\chi))$ and then,

$$\Sigma_S = \int_0^{\chi_D} \sum_i eZ_i(\beta_i^S(\chi)C_i^S(\chi) - \beta_i^f C_i^f)\,d\chi. \quad (13)$$

The difference between the ionic mobilities in the EDL and in the free electrolyte arises from the difference between the ionic strength in the EDL [which is given by: $I_S(\chi) \equiv (1/2)\sum_i Z_i^2 C_i^S(\chi)$] and in the free electrolyte (Eq. (8)) (because ionic mobilities are sensitive to the ionic strength, e.g., MacInnes 1951). In the rest of the paper, we will consider that the ionic mobilities in the EDL are constant (and should be evaluated at the mean ionic strength of the EDL), but at least, in first order is possible to equate ionic mobilities in the EDL and in the free electrolyte. The specific surface conductance can be also written by the sum of its ionic contributions: $\Sigma_S = e\sum_i (Z_i\Sigma_i^S)$, where the ionic contributions Σ_i^S are thus defined by,

$$\Sigma_i^S \equiv \int_0^{\chi_D} (\beta_i^S(\chi)C_i^S(\chi) - \beta_i^f C_i^f)\,d\chi. \quad (14)$$

Note that the ionic contributions Σ_i^S, such as defined, are dimensionally different from Σ_S. Using Eqs (4), (10), and (13), the specific surface conductance is given by,

$$\Sigma_S = 2\chi_d\left(\sum_i (eZ_i)\beta_i^S C_i^f \right.$$
$$\left. \times \exp\left((\pm 1)Z_i \frac{\tilde{\varphi}_d}{2}\right) - \sigma_f\right). \quad (15)$$

Following Pride (1994), Revil & Glover (1997) show that if the porous medium is submitted to an external electrical field, E, an electro-osmotic contribution must also be accounted in the total specific surface conductance. The electro-osmotic surface conductance is due to a convective electrical current in the EDL induced by a macroscopic electrical field (spontaneous electrical field generated by a concentration gradient, a fluid pressure gradient or a temperature gradient, or due to the applied electric field in the electrical conductivity problem). The electrical force acting on the excess of charged ions in the EDL is transmitted by friction to the fluid in the interconnected pore space. Despite this fact, Revil & Glover (1997) show that the problem, including electro-osmosis, is similar to the electrical problem considered previously if the classic ionic mobilities β_i, in the formulation

for Σ_S, are replaced by effective mobilities B_i defined by,

$$B_i \equiv \beta_i + \frac{2\varepsilon_f k_b T}{\eta_f eZ_i}, \quad (16)$$

where ε_f is the fluid dielectric permittivity (in $F\,m^{-1}$), η_f is the fluid dynamic viscosity ($kg\,m^{-1}\,s^{-1}$, $\eta_f \approx 10^{-3}$ Pa s for water at 25°C), T is the temperature in Kelvin, k_b is the Boltzmann's constant ($1.381 \cdot 10^{-23}\,J\,K^{-1}$), and Z_i the valence of ion (i). An evaluation of the magnitude of the two terms of Eq. (16) show that the electro-osmotic effect is appreciable (10–30% of the overall mobility), and therefore should not be neglected.

The electro-neutrality requirement between the surface charge density and the diffuse layer density is given, in absence of charge in the Stern layer (i.e., in absence of specific adsorption on the mineral surface), by,

$$Q_S^0 + Q_S = 0, \quad (17)$$

where Q_S^0 is given by Eq. (2), and Q_S by Eq. (11). Equation (17) relates surface and diffuse layer phenomena, whereas the link between diffuse layer and bulk electrolyte parameters is given by the Boltzmann distribution for the ionic concentrations. Equation (17) can be used together with Eqs (2) and (11) to determine the Stern plane potential as a function of pH and pore fluid concentrations. We give an example of such a derivation in the next sub-section.

Equations (11) and (15) indicate that the final result, in terms of diffuse layer charge and electrical conductivity perturbation, is equivalent to emptying a region of thickness $2\chi_d$ of ions possessing the same charge as surface sites. Consequently, the 'equivalent thickness' of the diffuse layer is equal to $2\chi_d$, and is a salinity-dependent parameter. The 'free porosity', ϕ_f, which is the fractional porosity associated with the 'free electrolyte' is therefore given by,

$$\phi_f = \phi(1 - \theta), \quad (18)$$

$$\theta \equiv \frac{2\chi_d S}{V_p} = 2\frac{S}{V_p}\sqrt{\frac{\varepsilon_f k_b T}{2e^2 I_f}}, \quad (19)$$

where V_p is the interconnected pore volume and S the interface area of the mineral–water interface. When the double layer reaches its minimum thickness, the counter-ions are concentrated in the Helmoltz layer. Consequently, the Helmoltz layer thickness constitutes the lower limit of the diffuse layer thickness. It can be computed from the hydration diameter of the counter-ions present at the surface: $\chi_H = 2r_i^{aq}$, where r_i^{aq} is the mean hydration radius of the counter-ions. To account for this effect, Clavier et al. (1984) introduced a 'double-layer expansion factor' α_d,

$$\alpha_d \equiv 2\chi_d/\chi_H \geq 1. \qquad (20)$$

The parameter α_d decreases when the ionic strength of the electrolyte increases. The minimum value of the double-layer expansion factor is that when the diffuse layer thickness is minimum and equals χ_H. Note, the factor 2 in the present equation compared with Eq. (9) of Clavier et al. (1984). Equation (20) leads us to introducing the notion of 'EDL high-salinity domain'. This domain is defined by an electrolyte concentration greater than a critical concentration defined by the equality between $2\chi_d$ and χ_H. Consequently, for a symmetric electrolyte, this critical concentration is given by $C_f^{crit} \equiv 2\varepsilon_f k_b T/(e^2 Z^2 \chi_H^2)$. Approximately, $C_f^{crit} \approx 0.729 \, \mathrm{mol \, l^{-1}}$ for a NaCl solution at 25°C, which corresponds to a conductivity of $6.3 \, \mathrm{S \, m^{-1}}$. Consequently, the term $2\chi_d$ in Eqs (11) and (15) can be replaced by $\alpha_d \chi_H$ to compute electrical properties of the EDL in a large salinity domain.

Application to a quartz-electrolyte system

Surface reactions. The determination of the Stern potential, φ_d, is the key that enables EDL properties to be calculated. The mineral surface potential distribution in the EDL, φ_d, is related to the surface charge density, Q_S, (Eq. (11)), as well as the specific surface conductance, Σ_S, (Eq. (15)). The Stern plane potential depends on the specific ionization reactions of the mineral surface under consideration. Consequently, to make further progress, we must specialize. Consider a quartz surface covered by silanol groups: >Si-OH, where > refers to the mineral lattice, the chemical reactions by which a surface site can become positively or negatively charged can be written by (e.g., Yates & Healy 1976),

$$> \mathrm{SiOH} \overset{K_{(-)}}{\rightleftharpoons} > \mathrm{SiO^-} + \mathrm{H^+}, \qquad (21)$$

$$> \mathrm{SiOH} + \mathrm{H^+} \overset{K_{(+)}}{\rightleftharpoons} > \mathrm{SiOH_2^+}, \qquad (22)$$

where $K_{(+)}$ and $K_{(-)}$ are the intrinsic equilibrium constants of these reactions. In Eqs (21) and (22), H$^+$, and consequently OH$^-$ due to the water dissociation, are the 'potential-determining ions' and the surface charge depends on

the proton concentration close to the surface (function of pH and Stern plane potential due to the Boltzmann equation). These chemical reactions lead to three species on the surface >SiOH, >SiOH$_2^+$ and >SiO$^-$. As previously defined (see Eq. (1)), Γ_S^0 is the total surface site density and, therefore, a conservation equation can be written by,

$$\Gamma_S^0 = \Gamma_{\mathrm{SiOH}}^0 + \Gamma_{\mathrm{SiO^-}}^0 + \Gamma_{\mathrm{SiOH_2^+}}^0, \qquad (23)$$

where Γ_i^0 are the concentration of surface groups (i) (in m^{-2}), see Eq. (1). The application of mass action law to Eqs (21) and (22) gives,

$$K_{(-)} = \frac{\Gamma_{\mathrm{SiO^-}}^0 C_{\mathrm{H^+}}^0}{\Gamma_{\mathrm{SiOH}}^0}, \qquad (24)$$

$$K_{(+)} = \frac{\Gamma_{\mathrm{SiOH_2^+}}^0}{\Gamma_{\mathrm{SiOH}}^0 C_{\mathrm{H^+}}^0}, \qquad (25)$$

where $C_{\mathrm{H^+}}^0$ is the concentration of hydrogen ions on the mineral surface (i.e., for $\chi = 0$). Using the Boltzmann distribution (Eq. (4)), $C_{\mathrm{H^+}}^0$ can be related to the concentration of H$^+$ ions in the free electrolyte, $C_{\mathrm{H^+}}^f$, and the Stern potential, φ_d, by

$$C_{\mathrm{H^+}}^0 = C_{\mathrm{H^+}}^f \exp(\tilde{\varphi}_d). \qquad (26)$$

The fractional availabilities of positive and negative surface site, defined by (see p. 254),

$$\Omega_{(+)}^0 = \Gamma_{\mathrm{SiOH_2^+}}^0/\Gamma_S^0, \qquad (27)$$

$$\Omega_{(-)}^0 = \Gamma_{\mathrm{SiO^-}}^0/\Gamma_S^0, \qquad (28)$$

are given from Eqs (23), (24), (25), (27), and (28) (Revil & Glover 1997) by,

$$\Omega_{(+)}^0 = \frac{K_{(+)} C_{\mathrm{H^+}}^0}{1 + K_{(+)} C_{\mathrm{H^+}}^0 + \dfrac{K_{(-)}}{C_{\mathrm{H^+}}^0}}, \qquad (29)$$

$$\Omega_{(-)}^0 = \frac{(K_{(-)}/C_{\mathrm{H^+}}^0)}{1 + K_{(+)} C_{\mathrm{H^+}}^0 + \dfrac{K_{(-)}}{C_{\mathrm{H^+}}^0}}. \qquad (30)$$

The fractional surface site density of neutral groups, Ω_{SiOH}^0, is given by: $\Omega_{\mathrm{SiOH}}^0 = 1 - \Omega_{(+)}^0 - \Omega_{(-)}^0$ (from Eq. 23). Equation (2) allows us to determine (with $Z_{(\pm)}^0 = 1$ in this case for >SiO$^-$ and >SiOH$_2^+$ groups, and 0 for >SiOH group) the electrical charge density of the mineral surface,

$$Q_S^0 = e\Gamma_S^0(\Omega_{(+)}^0 - \Omega_{(-)}^0)$$

$$= e\Gamma_S^0 \frac{K_{(+)}C_{H^+}^0 - \dfrac{K_{(-)}}{C_{H^+}^0}}{1 + K_{(+)}C_{H^+}^0 + \dfrac{K_{(-)}}{C_{H^+}^0}}. \qquad (31)$$

The condition called the point of zero charge (noted as PZC hereinafter) is defined as the thermodynamical condition at which the macroscopic effects of mineral charged surface groups cancel each other out. Consequently, the mineral surface charge is considered to be zero at the PZC, which implies no EDL and no surface electrical conduction. It follows from the previous equation that,

$$(C_{H^+}^f)_{pzc} = \sqrt{K_{(-)}/K_{(+)}}, \qquad (32)$$

As $pH = -\log[C_{H^+}^f]$, the pH corresponding to the PZC is,

$$pH(pzc) = -\frac{1}{2}\log\left(\frac{K_{(-)}}{K_{(+)}}\right). \qquad (33)$$

This last relationship gives the aqueous concentration of H^+ needed to produce a surface with an average neutral charge over is entire area. We have $\Gamma_{SiOH_2^+}^0 > \Gamma_{SiO^-}^0$ for $pH < pH(pzc)$ resulting in a positive surface charge, and $\Gamma_{SiOH_2^+}^0 < \Gamma_{SiO^-}^0$ for $pH > pH(pzc)$ resulting in a negative surface charge. Equation (33) is equivalent to the relationship derived by Glover *et al.* (1994). For quartz, taking the values of the equilibrium constants (e.g., Revil & Glover 1997), Eq. (33) gives $pH(pzc) = 3$.

Electrolyte reactions and diffuse layer densities. To control the pH we consider the following chemical reactions in the particular case of the presence of a sodium chloride salt: $NaCl \rightarrow Na^+ + Cl^-$, $HCl \rightarrow H^+ + Cl^-$ ($pH < 7$), $NaOH \rightarrow Na^+ + OH^-$ ($pH > 7$). The electrolyte concentration is noted C_f. We assume a complete dissociation for NaCl, HCl and NaOH and we assume the ideality of the solution (i.e., ionic concentration and ionic activity can be equated). The water dissociation reaction is: $H_2O \rightleftharpoons OH^- + H^+$, and the dissociation constant of water K_w is close to $10^{-13.8}$ at $25°C$ ($K_w = C_{H^+}^f C_{OH^-}^f$ if the water activity can be assumed to be 1). We also assume that no HCl is added when $pH > 7$ and that no NaOH is added when $pH < 7$ (of course, one could always

have both present and arrive at a given pH with an effect on the salt-ion concentrations, Na^+ and Cl^-). Consequently, for an acidic pH, the ionic concentrations in the free electrolyte are

$$C_{Na^+}^f = C_f, \qquad (34)$$

$$C_{Cl^-}^f = C_f + C_a, \qquad (35)$$

$$C_{H^+}^f = 10^3 N\, 10^{-pH} = C_a, \qquad (36)$$

$$C_{OH^-}^f = 10^3 N\, 10^{pH-pK_w}, \qquad (37)$$

where N is the Avogadro's number ($6.02 \cdot 10^{23}\ mol^{-1}$), $pK_w \equiv -\log_{10} K_w$, C_a (in m^{-3}) is the HCl concentration, C_f is the salt concentration. For a basic pH, ionic concentrations in the free electrolyte are given by

$$C_{Na^+}^f = C_b + C_f, \qquad (38)$$

$$C_{Cl^-}^f = C_f, \qquad (39)$$

$$C_{H^+}^f = 10^3 N\, 10^{-pH}, \qquad (40)$$

$$C_{OH^-}^f = 10^3 N\, 10^{pH-pK_w} = C_b, \qquad (41)$$

where C_b (in m^{-3}) is the NaOH concentration. The fractional occupancy of ions in EDL are defined by,

$$\Omega_i^S \equiv \frac{\displaystyle\int_0^{\chi_D} C_i^S(\chi)\,d\chi}{\displaystyle\sum_j \int_0^{\chi_D} C_j^S(\chi)\,d\chi}$$

$$= \frac{C_i^f e^{(\pm 1)Z_i(\bar\varphi_d/2)}}{\displaystyle\sum_j C_j^f e^{(\pm 1)Z_j(\bar\varphi_d/2)}}. \qquad (42)$$

This second equation is obtained using the Pride's approximation (Eq. (10)) and the Boltzmann's distribution for the EDL concentrations (Eq. (4)). Using the electroneutrality equation in the free electrolyte,

$$C_{Cl^-}^f + C_{OH^-}^f = C_{Na^+}^f + C_{H^+}^f, \qquad (43)$$

the fractional occupancy of ions in the EDL can be easily determined by,

$$\Omega_{H^+}^S = \frac{C_{H^+}^f \exp(\tilde\varphi_d/2)}{2(C_{Na^+}^f + C_{H^+}^f)\cosh\dfrac{\tilde\varphi_d}{2}}, \qquad (44)$$

$$\Omega_{Na^+}^S = \frac{C_{Na^+}^f \exp(\tilde\varphi_d/2)}{2(C_{Na^+}^f + C_{H^+}^f)\cosh\dfrac{\tilde\varphi_d}{2}}, \qquad (45)$$

$$\Omega_{OH^-}^S = \frac{C_{OH^-}^f \exp(-\tilde{\varphi}_d/2)}{2(C_{Na^+}^f + C_{H^+}^f)\cosh\dfrac{\tilde{\varphi}_d}{2}}, \qquad (46)$$

$$\Omega_{Cl^-}^S = \frac{C_{Cl^-}^f \exp(-\tilde{\varphi}_d/2)}{2(C_{Na^+}^f + C_{H^+}^f)\cosh\dfrac{\tilde{\varphi}_d}{2}}. \qquad (47)$$

Consequently, the fractional occupancies of ions in the EDL are a function of the ionic concentrations in the free electrolyte and the Stern potential.

Stern plane potential determination. The intensity and the sign of the surface charge are not only dependent on the potential determining ions (H^+ and OH^-), but also on the ionic strength of the free electrolyte. The process is not simple because surface reactions are stimulated not only by the pH in the free electrolyte, but also by the φ_d-potential in a feed-back process (because the sign of the φ_d-potential also controls the fractional ionic occupancies in the vicinity of the mineral surface). The surface charge density is obtained using Eqs (2), (29), and (30) to give the following equation,

$$Q_S^0 = -e\Gamma_S^0$$

$$\times \frac{1 - \dfrac{K_{(+)}}{K_{(-)}}10^{-2pH}e^{2\tilde{\varphi}_d}}{1 + \dfrac{K_{(+)}}{K_{(-)}}10^{-2pH}e^{2\tilde{\varphi}_d} + \dfrac{1}{K_{(-)}}10^{-pH}e^{\tilde{\varphi}_d}}.$$

$$(48)$$

The diffuse layer charge density Q_S (also in $C\,m^{-2}$) is obtained from Eq. (11) as,

$$Q_S = 2e\chi_d[-(C_{Cl^-}^f + C_{OH^-}^f)e^{-(\tilde{\varphi}_d/2)}$$
$$+ (C_{Na^+}^f + C_{H^+}^f)e^{(\tilde{\varphi}_d/2)}]. \qquad (49)$$

Using the electro-neutrality requirement of charge in the free electrolyte, Eq. (41), and the Debye screening length (see Eq. (7): $1/\chi_d^2 = 2e^2(C_{Na^+}^f + C_{H^+}^f)/(\varepsilon_f k_b T)$), gives Q_S as for $(0 < pH < 7)$,

$$Q_S = \sqrt{8\varepsilon_f k_b T N\,10^3(C_f + 10^{-pH})}$$

$$\times \sinh\left(\frac{\tilde{\varphi}_d}{2}\right), \qquad (50)$$

(where $C_f + 10^{-pH}$ must be replaced by $C_f + 10^{ph-pK_w}$ for pH greater than 7 to account for the modification of the ionic strength due to

OH^- ion concentration). Equation (50) is similar to the Gouy–Chapman equation between the φ_d-potential and the diffuse layer charge density Q_S (e.g., Fuerstenau 1983).

The Stern plane potential, φ_d, must satisfy the electro-neutrality requirement between the surface charge and the diffuse layer charge in equilibrium conditions. Consequently, as a direct consequence of Eqs. (17), (48), and (50), the Stern plane potential must satisfy the following equation (recall that $\tilde{\varphi}_d \equiv -e\varphi_d/k_b T$),

$$F[X] \equiv \frac{\eta}{2}\sqrt{C_f + 10^{-pH} + 10^{pH-pK_w}}\left(X - \frac{1}{X}\right)$$

$$\times \left(1 + \delta 10^{-2pH}X^4 + \frac{1}{K_{(-)}}10^{-pH}X^2\right)$$

$$+ \delta 10^{-2pH}X^4 - 1 = 0 \qquad (51)$$

$$\text{where} \begin{cases} \eta \equiv \sqrt{8\varepsilon_f k_b T N 10^3}/(e\Gamma_S^0) \\ \delta \equiv K_{(+)}/K_{(-)} \end{cases}$$

and $X \equiv \exp(\tilde{\varphi}_d/2)$. The solution of Eq. (51) (for the direct problem pH and C_f are the input parameters, and φ_d is the output parameter) gives the relationship between the NaCl concentration, the pH and the Stern plane potential. For a realistic range of X-values, the equation $F(X) = 0$ has only one root (Revil & Glover 1997), and consequently, the φ_d-potential can be found using Eq. (51).

Discussion. The fractional site occupancies of a quartz surface can be represented by parametric diagrams for each species. As seen in Fig. 2a, for increasing pH up to 5, an increasingly more significant part of the surface consists of negative >Si-O$^-$ sites. Positive sites are excluded from the surface in this pH-domain. The importance of the fraction of neutral sites should also be noted. In Fig. 2b, we give a comparison with the Brady's experimental data (1992), based on potentiometric titration of SiO$_2$ surfaces at 25°C. For pH = 2–4, the surface contains essentially neutral sites (Fig. 2a), and consequently the φ_d-potential and the surface charge density are very low for such pH (Figs 3 and 4). Between pH 4.5 and pH 3 the nature of the surface changes with >SiOH$_2^+$ site replacing the >Si-O$^-$ sites. At pH = pH(PZC), the surface is occupied by equal concentrations of negative and positive sites.

Revil & Glover (1997) show that the diffuse layer is dominated by sodium ions around pH 7 but chloride ions are not always negligible. At progressively lower pH values, the surface

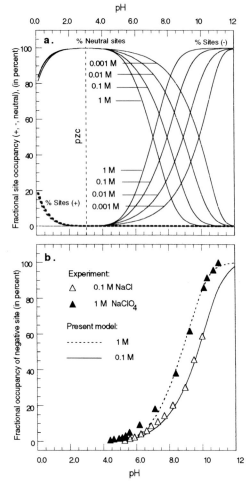

Fig. 2. (a) Fractional site occupancies (in percent) as
a function of pH and electrolyte concentration
(moles/litre), ($T = 25°C$, pH(pzc) = 3, $K_{(-)} = 10^{-6.3}$,
1.5 sites nm^{-2}), pzc: point of zero charge.
(b) Comparison with Brady's experimental results
(Brady 1992). Curves are the result of the present
model (surface site density: $1.00 \, \text{mol kg}^{-1}$,
$K_{(-)} = 10^{-7.5}$).

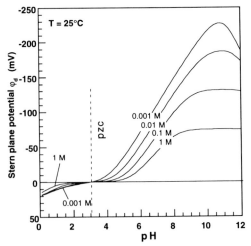

Fig. 3. Stern plane potential φ_d versus pH at $T = 25°C$
for different electrolyte concentrations (pzc: point of
zero charge).

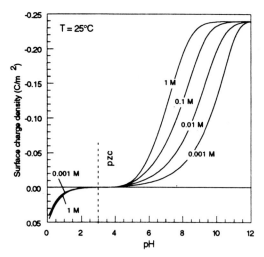

Fig. 4. Surface charge density Q_0^S (C m^{-2}) versus pH
for different electrolyte concentrations ($T = 25°C$) (pzc:
point of zero charge).

becomes positively charged and the EDL contains H$^+$, Cl$^-$, and Na$^+$ in comparable proportions. The little competition between Cl$^-$ and
OH$^-$ ions should be noted: the availability of
OH$^-$ in solution is small at those pH for which
positive charged surface sites exist in significant
quantity.

The variation of the Stern plane potential with
the pH and the electrolyte concentration is given
in Fig. 3. In practice, the Stern plane potential is
not measurable experimentally, however the so-
called 'shear plane electrical potential', ζ, gives a

precise measurement of the variation of φ_d with
ionic concentration, the pH and the temperature
(e.g., Ishido & Mizutani 1981). In the case of the
streaming potential (due to the motion of one
part of the diffuse layer induced by a pore fluid
differential pressure), the ζ-potential is the
electrical potential at the slipping plane or
shear plane (i.e., the potential within the
double layer at the zero velocity surface).
Although the ζ-potential is undoubtedly an
important parameter, there is a problem knowing the location of the shear plane with respect

to the surface in the 'thin and flat EDL assumption'. We define this distance as χ_ζ, and therefore from Eq. (6),

$$\zeta \equiv \varphi(\chi_\zeta) \approx \varphi_d \exp(-\chi_\zeta/\chi_d). \qquad (52)$$

In Fig. 5, the Stern plane potential and the ζ-potential are plotted against the electrolyte concentration, together with some experimental data for a quartz-KCl system (Scales *et al.* 1990). Because we have no information about the exact value of χ_ζ, we use χ_ζ as an adjustable parameter and choose its value so that the theoretical ζ–pH relationship fits the experimental data. The value of χ_ζ derived from this procedure is 2.4 Å.

In Fig. 6, the specific surface conductance is estimated for pH 7, 8, and 10, and for electrolyte concentration between 10^{-5}–$1 \, \text{mol} \, l^{-1}$. The specific surface conductance decreases at high salinity (0.5–$1 \, \text{mol} \, l^{-1}$) due to a decrease in the difference between the average electrical conductivity in the EDL and the electrolyte conductivity of the 'free electrolyte' (recall that Σ_S is the difference between the local conductivity and the electrolyte conductivity integrated over the EDL thickness, see Eq. (15)). At low salinity, the second term of Eq. (15) becomes negligible, and consequently the specific surface conductance depends indirectly on the mineral

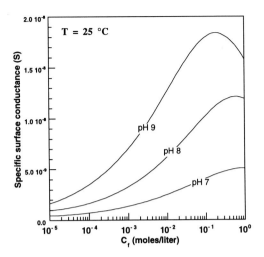

Fig. 6. Specific surface conductance (S) versus electrolyte concentration $(\text{mol} \, l^{-1})$ pH $= 7, 8$ and 9 and $T = 25°C$.

surface charge density, Q_S^0, which decreases when the electrolyte concentration decreases. Consequently, the specific surface conductance decreases with the electrolyte concentration at low salinity. The result of these remarks is that we can say, heuristically, that when the surface is not completely saturated with negative sites the specific surface conductance (and therefore the surface conduction) should be very low at very low electrolyte concentrations, as well as at high electrolyte concentrations, with a peak in surface conduction occurring 0.1–$1 \, \text{moles} \, l^{-1}$ depending on the surface site density (Fig. 6). Note that surface conduction in quartz is low at pH 7 due to a significant fraction of sites being neutral at this pH.

Electrical conductivity in porous media

Effective conductivity

The effects of surface conductivity, microstructure and diffusion processes on porous media were studied thoroughly in the DC-frequency limit by Johnson, Plona, and Kojima in 1986 (called JPK model hereinafter) and related works (Johnson & Sen 1988; Schwartz & Sen 1988; Schwartz *et al.* 1989; Johnson & Schwartz 1989; Sen 1989; Sen & Goode 1992). In the JPK model, the porous media is composed of an inter-connected pore volume saturated by a binary electrolyte, and an insulating phase called the matrix. The matrix is composed of

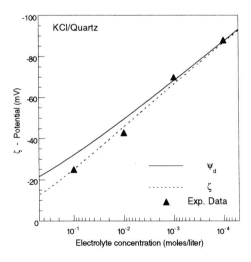

Fig. 5. Zeta potential (mV) versus electrolyte concentration $(\text{mol} \, l^{-1})$. Experimental data are from Scales *et al.* (1990) (fused silica, $T = 20 \pm 1°C$, pH $= 5.8 \pm 1$, KCl electrolyte). The heavy line and the dotted line are computed for the ψ_d-potential and the ζ-potential respectively (parameters used: pH $= 5.9$, 1.5 sites nm^{-2}, $K_{(-)} = 10^{-6.3}$, pH(pzc) $= 3$, $\chi_S = 2.4 \, \text{Å}$).

mineral grains and the non-connected porosity. The surface of grains which compose the matrix are typically charged and the counter-ions required by macroscopic electrical neutrality form within a thin diffuse layer around them (see above). Surface conduction within this electrical diffuse layer can contribute substantially to the effective electrical conductivity of the porous medium. In the JPK model, surface conduction is accounted through the specific surface conductance Σ_S introduced in the previous section, but the dependence of this parameter upon the salinity of the pore-fluid is not addressed.

One of the conclusions of the JPK model is that the electrical rock conductivity is described by four well-defined micro-geometrical parameters characteristic of the porous medium. The relevant quantities are (a) the pore space formation factor, F, which is related to the pore space tortuosity by,

$$\tau_V^2 = F\phi, \qquad (53)$$

(where ϕ is the fractional interconnected porosity), (b) the surface formation factor, f, which is related to the surface tortuosity by,

$$\tau_S^2 = f\phi(S/V_p), \qquad (54)$$

and (c) parameters Λ and λ, which measure, respectively, the relevant dimensions associated with transport in the interconnected pore volume and mineral surface (actually the EDL associated with the mineral surface charge). These parameters are defined by the following equations:

$$\frac{1}{F} = \frac{1}{V} \int |e_b|^2 \, dV_p, \qquad (55)$$

$$\frac{1}{f} = \frac{1}{V} \int |e_S|^2 \, dS, \qquad (56)$$

$$\frac{2}{\Lambda} = \frac{\int |e_b|^2 \, dS}{\int |e_b|^2 \, dV_p}, \qquad (57)$$

$$\frac{2}{\lambda} = \frac{\int |e_S|^2 \, dS}{\int |e_S|^2 \, dV_p}, \qquad (58)$$

where

$$e_b \equiv -\frac{L}{\Delta\Psi} \nabla\psi_b \quad \text{and} \quad e_S \equiv -\frac{L}{\Delta\Psi} \nabla\psi_S.$$

In these equations, V is the volume of the porous medium, $\nabla\psi/L$ is the nominal macroscopic electrical potential gradient through the porous medium, ψ_b is the local potential in the interconnected pore space due to the external electric field in the pore space in absence of surface conduction, and ψ_S is the local potential in the interconnected pore space when bulk conduction in the free electrolyte can be neglected by comparison with surface electrical conduction in the EDL. It should be emphasized that the local potential and ψ_S should not be confused with the local potential in the EDL, $\varphi(\chi)$, due to the pore-matrix charge, and introduced in the previous section. The differentials dS and dV_p denote integration over the interconnected pore surface and inter-connected pore volume respectively.

The presence of the power 2 in the tortuosity expression, Eq. (53), has often been a source of confusion (e.g., in Stoessel & Hanor 1975). It should be noted that this relation is a consequence of a capillary model, but can be generalized to a large range of porous media (Bergman *et al.* 1975). The inverse of the formation factor appears to be a measure of the effective interconnected porosity. The formation factor is a scale invariant material properties. The norm of the local electric field $|e_b|^2$ acts as a weighting function, giving less weight to poorly conducting pores than to highly conducting pores. For example $|e_b|^2$ vanishes in dead-ends, as a consequence of the continuity equation, thus deads-ends are not taken into account by F. The microstructural parameters Λ and λ are characteristic pore-size dimensions associated with pore and surface transport respectively (e.g., Schwartz *et al.* 1989). For an arbitrary porous medium, the inequality $\lambda \geq \Lambda$ is always satisfied (Revil & Glover 1997). The length scale Λ can be interpreted as an effective pore radius for transport in the interconnected pore space volume (Λ is not rigorously a geometrical parameter, however, it approximates the radius of the narrow throats that control the transport, e.g., Clennel, this volume). The influence of surface properties on bulk properties depends on the throats size which control transport properties relative to the EDL thickness. Consequently, the parameter θ_d defined by,

$$\theta_d \equiv \frac{2\chi_d}{\Lambda} = \frac{2}{\Lambda} \sqrt{\frac{\varepsilon_f k_b T}{2e^2 I_f}}, \qquad (59)$$

should be considered as a key parameter to evaluate, for a given porous material, if surface properties can be neglected ($\theta_d \approx 0$), or not

$(\theta_d > 0.1)$. It should be noted that if the effective pressure increases, and for reversible deformations, Λ is expected to decrease, so the parameter θ_d is expected to increase. Consequently, a porous rock can exhibit few surface effects at the atmospheric pressure, but can show strong surface effects in 'natural conditions' of pressure and temperature. In this order of idea, the 'anomalous' low electrical resistivity usually observed in the lower part of the crust could be due to a 'surface conductivity' phenomenon.

In absence of electrical conduction in the EDL, the effective electrical conductivity, σ_c, is given by,

$$\sigma_c = \frac{1}{F}\sigma_f \qquad (60)$$

where σ_f is the electrical conductivity of the electrolyte solution in the interconnected pore space. The total rate of energy dissipated by the Joule effect in the porous medium by cations and anions is written as $\dot{W}_i = \sigma_{c,i}V|\nabla\Psi|^2$ (where $\sigma_{c,i}$ is the effective conductivity contribution of the porous medium due to ionic species (i)). We introduce the fraction of electrical current transported by ionic species (i), respectively in the free electrolyte and in the diffuse layer, by $t_i^{f,hf}$ and $t_i^{S,hf}$. These parameters are called the Hittorf or transport numbers for the free electrolyte and the diffuse layer, and are rigorously defined in the next section. Consequently, by equating $\dot{W}_i$ to the sum of the local dissipations due to electrical conduction in the diffuse layer and in the free electrolyte by species (i), we obtain,

$$\sigma_{c,i} = \frac{1}{V|\nabla\Psi|^2}\left(\int |\nabla\psi|^2 t_i^{f,hf}\sigma_f \, dV_p \right.$$
$$\left. + \int |\nabla\psi|^2 t_i^{S,hf}\Sigma_S \, dS\right), \quad (61)$$

where Σ_S is the specific surface conductance defined in the first section, ψ is the local potential in the conducting phase due to the external electric field ($\psi \to \psi_b$ in absence of surface conduction, and $\psi \to \psi_S$ when bulk conduction in the free electrolyte can be neglected by comparison with surface electrical conduction in the EDL). It should be noted that Eq. (61) must be true for all salinities, except maybe for very low salinities when the diffuse layer thickness is as important as Λ, and is independent of any relationship between the specific surface conductance and the electrolyte concentration. However, Eq. (61) cannot be used directly because the local potential ψ is not known.

Ionic high and low salinity limits (noted I.H.S. and I.L.S. limits) are defined respectively for each ionic conductivity contribution by

$$\text{I.H.S.:} \quad t_i^{f,hf}\sigma_f > \frac{1}{\Lambda}t_i^{S,hf}\Sigma_S, \qquad (62)$$

$$\text{I.L.S.:} \quad t_i^{f,hf}\sigma_f < \frac{2}{\lambda}t_i^{S,hf}\Sigma_S. \qquad (63)$$

It should be noted that the transition domain between the high and low salinity domains is different for each ionic species contribution. For the effective conductivity, high and low salinity limits are defined by,

$$\text{H.S.:} \quad \sigma_f > \frac{2}{\Lambda}\Sigma_S, \qquad (64)$$

$$\text{L.S.:} \quad \sigma_f < \frac{2}{\lambda}\Sigma_S. \qquad (65)$$

Consequently H.S. and L.S. domains for the rock conductivity are not similar to high and low salinity domains for the different ionic contributions. Another point is that the flux of electrical current concentrates in regions of highest conductivity. Thus, at low salinity, the flux of electrical current concentrates more in the EDL near the mineral surface, and, as the salinity increases, it spreads increasingly into the pore space. Because the distribution of local conductivity changes with the concentration (as a result of the different dependence of the specific surface conductance and the electrolyte conductivity with the ionic concentrations), the local potential distribution ψ is also salinity dependent. The electrical potential distribution ψ_S is the local electrical potential distribution in the interconnected pore space (due to $\Delta\Psi$) when the surface conduction is dominant. The difference between the ψ_S and ψ_b-distributions arises from two effects. First, the roughness of the pore surface imposes that the current lines for bulk and surface conductions are not parallel, hence causing different behaviors between high and low salinity limits. However, even if the current flux lines are locally parallel everywhere, because of pore-scale heterogeneity, the distribution of current in the porous medium is different for bulk and surface conduction leading to the same effect (Bernabé & Revil 1995). Throughout this paper we consider that surface conduction takes place in the EDL, and because the diffuse layer thickness grows as the inverse square root of the electrolyte salinity, it appears to be a problem to define the conditions in which the ψ_S-distribution occurs. This is because the analysis in the first section is valid only for thin EDL. Consequently, we should have,

$$\psi(r) \xrightarrow[\text{H.S.}]{} \psi_b(r), \tag{66}$$

$$\psi(r) \xrightarrow[\substack{\text{L.S.}\\ \{\chi_D \ll \Lambda}]{} \psi_S(r), \tag{67}$$

where χ_D is the total thickness of the EDL (actually in the present problem, the size of the perturbed conductivity layer in the vicinity of the mineral surface, i.e. $2\chi_d$). We point out that the two conditions: (a) L.S. domain (Eq. 65), (b) $\chi_D \ll \Lambda$, allow us to determine a salinity domain where $\psi \to \psi_S$.

Equation (61) allows the derivation of two linear equations true in high and low salinity limits (H.S. and L.S. respectively) which represent physically justified equations deduced by a perturbation method (e.g., Schwartz *et al.* 1989, and references therein),

$$\sigma_{c,i} = \frac{1}{F}\left(t_i^{f,hf}\sigma_f + \frac{2}{\Lambda} t_i^{S,hf}\Sigma_S\right), \quad \text{H.S.} \tag{68}$$

$$\sigma_{c,i} = \frac{1}{f}\left(t_i^{S,hf}\Sigma_S + \frac{\lambda}{2} t_i^{f,hf}\sigma_f\right). \quad \text{L.S.} \tag{69}$$

Considering the complete range of salinities, the rock ionic conductivity contribution, for cations or anions, is some function, G, of the ionic contributions in the free electrolyte, $t_i^{f,hf}\sigma_f$, and in the diffuse layer, $t_i^{S,hf}\Sigma_S$. JPK connects H.S. and L.S. salinity limits with a smooth function (a Padé approximation, i.e., a ratio of two polynomials). Consequently, the ith ionic contribution to the rock conductivity is obtained by,

$$\sigma_{c,i} = G[t_i^{S,hf}\Sigma_S; t_i^{f,hf}\sigma_f]$$

$$= t_i^{f,hf}\sigma_f \tilde{G}\left[\frac{t_i^{S,hf}\Sigma_S}{t_i^{f,hf}\sigma_f}\right], \tag{70}$$

where

$$\tilde{G}[X] = \frac{b + cX + dX^2}{1 + aX}, \tag{71}$$

and

$$\begin{cases} a = \dfrac{2/(\Lambda F) - 1/f}{\lambda/2f - 1/F} \\ b = 1/F \end{cases} \quad \begin{cases} c = \dfrac{1 - \lambda/\Lambda}{f - \lambda F/2} \\ d = a/f. \end{cases}$$

In numerical simulations, the Padé functional gives a good approximation for the intermediate conductivities (Schwartz *et al.* 1989; Bernabé &

Revil 1995). The total rock conductivity, σ_c is the sum of all rock ionic conductivity contributions,

$$\sigma_c = \sum_i \sigma_{c,i}. \tag{72}$$

It should be emphasized again that all the previous equations are given under the 'thin and flat EDL assumption' described in the first section.

A high salinity (i.e., in the H.S. domain previously defined, e.g., $0.1\,\text{mol}\,\text{l}^{-1}$ NaCl in a clayey sandstone) a linear equation between the effective electrical conductivity and the electrolyte conductivity can be written from the previous analysis under the form,

$$\sigma_c = \frac{1}{F}\left(\sigma_f + \frac{1}{\Lambda}\Sigma_S\right). \tag{73}$$

As discussed by Sen (1989), this equation resembles the empirical relationship given by Waxman & Smits (1968) for clayey sands and sandstones,

$$\sigma_c = \frac{1}{F}\left(\sigma_f + \frac{1}{\Lambda} BQ_V\right). \tag{74}$$

In Eq. (74), B is the equivalent mobility of the counterions in the Helmoltz layer which is the lower limit of the EDL in the 'EDL high salinity domain' (see Eq. (20) and the discussion at the end of the first section), and Q_V, the 'excess charge' per unit pore volume is defined, from the 'cation exchange capacity' (CEC), by

$$Q_V \equiv \rho_m \frac{1 - \phi}{\phi} \text{CEC} \tag{75}$$

where ρ_m is the matrix density (in $\text{g}\,\text{m}^{-3}$). The CEC (in $\text{mol}\,\text{kg}^{-3}$ of matrix weight) indicates the surface charge of a given quantity of clayey rock. A comparison between Eqs (73) and (74), indicates an obvious similarity between $2\Sigma_S/\Lambda$ and BQ_V. We will use such a similarity further in the paper. Implicitly, the assumptions behind this similarity are the following: (a) the distinction between $2/\Lambda$ and V_p/S is ignored, (b) the size of the EDL (or the Helmoltz layer) can be neglected by comparison to Λ. Several mechanism are responsible for the clay surface charge (e.g., Fripiat *et al.* 1971; Clark-Monks & Ellis 1973), and include: (1) isomorphous substitution of high valence cation with one of lower valence (e.g., substitution of Al^{3+} for Si^{4+} in the tetrahedral layer); such a phenomenon is an important feature of 2:1 clay mineral structure, in particular, aluminium substitutes for silicon in the tetrahedral sheet (vermiculite) and magnesium substitutes for aluminium in the octahedral sheet (montmorillonite); (2) broken bonds along the edges of clay platelets, (3) amphoteric reactions (i.e., acid/base reactions like Eqs (21) and (22)) of exposed hydroxyl groups (silanol and aluminol groups) singly and doubly coordinated, (4) under certain conditions, structural cations other than H^+ could become exchangeable (e.g., at low pH value, Al^{3+} ions could move to exchange positions). This situation leads to an excess charge on clay surface. exchangeable ions attach to the clay surface, either by electrostatic or chemical interactions, and balance the surface charge. The CEC of the clay, usually expressed in milliequivalents per unit weight of clay (meq g^{-1}, i.e., 10^{-3} mol g^{-1}), is a measure of the concentration of the Exchangeable cations. In a 1:1 clay-structure, isomorphous cation substitutions are unfrequent, the CEC is quite low and almost entirely due to hydroxyl edge sites. In a 2:1 clay-structure, edge sites could represent an insignificant contribution to the total CEC (Lipsicas 1984). The specific surface area (in m^2 g^{-1}) and the total CEC appear to be dependent parameters. On a log–log plot, various experimental data corresponding to 1:1 and 2:1 clays falls on a single straight line indicating direct proportionality between these two parameters (Fig. 7). Consequently, the equivalent surface charge density for clay minerals is given by,

$$\Gamma_S^0 = \frac{1}{e}\frac{V_p}{S}\,Q_V \approx 1\text{--}3 \text{ sites nm}^{-2}. \qquad (76)$$

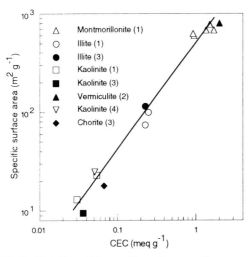

Fig. 7. Variation of the specific surface area (in m^2 g^{-1}) with the CEC (in meq g^{-1}) for various clay minerals. The line indicates a surface charge density of 1.5 charge nm^{-2}. Experimental data are from: (1) Patchett (1975), (2) Lipsicas (1984), (3) Zundel & Siffert (1985), (4) Lockhart (1980).

Consequently, it appears that clays with important structural differences show a relatively similar effective total surface charge density, even if, as said before, the structural and surface charge generation properties of clay minerals are quite different from one to another.

Hittorf transport numbers

The free electrolyte conductivity is given by $\sigma_f = \sum_i (eZ_i\beta_i^f C_i^f)$ where C_i^f is the ionic concentration of (i) in the free electrolyte, β_i^f the ionic mobility of (i) in the free electrolyte, and Z_i the valence of (i). Transference and Hittorf numbers of species (i) (respectively t_i^f and $t_i^{f,hf}$) in the bulk fluid are defined by (e.g., MacInnes 1961),

$$\begin{cases} t_i^f \equiv \dfrac{(\pm 1)\beta_i^f eC_i^f}{\sum_j \beta_j^f eZ_j C_j^f} = \dfrac{(\pm 1)\beta_i^f eC_i^f}{\sigma_f} & (77) \\[4mm] t_i^{f,hf} \equiv (\pm 1)Z_i t_i^f \quad \text{and} \quad \sum_i (t_i^{f,hf}) = 1 & (78) \end{cases}$$

where $(\pm 1)eZ_i$ is the charge of the ion of type (i). The parameter $t_i^{f,hf}$ represents the ratio between the electrical current due to species (i) divided by the total current in the free electrolyte (e.g., for a NaCl solution, $t_{Na}^{f,hf} \approx 0.38$ at 25°C, i.e., 38% of the electrical current is transported by sodium). Surface transference and Hittorf numbers of species (i) (respectively t_i^S and $t_i^{S,hf}$) can now be defined by analogy with t_i^f and $t_i^{f,hf}$,

$$\begin{cases} t_i^S \equiv \dfrac{(\pm 1)e\Sigma_i^S}{\sum_j eZ_j\Sigma_j^S} = \dfrac{(\pm 1)e\Sigma_i^S}{\Sigma_S} & (79) \\[4mm] t_i^{S,hf} \equiv (\pm 1)Z_i t_i^S \quad \text{and} \quad \sum_i (t_i^{S,hf}) = 1 & (80) \end{cases}$$

where $t_i^{S,hf}$ represents the difference of the relative fraction of the electrical current due to ionic species (i) between the EDL and the free electrolyte. We can establish a relationship between the surface Hittorf numbers and the Stern plane potential, φ_d, defined in the first section. Using Eqs (10), (14), (79), and (80), we have,

$$t_i^{S,hf} = \frac{C_i^f\left(\beta_i^S \exp\left((\pm 1)Z_i\dfrac{\tilde{\varphi}_d}{2}\right) - \beta_i^f\right)}{\sum_j C_j^f\left(\beta_j^S \exp\left((\pm 1)Z_i\dfrac{\tilde{\varphi}_d}{2}\right) - \beta_j^f\right)}. \qquad (81)$$

Macroscopic Hittorf numbers can also be defined for the whole porous medium. By analogy with the local Hittorf numbers, we have

$$T_i^{hf} \equiv \frac{\sigma_{c,i}}{\sum_i \sigma_{c,i}} = \frac{\sigma_{c,i}}{\sigma_c} \qquad (82)$$

(and consequently, $\sum_i (T_i^{hf}) = 1$). The parameters T_i^{hf} represent the electrical current due to ions of type (i) in the porous medium divided by the total electrical current of all species. The relations between the macroscopic Hittorf numbers and the microstructure are given via the Eqs (70) and (72). In Fig. 8, we give the variation of the effective Hittorf numbers with the salinity in the case where the interconnected pore volume is saturated by a binary symmetric electrolyte.

Relation between structural parameters

For an arbitrary porous medium, there is no relation between the electrical formation factor F and the porosity ϕ because F corresponds to a dynamical property whereas ϕ is just a geometrical one. However, Hashin & Shrikman (1962) have shown from a variational method that the following inequality (in the isotropic case) is always satisfied,

$$F \geq \frac{3 - \phi}{2\phi}. \qquad (83)$$

As an illustration, we test with success in Fig. 9 this equation on two experimental data sets obtained on limestone and basalt samples. However, natural porous media do not have arbitrary porous structure. For special types of micro-geometry, rigorous relationships between the electrical formation factor and the porosity may be found. For example, for granular materials with spherical grains, Sen *et al.* (1981) established that,

$$F = \phi^{-3/2}. \qquad (84)$$

Figure 10 illustrates such an expression for this particular class of porous media. Bernabé & Revil (1995) and Revil & Glover (1997) also give two inequalities based upon energetical considerations between the four microstructural parameters introduced by JPK for arbitrary porous media,

$$\frac{1}{F} \leq \frac{\lambda}{2f} \quad \text{and} \quad \frac{2}{F\Lambda} \geq \frac{1}{f} \qquad (85)$$

(and consequently, $\lambda \geq \Lambda$). In the particular case of granular material with spherical grains, Johnson & Sen (1988) derived the following relationship between Λ, the porosity, ϕ, and the mean grain diameter, d,

$$\Lambda = \frac{2\phi d}{9(1 - \phi)}. \qquad (86)$$

Theoretical and numerical simulations, and experimental evidence show that permeability,

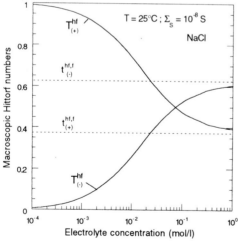

Fig. 8. Variation of the macroscopic or effective Hittorf numbers with the electrolyte concentration (mol l^{-1}). For the computation, we have taken a constant specific surface conductance and the following microstructural parameters values: $F = 10$; $f = 1\,\mu m$; $\Lambda = 0.1\,\mu m$; $\lambda = 1\,\mu m$.

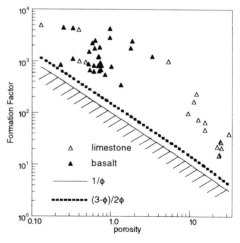

Fig. 9. Electrical formation factor versus porosity for two different rocks (limestone and basalt). The samples used here present a large variety of microstructure.

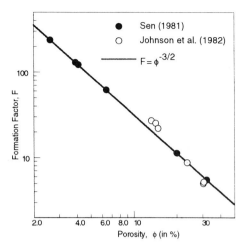

Fig. 10. Electrical formation factor versus porosity for granular porous media with spherical grains. The prediction of the model of Sen *et al.* (1981) is also given.

k, is closely related to the formation factor, F, and the characteristic length Λ by

$$k \approx a\Lambda^2/F, \qquad (87)$$

where a is a constant parameter (Johnson *et al.* 1986; Avellenada & Torquato 1991; Bernabé & Revil 1995). Figure 11 shows a comparison between measured and computed permeability estimated from electrical conductivity measurements on shaley sandstone. The relatively large dispersion suggests that the equations developed

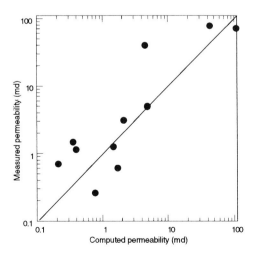

Fig. 11. Measured permeability versus computed permeability from electrical conductivity measurements (experimental data are from Vinegar & Waxman, 1984).

for the electrical conductivity are too simple to account completely for the highly complex structure of shaley rocks.

Membrane potential and ionic diffusion

Junction potential and ionic diffusion in an electrolyte

The self-diffusion coefficients, $\mathbf{D}^f_{(\pm)}$, of positively and negatively charged carriers describe the spreading of isotope-marked ions due to random collisions (i.e., internal forces), in the absence of concentration gradients. Ionic mobility, $\beta^f_{(\pm)}$, is the drift of ions due to external forces such as electrical potential gradients. Self-diffusion coefficient and mobility are related to each other by the classical Nernst-Einstein relationship $\mathbf{D}^f_{(\pm)} \equiv k_b T \beta^f_{(\pm)}/(Z_{(\pm)}e)$. Onsager & Fuoss (1932) have shown that in a concentration gradient the effective fluid diffusion coefficient is equal to the fluid self-diffusion coefficients (or so-called 'tracer diffusion coefficient') corrected by an additional term arising from cross-coupling of the ionic-fluxes. When a salt diffuses, the separate ions formed in solution cannot diffuse independently, although their mobilities may be different. When one ion type starts to move ahead to the other because of its higher mobility, the resultant separation of charge produces an electric field which retards the fast ion and accelerates the slow ion, so that both move at the same rate. This electric field appears as the liquid-junction potential when this potential gradient is measured between two different solutions.

For a binary electrolyte, the solution concentration is defined by $C_f = C^f_{(+)}/\nu_{(+)} = C^f_{(-)}/\nu_{(-)}$, where $\nu_{(\pm)}$ are the stoichiometric coefficients (i.e., the numbers of cations and anions produced by one molecule of electrolyte), and $C^f_{(\pm)}$ the ionic concentrations of cations and anions. The electro-neutrality equation, $Z_{(+)}C^f_{(+)} = Z_{(-)}C^f_{(-)}$, requires $\nu_{(+)}Z_{(+)} = \nu_{(-)}Z_{(-)}$. In an ideal electrolyte, in the presence of ionic concentration gradients, the ionic diffusion coefficients $D^f_{(\pm)}$ are different of the self-diffusion coefficient, $\mathbf{D}^f_{(\pm)}$ (e.g., McDuff & Ellis 1979),

$$D^f_{(\pm)} = \mathbf{D}^f_{(\pm)}\left(1 + (\pm 1)\frac{eZ_{(\pm)}}{k_b T}\frac{\mathrm{d}\psi_m}{\mathrm{d}\ln C_f}\right), \quad (88)$$

where ψ_m is the liquid junction potential in the electrolytic solution due to the concentration gradients. This potential is given by (e.g., MacInnes 1961, Eq. (4) Chap. 13, p. 222),

$$\nabla \psi_m = -\frac{k_b T}{e} \left(\frac{t_{(+)}^{f,hf}}{Z_{(+)}} - \frac{t_{(-)}^{f,hf}}{Z_{(-)}} \right) \nabla \ln C_f, \quad (89)$$

where $t_{(\pm)}^{f,hf}$ are the Hittorf numbers defined by Eqs (77) and (78) (note that $t_{(-)}^{f,hf} + t_{(+)}^{f,hf} \equiv 1$). The ionic diffusion coefficients of Eq. (88) are defined by the Fick's law:

$$j_{(\pm)}^d = -D_{(\pm)}^f \nabla C_{(\pm)}^f = \nu_{(\pm)} j_d,$$

$$j_d = -D_f \nabla C_f,$$

where $j_{(\pm)}^d$ is the diffusion flux of cations and anions, j_d is the salt diffusion flux, and D_f is the salt (or mutual) diffusion coefficient. It follows that $D_f = D_{(+)} = D_{(-)}$ in order to satisfy the disappearance of any electrical current in absence of an external electric field. Finally, the combination of Eqs (88) and (89) gives the fluid diffusion coefficient of the binary electrolyte,

$$D_f = D_{(\pm)}^f = \frac{k_b T}{e} \frac{\beta_{(+)}^f \beta_{(-)}^f}{\beta_{(+)}^f + \beta_{(-)}^f}$$

$$\times \left(\frac{Z_{(+)} + Z_{(-)}}{Z_{(+)} Z_{(-)}} \right), \quad (90)$$

This last equation is known as the Nernst–Hartley equation. The electrical potential gradient would increase the flux of ions carrying a charge of one polarity and, at the same time, decrease the flux of opposite charged ions in order to verify the electroneutrality requirement. Consequently, electrical ionic interactions significantly modify the fluxes of diffusing ions and produce a large difference between the self-diffusion coefficient and the effective diffusion coefficient in a concentration field. As shown hereinafter, this effect is enhanced in porous media with an electrical diffuse layer at the pore-matrix interface.

Membrane potential

When two electrolyte solutions of different concentrations, at the same temperature and pressure, are separated by a porous medium in which an EDL exists at the pore-matrix interface, a voltage drop $\Delta \psi_m$, known as 'membrane potential', develops across the porous medium (e.g., Sen 1989, and references therein). The membrane potential arises from two contributions that are called the 'exclusion' and the 'diffusion' potentials (e.g., Westermann-Clark & Christoforou 1986). The first contribution arises

because of the charge of the mineral surface, Q_S^0, described in the first section. As shown in this section, the mineral surface charge repels ions whose charge has the same sign as that of the dominant charged surface mineral group (and called the 'coins'), and attracts ions whose charge has the opposite sign (see Eq. (4)), which are called the 'counterions'. There is therefore an excess of counterions in the pore and an exclusion (a depletion) of coins. As seen in the first section, this phenomenon is restricted to the so-called EDL. In the presence of an electrolyte concentration gradient across the porous medium, the counterions would diffuse through the inter-connected pore space to the low concentration side of the porous medium, but since they are partially excluded from pores, the coins would build up at the high concentration entrance to the pores. The separation of charge would result in an electrical potential gradient that would retard counterions and help coins through the interconnected pore space. This potential gradient, called the 'exclusion potential' prevents the buildup of ions through the porous medium.

The second contribution to the potential membrane arises because the ions of an electrolyte have different mobilities. The mobility of the aqueous chloride ion, for example, is higher than that of the aqueous sodium ion (and it result that the Hittorf number of sodium ion is lower than the Hittorf number of chloride ion in an NaCl solution). When a porous medium is placed between electrolyte solutions with different concentrations, the ions of the electrolyte diffuses from the solution with the higher concentration to that with the lower concentration. Since the mobilities of the ions of an electrolyte differ in general, one type of ion diffuses more rapidly from the high concentration side, through the porous medium, to the low concentration side. Meanwhile, the less mobile ions are left behind on the high concentration side. Hence, a separation of charge across the porous medium would occur because of the difference in ionic mobilities. This type of separation of charge would cause an electrical potential gradient. However, the solutions on each side of the porous medium are not connected except through the porous medium, so this potential gradient develops in the absence of an electrical current in order to help the less mobile ions across the porous medium and to retard the more mobile ions. This 'diffusion potential gradient' would arise across the porous medium even in absence of charge on the mineral surface (e.g., in the PZC condition described in the first section), and is completely similar to the 'junction potential' defined in the previous sub-section (Eq. (89)). The 'diffusion or junction potential' can be generalized from Eq. (89) to a multi-ionic electrolyte by,

$$\nabla \psi_m^{dif} = -\frac{k_b T}{e}\left[\sum_i (\pm 1)\, \frac{t_i^{hf}}{Z_i}\, \nabla \ln C_i^f\right], \quad (91)$$

Sen (1989) gave a model for membrane potential in porous media saturated by a binary symmetric electrolyte. Sen's model (1989) can be easily generalized for a multi-ionic electrolyte, and consequently the 'membrane potential' can be written by,

$$\nabla \psi_m = -\frac{k_b T}{e}\left[\sum_i (\pm 1)\, \frac{T_i^{hf}}{Z_i}\, \nabla \ln C_i^f\right], \quad (92)$$

where T_i^{hf} are the macroscopic Hittorf numbers defined by Eq. (82). One of the differences between the present model and the Sen's model, is that the electro-osmotic contribution to ionic current density in the interconnected pore space is here *ipso facto* considered (through the link between the macroscopic Hittorf numbers, T_i^{hf}, the specific ionic conductances, Σ_i^S, given by Eq. (15), and the comments around Eq. (16)). In the special case where the surface mineral charge is zero (and consequently, $\varphi_d = 0$, and hence, $\Sigma_i^S = 0$), we have from Eqs (68), (72), and (82),

$$\lim_{Q_S^0 \to 0} T_i^{hf} = t_i^{f,hf}, \quad (93)$$

where $t_i^{f,hf}$ are the Hittorf numbers of ions in the free electrolyte defined by Eqs (77) and (78). Consequently, we can check that in the absence of charge at the mineral-water interface, the 'membrane potential' reduces to the 'diffusion or junction potential',

$$\lim_{Q_S^0 \to 0} \nabla \psi_m = \nabla \psi_m^{dif}. \quad (94)$$

Consequently, the 'exclusion potential' can be written from the previous statements by,

$$\nabla \psi_m^{exc} = -\frac{k_b T}{e}$$
$$\times \left[\sum_i (\pm 1)\, \frac{T_i^{hf} - t_i^{f,hf}}{Z_i}\, \nabla \ln C_i^f\right], \quad (95)$$

In Fig. 12, we give the values of the effective Hittorf number of cation deduced from membrane potential experimental data of Thomas (1976) in the case of clean and clayey sandstones saturated with a sodium chloride solution. These values are plotted as a function of the Q_V

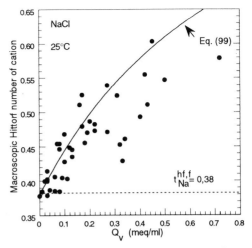

Fig. 12. Macroscopic Hittorf number of cation for clean and shaley sandstones deduced from Thomas' membrane potential measurements (the geometrical mean salt concentration is 0.35 mol l^{-1}). The solid line is computed from Eq. (99) with $B = 4$ (S m^{-1})/ (meq ml^{-1}), and $t_{Na}^{f,hf} = 0.38$ (see Thomas 1976).

parameter which is computed from CEC measurements by Eq. (75), e.g., Waxman & Smits (1968). The Q_V parameter can be considered, in first approximation, as a shaliness factor. From Eq. (82), and the similitude between $2\Sigma_S/\Lambda$ and BQ_V, the macroscopic Hittorf numbers can be put under an empirical form for clayey rocks,

$$T_{(\pm)}^{hf} = \sigma_{c,(\pm)}/\sigma_c \quad (96)$$

$$\sigma_{c,(-)} = \frac{1}{F}\left(t_{(-)}^{f,hf}\sigma_f\right) = \frac{1}{F}\left(1 - t_{(+)}^{f,hf}\right)\sigma_f \quad (97)$$

$$\sigma_{c,(+)} = \frac{1}{F}\left(t_{(+)}^{f,hf}\sigma_f + \frac{2}{\Lambda}\Sigma_S\right)$$

$$\approx \frac{1}{F}\left(t_{(+)}^{f,hf}\sigma_f + BQ_V\right) \quad (98)$$

and consequently,

$$T_{(+)}^{hf} = 1 - T_{(-)}^{hf} = \frac{t_{(+)}^{f,hf}\sigma_f + BQ_V}{\sigma_f + BQ_V}. \quad (99)$$

This last equation is similar to Eq. (5b) given by Clavier *et al.* (1984). A comparison between experimental data and Eq. (99) is given in Fig. 12. The more the porous media is filled with clay minerals, the larger $T_{(+)}^{hf}$ is (for a 'perfect shale' which can be considered as a perfect membrane for anions, we would have: $T_{(+)}^{hf} = 1$ and consequently, $T^{hf}_{(-)} = 0$). For a binary electrolyte, the membrane potential of a perfect shale and its associated exclusion potential are given by,

$$\nabla \psi_m = -\frac{k_b T}{e} \nabla \ln C_f \tag{100a}$$

$$\nabla \psi_m^{exc} = -\frac{2k_b T}{e} t^{f,hf}_{(-)} \nabla \ln C_f. \tag{100b}$$

Taking a ten-fold concentration gradient, $C^f_I = 0.1 \, \mathrm{mol \, l^{-1}}$, and $C^f_{II} = 1 \, \mathrm{mol \, l^{-1}}$, and a distance L between the two sides of $1 \, \mathrm{m}$, the electrical field for a perfect membrane is: $E_m = -\Delta \psi_m^{II-I}/L$ and is equal to $59 \, \mathrm{mV \, m^{-1}}$ at $25°C$, and $74 \, \mathrm{mV \, m^{-1}}$ at $100°C$.

Effective diffusion coefficient

Traditionally, the rock diffusion coefficient, D, is computed from the pore fluid diffusion coefficient, D_f, corrected for bulk tortuosity (e.g., Li & Gregory 1974). The bulk tortuosity can be derived from electrical conductivity measurements by Eq. (53), e.g., McDuff & Ellis (1979). The existence of such relationships result from the analogy between the ions transport under an electrical field and under a concentration gradient (such analogy is valid only in absence of surface charge and its associated EDL at the mineral–water interface). Consequently, in steady-state conditions and in absence of mineral surface electrostatic phenomena,

$$D = \frac{D_f}{F\phi}. \tag{101}$$

The effect of an EDL at the mineral water interface on ionic diffusion currents have been studied by Revil *et al.* (1996). They found that the effective diffusion coefficient of a binary non-symmetric electrolyte is given as the product on the effective diffusion coefficient without surface effect and a correction term due to 'surface' diffusion in the EDL,

$$D = \frac{D_f}{F\phi} \xi, \tag{102}$$

where the correction term ξ is a dimensionless parameter defined as,

$$\xi \equiv F \frac{\tilde{G}_{(+)} \tilde{G}_{(-)}}{t^{f,hf}_{(+)} \tilde{G}_{(+)} + t^{f,hf}_{(-)} \tilde{G}_{(-)}}. \tag{103}$$

The notation $\tilde{G}_{(\pm)}$ is used in the following sense $\tilde{G}_{(\pm)} \sim \tilde{G}[t^{S,hf}_{(\pm)} \Sigma_S / t^{f,hf}_{(\pm)} \sigma_f]$, where $\tilde{G}[X]$ is defined in the second section, and F is the pore volume

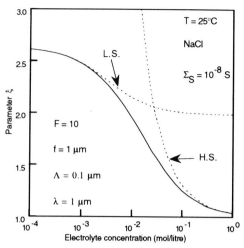

Fig. 13. Parameter ξ versus electrolyte concentration (in $\mathrm{mol \, l^{-1}}$). Low and high salinity asymptotes are represented as dotted lines, the Padé approximation for the electrical ionic contributions to the rock conductivity is drawn as a plain line.

formation factor. Consequently, the presence of the electrical diffuse layer can cause an amplification of the effective diffusion coefficient D depending on the electrolyte concentration in relation with the four fundamental microgeometrical parameters given in Section II. In the case of a negative surface charge, we give in Fig. 13 the variation of ξ versus the electrolyte concentration for a NaCl solution ($t^{f,hf}_{Na} = 0.38$). In this figure we assumed that the specific surface conductance is a constant ($\Sigma_S = 10^{-8} \, \mathrm{S}$). Note that this assumption is not needed in the model presented here and Σ_S could be salinity dependent. As shown in Fig. 13 and in Revil *et al.* (1996), ξ obeys the following inequality: $1 \le \xi \le 1/t^{f,hf}_{(+)}$. We will show in a future work that it is possible to include in the present derivation some specific adsorption phenomena, in the Stern layer, of the ions from the free electrolyte. Such adsorption can be account, for example for Na^+, Ca^{2+}, Cl^- on a quartz surface, by the following equations

$$> SiO^- + Na^+ \rightleftarrows \, > SiO^- Na^+ \tag{104}$$

$$> SiO^- + Ca^{2+} \rightleftarrows \, > SiO^- Ca^{2+} \tag{105}$$

$$> SiOH_2^+ + Cl^- \rightleftarrows \, > SiOH_2^+ Cl^-. \tag{105}$$

The concentration variation due to these reactions acts as a source term in the conservation equation for ionic diffusion flux.

Streaming potential

Electrokinetic phenomena are defined as arising from the relative motion between a charged surface and its associate double layer. A hydraulic flow in the interconnected pore space drags one part of the diffuse layer, and consequently is responsible of a convective electrical current density. To prevent any buildup of ions in the porous medium, this convective electrical current is counterbalanced by an electrical conductive current. It results in an electrical potential gradient called the 'streaming potential gradient' or 'streaming electrical field'. This potential gradient is given in the DC frequency limit and for laminar fluid flow by Pride (1994),

$$\nabla \psi_p = \frac{\varepsilon_f \zeta}{\eta_f \left(\sigma_f + \dfrac{2}{\Lambda} \Sigma_S \right)} \nabla p \qquad (107)$$

where η_f is the dynamic viscosity of the fluid saturating the pore space in Pa s, and the other parameters are as described in the first and second sections (ζ by Eq. (52), Λ by Eq. (57), Σ_S by Eqs (15) and (16)). Equation (103) gives a better understanding of the 'modified Helmoltz–Smoluchowski equation' of Rutgers (1940). Indeed, Rutgers measured the electrokinetic potential of capillaries of different diameters with very dilute solutions. He found an equation equivalent to Eq. (107) where $\Lambda = r$ (r being the capillary radius). A similar equation can also be found in Watillon & De Becker (1970), and in Ishido & Mizutani (1981). Using the similarity between $2\Sigma_S/\Lambda$ and BQ_V discussed on p. 264, Eq. (107) can be rewritten under an empirical form,

$$\nabla \psi_p = \frac{\varepsilon_f \zeta}{\eta_f (\sigma_f + BQ_V)} \qquad (108)$$

more suitable for practical applications. This equation can be used to compute fluid over-pressure in formations around a well from the spontaneous potential recorded on the wall of the well (if the other spontaneous potential source, like the membrane potential, can be considered as negligible or can be removed to the main signal). If we are concerned with situations in which the electrical surface conduction can be neglected by comparison with electric conduction in the free electrolyte, Eq. (103) reduced to the classical Helmoltz–Schmoluchowski equation (Overbeek 1952; Ishido & Mizutani 1981),

$$C_p = \frac{\mathrm{d}\psi_p}{\mathrm{d}p} = \frac{\varepsilon_f \zeta}{\eta_f \sigma_f} \qquad (109)$$

which is independent of any microstructural parameter. Electrokinetic experiments described in the literature are often oriented toward the determination of ζ using this last equation for various liquid–solid systems.

Thermoelectric potential

Thermochemical effect

When two electrolyte solutions at different temperatures, but same ionic concentrations are separated by a porous medium, a voltage drop, ψ_T, known as the thermoelectric potential, develops across the porous medium. Marshall & Madden (1959) presented some laboratory experimental results of the thermoelectric coupling coefficient (the ratio of the electrical potential drop to the temperature difference) with an average value of about $+0.3\,\mathrm{mV}\,^\circ\mathrm{C}^{-1}$. Corwin & Hoover (1979) obtained from previous studies, for a variety of rock types, a range of the thermoelectric coupling coefficient: -0.25 to $+1.5\,\mathrm{mV}\,^\circ\mathrm{C}^{-1}$. For some clayey sandstone and shale taken from a geothermal area, Fitterman & Corwin (1982) measured a coupling coefficient ranging from $+0.01$ to $+0.18\,\mathrm{mV}\,^\circ\mathrm{C}^{-1}$.

The model of Marshall & Madden (1959) can be easily generalized and integrated to the present work to compute, for a multi-ionic electrolyte, the thermoelectric potential as a function of the temperature gradient,

$$\nabla \psi_T = -\frac{1}{e}$$
$$\times \left[\sum_i (\pm 1) \frac{T_i^{hf}}{Z_i} \left(\frac{Q_i^f}{T} - S_i^f \right) \right] \nabla T, \quad (110)$$

where S_i^f are the partial entropy per ion, and Q_i^f are the ionic heat of transport per ion. As the effective Hittorf numbers are dependent on the microstructure and the mineral surface properties, the thermo-electric coupling coefficient, $C_T = \mathrm{d}\psi_T/\mathrm{d}T$, is therefore a function of these properties. The coupling factor C_T is also temperature dependent because: (a) effective Hittorf numbers are temperature dependent, (b) due to the term Q_i^f/T. The temperature dependence of the effective Hittorf numbers can be computed from Eqs (70), (71), and (82) if the temperature dependence of the electrolyte conductivity and specific surface conductance are known (the temperature dependence of the local Hittorf numbers can be neglected, e.g., Thomas 1976). The dependence of these parameters can be written by a linear form (e.g., Revil *et al.* 1996),

$$\sigma_f(I_f,T) = \sigma_f(I_f,T_0)[1 + \alpha_f(T - T_0)], \quad (111)$$

$$\Sigma_S(I_f,T) = \Sigma_S(I_f,T_0)[1 + \alpha_S(T - T_0)], \quad (112)$$

where $\alpha_f = 0.023 \pm 0.001°C^{-1}$ for a NaCl electrolyte, and $\alpha_S \approx 0.040 \pm 0.010°C^{-1}$. We consider now a binary symmetric 1:1 electrolyte (like NaCl). The thermoelectric coupling coefficient C_T is maximum for a perfect membrane (e.g., the perfect shale of Clavier *et al.* 1984: $T^{hf}_{(+)} = 1$ and consequently $T^{hf}_{(-)} = 0$), and for high temperature (such as $Q^f_{(+)}/T \ll S^f_{(+)}$). Consequently such potential is given by,

$$C_T = \frac{1}{e} S^f_{(+)}. \quad (113)$$

For a sodium chloride solution, this last equation gives: $C_T \approx +0.6\,mV\,°C^{-1}$. In absence of charge at the mineral-water interface, the coupling coefficient can be estimated from Eqs (93) and (110),

$$C_T = -\frac{1}{e}\left[\frac{t^{f,hf}_{(+)}}{Z_{(+)}}\left(\frac{Q^f_{(+)}}{T} - S^f_{(+)}\right)\right.$$
$$\left. -\frac{t^{f,hf}_{(-)}}{Z_{(-)}}\left(\frac{Q^f_{(-)}}{T} - Sf_{(-)}\right)\right]. \quad (114)$$

For a sodium chloride solution, this last equation gives approximately $C_T \approx -0.1\,mV\,°C^{-1}$. Consequently, the present model is compared favorably with the range of the experimental results described in the beginning of this sub-section.

Thermoelectrokinetic effect

Zablocki (1976) noted a spontaneous potential anomaly of +2.3 V over the Kilauea geothermal area in Hawaii. Anderson & Johnson (1976) reported another large value around +1.1 V for the Long Valley geothermal area (in California), and Fitterman & Corwin (1982) found a dipolar anomaly of ±80 mV in the Cerro Prieto geothermal field (Baja California, Mexico). It is quite clear that such anomalies cannot be explained by the previous process. Ishido & Mizutani (1981), and Morgan *et al.* (1989) suggest that the best candidate to explain them could be streaming potential related to fluid flow forming convection cells. The net fluid pressure

increase caused by heating a fixed-volume fluid filled pore can be related to the isobaric coefficient of thermal expansion, α_f, and the isothermal coefficient of compressibility, β_f, by,

$$dp = \frac{\alpha_f}{\beta_f} dT \quad (115)$$

and the ratio α_f/β_f ranges from about 1 to 18 bar °C^{-1} over the conditions encountered in hydrothermal systems (Norton 1984). In the interconnected pore volume, this fluid pressure is dissipated by fluid flow. The thermo-electrokinetic coupling coefficient can be estimated by,

$$C_T \equiv \frac{d\psi_T}{dT} = \frac{d\psi_T}{dp}\frac{dp}{dT} = C_p\frac{\alpha_f}{\beta_f} \quad (116)$$

where C_p is the streaming potential coupling coefficient given from Eq. (108) by,

$$C_p \equiv \frac{d\psi_p}{dp} = \frac{\varepsilon_f\zeta}{\eta_f\left(\sigma_f + \frac{2}{\Lambda}\Sigma_S\right)}. \quad (117)$$

The order of magnitude of such effect is $C_T \approx 5{-}100\,mV\,°C^{-1}$, and consequently, hydrothermal convection could be a very efficient mechanism to produce spontaneous electrical potential anomalies. Consequently, measurements of spontaneous potential anomalies may provide an efficient means of monitoring subsurface flow and give some insight into the rate and direction of fluid flow in an hydrothermal area.

This paper can be seen as a first step to produce a coherent model of the transport and coupling electrical properties in a porous medium saturated by a multi-ionic electrolyte. In such systems, we have shown that some macroscopic properties (electrical conductivity, effective diffusion coefficient and spontaneous potential coupling coefficients) are dependent on (1) four intrinsic micro-geometrical parameters, (2) mineral surface properties, and (3) properties of the electrolyte. The micro-geometrical parameters have been identified to the parameters introduced in Johnson *et al.* (1986) and related works, and a methodology is given to compute mineral surface properties, whereas electrolyte properties are well known. The influence of the effective pressure on the properties discussed in the present paper will be investigated in a future work. Such a link is important in many fields (including the oil industry) because it is certain that the ratio of surface properties/bulk properties should increase with depth as a result of the following: (a) a more important activation energy for surface processes than for bulk processes (e.g., Eq. (111) by comparison with Eq. (112)) and because the temperature increases with depth, (b) as a direct consequence of the decrease of the ratio between the pore volume and the mineral surface area (as a consequence of mechanical and diagenetic compactions).

References

ANDERSON, L. A. & JOHNSON, G. R. 1976. Application of the self-potential method to geothermal exploration in Long Valley, California. *Journal of Geophysical Research*, **81**, 1527–1532.

ANTRAYGUES, P. & AUBERT, M. 1993. Self potential generated by two-phase flow in a porous medium: experimental study and volcanological applications. *Journal of Geophysical Research*, **98**, 22 273–22 281.

AVELLENADA, M. & TORQUATO, S. 1991. Rigorous link between fluid permeability, electrical conductivity and relaxation times for transport in porous media. *Physics of Fluids. A. Fluid Dynamics*, **3**, 2529–2540.

BERNABÉ, Y. & REVIL, A. 1995. Pore-scale heterogeneity, energy dissipation and the transport properties of rocks. *Geophysical Research Letters*, **22**, 1529–1532.

BERGMAN, D. J., HALPERIN, B. I. & HOHENBERG, P. C. 1975. Hydrodynamic theory applied to fourth sound in a moving superfluid. *Physical Review*, **B11**, 4253–4263.

BOCKRIS, J. O'M. 1973. *Electrochemistry*. University Park Press, London.

BRADY, P. V. 1992. Silica surface chemistry at elevated temperatures. *Geochimica et Cosmochimica Acta*, **56**, 2941–2946.

CLARK-MONKS, C. & ELLIS, B. 1973. The characterization of anomalous adsorption sites on silica surfaces. *Journal of Colloid and Interface Science*, **44**, 37–49.

CLAVIER, C., COATES, G. & DUMANOIR, J. 1984. The theoretical and experimental bases for the dual-water mode for the interpretation of shaly sands. *Society of Petroleum Engineers Journal*, **24**, 153–169.

CORWIN, R. F. & HOOVER, D. B. 1979. The self-potential method in geothermal exploration. *Geophysics*, **44**, 226–245.

DOLL, H. G. 1949. The SP log: theoretical analysis and principles of interpretation. *Transactions AIME [American Institute of Mining, Metallurgical, and Petroleum Engineers]*, **179**, 146–185.

FITTERMAN, D. V. & CORWIN, R. F. 1982. Inversion of self-potential data from the Cerro-Prieto geothermal field, Mexico. *Geophysics*, **47**, 938–945.

FIXMAN, M. 1980. Charged macromolecules in external fields. I. The sphere. *Journal of Chemical Physics*, **72**, 5177–5186.

FRIPIAT, J., CHAUSSIDON, J. & JELLI, A. 1971. *Chimie-Physique des Phénomènes de Surface, Application aux Oxides et aux Silicates*. Masson et Cie.

FUERSTENAU, D. W. 1983. Adsorption and electrical double layer phenomena at mineral-water interfaces. *In*: JOHNSON, D. L. & SEN, P. N. (eds) *Physics and Chemistry of Porous Media*. AIP Conference Proceedings, New York, 209–223.

GLOVER, P. W J., MEREDITH, P. G., SAMMONDS, P. R. & MURRELL, S. A. F. 1994. Ionic Surface Electrical Conductivity on Sandstone. *Journal of Geophysical Research*, **99**, 21 635–21 650.

HASHIN, Z. & SHTRIKMAN, S. 1962. A variational approach to the theory of the effective magnetic permeability of multiphase materials. *Journal of Applied Physics*, **33**, 3125–3131.

HAYES, K. F., PAPELIS, C. & LECKIE, J. O. 1988. Modeling ionic strength effects on anion adsorption at hydrous oxide/solution interfaces. *Journal of Colloid and Interface Science*, **125**, 717–726.

ILLER, R. K. 1979. *The Chemistry of Silica: Solubility, Polymerization Colloid and Surface Properties, and Biochemistry*, Wiley and Sons, New York.

ISHIDO, T. & MIZUTANI, H. 1981. Experimental and theoretical basis of electrokinetic phenomena in rock-water systems and its applications to geophysics. *Journal of Geophysical Research*, **86**, 1763–1775.

JOHNSON, D. L. & SCHWARTZ, L. M. 1989. Unified theory of geometrical effects in transport properties of porous media. *In*: *Proceedings of the SPWLA 30th Annual Logging Symposium*, Paper E.

— & SEN, P. N. 1988. Dependence of the conductivity of a porous medium on electrolyte conductivity. *Physical Reviews*, **B37**, 3052–3510.

——, PLONA, T. J. & KOJIMA, H. 1986. Probing Porous Media with 1st Sound, 2nd Sound, 4th Sound and 3rd Sound, in Physics and chemistry of porous media-II. *In*: BANAVAR, J. R., KOPLIK, J. & WINKLER, K. W. (eds) *Proceedings of the Second International Symposium on the Physics and Chemistry of Porous Media*. AIP Conference Proceedings, **154**, 243–277.

——, SCALA, C., PASIERB, F. & KOJIMA, H. 1982. Tortuosity and acoustic slow waves. *Physics Reviews Letters*, **49**, 1840–1844.

LANDAU, L. D. & LIFSHITZ, E. M. 1980. *Statistical Physics, Part 1* (3rd Edition). Pergamon Press, Inc.

LI, Y.-H. & GREGORY, S. 1974. Diffusion of ions in sea water and deep-sea sediments. *Geochimica et Cosmochimica Acta*, **38**, 703–714.

LIPSICAS, M. 1984. Molecular and surface interactions in clay intercalates. *In*: JOHNSON, D. L. & SEN, P. N. (eds) *Physics and Chemistry of Porous Media*. American Institute of Physics, New York, 191–202.

LOCKHART, N. C. 1980. Electrical properties and the surface characteristics and structure of clays. II. Kaolinite. A nonswelling clays. *Journal of Colloid and Interface Science*, **74**, 520–529.

LYKLEMA, J. 1985. Electrochimie interfaciale des oxydes insolubles. *In*: *Solid–Liquid Interactions in Porous Media*. Editions Technip, 93–108.

McDuff, R. E. & Ellis, R. A. 1979. Determining diffusion coefficients in marine sediments: a laboratory study of the validity of resistivity techniques. *American Journal of Science*, **279**, 666–675.

MacInnes, D. A. 1961. *The Principles of Electrochemistry* (revised edition). Dover Publications, New York.

Marshall, D. J. & Madden, T. R. 1959. Induced polarization, a study of its causes. *Geophysics*, **24**, 790–816.

Mizutani, H. & Ishido, T. 1976. A new interpretation of magnetic field variation associated with the Matsushiro earthquakes. *Journal of Geomagnetism and Geoelectricity*, **28**, 179–188.

Morgan, F. D., Williams, E. R. & Madden, T. R. 1989. Streaming potential properties of Westerly granite with applications. *Journal of Geophysical Research*, **94**, 12 449–12 461.

Norton, D. L. 1984. Theory of hydrothermal systems. *Earth Planet*, **12**, 155–177.

Onsager, L. & Fuoss, R. M. 1932. Irreversible processes in electrolytes. Diffusion, conductance, and viscous flow in arbitrary mixtures of strong electrolyte. *Journal of Physical Chemistry*, **36**, 2689–2778.

Overbeek, J. Th. 1952. *Colloid Science*. Elsevier, New York.

Patchett, J. G. 1975. An investigation of shale conductivity. *Transactions of SPWLA 16th Annual Logging Symposium, New Orleans*, Paper V.

Pezard, P. A. 1990. Electrical properties of mid-ocean ridge basalt and implications for the structure of the oceanic crust in Hole 504B. *Journal of Geophysical Research*, **95**, 9237–9264.

Pride, S. 1994. Governing equations for the coupled electromagnetics and acoustics of porous media. *Physical Review*, **B50**, 15 678–15 696.

—— & Morgan, F. D. 1991. Electrokinetic dissipation induced by seismic waves. *Geophysics*, **56**, 914–925.

Revil, A. & Glover, P. W. J. 1997. Theory of ionic surface electrical conduction in porous media. *Physical Review*, in press.

——, Darot, M. & Pezard, P. A. 1996. Influence of the electrical diffuse layer and microgeometry on the ionic diffusion coefficient in porous media. *Geophysical Research Letters*, **22**, 1529–1532.

——, ——, —— & Becker, K. Electrical conduction in oceanic dikes, DSDP-ODP Hole 504B. *In*: Alt, J. C., Kinoshita, H. *et al. Proceedings of the Ocean Drilling Program Scientific Results*, **148**. ODP, College Station, TX, 297–305.

Rutgers, A. J. 1940. Streaming potentials and surface conduction. *Transactions of the Faraday Society*, **36**, 69–80.

Sen, P. N. 1981. Relation of certain geometrical features to the dielectric anomaly of rocks. *Geophysics*, **46**, 1714–1720.

——1989. Unified model of conductivity and membrane potential of porous media. *Physical Review*, **B39**, 9508–9517.

—— & Goode, P. A. 1992. Influence of temperature on electrical conductivity on shaly sands. *Geophysics*, **57**, 89–96.

——, Scala, C. & Cohen, M. H. 1981. Self-similar model for sedimentary rocks with application to the dielectric constant of fused glass beads. *Geophysics*, **46**, 781–795.

Scales, P. J., Grieser, F. & Healy, T. W. 1990. Electrokinetics of the muscovite mica-aqueous solution interface. *Langmuir*, **6**, 582–589.

Schwartz, L. M. & Sen, P. N. 1988. Electrolyte conduction in partially saturated shaly formations. *Annual Fall Technology Conference and Exhibition, Society of Petroleum Engineers, AIME*, SPE paper 18131.

——, —— & Johnson, D. L. 1989. Influence of rough surfaces on electrolytic conduction in porous media. *Physical Review*, **B40**, 2450–2457.

Stoessel, R. K. & Hanor, J. S. 1975. A nonsteady state method for determining diffusion coefficients in porous media. *Journal of Geophysical Research*, **80**, 4979–4982.

Stipp, S. L. & Hochella, M. F. Jr. 1991. Structure and bonding environments at the calcite surface as observed with X-ray photoelectron spectroscopy (XPS) and low energy electron diffraction (LEED). *Geochimica et Cosmochimica Acta*, **55**, 1723–1736.

Thomas, E. C. 1976. The determination of Q_V from membrane potential measurements on shaly sands. *Journal of Petroleum Technology*, **259**, 1087–1096.

Van cappellen, P., Charlet, L., Stumm, W. & Wersin, P. 1993. A surface complexation model of the carbonate mineral-aqueous solution interface. *Geochimica et Cosmochimica Acta*, **57**, 3505–3518.

Vinegar, H. J. & Waxman, M. H. 1984. Induced polarization of shaly sands. *Geophysics*, **49**, 1267–1287.

Watillon, A. & De Backer, R. 1970. Potentiel d'écoulement, courant d'écoulement et conductance de surface à l'interface eau–verre. *Journal of Electroanalytical Chemistry*, **25**, 181–196.

Waxman, M. H. & Smits, L. J. M. 1968. Electrical conductivities in oil-bearing shaly sands. *Society of Petroleum Engineers Journal*, **8**, 107–122.

Westerman-Clark, G. B. & Christoforou, C. C. 1986. The exclusion-diffusion potential in charged porous membranes. *Journal of Electroanalytical Chemistry*, **198**, 213–231.

Wyllie, M. R. J. 1949. A quantitative analysis of the electrochemical component of the SP curve. *Trans-actions AIME [American Institute of Mining, Metallurgical, and Petroleum Engineers]*, **186**, 17–26.

Yates, D . E. & Healy, T. W. 1976. The structure of the silica/electrolyte interface. *Journal of Colloid and Interface Science*, **55**, 9–19.

Zablocki, C. J. 1976. Mapping thermal anomalies on an active volcano by the self-potential method, Kilauea, Hawaii. *In: Proceedings of the 2nd U.N. Symposium on the Development and Use of Geothermal Resources*, **2**. US Government Printing Office, Washington DC, 1299–1309.

ZHANG, Z. Z., SPARKS, D. L. & SCRIVNER, N. C. 1994. Characterization and modeling of the Al-oxide/aqueous solution interface. I. Measurement of electrostatic potential at the origin of the diffuse layer using negative adsorption of Na^+ ions. *Journal of Colloid and Interface Science*, **162**, 244–251.

ZUNDEL, J. P. & SIFFERT, B. 1985. Mécanisme de rétention de l'octylbenzene sulfonate de sodium sur les minéraux argileux. *In: Solid–Liquid Interactions in Porous Media*. Editions Technip, 447–462.

Fractal geometry, porosity and complex resistivity: from rough pore interfaces to hand specimens

BRÍGIDA RAMATI P. DA ROCHA[1] & TAREK M. HABASHY[2]

[1] *Federal University of Pará, Belém, Pará, Brazil*
[2] *Schlumberger-Doll Research, Ridgefield, CT, USA*

Abstract: We propose a new model to interpret the electrical behavior of rocks containing metallic or clay particles. This new model encompasses some of the other commonly used models as special cases.

This model is a generalization of two models, one developed by Dias (Journal of Geophysical Research 77, 4945–4956, 1972) and another by Pelton *et al.* (*Geophysics* 43, 589–609, 1978). Its circuit analog includes an impedance $K(i\omega)^{-\eta}$ which simulates the effects of the fractal rough pore interfaces between the conductive grains (metallic or clay minerals which are blocking the pore paths) and the electrolyte. This generalized Warburg impedance is in series with the resistance of the blocking grains and both are shunted by the double layer capacitance. This combination is in series with the resistance of the electrolyte in the blocked pore passages. The unblocked pore paths are represented by a resistance which corresponds to the normal DC resistivity of the rock. The parallel combination of this resistance with the 'bulk' sample capacitance is finally connected in parallel to the rest of the above-mentioned circuit.

The model was tested over a wide range of frequencies against experimental data obtained for amplitude and phase of resistivity or conductivity as well as for the complex dielectric constant. The samples studied are those of sedimentary, metamorphic and igneous rocks.

Rocks are formed by a makeup of minerals, a porous network and a variable fluid quantity (air, water, oil or gas). Their properties are, therefore, basically determined by: (1) the mineralogical association constituting the rock (matrix); (2) the spatial distribution of the constituent minerals in the solid rock (texture); (3) the total pore space (porosity) and the geometric distribution of pores; (4) the type and concentration of fluids present in the pore space; (5) the presence of conductive minerals (clays and metallic minerals) contrasting with the matrix which is normally of high resistivity; (6) the distribution pattern of these minerals (veins, disseminated, massive); (7) the grain-size distribution of the conductive minerals.

Rock forming minerals are, in the majority, silicates having high resistivities (10^6 to 10^{14} Ωm). In crustal rocks the pore space is normally filled with electrolytic solutions of electrical resistivity much lower than silicates (10^2 to 10 Ωm). Electrical conduction in rocks is, therefore, mainly via the ionic solutions. Only when the rock matrix is formed by conductive minerals (e.g., metallic minerals or clays) will the conduction current through the matrix be appreciable. The parameters controlling the electrical current flux in the solutions include, among others, the density of ions, ionic valency, mobility and temperature.

Some minerals like metallic particles have a considerable concentration of free electrons by volume, and in the equilibrium state they are concentrated near the mineral grain surface. When the mineral grain is in contact with an ionic solution an electrically charged double layer is responsible for the low frequency rock polarization phenomena. Otherwise, some minerals, specially the clays, permit the substitution of metallic ions in their structure by others of lower valency (for example Al^{+++} by Mg^{++}), generating an electrically negative charged surface in the clay, which will also generate an electrically charged double layer when in contact with the ionic solution. For this phenomenon to be important one needs a large number of metallic or clay grains disseminated in the volume. The intensity of this polarization will depend on the content and distribution of conductive minerals. The more disseminated these minerals are, the greater is this polarization effect.

Warburg (1899) proposed an inverse square root frequency dependence for the polarization observed at the electrode-electrolyte interfaces. Recently, Pfeifer & Avnir (1983) demonstrated that this type of polarization obeys a more general frequency dependence due to the rough electrode surfaces. This general relation is of the form $K(i\omega)^{-\eta}$, where K is the diffusivity of the charged ions at the fractal porous surface and η is related to the fractal geometry of the medium. One of the main characteristics of this type of low frequency polarization is the constant phase angle (CPA) response.

From Lovell, M. A. & Harvey, P. K. (eds), 1997, *Developments in Petrophysics*, Geological Society Special Publication No. 122, pp. 277–286.

Katz & Thompson (1985) measured the fractal dimensions of the porous surfaces in sandstones and showed their self-similarity over three to four orders of magnitude. Liu (1985), using a self-similar Cantor bar to model a rough interface between an electrode and the electrolyte, obtained a response similar to the CPA response. This frequency response depends on the fractal dimension of the interface which can be measured under a microscope. Kaplan & Gray (1985) applied the Liu model for a random fractal surface and derived a relationship between the fractal parameters and the ac response of the interface. They were able to show that the exponent of the average admittance is directly related to the fractal dimension which is the same result obtained by Liu for the regular fractal. Liu et al. (1986) proposed that the CPA originates from the fractal geometry of the rough interface. They used the impedance of the model of an interface composed of a random combination of Cantor block type models to prove that the CPA behavior is caused by the fractal geometry of the rough pore interfaces. They also suggested an experiment that would enable the general verification of their speculation.

Wong et al. (1986), studying the fractal roughness of sandstones and shales, demonstrated that the clays are responsible for the CPA behavior in this type of rock. Kaplan et al. (1987), using a self-affine Cantor block model for the electrolyte surface at the metal-electrolyte interface were able to prove that the frequency exponent (η) of the impedance in the CPA element $Z(w) = (jw)^{-\eta}$ depends on the definition of the fractal dimension. Krohn (1988a, b) used SEM and optical measurements in thin sections to calculate the fractal nature of the surfaces in sandstones, shales and carbonates. Ruffet et al. (1991) were able to demonstrate a relationship between the complex conductivity and the fractal nature of porosity in sandstones and shales.

In this paper, we hypothesize that this fractal behavior observed in electrodes and simple type rocks could be extended to other more complex rocks. We introduce a new model for the complex conductivity of rocks which includes such a fractal behavior and we test the model against experimental data collected for sedimentary, metamorphic and igneous rocks.

Complex conductivity and dielectric constant models

On a macroscopic scale the electrical conductivity (σ) (or its inverse function, the electrical resistivity, ρ) and the dielectric constant (k) are the constitutive parameters related to the mean electrical behavior of a medium.

We define the electrical conductivity as the constitutive parameter that relates the electrical current density ($\vec{J}$), caused by free charges present in the medium moving under the influence of an applied electric field ($\vec{E}$), and the intensity of this field. The constitutive parameter that relates the electric flux density ($\vec{D}$) and the electric field ($\vec{E}$) gives the electrical permittivity of the material. Hence,

$$\vec{J} = \sigma\vec{E} \tag{1}$$
$$\vec{D} = \epsilon\vec{E}. \tag{2}$$

In rocks, conduction of electricity normally involves transport mechanisms (such as ionic diffusion and ionic replacement in the neighbouring solid–liquid interface) introducing a phase difference between the transport current density and the electric field. Conduction also undergoes total or partial ionic retention processes (total or partial ionic adsorption capable of producing electrical double layers with a fixed positive diffuse layer structure), resulting again in an additional phase difference between the current density and the applied electrical field (Marshall & Madden 1959). As a consequence, the electrical conductivity and dielectric constant are, in general, complex quantities. This is usually represented as

$$\sigma = \sigma' + i\sigma'' \tag{3}$$
$$\epsilon = \epsilon' - i\epsilon''. \tag{4}$$

which results, after grouping the real and imaginary parts, in the following form for the first Maxwell's Equation (Ampère's Law) in a medium without internal sources:

$$\nabla \times \vec{H} = [(\sigma' + \omega\epsilon'') + i(\sigma'' + \omega\epsilon')]\vec{E}. \tag{5}$$

The real part, given by $(\sigma' + \omega\epsilon'')$, is responsible for the in-phase component of the observed current, whereas the imaginary part, given by $(\sigma'' + \omega\epsilon')$, is responsible for the quadrature component (Fraser et al. 1964; Dias 1968, 1972; Fuller & Ward 1970).

Many models have been proposed to describe the electrical behaviour of rocks, each of which attempted to explain a particular feature in some restricted frequency interval. For demonstrative purposes, we show in Fig. 1 some typical electrical data by Mahan et al. (1986) and Garrouch & Sharma (1994). In Fig. 2 we show the Argand diagram for three widely used models: Debye, Cole–Cole and Davidson–Cole.

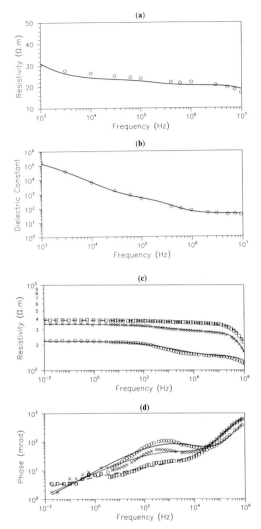

Fig. 1. Some typical complex resistivity and dielectric curves. (**a, b**) Experimental complex dielectric data from Garrouch & Sharma, 1994. (**c, d**) Experimental complex resistivity data from Mahan *et al.* 1986.

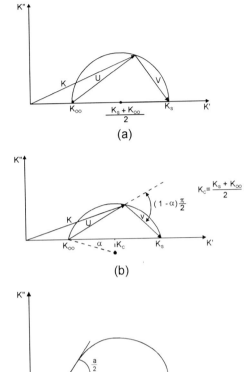

Fig. 2. Argand diagrams for: (**a**) Debye model, (**b**) Cole–Cole model, (**c**) Davidson–Cole model.

In Fig. 2a, we show the Debye model (1929) which was originally proposed to explain the behaviour of the dielectric constant of some polar liquids at very high frequencies. The specific features of this model are: it has a single relaxation time which is easily observed on the phase spectrum, a single dispersion in the dielectric constant, and a semicircular relationship between the real and imaginary parts of the dielectric constant on the Argand diagram. The Debye model is associated with the polarization of electrons on the atomic scale and its characteristic relaxation time is of the order of 10^{-15} s. In a macroscopic scale this model could

be represented by a parallel combination of a resistor and a capacitor, where the relaxation time is given by the time constant of the circuit. The dielectric constant as represented by the Debye model is given by

$$k = k_\infty + \frac{k_o - k_\infty}{1 + i\omega\tau} \qquad (6)$$

where k_∞ and k_o are the high and low frequency limits of the dielectric constant.

In Fig. 2b, we show the Argand diagram for the Cole–Cole model (1941), which was originally proposed as an empirical model to explain a molecular polarization effect in the dielectric data of some liquids. The characteristic relaxation time of the Cole–Cole model is in the order of 10^{-14} to 10^{-12} s. The dielectric constant as represented by this model is given by

$$k - k_\infty = \frac{k_o - k_\infty}{1 + (i\omega\tau)^{1-\alpha}} \qquad (7)$$

where $0 \leq \alpha \leq 1$ is a parameter characteristic of each particular material. Its value determines the angle formed between the radius of the semi-circle and the real axis in the Argand diagram at the high frequency limit.

Davidson & Cole (1951) extended this model to explain dielectric data observed in liquids such as glycerol, propylene–glycol and n-propanol. The dielectric constant as represented by this model is given by

$$k - k_\infty = \frac{k_o - k_\infty}{(1 + i\omega\tau)^a} \qquad (8)$$

where $0 \leq a \leq 1$. In this case, the value of α determines the angle at which the arc in the Argand diagram (shown in Fig. 2c) makes with the real axis at the intersection point at the high frequency limit.

Dias (1968, 1972) proposed a conductivity model which included a Warburg impedance element to explain the behaviour of the polarizable double layer formed between the rock electrolyte and the metallic mineral grains. This model is capable of explaining many polarization curves; however, when divided by $i\omega$ the corresponding dielectric constant goes to zero at high frequencies. Moreover, in the low frequency range this model is incapable of explaining the CPA behavior and the varying asymptotic behavior of the phase spectrum of the resistivity curves as a function of the rock type.

In 1977, Pelton proposed a model similar to Dias's model where he replaced the Warburg's impedance by a more general nonlinear diffusion-type element $(i\omega\tau)^{-c}$. Recently, the exponent c was interpreted as being the result of the fractal nature of the rough surfaces of the interfaces between an electrolyte and the metallic or clay particles. Besides, Pelton's model did not include the capacitance of the double layer and its associated resistance. The lack of these two elements makes Pelton's model incapable of explaining more complex electrical responses in rocks.

Pelton *et al.* (1978) used an expression similar to that developed by Cole & Cole (1941) to explain the resistivity data obtained both in the laboratory and in the field. This model was not physically justified and, furthermore, its behaviour is not sufficiently adequate to explain the complex resistivity data obtained in the laboratory.

A new model to explain the electrical behaviour of rocks

In this paper, we propose a new model to describe the electric behavior of rocks. This model is represented by the equivalent analogue circuit of Fig. 3 accounting for the various mechanisms involved in the conduction and polarization processes that are present in a rock.

The proposed model includes a fractal rough surface impedance element $K(i\omega)^{-\eta}$ analogous to the nonlinear element of the circuit proposed by Dias (1968, 1972). This new model also includes a capacitance associated with the bulk capacitance of the material. In the absence of this capacitive element, the dielectric constant will have a zero value at high frequencies. The pore charged double layer consists of a pure resistive term (resulting from loss of energy due to collision of the free carriers during their motion across the charged double layer) and a capacitive term (caused by oscillations of the bound charges in the double layer).

At high frequencies, another dispersion phenomenon caused by electronic, molecular and ionic polarization becomes important. This is represented by the bulk capacitance of the material which is directly dependent on the amount of mineral or fluid present in the rock that has the higher dielectric constant. Since the

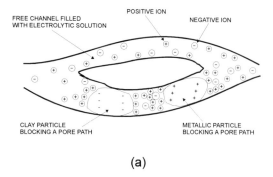

(a)

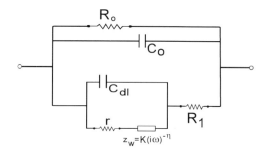

(b)

Fig. 3. (**a**) Basic cell of the electrical conduction in rocks. (**b**) Equivalent analogue circuit of the mean behaviour of the medium.

water is the rock's constituent with the higher dielectric constant for frequencies below 10^{14} Hz, it will normally be the most important element influencing this high frequency polarization. Consequently, the porosity of the rock will be the most important factor. This polarization will also be influenced by the ionic content of the water (Johnson & Sen 1988).

Across the whole frequency spectrum, a rock could be electrically characterized by two principal paths, one representing the free porous channel and the other representing the blocked porous channel. The free channel will represent the electrical paths in the rock which is free of clay or metallic particles, while the other will represent the paths with occurrence of these minerals.

Representing the time dependence of the electric field as e^{iwt}, the expression we propose for the complex electrical resistivity ρ is defined as:

$$Z = \frac{R_o}{1 + iw\tau_o}\left[1 - m\left(1 - \frac{1}{1 + \frac{1}{\delta_1 + \delta_2}(1 + u)}\right)\right]$$

$$m = \frac{R_o}{R_1 + R_o}, \quad \text{chargeability (Seigel 1959)}$$

$$\tau_1 = rC_{dl}, \quad \text{double layer relaxation time}$$

$$\tau_2 = R_o C_o, \quad \text{sample relaxation time}$$

$$\delta_r = \frac{r}{R_o}, \quad \text{grain percent resistivity}$$

$$\delta_1 = \frac{r}{R_1 + R_o} = m\delta_r$$

$$\delta_2 = \frac{K(iw)^{-\eta}}{R_1 + R_o} = \frac{m}{R_o}K(iw)^{-\eta}$$

$$u = iw\tau\left(1 + \frac{\delta_2}{\delta_1}\right)$$

with $Z = g\rho$, $R_o = g_0\rho_0$, $R_1 = g_1\rho_1$, $C_0 = g_{co}\epsilon$, $C_{dl} = g_{dl}\epsilon_{dl}$. The factors g, g_0, g_1, g_{co} and g_{dl} represent geometric factors given as a function of S_i/d_i, where S_i is the cross-section area and d_i is the corresponding electric length associated with the ith element in the current's path.

The physical meaning of the model parameters could be explained as follows.

(a) ρ_0: is the DC resistivity of the material which is closely related to the porosity of the rock. Archie's Law is one example of the relationship between porosity and the electrolyte resistivity of the material (Archie 1942).

(b) m is the chargeability parameter, first introduced by Seigel (1959), which relates the low and high frequency asymptotes of the rock resistivity:

$$m = \frac{\rho_0 - \rho_\infty}{\rho_0} = \frac{\rho_0}{\rho_0 + \rho_1}. \tag{9}$$

The relationship of m to the petrophysics of the rock is not very well established since it is strongly influenced by the rock's texture.

(c) $\tau_1 = rC_{dl}$ is the relaxation time constant related to the double layer oscillations (Rocha et al, 1991). It is related to the grain size and type of the blocking minerals (normally metallic minerals or clay particles).

(d) K is the diffusivity of the charged ions in the electrolyte. It depends mainly on the type and concentration of ions present in the electrolyte.

(e) η. This parameter is directly related to the fractal geometry of the medium and is determined by the type and distribution of the mineral causing the low frequency polarization.

(f) $\tau_2 = R_0 C_0$ is the time constant associated with the material as a whole. It depends on the rock fabric, the matrix properties and the total amount of water present in the rock.

(g) $\delta_r = r/R_0$ is a secondary parameter which relates the resistivity of the conductive grains with the DC resistivity value of the rock. Its value, for example, will be lower than unity for very good conductive grains and larger than unity for oxides.

It is instructive to note some particular values of the model parameters.

$m = 1$ this will occur when $R_1 \ll R_0$; this could occur in some situations like a rock filled with oil and water occupying the non-free paths or a rock presenting vein type mineralization.

$K = 0$: a very dried rock sample or a lunar rock.

$C_{dl} = 0, \tau = rC_{dl} = 0$ this could occur when $C_{dl} = 0$, that is, very low block grain surfaces, caused by non-blocking pores; or a $r = 0$, which is a very conductive grain, like a metallic particle. This is an important case. It is the Pelton's model, if we forget the very high frequency dispersion.

The new model response

The whole spectrum (10^{-6} to 10^{12} Hz) of the model can be observed in Fig. 4. Its most

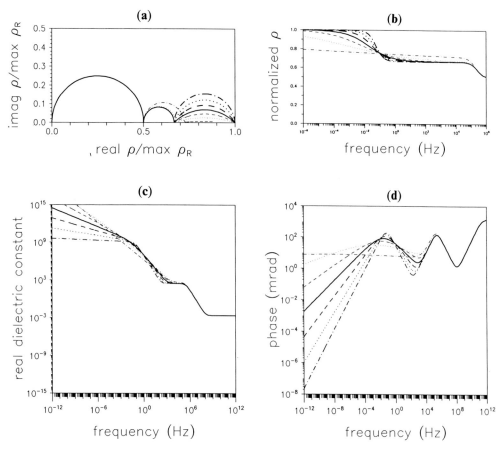

Fig. 4. The new model response, frequency interval of 10^{-6} to 10^{12} Hz, fixing $\rho_o = 100\,\Omega\text{m}$; $m = 0.5$; $\tau = 10^{-6}$ s; $\delta_r = 1.0$; $K = 100.0$; $\tau_0 = 10^{-12}$ s and curves n. 1, 2, 3 4, 5, 6, and 7, respectively to $\eta = 0.05$; 0.20; 0.35; 0.5; 0.65; 0.80 and 0.95 (**a**) Argand diagram for normalized resistivity; (**b**) normalized resistivity; (**c**) real dielectric constant (**d**) phase.

representative characteristics are the three clear relaxation processes shown in the Argand diagram: one for very high frequencies, which could be described as a Debye-type equation, related to the whole dielectric, dipolar polarization of the high resistivity rock matrix, and its total water polarization; another for the medium frequencies relaxation related to the interfacial capacitance of the conductive mineral grains; and the third for the low frequencies related to the fractal geometry of the rough grain surfaces.

Figure 4d shows the η effect for the phase curves when we can observe the CPA behavior. The lower the η value the lower the inclination of the Phase curves. A very interesting characteristic of the η effect can also be observed in the dielectric constant (Figure 4c). In this case the dielectric constant can also

be used to determine this parameter. It is important to note that even when the resistivity curve does not show any variation, the phase and dielectric constant curves permit the η determination.

Looking at the asymptotic behaviour of the model we observe that:

$$Z(w) = R_o \left[1 - \frac{R_o}{K}(iw)^\eta \right],$$

for low frequencies   (10)

$$Z(w) = R_o \left[1 - m \left\{ 1 - \frac{\delta_1}{iw\tau} \right\} \right],$$

for high frequencies   (11)

Since $w = 2\pi f$ and defining the $AMP(w)$ and $\phi(w)$ as the amplitude and phase of the impedance function, then:

In the low frequency limit:

$$\lim_{w \to 0} \frac{d\{\log \phi\}}{d\{\log f\}} = \eta \text{ and } \lim_{w \to 0} AMP(\omega) = R_o.$$

In the high frequency limit:

$$\lim_{w \to \infty} AMP(w) = R_o(1 - m)$$

and

$$\lim_{w \to \infty} \frac{d\{\log \phi\}}{d\{\log f\}} = 1$$

Results

It is possible to use all the models presented to fit the experimental data. Each of them will give us some parameters that need to be interpreted in a petrophysical way.

One can use Debye's model to explain the electrical behaviour of monocrystals to poly-crystalline rocks with very low porosities. Electronic, atomic and molecular polarizations are the mechanisms responsible for this relaxation.

The inclusion of conductive minerals will cause responses similar to the Cole–Cole and Davidson–Cole models. Here the Maxwell–Wagner polarization is one of the important mechanisms of polarization.

When water is introduced three situations can occur.

(1) A non-interactive situation, like quartz sand grains with pure water. This situation is found in one of the samples (27) of the experiments carried out by Mahan *et al.* (1986). In this case it is possible to interpret the porosity, the electrolyte conductivity and the response of the conductivity spectrum. Archie's law is valid in this situation. In the high frequency range the effect of water is equivalent to a high resistivity phase without any interaction with the solid phase. In this case the electromagnetic propagation tool (the EPT™ made by Schlumberger) can provide the water content from the dielectric constant. The parameters obtained for Mahan *et al.* data are a good example of the above interpretation.

(2) An interactive situation. In this case, Archie's law is not valid. The electrochemical polarization becomes important in this situation.

(3) A multiphase interactive situation: clay minerals + electrolyte + minerals.

However, all the above cases can be analysed using the new model which comprises all the other models as particular cases.

We used a random search method to obtain the best simultaneous amplitude and phase (or complex dielectric constant/resistivity or complex conductivity) fitting parameters using the model and a cost function given by the logarithm of the ratio between the calculated and the observed values.

Table 1. *Mineralogial description of the samples from Klein & Sill (1982), Scott et al. (1967) and Garrouch & Sharma (1994)*

Sample	Mineralogical description
A	Single pyrite electrode
B	Artificial glass beads matrix containing disseminated pyrite (10%)
C	Artificial glass beads matrix containing disseminated Ca-montimorilonitic (3%)
G1	Shaly sand saturated with 20 000 ppm NaCl brine
S1	Mancos' shale (carbonaceous sill) with pyrite, water content 3.8%

Table 2. *Mineralogical description of the samples from Olhoeft (1985)*

Sample	Description
D1	Barren rock saturated with $0.05\,\Omega$m KCl solution
D2	Finely layered silty sandstone containing carbonaceous matter and shale with its natural water
D3	Mineralized sandstone saturated with $0.001\,\text{mol}\,l^{-1}$ KCl solution
D4	Graphite in a three electrode corrosion cell with $0.001\,\text{mol}\,l^{-1}$ KCl solution
D5	Nicolite in a three electrode corrosion cell with $0.001\,\text{mol}\,l^{-1}$ KCl solution
D6	Massive amorphous graphite saturated with a $0.001\,\text{mol}\,l^{-1}$ KCl solution
D7	Disseminated graphite in schist with a $0.001\,\text{mol}\,l^{-1}$ KCl solution
D8	Standard montmorilonite bentonite (API-26) composed of 3 parts $0.001\,\text{mol}\,l^{-1}$ KCl solution to 2 parts dry clay by weight
D9	Uncontaminated montimorilonitic soil
D10	Organic-waste-contaminated montimorilonitic soil
D11	Uncontaminated smectite soil
D12	Organic-waste-contaminated smectite soil
D13	Shaly sandstone from a producing zone of an oily well measured in nitrogen atmosphere
D14	Shaly sandstone from a producing zone of an oil well measured in air atmosphere

Table 3. *Mineralogical description of the synthetic samples presented by Mahan* et al. *(1986)*

Sample	Cp (%)	r_m (μm)	r_{qz} (μm)	ϕ (%)	ρ_e (Ωm)
M1	6.5	150–125	≤53	34.30	15.0
M2	6.5	150–125	53	34.30	52.1
M3	6.5	150–125	≤53	34.30	105.0
M4	9.0	53–45	≤53	20.70	16.8
M5	9.0	53–45	≤53	20.70	113.6
M6	9.0	53–45	≤53	20.70	475.0
M7	6.0	32–27	32–27	40.49	108.0
M8	3.0	32–27	32–27	43.28	108.0
M9	0.6	32–27	32–27	40.84	108.0
M10	9.0	53–45	≤53	20.70	108.0
M11	0.6	53–45	≤53	41.24	108.0

The samples are made of a quartz matrix and disseminated chalcopyrite particles varying sulfide grain size (r_m), matrix grain size (r_{qz}), porosily (ϕ) and electrolyte resistivity (NaCl); Cp = Chalcopyrite.

The model was tested over more than 200 complex resistivity and dielectric data of many authors.

The results of this fitting to the data of Scott *et al.* (1967), Klein & Sill (1982), Olhoeft (1985), Mahan *et al.* (1986) and Garrouch & Sharma (1994) are presented in Tables 4, 5, 6 and 7 respectively. Tables 1, 2 and 3 present the mineralogical description of the samples.

Examining the parameters we obtained for the Klein & Sill data, we could observe that m is very close to unity for a single pyrite interface and that η had a greater value associated with single and large interfaces as opposed to the case of the commonly occurring grain interfaces. For disseminated pyrite, m is close to 0.5: both single electrode and disseminated pyrite present a very low δ_r as we expected. For the clays plus glass beads samples the lower η represents the more 'fractal' characteristic of the clay minerals. We also observe a very low m showing that $\rho_1 \gg \rho_0$, indicating that pores are blocked by the clays.

In Olhoeft's (1985) data one of the most interesting sets of parameters is the one we obtained for a single graphite sample. In this case, the value of δ_r was found to be very close to zero.

Conclusions

We have developed a new model capable of explaining the complex resistivity/dielectric data over a wide frequency range with physically acceptable values for the model parameters. Our investigation introduces a new method for obtaining valuable information about the electrical behaviour of rocks.

The introduction of the fractal roughness factor permits the investigation of the texture

Table 4. *Model parameters for data of Klein & Sill (1982)*

Sample	ρ_0 (Ωm)	m	τ_1 (μs)	δ_r	K ($s^{-\eta}$)	η	τ_2 (s)
A	4.84	0.99998	4.499	0.001	0.7792	0.701	1.0×10^{-16}
B	53.69	0.437	1.874	0.001	2478	0.692	1.0×10^{-13}
C	10.55	0.095	99.99	2.481	256.7	0.466	1.0×10^{-10}

Table 5. *Model parameters for data of Olhoeft (1985)*

Sample	ρ_0 (Ωm)	m	τ_1 (μs)	δ_r	K ($s^{-\eta}$)	η	τ_2 (s)
D1	1.41	0.550	0.11×10^{-9}	8.4	52.64	0.05	0.171×10^{-12}
D2	49.99	0.512	0.34×10^{-6}	0.159	17.07	0.30	0.101×10^{-8}
D3	14.99	0.313	0.43×10^{-6}	0.81	30.60	0.37	1.08×10^{-12}
D4	49.78	0.957	0.49×10^{-3}	8.4	52.61	0.05	0.171×10^{-12}
D5	27.07	0.9899	0.20×10^{-4}	0.063	16.10	0.68	0.165×10^{-12}
D6	1500	0.9899	0.45×10^{-5}	0.041	94.72	0.31	0.113×10^{-14}
D7	16.15	0.9999	0.67×10^{-11}	0.5×10^{-10}	4517	0.11	0.141×10^{-14}
D8	2.991	0.685	0.34×10^{-7}	0.102	1.17	0.05	0.50×10^{-13}
D9	2.991	0.685	0.11×10^{-9}	8.4	52.61	0.05	0.17×10^{-12}
D10	4.999	0.76	0.16×10^{-6}	0.201	0.348	0.21	0.21×10^{-9}
D11	7.887	0.61	0.14×10^{-6}	1.757	240.7	0.38	0.11×10^{-11}
D12	5.687	0.9899	0.11×10^{-9}	0.003	275.2	0.12	0.654×10^{-14}
D13	42.56	0.12	0.70×10^{-8}	1.305	1760	0.29	0.10×10^{-16}
D14	5.687	0.9899	0.10×10^{-9}	0.003	275.2	0.12	0.65×10^{-14}

Table 6. *Model parameters for data of Mahan* et al. *(1986)*

Sample	ρ_0 (Ωm)	m	τ_1 (μs)	δ_r	K ($10^4\,s^{-\eta}$)	η	τ_2 (s)
M1	150.0	0.531	67.8	1.17	13.0	0.638	1×10^{-10}
M2	98.46	0.559	0.433	1.941	1.67	0.584	0.25×10^{-15}
M3	335.62	0.778	1.01	2.193	2.08	0.565	0.55×10^{-11}
M4	70.01	0.610	0.148	0.613	5.21	0.640	0.46×10^{-15}
M5	172.7	0.819	0.210	1.141	2.61	0.505	0.12×10^{-11}
M6	689.94	0.896	0.903	1.365	7.73	0.561	0.23×10^{-10}
M7	220.0	0.983	0.230	1.818	3.59	0.468	0.24×10^{-11}
M8	349.84	0.835	0.820	3.082	4.72	0.391	0.16×10^{-12}
M9	394.25	0.849	0.36	1.654	2.47	0.185	0.14×10^{-12}
M10	168.99	0.766	0.229	1.135	2.81	0.517	0.13×10^{-11}
M11	370.65	0.792	1.56	5.932	3.58	0.309	0.54×10^{-11}

Table 7. *Model parameters for data of Scott* et al *(1967) and Garrouch & Sharma (1994)*

Sample	ρ_0 (Ωm)	m	τ_1 (ns)	δ_r	K ($s^{-\eta}$)	η	τ_2 (ns)
G1	35.2	0.662	32.04	1.757	176.4	0.257	0.0015
S1	0.006	0.422	992.7	15.44	325.7	0.416	0.0073

of rocks which is a very important factor in explaining their electrical properties.

We showed that the amplitude of resistivity alone is not capable of showing the effect of the rough pore interfaces. On the other hand, the phase curves provide a reliable estimate of the fractal roughness factor, η. For sedimentary rocks this is a very important result (Snyder et al. 1977, 1981).

We have also shown that the chargeability factor m, defined as $R_o/(R_0 + R_1)$, does not seem to have any special physical meaning, despite of its broad use. A better parameter would be the resistivity ratio R_0/R_1 which appears in the high frequency asymptote and the resistivity ratio R_0/r which is needed to express the low frequency asymptote.

The authors thank Schlumberger-Doll Research and Federal University of Para for their support.

References

ARCHIE, G. E. 1942. The electrical resistivity log as an aid in determining some reservoir characteristics. *AIME Petroleum Technology*, 1–8.

COLE, K. S. & COLE, R. H. 1941. Dispersion and Absorption in Dielectrics, *Journal of Chemical Physics*, **9**, 341–351.

DAVIDSON, D. W. & COLE, R. H. 1951. Dielectric Relaxation in Glycerol, Propylene Glycol and n-Propanol, *Journal of Chemical Physics*, **19**, 1484–1490.

DEBYE, P. 1929. *Polar Molecules*. The Chemical Catalog Co. Inc. Also (1945) Dover Publications.

DIAS, C. A. 1968. *A Non-Grounded Method for Measuring Electrical Induced Polarization and Conductivity*, PhD Dissertation, Univ. California, Berkeley/USA.

——1972. Analytical Model for a Polarizable Medium at Radio and Lower Frequencies, *Journal of Geophysical Research*, **77**, 4945–4956.

FRASER, D. C., KEEVIL, JR., N. B. & WARD, S. H. 1964, Conductivity Spectra of Rocks from the Craigmont Ore Environment. *Geophysics*, **29**, 832–847.

FULLER, B. D. & WARD, S. H. 1970. Linear System Description of the Electrical Parameters of Rocks, *IEEE Transaction Geosciences and Electronics*, **GE 8**, 7–18.

GARROUCH, A. A. & SHARMA, M. M. 1994. The influence of clay content, salinity, stress, and wettability on the dielectric properties of brine-saturated rocks: 10 Hz to 10 MHz, *Geophysics*, **59**, 909–917.

JOHNSON, D. L. & SEN, P. N. 1988. Dependence of the conductivity of a porous medium on electrolyte conductivity. *Physical Review B*, **37**, 3502–3510.

KAPLAN, T. & GRAY, L. J. 1985. Effect of disorder on a fractal model for the ac response of a rough interface. *Physical Review Bulletin of the American Physical Society*, **32**, 7360–7366.

——, —— & LIU, S. H. 1987. Self-affine fractal model for a metal-electrolyte interface. *Physical Review Bulletin of the American Physical Society*, **35**, 5379–5381.

KATZ, A. J. & THOMPSON, A. H. 1985. Fractal sandstone pores: Implications for conductivity and pore formation. *Physical Review Letters*, **54**, 1325–1328.

KLEIN, J. D. & SILL, W. R. 1982. Electrical Properties of Artificial Clay-bearing Sandstones. *Geophysics*, **47**, 1593–1605.

KROHN, C. E. 1988*a*. Sandstones Fractal and Euclidean Pore Volume Distributions. *Journal of Geophysical Research*, **93(B4)**, 3286–3296.

——1988*b*. Fractal Measurements of Sandstones, Shales and Carbonates. *Journal of Geophysical Research*, **93(B4)**, 3297–3305.

LIU, S. H. 1985. Fractal model for the ac response of a rough interface. *Physics Review Letters*, **55**, 529–532.

——, KAPLAN, T. & GRAY, L. J. 1986. Theory of the ac response of rough interfaces. *In*: PIETRONERO, L. & TOSATTI, E. (eds) *Fractals in Physics*. Elsevier, 383–389.

MARSHALL, D. J. & MADDEN, T. R. 1959. Induced polarization : a study of its causes. *Geophysics*, **24**, 790–816.

MAHAN, M. K., REDMAN, J. D. & STRANGWAY, D. W. 1986. Complex resistivity of synthetic sulphide bearing rocks, *Geophysical Prospecting*, **34**, 743–768.

OLHOEFT, G. R. 1985. Low Frequency Electrical Properties. *Geophysics*, **50**, 2492–2503.

PELTON, W. H. 1977. *Interpretation of induced polarization and resistivity data*. PhD Thesis, Univ. of Utah, Salt Lake City.

——, WARD, S. H., HALLOF, P. G., SILL, W. R. & NELSON, P. H. 1978. Mineral Discrimination and Removal of Inductive Coupling with Multifrequency IP. *Geophysics*, **43**, 588–609.

PFEIFER, P. & AVNIR, D. 1983. Chemistry in non-integer dimensions between two and three. I. Fractal theory of heterogeneous surfaces. *Journal of Chemical Physics*, **79**, 3558–3565.

ROCHA, B. R. P., LEÃO, J. W. D. & DIAS, C., A. 1991. Ajuste automático de condutividade complexa para descrever o comportamento elétrico de rochas a baixas frequências, *In*: *Anais do II Congresso Internacional de Geofísica*, setembro de 1991, 260–265.

RUFFET, C., GUENGUEN, Y. & DAROT, M. 1991. Complex conductivity measurements and fractal nature of porosity. *Geophysics*, **56**, 758–768.

SEIGEL, H. O. 1959. A theory for induced polarization effects (for step Excitation Function). *In*: WAIT, J. R. (Ed.) *Overvoltage Research and Geophysical Applications*. Pergamon Press Inc., 4–21.

SCOTT, J. H., CARROLL, R. D. & CUNNINGHAM, D. R. 1967. Dielectric constant and electrical conductivity measurements of moist rock: A new laboratory method. *Journal of Geophysical Research*, **72**, 5101–5115.

SNYDER, D. D., MERKEL, R. H. & WILLIAMS, J. T. 1977. Complex formation resistivity – the forgotten half of the resistivity log. *SPWLA, 18th Ann. Logg. Symp.*, June 5–8 1977.

——, KOLVOORD, R. W., FRANGOS, W., BAJWA, Y., FLEMING, D. B. & TASOI, M. T. 1981. Exploration for petroleum using complex resistivity Measurements. *In*: *Advances in Induced Polarization and Complex Resistivity*, Univ. Arizona, Jan. 5–7 1981.

WARBURG, E. 1899. On the behavior of so-called unpolarizable electrodes with respect to alternating current. *Annalen der Physik und Chemie*, **3**, 493–499.

WONG, P., HOWARD, J. & LIN, J. 1986. Surface Roughening and the Fractal Nature of Rocks. *Physical Review Letters*, **57**, 637–640.

Fractal geometry, porosity and complex resistivity: from hand specimen to field data

BRÍGIDA RAMATI P. DA ROCHA[1] & TAREK M. HABASHY[2]

[1] *Federal University of Pará, Departamento de Engenharia Elérica, CP 8619, CEP 66075-900 Belém, Paré, Brazil*
[2] *Schlumberger-Doll Research, Old Quarry Road, Ridgefield, CT, USA 06877-4108*

Abstract: In the induced polarization (IP) method of geophysical prospecting, the frequency dependence of the constitutive parameters (conductivity and permittivity) of rocks is exploited to distinguish between different types of rocks with different types of mineralization. Unavoidably, the field data measured using either a four electrode array or an inductive type setup employing ungrounded loops will not directly provide the induced polarization properties of the geological medium to the exclusion of the 'electromagnetic coupling'. In fact one can not measure one effect without the other. Inference of the medium constitutive parameters from field data is often masked and confused by the purely electromagnetic effects that would be present even when the medium is non-dispersive.

In this paper we present a framework by which one can separate the electrochemical effects from the electromagnetic effects which are usually intertwined in a complicated manner in the measured field data. We formulate the problem in a fairly general way without introducing the customary quasi-static approximation. The model we use is that of a layered earth with one of the layers representing a polarizable medium and the rest of the layers are non-polarizable. The effective complex resistivity of the polarizable medium, which is described in a companion paper, is given by a generalized Cole–Cole type resistive network.

Conrad Schlumberger at the beginning of this century measured the ground electrical resistivity and observed that after the current was switched off, a small potential difference could still be measured during some time. He observed that this effect increases over highly conductive materials buried near the surface and reasoned that this type of polarization could be used to locate either a metallic mass or a highly conductive ore. He established a new geophysical survey method by sending a direct current into the ground, interrupting the current and measuring the polarization effect for prospecting metallic conductors like pyrite, chalcopyrite, galena or graphite. The technique for measuring the voltage after interrupting the input current was employed during the second world war to map mines and submarines in the sea, and after the war it was used to locate mineral deposits of copper and other strategic metals (Collet 1989).

Many researchers contributed to the development of the method. Seigel (1959) introduced the 'chargeability' concept associated with the variation of the observed voltage in time domain measurements. This method has evolved from direct current measurements (time domain) to alternating current measurements (frequency domain) where the polarization effect is observed in the variation of the apparent resistivity with the frequency. Wait (1959)

introduced the PFE concept to measure the polarization effect using the variation of resistivity between two frequencies. Madden & Cantwell (1967), and Marshall & Madden (1959) established the basis of the IP phenomena introducing the concepts of electrode polarization, membrane polarization and metal factor. Vacquier *et al.* (1957) applied the method to ground water research and Daknov *et al.* (1967) applied it in well logging. Komarov & Kotov (1968) showed the possibility of using the induced polarization measurements to determine the formation permeability, and the difficulties of obtaining good results caused by the non unique relations between porosity and permeability.

The method evolved from two single-frequency measurements to multi-frequency measurements and is currently used in metallic mineral prospecting, ground water investigations, oil prospecting and environmental studies, as well as to the crosshole tomography (Iseki & Shima, 1991).

One of the problems in using this method is separating the effect due to the intrinsic variation of the ground resistivity with the frequency and the inductive coupling caused by the electromagnetic interaction between the measuring array and the ground.

In this paper we use a new model for the intrinsic electrical properties of the medium and

From Lovell, M. A. & Harvey, P. K. (eds), 1997, *Developments in Petrophysics*, Geological Society Special Publication No. 122, pp. 287–297.

analyse the induced polarization response of the ground. It is shown that this new model is capable of explaining the field data and also provides a good tool for rock properties determination.

The IP response over a layered Earth

Figure 1 presents a collinear dipole–dipole configuration used to measure the ground resistivity over an n-layered Earth. A current is introduced into the ground through electrodes A and B and the voltage is measured between the electrodes M and N. In the normal dipole–dipole configuration the spacing between the current and potential electrodes is fixed and the distance between the two couples of electrodes is a multiple of the dipole length (normally from 1 to 7 times)

To calculate the voltage measured by the receiver electrodes we will solve the electromagnetic problem for the dipole-dipole configuration. Starting from Maxwell's equations, and assuming an $e^{-i\omega t}$ time dependence

$$\nabla \times \bar{E} = i\omega\mu\bar{H} \qquad (1)$$

$$\nabla \times \bar{H} = -i\omega\epsilon\bar{E} + \sigma\bar{E} + \bar{J}_s \qquad (2)$$

where $\bar{J}_s$ is the current impressed in the medium.

Using the complex conductivity by combining the conductivity (σ) and displacement factor ($i\omega\epsilon$), we could write for the current density J

$$\bar{J} = \sigma^*\bar{E} \qquad (3)$$

and equation (2) could be written as

$$\nabla \times \bar{H} = \bar{J} + \bar{J}_s. \qquad (4)$$

The complex conductivity σ^* is given by the model presented in a companion paper (Rocha & Habashy, this volume) and given by

$$Z = \frac{\rho_o}{1 + i\omega\tau_o}$$

$$\times \left[1 - m \left(1 - \frac{1}{1 + \dfrac{1}{\delta_1 + \delta_2}(1 + u)} \right) \right] \qquad (5)$$

with $m = R_o/(R_1 + R_o)$ being the chargeability (Seigel 1959); $\tau_1 = rC_{dl}$, the double layer relaxation time (Rocha et al. 1991); $\tau_2 = R_oC_o$, the sample relaxation time; $\delta_r = r/R_o$, the grain percent resistivity; $\delta_1 = r/(R_1 + R_o) = m\delta_r$ and $\delta_2 = K(i\omega)^{-\eta}/(R_1 + R_o) = (m/R_o)K(i\omega)^{-\eta}$; $u = i\omega\tau(1 + \delta_2/\delta_1)$.

In a homogeneous unbounded medium, the electric field is given by

$$\bar{E}(\bar{r}) = i\omega\mu \int d\bar{r}' \frac{e^{ik|\bar{r}-\bar{r}'|}}{4\pi|\bar{r} - \bar{r}'|} \bar{J}_s(\bar{r}')$$

$$+ \frac{1}{\sigma^*} \nabla \int d\bar{r}' \frac{e^{ik|\bar{r}-\bar{r}'|}}{4\pi|\bar{r} - \bar{r}'|} \nabla' \cdot \bar{J}_s(\bar{r}'). \qquad (6)$$

For a dipole pointing along the y axis, located at $x = 0$, $z = z_0$ and of length L which extends from $y = y_1$ to $y = y_2$ ($y_2 > y_1$ with $y_2 - y_1 = L$), its volumetric current density is given by

$$\bar{J}_s(\bar{r}) = \hat{y}I\delta(x)\delta(z - z_o)$$

$$\times [H(y - y_1) - H(y - y_2)] \qquad (7)$$

where $H(.)$ is the Heaviside step function. Note that for an infinitesimal dipole, where

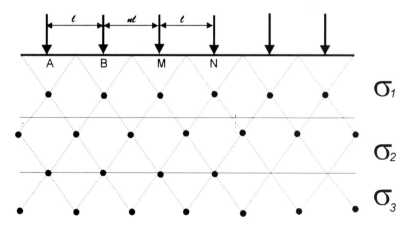

Fig. 1. Geometry of the dipole–dipole collinear configuration.

$y_2 - y_1 = L \to 0$, its volume current density is given by

$$\bar{J}_s(\bar{r}) = \hat{y}IL\delta(x)\delta(z - z_o)$$

$$\times \lim_{L \to 0} \left\{ \frac{H(y - y_1) - H(y - y_2)}{L} \right\} \quad (8)$$

$$= \hat{y}IL\delta(x)\delta(z - z_o)\delta(y - y_o)$$

$$= \hat{y}IL\delta(\bar{r} - \bar{r}_o) \quad (9)$$

where

$$y_o = y_1 = y_2 \quad (10)$$

$$\bar{r}_o = \hat{y}y_o + \hat{z}z_o. \quad (11)$$

The charge accumulated at the ends of the dipole, given by equation (7) is given by

$$\nabla \cdot \bar{J}_s(\bar{r}) = I\delta(x)\delta(z - z_o)$$

$$\times \frac{d}{dy}[H(y - y_1) - H(y - y_2)] \quad (12)$$

$$= I\delta(x)\delta(z - z_o)$$

$$\times [\delta(y - y_1) - \delta(y - y_2)] \quad (13)$$

$$= I\delta(\bar{r} - \bar{r}_1) - I\delta(\bar{r} - \bar{r}_2) \quad (14)$$

where

$$\bar{r}_1 = \hat{y}y_1 + \hat{z}z_o \quad (15)$$

$$\bar{r}_2 = \hat{y}y_2 + \hat{z}z_o. \quad (16)$$

Equation (14); represents two point current electrodes of opposite polarities, $I\delta(\bar{r} - \bar{r}_1)$ representing the injected current (measured current) and $-I\delta(\bar{r} - \bar{r}_2)$ representing the return current, located at a distance L away from the injected current.

Substituting equations (7) and (14) in (6), we obtain

$$\bar{E}(\bar{r}) = \hat{y} \, i\omega\mu \frac{I}{4\pi} \int_{y_1}^{y_2} dy' \frac{1}{R} e^{ikR} + \frac{I}{4\pi\sigma^*}$$

$$\times \nabla \left[\frac{1}{R_1} e^{ikR_1} - \frac{1}{R_2} e^{ikR_2} \right] \quad (17)$$

where

$$R^2 = x^2 + (y - y')^2 + (z - z_o)^2 \quad (18)$$

$$R_1^2 = x^2 + (y - y_1)^2 + (z - z_o)^2 \quad (19)$$

$$R_2^2 = x^2 + (y - y_2)^2 + (z - z_o)^2 \quad (20)$$

where R_1 is the distance of the observation point from the measure current electrode located at $\bar{r}_1$ and R_2 is the distance from the return current electrode located at $\bar{r}_2$.

Assuming the length of the dipole L to be small compared to the wavelength, the expression for the electric field of equation (17) can be approximated by

$$\bar{E}(\bar{r}) = \frac{I}{4\pi\sigma^*} \nabla \left[\frac{1}{R_1} e^{ikR_1} - \frac{1}{R_2} e^{ikR_2} \right] \quad (21)$$

which is the electric field due to the two point current electrodes at $\bar{r}_1$ and $\bar{r}_2$. The associated potential is given by:

$$V(\bar{r}) = -\frac{I}{\sigma^*} \left[\frac{e^{ikR_1}}{4\pi R_1} - \frac{e^{ikR_2}}{4\pi R_2} \right]. \quad (22)$$

For the dipole–dipole configuration, where the receiver is a dipole which is collinear with the source dipole and extends from y_{1r} to y_{2r} the potential difference measured by the receiver dipole is given by:

$$V_r \equiv V_2 - V_1$$

$$= \frac{I}{\sigma^*} \left[\frac{e^{ik(y_2 - y_{2r})}}{4\pi(y_2 - y_{2r})} - \frac{e^{ik(y_{2r} - y_1)}}{4\pi(y_{2r} - y_1)} \right.$$

$$\left. - \frac{e^{ik(y_2 - y_{1r})}}{4\pi(y_2 - y_{1r})} + \frac{e^{ik(y_{1r} - y_1)}}{4\pi(y_{1r} - y_1)} \right]. \quad (23)$$

In the dipole–dipole configuration we call

$$y_{1r} - y_1 = y_{ll} \quad (24)$$

$$y_{2r} - y_1 = y_{lr} \quad (25)$$

$$y_2 - y_{2r} = y_{rr} \quad (26)$$

$$y_2 - y_{1r} = y_{rl} \quad (27)$$

we have

$$V_r = \frac{I}{\sigma^*} \left[\frac{e^{iky_{rr}}}{4\pi y_{rr}} - \frac{e^{iky_{lr}}}{4\pi y_{lr}} - \frac{e^{iky_{rl}}}{4\pi y_{rl}} + \frac{e^{iky_{ll}}}{4\pi y_{ll}} \right]. \quad (28)$$

Assuming the spacings y_{rr}, y_{lr}, y_{rl} and y_{ll} to be small compared to the wavelength, we can approximate the potential difference by

$$V_r = \frac{I}{4\pi\sigma^*} \left[\frac{1}{y_{rr}} - \frac{1}{y_{lr}} - \frac{1}{y_{rl}} + \frac{1}{y_{ll}} \right]. \quad (29)$$

Hence the apparent resistivity as measured by the dipole–dipole setup is given by

$$\rho_a = \frac{4\pi}{K} \frac{V_r}{I} \quad (30)$$

where K is a geometric factor given by

$$K = \frac{1}{\gamma_{rr}} - \frac{1}{\gamma_{lr}} - \frac{1}{\gamma_{rl}} + \frac{1}{\gamma_{ll}}. \tag{31}$$

Applying Sommerfeld's identity

$$\frac{e^{ik|r-r'|}}{4\pi|r-r'|} = \frac{i}{4\pi} \int_0^\infty dk_\rho \frac{k_\rho}{k_z}$$

$$\times J_o(k_\rho|\bar\rho - \bar\rho'|) e^{ik_z|z-z'|} \tag{32}$$

to equation (22) we obtain

$$V(\bar r) = \frac{iI}{4\pi\sigma^*} \left[\int_0^\infty dk_\rho \frac{k_\rho}{k_z} J_o(k_\rho\rho_2) e^{ik_z|z-z_o|} \right.$$

$$\left. - \int_0^\infty dk_\rho \frac{k_\rho}{k_z} J_o(k_\rho\rho_1) e^{ik_z|z-z_o|} \right] \tag{33}$$

where

$$k_z^2 + k_\rho^2 = k^2 = i\omega\mu\sigma^* \tag{34}$$

$$\rho_1^2 = x^2 + (y - y_1)^2 \tag{35}$$

$$\rho_2^2 = x^2 + (y - y_2)^2. \tag{36}$$

Equation (33) gives the potential distribution in a homogeneous unbounded medium.

If we assume that the electrodes are located in region (0) which has a conductivity of σ_o^*. A stratified layered medium occupies the half-space $z < 0$. The general distribution of the potential in region (0) (i.e., the upper half-space $z > 0$) is given by:

$$V(\bar r) = \frac{iI}{4\pi\sigma_o^*} \int_0^\infty dk_\rho \frac{k_\rho}{k_{0z}} \{J_o(k_\rho\rho_2) - J_o(k_\rho\rho_1)\}$$

$$\times [e^{ik_{0z}|z-z_o|} + A e^{ik_{0z}z} + B e^{-ik_{0z}z}] \tag{37}$$

where

$$k_{0z}^2 + k_\rho^2 = i\omega\mu\sigma_o^*. \tag{38}$$

Since the potential has to vanish when $z \to \infty$ (from the radiation condition), then $B = 0$. Applying the boundary condition at the surface of the layered medium (at $z = 0$) where the ratio of the reflected wave to the incident wave is given by the global reflection coefficient R_0, we obtain

$$\left. \frac{A e^{ik_{0z}z}}{e^{ik_{0z}(z_o-z)}} \right|_{z=0} = R_0 \tag{39}$$

where R_0 is the global reflection coefficient at the interface with the layered medium located at $z = 0$. From equation (39) we obtain

$$A = R_0 e^{ik_{0z}(z_o)}. \tag{40}$$

Hence

$$V(\bar r) = \frac{iI}{4\pi\sigma_o^*} \int_0^\infty dk_\rho \frac{k_\rho}{k_{0z}} [J_o(k_\rho\rho_2) - J_o(k_\rho\rho_1)]$$

$$\times [e^{ik_{0z}|z-z_o|} + R_0 e^{ik_{0z}(z+z_o)}] \tag{41}$$

where R_0 is obtained from the following recurrence relationship:

$$R_n = \frac{r_{n,n+1} + R_{n+1} e^{i2k_{(n+1)z}d_{n+1}}}{1 + r_{n,n+1} R_{n+1} e^{i2k_{(n+1)z}d_{n+1}}},$$

$$n = N - 1, \ N - 2, \dots, 0 \tag{42}$$

with the reflection coefficient at the lowermost layer:

$$R_N = 0 \tag{43}$$

and

$$r_{n,n+1} = \frac{\sigma_n^* k_{nz} - \sigma_{n+1}^* k_{(n+1)z}}{\sigma_n^* k_{nz} + \sigma_{n+1}^* k_{(n+1)z}} \tag{44}$$

$$k_{nz}^2 + k_\rho^2 = i\omega\mu\sigma_n^*. \tag{45}$$

For the dipole–dipole configuration, where the receiver is a dipole which is collinear with the source dipole and extends from y_{1r} to y_{2r}, where $y_2 > y_{2r} > y_{1r} > y_1$, the potential difference measured by the receiver dipole is given by:

$$V_r = \frac{iI}{4\pi\sigma_o^*} \int_0^\infty dk_\rho \frac{k_\rho}{k_{0z}} [J_o\{k_\rho(y_2 - y_{2r})\}$$

$$- J_o\{k_\rho(y_{2r} - y_1)\} - J_o\{k_\rho(y_2 - y_{1r})\}$$

$$+ J_o\{k_\rho(y_{1r} - y_1)\}](1 + R_0 e^{i2k_{0z}z_o}). \tag{46}$$

Using the relations (24) to (27) in equation (46) and assuming both the source and receiver dipoles to be located at the surface of the layered medium, i.e., $z = 0 = z_o$, the potential difference measured by the receiver dipole is thus given by:

$$V_r = \frac{iI}{4\pi\sigma_o^*} \int_0^\infty dk_\rho \frac{k_\rho}{k_{0z}} [J_o(k_\rho y_{rr}) - J_o(k_\rho y_{lr})$$

$$- J_o(k_\rho y_{rl}) + J_o(k_\rho y_{ll})](1 + R_0) \tag{47}$$

and the apparent resistivity is given by

$$\rho_a = \frac{i}{K\sigma_o^*} \int_0^\infty dk_\rho \frac{k_\rho}{k_{0z}} [J_o(k_\rho y_{rr}) - J_o(k_\rho y_{lr})$$

$$- J_o(k_\rho y_{rl}) + J_o(k_\rho y_{ll})](1 + R_0) \tag{48}$$

where σ_0^* is the complex conductivity of air given by

$$\sigma_0^* = -i\omega\epsilon_o \tag{49}$$

$$k_{0z}^2 + k_\rho^2 = \omega^2 \mu\epsilon_0 \tag{50}$$

$$R_0 = \frac{r_{0,1} + R_1 \, e^{i2k_{1z}d_1}}{1 + r_{0,1} R_1 \, e^{i2k_{1z}d_1}} \tag{51}$$

$$r_{0,1} = \frac{\sigma_0^* k_{0z} - \sigma_1^* k_{1z}}{\sigma_0^* k_{0z} + \sigma_1^* k_{1z}} \tag{52}$$

and

$$1 + R_0 = \frac{1 + R_1 \, e^{i2k_{1z}d_1}}{1 + r_{0,1} R_1 \, e^{i2k_{1z}d_1}} (1 + r_{0,1}) \tag{53}$$

$$1 + r_{0,1} = \frac{2\sigma_0^* k_{0z}}{\sigma_0^* k_{0z} + \sigma_1^* k_{1z}}$$

$$= \frac{\sigma_0^* \, k_{0z}}{\sigma_1^* \, k_{1z}} (1 + r_0) \tag{54}$$

where

$$r_0 = -r_{0,1} = \frac{\sigma_1^* k_{1z} - \sigma_0^* k_{0z}}{\sigma_1^* k_{1z} + \sigma_0^* k_{0z}} \tag{55}$$

Hence,

$$\rho_a = \frac{i}{K\sigma_1^*} \int_0^\infty dk_\rho \frac{k_\rho}{k_{1z}} [J_0(k_\rho y_{rr}) - J_0(k_\rho y_{lr})$$

$$- J_0(k_\rho y_{rl}) + J_0(k_\rho y_{ll})]$$

$$\times \frac{1 + R_1 \, e^{i2k_{1z}d_1}}{1 - r_0 R_1 \, e^{i2k_{1z}d_1}} (1 + r_0). \tag{56}$$

The response of a homogeneous polarizable Earth

The response of a polarizable half-space is presented in Fig. 2 which shows a great resemblance to those obtained for a hand specimen at low frequencies (Rocha & Habashy this volume). This is the reason why it is possible to use the same model for the electrical properties of an intrinsic polarizable medium to fit the field data.

Pelton et al (1978) used a Cole–Cole type function to explain their field data, but, as this kind of model is incapable of explaining the behaviour of all geological materials, they suggested that the second phase maximum present in the field data was caused by inductive coupling. It is demonstrated that this could be better explained by our proposed model. This result also justifies the use of this kind of

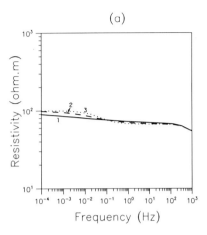

(a)

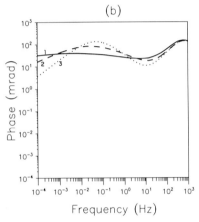

(b)

Fig. 2. The new model collinear dipole response, frequency interval of 10^{-4} to 10^3 Hz, fixing $\rho_o = 100\,\Omega \cdot \text{m}$; $m = 0.5$; $\tau = 10^{-6}$ s; $\delta_r = 1.0$; $K = 100.0$; $\tau_o = 10^{-12}$ s and a dipole separation of 1.0 m; curves n. 1, 2 and 3, respectively to $\eta = 0.25$, 0.50, 0.75 (**a**) amplitude of apparent resistivity; (**b**) phase.

function to obtain equivalent homogeneous layer parameters as has been used by Vanhala & Peltoniemi (1992) with Pelton's model. Pelton's model appear to be incapable of giving the important information concerning the texture of the rocks which is related to the fractal geometry present in the η parameter of the new model.

In order to test the applicability of the model we fitted field data obtained by Vanhala & Peltoniemi (1992). The data were collected over sulphide, oxide and graphitic gneiss deposits in Finland (Vanhala & Peltoniemi 1992). A standard Gauss–Newton search method was used to obtain the phase of the measured ground resistivity fitting parameters with a cost function given by the logarithm of the ratio between the calculated and the observed values.

Figure 3 presents experimental phase measured with a dipole–dipole array for Eko deposit (disseminated magnetite, net-textured Ni–Cu ore and graphite gneiss) and corresponding fitted curves. The same is shown in Fig. 4 to Kilvenjärvi chacopyrite and pyrrhotite deposits; Fig. 5 to Jokisivu–Aurum deposit and Fig. 6 shows the fitting to weakly disseminated gneiss from several deposits.

Table 1 shows a brief description of the deposits and the dipole array used in the measurements; Table 2 presents the parameters obtained for the published data.

Figure 7 shows a crossplot of η versus ρ_o where we could observe the possibility of distinguishing some deposits. Weakly disseminated gneiss have lower η and ρ_o while the higher grade deposits show a greater η. Figure 8 shows a crossplot of K versus η where the calcopirite are distinguished from other deposits (higher η and lower K). Figures 9, 10 and 11 show some crossplot that could not permit any distinction between the different deposits. Among these the chargeability

parameter does not show the capability of distinguishing between the different deposits.

Vanhala & Peltoniemi (1992) used Cole–Cole-type functions (Pelton *et al.* 1978) to fit the data and correlated with the mineralogical description and grain size of the deposits. They noted some disadvantages of using Cole–Cole-type functions mainly due to the discrepancies in the parameter values when using low frequency or high-frequency data. Since our proposed model is valid over a broader frequency range, this problem does not occur. Furthermore, with Cole–Cole-type function, unrealistically large relaxation times are usually obtained which vary from few seconds up to thousands of seconds. With this new model the relaxation time usually lie in the realistic range of nanoseconds to milliseconds.

The response of three-layer polarizable Earth

In order to test if the fractal parameters could be observed (and therefore measured), we

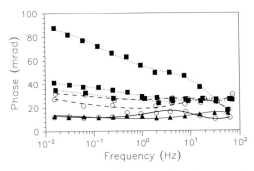

Fig. 3. Experimental and fitted curves for the phase of complex resistivity for Eko, Ni–Cu deposit (Vanhala & Peltoniemi 1992).

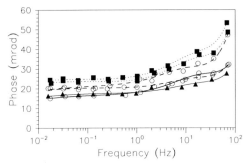

Fig. 5. Experimental and fitted curves for the phase of complex resistivity for Jokisivu, Au deposit (Vanhala & Peltoniemi 1992).

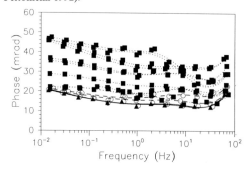

Fig. 4. Experimental and fitted curves for the phase of complex resistivity for Kilvenjärvi, chalcopyrite and pyrrhotite deposits (Vanhala & Peltoniemi 1992).

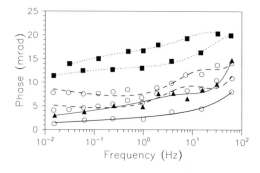

Fig. 6. Experimental and fitted curves for the phase of complex resistivity for weakly disseminated gneiss (Vanhala & Peltoniemi 1992).

Table 1. *Mineralization description of deposits (Vanhala & Peltoniemi 1992)*

Curve	Place/deposit type/mineralization	Dipole–dipole array
Fig. 3c-1	Eko Ni–Cu-deposit, disseminated magnetite	$a = 20$ m, $n = 2$
Fig. 3c-2	Eko Ni–Cu-deposit, disseminated magnetite	$a = 20$ m, $n = 2$
Fig. 3c-3	Eko Ni–Cu-deposit, net-texturized Ni–Cu ore	$a = 20$ m, $n = 2$
Fig. 3c-4	Eko Ni–Cu-deposit, net-texturized Ni–Cu ore	$a = 20$ m, $n = 2$
Fig. 3c-5	Eko Ni–Cu-deposit, net-texturized Ni–Cu ore	$a = 20$ m, $n = 2$
Fig. 3c-6	Eko Ni–Cu-deposit, graphitic gneiss	$a = 20$ m, $n = 2$
Fig. 3c-7	Eko Ni–Cu-deposit, graphitic gneiss	$a = 20$ m, $n = 2$
Fig. 3e-1	Kilvenjarvi chalcopyrite and pyrrhotite deposit	$a = 20$ m, $n = 4$
Fig. 3e-2	Kilvenjarvi chalcopyrite and pyrrhotite deposit	$a = 20$ m, $n = 2$
Fig. 3e-3	Kilvenjarvi chalcopyrite and pyrrhotite deposit	$a = 20$ m, $n = 4$
Fig. 3e-4	Kilvenjarvi chalcopyrite and pyrrhotite deposit	$a = 20$ m, $n = 4$
Fig. 3e-5	Kilvenjarvi chalcopyrite and pyrrhotite deposit	$a = 20$ m, $n = 4$
Fig. 3e-6	Kilvenjarvi chalcopyrite and pyrrhotite deposit	$a = 20$ m, $n = 4$
Fig. 3e-7	Kilvenjarvi chalcopyrite and pyrrhotite deposit	$a = 20$ m, $n = 2$
Fig. 3e-8	Kilvenjarvi chalcopyrite and pyrrhotite deposit	$a = 20$ m, $n = 4$
Fig. 3e-9	Kilvenjarvi chalcopyrite and pyrrhotite deposit	$a = 20$ m, $n = 4$
Fig. 3f-1	Jokisivu Au deposit, disseminated pyrrhotite	$a = 5$ m, $n = 2$
Fig. 3f-2	Jokisivu Au deposit, disseminated Au	$a = 5$ m, $n = 2$
Fig. 3f-3	Jokisivu Au deposit, disseminated Au	$a = 5$ m, $n = 2$
Fig. 3f-4	Jokisivu Au deposit, disseminated Au	$a = 5$ m, $n = 2$
Fig. 3f-5	Jokisivu Au deposit, disseminated pyrrhotite	$a = 5$ m, $n = 2$
Fig. 3f-6	Jokisivu Au deposit, disseminated Au	$a = 5$ m, $n = 2$
Fig. 4-1	Gneiss circundante deposits Eko, Pisamanieni, Kilvenjoki	
Fig. 4-2	Gneiss circundante deposits Eko, Pisamanieni, Kilvenjoki	
Fig. 4-3	Gneiss circundante deposits Eko, Pisamanieni, Kilvenjoki	
Fig. 4-4	Gneiss circundante deposits Eko, Pisamanieni, Kilvenjoki	
Fig. 4-5	Gneiss circundante deposits Eko, Pisamanieni, Kilvenjoki	
Fig. 4-6	Gneiss circundante deposits Eko, Pisamanieni, Kilvenjoki	

Table 2. *Model parameters for data of Vanhala & Peltoniemi (1992)*

Sample	ρ_o (μm)	m	τ_1 (ms)	δ_r (10^{-4})	K (10^{-3} s$^{-\eta}$)	η	τ_2
Fig. 3c-1	44.25	0.396	2.547	12.53	9.379	0.142	20.2 μs
Fig. 3c-2	58.66	0.497	0.500	11.02	7.799	0.131	0.55 ens
Fig. 3c-3	50.84	0.552	12.20	168.2	7.435	0.172	0.02 ps
Fig. 3c-4	28.93	0.302	2.547	92.06	30.34	0.236	47.76 μs
Fig. 3c-5	8.789	0.321	4.911	416.5	8.863	0.240	19.61 μs
Fig. 3c-6	6.978	0.308	0.002	1.23	8.219	0.278	34.63 μs
Fig. 3c-7	3.234	0.498	43.82	657.7	1.650	0.333	10.21 ps
Fig. 3e-1	12.80	0.309	19.04	188.7	4.766	0.231	37.1 μs
Fig. 3e-2	11.05	0.274	14.62	183.0	5.502	0.236	39.5 μs
Fig. 3e-3	7.745	0.253	0.132	4.421	7.353	0.212	39.8 μs
Fig. 3e-4	6.455	0.205	0.187	7.334	8.892	0.239	36.6 μs
Fig. 3e-5	4.506	0.174	0.175	17.60	12.25	0.313	27.7 μs
Fig. 3e-6	5.380	0.224	28.91	1 440	9.462	0.293	33.9 μs
Fig. 3e-7	3.752	0.140	155.6	15 630	16.25	0.424	37.8 μs
Fig. 3e-8	4.222	0.165	77.43	7 064	12.99	0.339	40.6 μs
Fig. 3e-9	6.481	0.283	29.31	1 104	8.173	0.279	56.7 μs
Fig. 3f-1	6.048	0.155	12.90	3 264	33.42	0.248	42.9 μs
Fig. 3f-2	3.040	0.155	0.0009	0.210	15.09	0.250	40.1 μs
Fig. 3f-3	3.497	0.182	0.0008	0.127	13.00	0.246	47.5 μs
Fig. 3f-4	3.257	0.188	0.0019	0.218	12.60	0.241	84.8 μs
Fig. 3f-5	4.451	0.209	29.45	3 615	13.81	0.267	79.5 μs
Fig. 3f-6	4.543	0.244	0.0040	0.421	11.10	0.219	89.2 μs
Fig. 4-1	0.2144	0.055	3.439e-10	0.130	22.35	0.185	1.00 fs
Fig. 4-2	0.7960	0.033	0,012	8.984	38.07	0.304	33.1 μs
Fig. 4-3	13.11	0.242	0.691	9.342	4.516	0.117	22.7 μs
Fig. 4-4	11.22	0.255	0.593	16.72	5.282	0.136	22.9 μs
Fig. 4-5	0.9126	0.146	6.006e-10	0.008	6.723	0.202	1.00 fs
Fig. 4-5	0.8545	0.142	9.675e-11	0.0007	7.332	0.253	1.00 fs

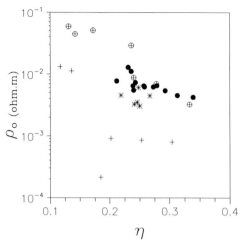

Fig. 7. DC resistivity (ρ_o) versus η for some mineral deposits. +, Weakly disseminated gneiss; $\oplus$, Eko (Ni–Cu) deposit; $\star$, Jokisivu (Au) deposit; •, Kilvenjärvi (chalcopyrite and pyrrhotite) deposit.

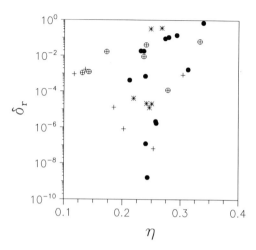

Fig. 9. δ_r versus η for some mineral deposits. +, Weakly disseminated gneiss; $\oplus$, Eko (Ni–Cu) deposit; $\star$, Jokisivu (Au) deposit; •, Kilvenjärvi (chalcopyrite and pyrrhotite) deposit.

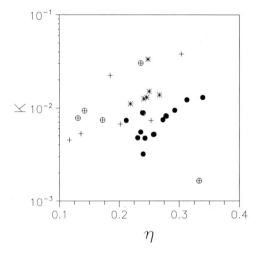

Fig. 8. K versus η for some mineral deposits. +, Weakly disseminated gneiss; $\oplus$, Eko (Ni–Cu) deposit; $\star$, Jokisivu (Au) deposit; •, Kilvenjärvi (chalcopyrite and pyrrhotite) deposit.

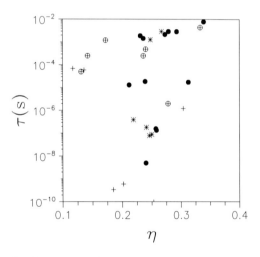

Fig. 10. τ versus η for some mineral deposits. +, Weakly disseminated gneiss; $\oplus$, Eko (Ni–Cu) deposit; $\star$, Jokisivu (Au) deposit; •, Kilvenjärvi (chalcopyrite and pyrrhotite) deposit.

calculated the response of a three-layer Earth, in which the second layer is a polarizable medium, with the intrinsic electrical properties given by the fractal complex resistivity (Equation (5)).

Figures 12, 13 and 14 present the response of the three-layer Earth for different thickness of the overburden, and different fractal parameters of the polarizable medium. We observe that the phase is mainly affected by the parameters of the polarizable layer while the amplitude is more dependent on the combined

layering. When the same ρ is used for all three layers and the overburden is thicker than the polarizable layer, the amplitude of the response will be the same as that of an unpolarizable half space. The value of the phase will be dependent on the layering, while the shape of the curve will be dependent on the fractal parameters. This indicates that it is possible to determine the parameters of the polarizable layer even in the presence of a thick overburden.

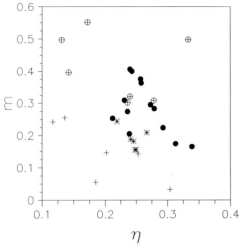

Fig. 11. Chargeability (*m*) versus η for some mineral deposits. +, Weakly disseminated gneiss; ⊕, Eko (Ni–Cu) deposit; ⋆, Jokisivu (Au) deposit; •, Kilvenjärvi (chalcopyrite and pyrrhotite) deposit.

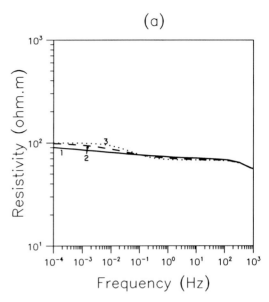

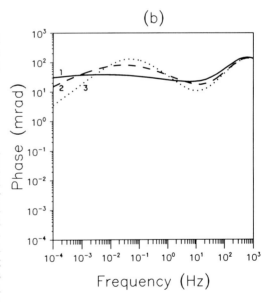

Fig. 12. The dipole–dipole resistivity calculated for a three-layer Earth where the polarizable layer is between two non-polarizable layers, for varying η. Frequency interval: 10^{-4} to 10^3 Hz. Dipole separation of 10 m. Parameters of the polarizable layer: $\rho_o = 100\,\Omega \cdot m$; $m = 0.5$; $\delta_r = 1$; $\tau = 10^{-6}$ s; $\tau_f = 10^{-3}$ s and $\tau_0 = 10^{-12}$ s; resistivity of the first and third layers $= 100\,\Omega \cdot m$; Thickness of the first layer: 1 m; (**a**) amplitude of the calculated apparent resistivity; (**b**) phase curve in log–log scale. Curves n. 1, 2 and 3, respectively to $\eta = 0.25$; 0.5; and 0.75.

From simulation results, it is also shown that the fractal frequency exponent of the polarizable layer dominates the phase frequency response at low frequencies.

Conclusions

In this paper we have presented a framework by which one can separate the electrochemical effects from the electromagnetic effects which are usually intertwined in a complicated manner in the measured field data. We have formulated the problem in a fairly general way without introducing the customary quasi-static approximation. The model that we use is that of a layered Earth with one of the layers representing a polarizable medium and the rest of the layers are non-polarizable. The effective complex resistivity of the polarizable medium is given by a new model which is described in a companion paper.

At low frequencies, the response of a polarizable medium embedded in a layered medium is similar to that of the polarizable medium in the absence of layering. The electromagnetic model of a layered earth combined with the new fractal model, described in the companion paper, enables one to delineate between the electromagnetic coupling and the true response of the polarizable medium.

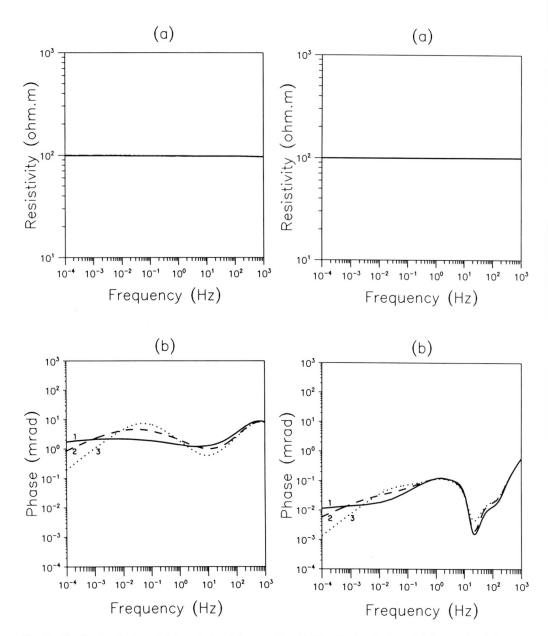

Fig. 13. The dipole–dipole resistivity calculated for a three-layer Earth where the polarizable layer is between two non-polarizable layers, for varying η. Frequency interval: 10^{-4} to 10^3 Hz. Dipole separation of 10 m. Parameters of the polarizable layer: $\rho_o = 100\,\Omega \cdot$ m; $m = 0.5$; $\delta_r = 1$; $\tau = 10^{-6}$ s; $\tau_f = 10^{-3}$ s and $\tau_0 = 10^{-12}$ s; resistivity of the first and third layers = $100\,\Omega \cdot$ m; Thickness of the first layer: 10 m; (**a**) amplitude of the calculated apparent resistivity; (**b**) phase curve in log–log scale. Curves n. 1, 2 and 3, respectively to $\eta = 0.25$; 0.5; and 0.75.

Fig. 14. The dipole–dipole resistivity calculated for a three-layer earth where the polarizable layer is between two non-polarizable layers, for varying η. Frequency interval: 10^{-4} to 10^3 Hz. Dipole separation of 10 m. Parameters of the polarizable layer: $\rho_o = 100\,\Omega \cdot$ m; $m = 0.5$; $\delta_r = 1$; $\tau = 10^{-6}$ s; $\tau_f = 10^{-3}$ s and $\tau_0 = 10^{-12}$ s; resistivity of the first and third layers = $100\,\Omega \cdot$ m. Thickness of the first layer: 100 m; (**a**) amplitude of the calculated apparent resistivity; (**b**) phase curve in log–log scale. Curves n. 1, 2 and 3, respectively to $\eta = 0.25$; 0.5; and 0.75.

The authors thank Schlumberger-Doll Research and Federal University of Para for their support.

References

COLLETT, L.S. 1989. History of the induced-polarization method. *In*: FINK, J.D. *ET AL.* (eds) *Induced Polarization: Applications and Case Histories.* SEG, 5–23.

DAKHNOV, V. N., LATISHOVA, M.G. & RYAPOLOV, V.A. 1967. Well logging by means of induced polarization (electrolytic logging). *The Log Analyst*, Nov–Dec 1967, 3–18.

ISEKI, S. & SHIMA, H. 1991. Induced polarization tomography: a crosshole imaging technique using chargeability and resistivity. *Proceedings of Annual Meeting*, SEG, 439–442.

KOMAROV, S.G. & KOTOV, P. T. 1968. On determining formation permeability from induced polarization. *The Log Analyst*, May–Jun, 1968: 12–17.

MADDEN, T. R. & CANTWELL, T. 1967. Induced polarization: a review. *In*: *Mining Geophysics*, 2, SEG, 373–400.

MARSHALL, D. J. & MADDEN, T. R. 1959. Induced polarization: a study of its causes. *Geophysics*, 24, 790–816.

PELTON, W. H., WARD, S. H., HALLOF, P. G., SILL, W.R. & NELSON, P. H. 1978. Mineral discrimination and removal of inductive coupling with multifrequency IP. *Geophysics*, 43, 588–609.

ROCHA, B. R. P. & HABASHY, T. M. 1997. Fractal geometry, porosity and complex resistivity: from rough pore interfaces to hand specimens. *This volume*.

——, LEÃO, J. W. D. & DIAS, C. A. 1991. Ajuste Automático de Condutividade Complexa para Descrever o Comportamento Elétrico de Rochas a Baixas Frequências. *Anais do II Congresso Internacional de Geofísica*, I, 260–265.

SEIGEL, H. O. 1959. Mathematical formulation and type curves for induced polarization. *Geophysics*, 24, 547–563

VANHALA, H. & PELTONIEMI, M. 1992. Spectral IP studies of Finnish ore prospects. *Geophysics*, 57, 1545–1555.

VACQUIER, V., HOLMES, C. R., KINTZINGER, P. R. & LAVERGNE, M. 1957. Prospecting for ground water by induced electrical polarization. *Geophysics*, 38, 49–60.

WAIT, J. R. 1959. The variable-frequency method. *In*: WAIT, J. R. (ed.) *Overvoltage Research and Geophysical Applications*. Pergamon Press, New York, 29–49.

Tortuosity: a guide through the maze

M. BEN CLENNELL

Department of Earth Sciences, University of Leeds, Leeds LS2 9JT, UK

Abstract: Despite its widespread use in petrophysics, tortuosity remains a poorly understood concept. Tortuosity can have various meanings when used by physicists, engineers or geologists to describe different transport processes taking place in a porous material. Values for geometrical, electrical, diffusional and hydraulic tortuosity are in general different from one another. Electrical tortuosity is defined in terms of conductivity whereas hydraulic tortuosity is usually defined geometrically, and diffusional tortuosity is typically computed from temporal changes in concentration. A better approach may be to define tortuosity in terms of the underlying flux of material or electrical current with respect to the forces which drive this flow.

Unsteady transport processes, including diffusion, can be described only by a population of tortuosities corresponding to the different flow paths taken by particles traversing the medium. In measurements of steady flow (e.g., those normally used to obtain resistivity or permeability), information about particle travel times is lost, and so the multiple values of tortuosity are homogenised. It can be shown that the maximum amount of information about pore structure is embedded in transport processes that combine advective and diffusive elements.

Most existing formulations of tortuosity are model-dependent, and cannot be correlated with independently measurable pore-structure properties. Nevertheless, tortuosity underpins the rigorous relationships between transport processes in rocks, and ties them with the underlying geometry and topology of their pore spaces. Tortuosity can be redefined in terms of the energetic efficiency of a flow process. The efficiency is related to the rate of entropy dissipation (or isothermally, energy dissipation) with respect to a simple, non-tortuous model medium using the postulates of non-equilibrium thermodynamics. Through Onsager's reciprocity relation for coupled flows it is possible to inter-relate efficiency for pairs of transport processes, and so go some way towards unifying tortuosity measures. In this way we can approach the goal of predicting the value of one transport parameter from measurements of another.

Tortuosity is a term used to describe the sinuosity and interconnectedness of the pore space as it affects transport processes through porous media. Tortuosity has no simple or universal definition: different measures of tortuosity are employed by geologists, engineers and chemists to describe the resistive and retarding effects of the pore structure on a range of conduction, advection and diffusion processes. In recent years a number of workers have reviewed the tortuosity concept critically. Foremost among these is Dullien (1992), whose monograph is recommended as an introduction to transport in porous media. This paper focuses on porous media whose properties are homogeneous and isotropic and only considers transport on the micro- to meso-scales (pore to core dimensions). Manifestations of tortuosity on larger scales, in fractured rocks, and where percolation properties become dominant are not covered here.

In the second section, tortuosity measures that have been applied to different transport processes are reviewed. Three areas are given detailed attention. Single-phase hydrodynamic flow at low Reynolds number (so-called creeping flow, where the ratio of inertial to viscous forces is small) is examined in the context of the measurement and prediction of permeability. Low frequency electrolytic conduction processes are described in relation to formation factor and Archie's Law. Diffusional flow processes are explored in the widest sense, including a consideration of dispersion and the analogy of random walks. Tortuosities derived for other transport processes, including magnetic resonance probes and multi-phase flows, are included for completeness but treated briefly.

In the third section, topological aspects of three-dimensional porous media are examined using network models. The long history of network models reveals their benefits and limitations for the understanding and solution of flow problems. We employ the following paradigm:

'what is found to be universally valid for simple pore structures must also apply to transport in real rocks.'

The fourth section uses tortuosity as a key to understand how diverse flow processes may make different use of the same pore space. An

From Lovell, M. A. & Harvey, P. K. (eds), 1997, *Developments in Petrophysics*, Geological Society Special Publication No. 122, pp. 299–344.

attempt is made to draw together recent advances in understanding from the physical sciences literature. Firstly, we explore the fundamental similarity between diffusion and electrical conduction. A consideration of the energy balance in transport processes leads to the concept of flow efficiency. This is followed by a discussion of characteristic length scales in porous media, and whether proposed measures have general validity. Finally, non-equilibrium thermodynamics is used to establish a direct relationship between electrical and hydrodynamic flows through an analysis of coupled transport.

The fifth section addresses whether tortuosity is a useful petrophysical parameter. Is tortuosity simply a fudge factor that modern physics has made obsolete, or is it a fundamental attribute of porous media that can be obscured, but not erased by emerging theories? In the concluding section the main findings of this review are summarised and some of the most promising approaches are highlighted. A wide range of material is covered in this paper, and so to aid the reader an annotated list of terms and symbols is provided in the Appendix.

Definitions of tortuosity

Geometrical tortuosity

Geometrical tortuosity can be defined as the ratio of the shortest path of connected points remaining in the fluid medium (i.e., continuously through the pores) that joins any two points defined by the vectors $\mathbf{r}_1$ and $\mathbf{r}_2$, with the straight line distance between them ($l_{\min}$). Geometrical tortuosity, τ_g, can be expressed mathematically as:

$$\tau_{g(\mathbf{r}_1,\mathbf{r}_2)} = \left[\frac{l_{\min(\mathbf{r}_1,\mathbf{r}_2)}}{|\mathbf{r}_1 - \mathbf{r}_2|}\right], \quad \mathbf{r}_2 \in f(\mathbf{r}_1) \quad (1)$$

(Adler 1992), where the pore region, occupied by the fluid, is denoted by 'f' (Fig. 1).

A more general definition of geometrical tortuosity can be found in the work of Russian engineers (Golin *et al.* 1992). The tortuosity coefficient is defined as the 'averaged ratio of pore length to the projection of the pores in the direction of flow'. This at once accepts that geometrical tortuosity is an averaged parameter, and that it is a structural characteristic of the medium, independent of any particular transport process.

The concept of a purely geometrical tortuosity factor has not found broad favour in the consideration of permeability, conductivity or

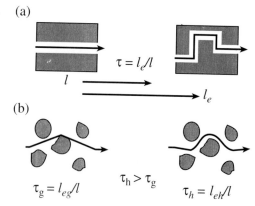

(a)

$$\tau = l_e/l$$

l_e

(b)

$$\tau_g = l_{eg}/l$$

$$\tau_h > \tau_g$$

$$\tau_h = l_{eh}/l$$

Fig. 1. (a) The most basic definition of tortuosity is the path length tortuosity, τ. This is the ratio of the length of the tortuous flow path l_e to the straight line length, l in the direction of flow. This is most easily visualized in a pore space consisting of single capillaries of constant width with no branching. **(b)** Geometrical tortuosity is defined as the shortest length between inflow and outflow points that avoids the solid obstacles, and this is realised as a zigzag path passing grains with close tangents. Carman's definition of hydraulic tortuosity differs from geometrical tortuosity in that it is the effective path length taken by the fluid (l_{eh}), rather than the shortest possible path (l_{eg}) which is considered. Fluid passes along a smoothed route through porous media so that τ_h is always greater than τ_g.

diffusivity. However, some workers, (e.g., Lee & Kozak 1987), have used a random-walk model to estimate tortuosity, and this can be considered to be a purely geometrical measure. This point is discussed further in the subsection on diffusional tortuosity.

Hydraulic tortuosity

Darcy's law. Darcy's law expresses the linear relationship between the forces available to drive fluid flow and the resultant volumetric flux when a gradient of fluid pressure, P, exists across a region of porous material. The rate of volumetric flow, Q, per unit sectional area, A, of the porous medium is proportional to the pressure gradient in the direction of flow (Bear 1972):

$$\frac{Q}{A} = -\frac{k}{\mu} \nabla(P + \rho gh). \quad (2)$$

The coefficient k is known as the permeability and has the dimensions of L^2. The parameter μ is the dynamic viscosity that resists the flow of the fluid, and ρ is its density, while h is the height above a reference level at which the hydrostatic pressure is defined as zero. The negative sign on the right-hand-side of equation (2) indicates that flow is in the direction of decreasing fluid pressure. The second term in parenthesis represents the work done by, or against, gravity (having acceleration g) if the flow is not horizontal (h is non-zero). The volumetric flux per unit area, Q/A, has the dimensions of LT^{-1} so will be referred to henceforth as the seepage velocity and given the vectorial symbol $\mathbf{u}$.

Capillary models of pore structure for hydraulic transport. Fluid flow in a straight cylindrical tube or capillary of radius r_0 can be described by the Hagen–Poiseuille equation:

$$Q = -\frac{\pi r_0^4}{8\mu}\frac{\Delta P}{l} \qquad (3)$$

where Q is the volumetric flow rate, μ is the dynamic viscosity of the fluid and ΔP is the macroscopic pressure difference between the two ends of the tube of length l. As the tube dimension decreases, the cross-sectional area decreases as the square of the radius, and the relative effect of viscous drag at the channel walls also increases in the same proportion. Thus overall fluid flow rate scales as the channel radius raised to the fourth power. In a porous medium consisting of an aggregate of equal sized grains, we could reason that the flow channels would be represented by a cluster of identical capillaries with a particular effective radius r_{eff}. The effect of reducing the grain size by a factor of two would be to reduce the flow velocity through each channel by a factor of 16. However, the number of flow channels would be increased by a factor of four, so the overall rate of flow, and so permeability, would scale with the square of the effective radius.

$$Q \propto r_{\text{eff}}^2 \quad \text{and} \quad k \propto r_{\text{eff}}^2. \qquad (4a,b)$$

This reasoning has led to many empirical models of permeability prediction for granular materials based on identification of the characteristic pore dimension. However, it was soon realized that knowledge of the grain size alone is not sufficient to predict the characteristic radius, and so permeability of the material. There is also a strong control exerted by the nature of the packing, which dictates the size and shape of the pores.

Kozeny's equation. In granular materials (such as sands), the pore surface area can be calculated from the size and shape of the constituent grains. On this basis, Kozeny (1927) introduced a semi-empirical equation to estimate the permeability of granular aggregates.

$$k = \frac{\phi r_h^2}{C_k} \qquad (5)$$

C_k is the 'Kozeny constant' and r_h is the hydraulic radius, defined as:

$$r_h = \frac{\text{cross-sectional area normal to flow}}{\text{'wetted perimeter' of the flow channels}}. \qquad (6)$$

This is the radius of the ideal circular capillary that would have the same hydraulic conductance as the granular material (see Van Brakel 1975, for details of the origin of this concept).

Kozeny assumed the pore space was equivalent to a bundle of parallel capillaries with a common hydraulic radius and cross-sectional shape described by r_h/C_k. He then introduced the concept that the real flow path could be tortuous, with an effective hydraulic path length l_{eh}. The permeability is then reduced by a factor of the sinuous length, l_{eh}, divided by the straight-line length in the direction of flow, l_{min}.

$$\tau_{\text{hK}} = \frac{l_{\text{eh}}}{l_{\text{min}}} \qquad (7)$$

where τ_{hk} is the hydraulic tortuosity of Kozeny (1927). His equation for permeability then becomes:

$$k = \frac{\phi r_h^2}{\beta \tau_{\text{hK}}}. \qquad (8)$$

In this way, the Kozeny constant is subdivided into a shape factor, β with a conservative value for a particular type of granular material, and a tortuosity factor, τ_{hK}, which represents the sinuous nature of the pore space. The shape factors for spheres and other regular particles can be calculated from estimates of their coefficients of hydrodynamic drag in Stokes flow (Happel & Brenner 1973).

Dupuit's relation. In considering the movement of fluid through a porous rock, it is usual to consider the flux in terms of seepage velocity $\mathbf{u}$ which is equal to the product of the intrinsic phase average velocity $\bar{\mathbf{v}}$, and the porosity, ϕ (Philips 1991).

$$\mathbf{u} = \bar{\mathbf{v}}\phi. \qquad (9)$$

The intrinsic phase average velocity is the average velocity at which a fluid particle is transported through the medium by steady laminar Newtonian viscous flow, and like $\mathbf{u}$ is a vector defined with reference to the macroscopic flow direction. Equation (8) is known as Dupuit's relation (following Carman 1956).

Carman (1937) pointed out that by introducing the tortuosity factor, Dupuit's relation had to be modified. This is because the interstitial velocity must increase if the external seepage velocity is to remain unchanged while the fluid traverses a longer path through the medium. The time taken for a fluid element to complete the path l_{eh} along the bent or twisted capillary at the faster interstitial velocity $(\mathbf{u}/\phi)(l_{eh}/l)$ must equal the measured time for the fluid to traverse the bulk sample length l at the seepage velocity $\mathbf{u}$. Equation (9) becomes:

$$\mathbf{u} = \bar{\mathbf{v}}\phi\,\frac{l}{l_{eh}}. \tag{10}$$

If we apply Darcy's law for one dimensional flow we find the seepage velocity is

$$\mathbf{u} = \left(\frac{l_{eh}}{l}\right)^2 \frac{\phi r_h^2}{\beta\mu}\frac{\Delta P}{l} \tag{11}$$

through each tortuous capillary. Therefore, the overall factor by which the flow rate is retarded in a bent capillary is proportional to the tortuosity squared (see also Epstein 1989).

The Kozeny–Carman equation. Carman (1937) modified Kozeny's equation to include a tortuosity factor that is the square of the ratio of the effective hydraulic path length, (l_{eh}), for fluid flow through the 'equivalent hydraulic channels' in a medium, to the straight-line distance through the medium in the direction of macroscopic flow. This measure has become known as the 'channel equivalent' hydraulic tortuosity factor (Dullien 1992), here denoted T_{hC}.

$$T_{hC} = \left(\frac{l_{eh}}{l}\right)^2 = \tau_{hK}^2. \tag{12}$$

A more mundane explanation of the length squared relationship is given by Dullien (1992), who reasons that if a bundle of capillaries are increased in length while the volume they occupy remains constant (porosity is conserved) then the number of flow channels must be reduced. This reduces to an argument of whether or not the flow paths fill space, a conjecture which is developed on p. 334. Carman himself accepted that this exponent of two was not rigorously derived but that experimental results suggested a value close to two for granular media and well sorted porous rocks.

Providing that a granular aggregate is not crushed or cemented, and that the grains do not interpenetrate or deform, the surface area is independent of the porosity, or tightness of packing of the material. However, the surface area per unit volume (known as the 'specific surface') will increase as the packing density increases. Thus, with knowledge of specific surface area, S_0, and porosity, the size of the 'equivalent hydraulic channels' can be estimated:

$$r_h = \frac{\phi}{S_0}. \tag{13}$$

Combining equations (8, 12) and (13) Kozeny's equation for permeability becomes:

$$k = \frac{\phi^3}{\beta\tau_{hK}^2(1-\phi)^2 S_0^2} \tag{13}$$

which is known as the Kozeny–Carman equation. This equation is only semi-empirical but has been found to be nearly exact for random packs of monodisperse spheres when results of experiments are compared with numerical simulations (Cancelliere *et al.* 1990). The Kozeny constant is about 4.8 in this case, while for reasonably well sorted and rounded sands, a value of about 5 is generally accepted. Estimated shape factors range in value from about 2 to 3 for typical granular materials (Carman 1956; Van Brakel 1975). The value of Kozeny hydraulic tortuosity, τ_{hK}, therefore ranges from about $\sqrt{2}$ to 2 in unconsolidated granular aggregates. In consolidated rocks and soils the value of τ_{hK}, ranges up to about five (Dullien 1992). Witt & Brauns (1982) suggest that there may be an upper limit on tortuosity in uncemented granular media of about 10, even when the grains are platy rather than rounded.

How is hydraulic tortuosity related to geometrical tortuosity? In general, hydraulic tortuosity and geometrical tortuosity are not the same. This is because the paths taken by fluid as it flows through the porous medium are not straight lines, or close tangents to the rock grains, but rather are smooth curves tending to follow the axes of the flow channels (Fig. 1). Also, as a result of viscous drag, fluid flow is more retarded at the channel walls than along channel axes, so not all paths are equally favoured. This matter is discussed in a mathematical context by Coussy (1995), and is expanded upon on p. 329.

Despite these obvious differences, some workers have attempted to estimate permeability using geometrical tortuosities (e.g., Arch & Maltman 1990; Shepard 1993). Both of these studies employed a 'ray-tracing' technique which tracks the shortest path, skirting obstacles, from one face of the material to the other. The advantage of such methods is their simplicity, but the results obtained cannot be used inter-changeably in a formulation such as the Kozeny–Carman equation.

Effect of pore-size distribution on hydraulic flow in capillary models. Van Brakel (1975) pointed out that capillary tube models force the Carman hydrodynamic tortuosity factor to be a function of the tube radius, and hence of grain size and porosity. A severe limitation on the Kozeny–Carman equation is the assumption that a single hydraulic radius represents the combined effects of the innumerable flow channels traversing the medium. This is the problem of 'parallel-type pore non-uniformities' explored by Dullien (1992) in his critique of capillary models; however Carman himself had realized that his approach is limited to the case of a nearly uniform pore size population. As the particle size distribution, and so pore size distribution broadens, the predicted permeabilities tend to underestimate the real value. Olsen (1962) concluded that this was the principal reason that the Kozeny–Carman relationship failed for fine-grained soils, where large inter-aggregate pores dominate the flow, but smaller intra-aggregate pores form the mode of the pore size population. It could be reasoned that in the case of a range of flow channel radii, a range of tortuosities will also pertain. Thus, if the flow channel radius cannot be homogenised to single value, then the tortuosity may not be expressible as a single factor. This conjecture is developed on p. 311.

Effects of serial pore non-uniformities on hydraulic flow in capillary models. Real pore spaces consist of flow paths with rough and convoluted walls, rather than cylindrical channels of uniform cross-section (Fig. 2). The effects of these constrictions and bulges is to reduce the overall efficiency of the flow channels, as energy is expended accelerating and decelerating the fluid, and in minor viscous losses as the streamlines diverge and converge. The retarding effects of constrictions have been referred to as kinematic tortuosity by Bernabé (1991).

Even in creeping flows where inertial terms in the governing Navier–Stokes equations can be neglected, the streamlines are forced to diverge from the direction of the pressure gradient, and so flow becomes less efficient than in a straight-walled capillary of equivalent (harmonic-mean) diameter. Schopper (1966) introduced a constrictivity factor separate from the geometrical component of tortuosity to account for flow divergence (see also Van Brakel 1975: Rink & Schopper 1976). Azzam & Dullien (1977) solved

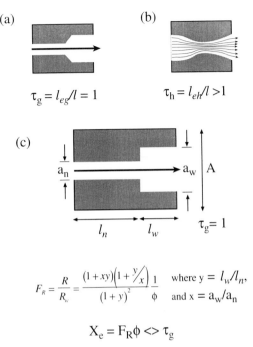

$$\tau_g = l_{eg}/l = 1 \qquad \tau_h = l_{eh}/l > 1$$

$$\tau_g = 1$$

$$F_R = \frac{R}{R_w} = \frac{(1+xy)\left(1+\dfrac{y}{x}\right)}{(1+y)^2}\frac{1}{\phi} \qquad \text{where } y = l_w/l_n,$$
$$\text{and } x = a_w/a_n$$

$$X_e = F_R\phi <> \tau_g$$

Fig. 2. Effect of changes in pore channel width on tortuosity. (**a**) The geometrical tortuosity is unaffected. (**b**) Fluid flow through a single constriction illustrating so-called 'kinematic' tortuosity due to convergence and divergence of flow lines. The effective hydrodynamic length, and so Carman hydraulic tortuosity is greater than in a channel of constant width. (**c**) The electrical tortuosity of a tube with a step change in diameter is not equal to the geometrical tortuosity (unity), but is a function of the ratio of the channel widths. From Dullien (1992).

the flow equations numerically for tubes with step changes in diameter, and so were able to calculate the effects of 'constrictedness' directly, at least in simple cases.

It will become evident in the section on electrical tortuosity that spreading and bunching of lines of flux also occurs during steady electrical conduction. Dullien (1992), used a 'constrictedness factor' to account for these 'serial-type pore non-uniformities'. The term 'kinematic' is therefore a misnomer, and the phenomenon is better called 'local flow divergence' unless momentum losses are significant. More importantly, it is not clear how the parallel components of sinuosity and constriction can ever be evaluated separately, except in an unrealistically simple model of the pore space of a rock. Indeed, Carman's definition of tortuosity takes both constrictedness and increased path length into account through his modification of Dupuit's relation.

Streamlines and tortuosity tensors. Dullien (1992) stated that:

'...the [hydraulic] tortuosity of a porous medium is a fundamental property of the streamlines, that is the lines of flux, in the conducting capillaries. It measures the deviation from the macroscopic flow direction in the fluid at every point.'

Carman (1956) attempted to visualize the streamlines in porous media flow by injecting dye into a glass bead pack. He found that the microscopic flow vectors diverged, on average, by an angle of about 45° from the macroscopic flow direction, and reasoned that the hydraulic tortuosity (τ_{hK}) could be estimated thus:

$$\tau_{hK} = \frac{l_{eh}}{l} \approx \frac{1}{\cos 45°} = \sqrt{2}; \quad \text{so} \quad T_{hC} \approx 2. \quad (15)$$

It may be that hydraulic tortuosity is determinate if we can somehow visualise the streamlines everywhere in a porous medium. For example, Northrup *et al.* (1992) and Peurrund *et al.* (1995) describe visualization methods using refractive-index matched media. However, from the results of Azzam & Dullien (1977), and subsequent calculations (Kostek *et al.* 1992; Saeger *et al.* 1995), it is evident that rapid changes in pore aperture, and blind pore spaces both produce recirculation features (eddies) even at very low Reynolds number. The presence of reconnected streamlines means that the notion of hydraulic tortuosity as a ratio of effective path lengths begins to lose its foundation (see p. 334). Nevertheless, the divergence vector envisaged by Carman can still be defined everywhere in the fluid (Fig. 3a & b).

Since at the continuum level this vector field varies from place to place in a continuous way, it is in theory possible to define a tensor that describes tortuosity in any pore space, including anisotropic and heterogeneous porous media. The single hydraulic tortuosity factor envisaged by Carman will therefore be the resultant of this tensor calculated over the sample volume, in the flow direction and for the given boundary conditions. Since the effect of this tortuosity field is to retard flow, some workers have found it convenient to define the tortuosity tensor so that it has a value less than one (the inverse of the Carman tortuosity factor). This is the approach followed by Bear (1972), who also showed that for a homogeneous, isotropic model of non-intersecting, straight capillaries, the tortuosity tensor produces an effective T_{hC} value of 3 in all directions (see also Freidman & Seaton 1996 and p. 318). This is rational, since

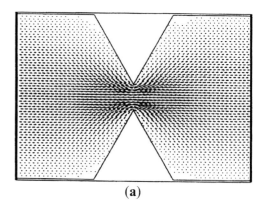

(a)

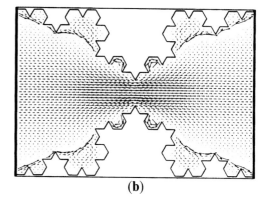

(b)

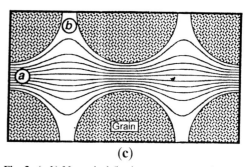

(c)

Fig. 3. (**a, b**) Numerical (lattice gas automaton) solutions to the hydrodynamic flow equations in idealized constricted pore systems with smooth (Euclidean surfaces) and with (fractal) surface roughness. The effect on surface roughness is minor, but flow is strongly concentrated in the throat regions in both instances. Regions of very low velocity recirculation (almost stagnant eddies) are produced in the re-entrant areas of the rough pore walls (the fluid velocity vectors in these regions are artificially magnified). Clearly, the effective hydraulic flow path is difficult to visualize in such a system. Reproduced with permission from Kostek *et al.* (1992). (**c**) Lines of electrical flux in a model pore space, each subtending 10% of total current, also show convergence and divergence, reducing the efficiency of flow. Reproduced with permission from Herrick & Kennedy (1994).

only a third of the capillary channels will carry flow in the desired direction.

Electrical tortuosity

Formation resistivity factor. The formation resistivity factor (F_R), of a porous medium is defined by the ratio of the conductivity of the porous medium saturated with an electrolyte to that of the free electrolyte (Archie 1942). It is commonly called simply the formation factor and in terms of measured conductivity (σ) or resistivity (ρ) it can be defined as:

$$F_R = \frac{\sigma_f}{\sigma_r} = \frac{\rho_r}{\rho_f}. \tag{16}$$

Where the subscripts f and r refer to the bulk electrolyte and bulk rock properties, respectively. Wyllie & Rose (1950) found that for electrical conduction, the conductivity of a sediment was inversely proportional to the porosity, so that the formation factor increased by a certain factor, X_e, with decreasing porosity.

$$F_R = \frac{X_e}{\phi}. \tag{17}$$

In the same way as we have already done for the Carman hydraulic tortuosity we can define electrical tortuosity in terms of an effective electrical path length l_{ee}:

$$\tau_{eW} = \frac{l_{ee}}{l}. \tag{18}$$

According to how equation (17) is derived, the 'generic' coefficient X_e becomes (l_{ee}/l) or $(l_{ee}/l)^2$. It appears that Wyllie & Rose (1950) redefined their electrical formation factor in order to retain parallelism with the existing Carman (1937) definition of tortuosity as a square of a ratio of lengths, i.e.,

$$\tau_{eCK}^2 = F_R\phi \quad \text{and} \quad \tau_{eW} = F_R\phi. \tag{19}$$

Winsauer *et al.* (1952) and Cornell & Katz (1953), used a different model of the pore structure to derive a relationship between formation factor and porosity, so that:

$$\tau_{eCK} = F_R\phi \quad \text{and so} \quad \tau_{eW}^2 = F_R\phi. \tag{20}$$

Wyllie never accepted this derivation of the conductivity equation (e.g., Wyllie & Gregory 1955), but the consensus is firmly on the side of equation (20) being correct (e.g., Walsh & Brace 1984). The discrepancy is simply a reflection of the different models: one pore space model conserves tube volume using the stereological

relationship between intercepted area and porosity while the other corrects for intercepted area of an inclined tube using an approximation that is exactly analogous to the modified Dupuit's relation used by Carman (1956). A truly rigorous derivation requires an assessment of the actual trajectories of electrical flow lines through the pore space. Regardless of these arguments, values of the factor X_e derived from formation factor measurements of rocks are known to range from about 1.3 to over 1000.

Effect of serial changes in pore diameter. Dullien (1992), following Owen (1952), modified the Wyllie electrical tortuosity to account for the effects of bulges and constrictions along the flow path (Figs 2c & 3c).

$$X_e = T_{eg} \cdot S' \tag{21}$$

where T_{eg} is the geometrical component of the electrical tortuosity and S', the electrical constrictedness factor, is the extra resistance (reduced flow efficiency) due to divergence of the flow lines in the bulges of the pore space. This modification is analogous to the separation of a geometrical tortuosity and a constrictedness factor in the case of hydrodynamic flow. Using this reduction, the electrical tortuosity is only equal to the geometrical tortuosity if all the flow channels are parallel-sided. Dullien showed that if this is not the case, it is easy to show that the calculated value for tortuosity varies very widely according to the model assumed for the pore space.

Archie's law. Archie (1942) found that the conductivity of a group of rocks belonging to a fairly uniform lithological unit, or of a similar type often varied systematically with the porosity and deduced an empirical relation, which has become known as Archie's 'law':

$$F_R = \phi^{-m}. \tag{22}$$

For a group of such rocks that are fully saturated, a logarithmic plot of formation factor, F_R, against porosity, ϕ produces a straight line with negative slope, m that generally lies between 1.3 and 3.0 (Parkhomenko 1967; Coates & Dumanoir 1973). The higher the value of m the greater is the increase in resistivity with decreasing porosity and this was found to correlate roughly with the degree of cementation of the rock formation and so is sometime known as the 'cementation exponent'. The important distinction between a global value of m equal to about 2 for a plot of F_R against ϕ for all rocks, and local values of m specific to sub-populations is discussed in an excellent paper by Madden (1976). Worthington (1993) demonstrates the importance of pore fluid salinity and pore surface conduction in the analysis of formation factor trends.

It can be shown that for a pore space consisting of uncorrelated pore and solid regions, random loss of porosity will result in a decrease in electrical conductivity such that m is equal to 2 (Dullien 1992). Higher values of m therefore indicate preferential loss of pore space important for conduction, while lower values of m can be explained by reduction in pore space that occurs predominantly in regions bypassed by flow (i.e., loss of dead-end porosity). Etris (1991) combined Archie's law, and the idea that the majority of the potential drop across the samples occurs in neck or throat regions of reduced cross-sectional area. From a quantitative study of rock thin-sections he deduced the following approximate relationship:

$$m \approx \frac{\log(\text{throat area})}{\log(\text{pore body area})}. \qquad (23)$$

This implies that the larger is the value of m the smaller are throats compared with pore bodies, and so the flow paths are less efficient (see p. 327).

Having examined some of the consequences of different mechanisms of porosity removal for the value of m, we can tie these in with changes in tortuosity effected by the same process. Combining equations [17] and [22] we get:

$$X_e = \phi^{(1-m)}. \qquad (24)$$

Electrical tortuosity and m are inextricably related, but m being a power in ϕ varies over a much smaller numerical range.

Modified Archie's law. Winsauer *et al.* (1952) introduced the modified Archie's law to account for the non-zero offset of the formation factor-porosity relationship that they observed.

$$F_R = a\phi^{-m}. \qquad (25)$$

The multiplication factor, a shows some variation between different lithologies, but remains at order unity, typically in the range 0.6 to 1.3 for marine sediments (Parkhomenko 1967). The factor a represents a combination of several second-order textural factors that affect the geometry of the pore structure, and it is sometimes considered that a is a type of tortuosity factor (e.g., Katsube *et al.* 1991). However, in equation (25), the electrical tortuosity is embedded in both a and m (Schopper 1966).

$$X_e = a\phi^{(1-m)}. \qquad (26)$$

The modified Archie equation has not found universal acceptance, largely because it does not satisfy the condition $F_R = 1.0$ when $\phi = 1.0$ except in the trivial case of $a = 1$. Nevertheless, given that porosities outside the range 0.1 to 0.6 are uncommon, we can use a empirically. It is also found that if we have a population of more than one type of particle, Archie's law applying to each, the overall law is like equation (25). Recently Perez-Rosales (1982) derived a generalized Archie's law from first principles, which unlike the Winsauer derivation, is valid for all porosity values:

$$F_R = 1 + G(\phi^{-m} - 1) \qquad (27)$$

here, G is a geometry factor somewhat different from a.

Corrected and 'true' electrical tortuosity
Katsube *et al.* (1991) presented a detailed discussion of electrical tortuosities. They measured the formation factor of a number of deeply buried shales from the Scotian shelf. They found the formation factor, corrected for surface conduction effects ranged from 140 to over 17 000. The porosity was determined using the mass balance method on samples saturated with water, and then dried and re-weighed. This they termed the 'effective' porosity (here we use wet-dry porosity ϕ_{wd}), which ranged from 0.008 to 0.09. Accordingly, the uncorrected electrical tortuosity factor X_e ranges from 11.9 to 141. Katsube *et al.* (1991) considered these values to be apparent tortuosities. They attempted to remove the effects of 'pocket porosity', ϕ_p, which they define as 'dead-end pore space and enlargements along the flow path', from the determination in order to arrive at a 'true tortuosity', τ_{etK}. They used the following equation

$$\phi_{wd} = \phi_p + \frac{b\tau_{etK}^2}{F_R}. \qquad (28)$$

The coefficient b is equal to 3/2 for isotropic material. Katsube *et al.* (1991) regarded their 'true' tortuosity as a better measure of the actual sinuousness of the pore channels traversing the rock. A mean value for τ_{etK} of about 3.3 was found for the eleven shale samples.

True tortuosity and pocket porosity can only be calculated for a population of samples, as it is assumed that the underlying geometry of the pore structure remains constant with changing porosity. Given that flow channels will undoubtedly change in shape as well as size during any feasible mechanism of porosity reduction, this assumption appears to be questionable.

Tortuosity in diffusion and dispersion

Traditional concepts of tortuosity used in petrophysics focus on hydraulic and electrical conduction. In this section it will be shown that the consideration of a spectrum of diffusional processes enables tortuosity to be understood more fully for all transport processes.

Molecular diffusion and Fick's laws. When the concentration of a component varies from place to place within a fluid mixture, the random motion of molecules tends to transport that component from areas of high to low concentration, so that over time the system becomes homogenised with respect to the concentration of each mobile component. The transport of matter in this way is known as molecular diffusion. Pure diffusion in the absence of advection and chemical interaction, i.e., the transport of an ideal tracer, can be described by Fick's laws of diffusion (see, e.g., Skelland 1974; Berner 1980; Shackelford & Daniel 1991. Crank 1975 gives a mathematical treatment).

Fick's first law states that the steady-state flux density q_D, of a mobile component is proportional to the spatial gradient of its concentration (c) in the fluid.

$$q_D = -D\nabla c. \qquad (29)$$

The negative sign indicates that flow of matter is in the direction of declining concentration. If q_D is measured as mass per unit area of sample per unit time, then c is measured in units of mass of tracer per unit volume of fluid. Alternatively, molar measures can be used for concentration and flux.

Fick's second law refers to the rate of change with time of concentration at points in the medium.

$$\frac{\partial c}{\partial t} = D\nabla^2 c. \qquad (30)$$

In both cases D is the diffusion coefficient or diffusivity, having the dimensions $L^2 T^{-1}$. Fick's laws are strictly valid for ideal tracers, and this requires a high level of dilution for real materials. Inside porous media, Fick's laws apply only where surface effects are negligible, so that molecular diffusion controls transport. The 'points' at which concentration is defined

are volumes in the porous medium that are large enough for the continuum approximation to be valid. In practice this means regions rather larger than the characteristic pore dimension.

Tortuosity for molecular diffusion in porous media. In a bulk solution consisting of a non-reactive solvent fluid and non-interacting solute components (a passive tracer), the molecular diffusivity (D_{bulk}) is controlled by the physical properties of the fluid and tracer, and the pressure and temperature of the system. Inside a porous medium under the same conditions, molecular diffusion is inhibited by the presence of boundary walls. Therefore the effective diffusivity (D_{por}) is reduced with respect to the bulk solution. There are many ways in which this retardation has been accounted for, most workers resorting to a tortuosity coefficient of one type or other. The terminology found in the literature is not consistent either within or between the various scientific disciplines, and so the remainder of this section tries to reconcile the various definitions that may be encountered.

The ratio of the effective to bulk diffusivity, has frequently been termed the diffusional tortuosity; and is usually given the symbol θ.

$$\theta_1 = \frac{D_{por}}{D_{bulk}}. \qquad (31a)$$

This definition, apparently introduced by Satterfield & Sherwood (1963; see Greenkorn 1983), remains the most prevalent in current engineering literature. The parameter θ_1 encompasses the effects of all geometrical and chemical interactions on the apparent diffusional constant relative to that found in free solution. Since it has a value less than one we will refer to θ, henceforth as a reciprocal tortuosity. Note that several workers, for example Berner (1980), use the symbol theta for a path length tortuosity analogous to that of Cornell & Katz. To avoid confusion while remaining consistent with our earlier usage, we can replace theta with an 'X' factor for diffusion (see Appendix):

$$\frac{D_{bulk}}{D_{por}} \equiv \theta_1 \equiv \frac{1}{X_{d1}}. \qquad (31b)$$

Another common approach has been to define reciprocal diffusional tortuosity to exclude measured porosity, so that only the effects of sinuosity and dead end-pore space are considered.

$$\theta_2 = \frac{D_{por}}{\phi D_{bulk}} \equiv \frac{1}{X_{d2}}. \qquad (31c)$$

X_{d2} appears at first to be directly analogous to the reciprocal of the electrical retardation factor X_e (Van Brakel 1975; Epstein 1989). We will examine this assumption on p. 322.

The final class of reciprocal tortuosities that the reader may encounter are formulated to exclude effective porosity, so constrictedness and blind pore space are accounted separately.

$$\theta_3 = \frac{D_{por}}{D_{bulk}\phi_{eff}} \equiv \frac{1}{X_{d3}}. \tag{31d}$$

This definition apparently corresponds to the reciprocal of the 'real' electrical tortuosity of Katsube *et al.* (1991).

Gas diffusion. So far we have considered only molecular diffusion, in which the tracer particle motion is random and uncorrelated, being dictated by interparticle collisions that conserve momentum. Under these circumstances, interactions with the pore walls are unimportant, except that they confine the molecular motion geometrically. Gradients of concentration, and so transport vectors, are not strongly influenced by the presence of walls, and so at the continuum scale, Fick's laws are obeyed. The molecular diffusion paradigm would apply, for example, to a dilute solution of a passive tracer in a liquid solvent at rest that is confined in a chemically inert porous medium whose pore structure is homogeneous and isotropic. If we are to treat more general diffusive transport, we must examine processes at the molecular scale. This is especially important for the case of gas diffusion, where we must consider the roles of both intermolecular and molecule wall-collisions in some detail in order to understand the macroscopic phenomena observed in nature (see Cunningham & Williams 1980).

According to the kinetic theory, the pressure of a gas is defined by the concentration of particles, and their velocities, which increase with increasing temperature. The internal pressure of a gas is a reflection of the mutual interaction of particles as they collide with each other and with the walls of the vessel (or the pore walls within a porous medium). Collisions between gas molecules can be considered completely elastic: momentum and kinetic energy are conserved in the gas as a whole at the collision points, and so throughout the body of the gas. When a molecule collides with a vessel or pore wall, total momentum is still conserved, but only if the collision is elastic and specular will no momentum be transferred from the fluid to the solid body of the enclosing medium. Specular collisions, where the angle of molecule incidence

and the angle of reflection are equal, occur when the walls are smooth on a length-scale comparable with the gas molecule dimension, and when the interaction time is essentially infinitesimal. If the walls are rough, or the particles reside for some time on the pore walls before travelling onwards in an arbitrary direction, then the collisions become diffuse, with the particles being scatted in all directions (Cunningham & Williams 1980). Under such circumstances, a stream of molecules travelling in a preferred direction will be retarded by multiple wall collisions, as on average some forwards momentum is lost for each increment of forwards travel (Fig. 4). This is true whether the flow is liquid or gas, or driven by a compositional gradient or a gradient of pressure.

Consider a mixture of two gases under isobaric conditions, but with a gradient in composition existing across a sample of a porous medium. In a high pressure mixture, the average distance molecules travel between collisions (or mean path length) will be very small. Therefore most collisions will occur between particles, and wall interactions will be relatively infrequent. The

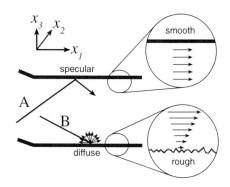

Fig. 4. Effect of pore wall roughness on motion of diffusing particles. In case of an atomically smooth wall, encountered by particle A (top inset), the collisions are specular and no momentum in the x_1 direction is lost, thus the net flow velocity profile is uniform and shows no reduction in speed at the walls. In the general case of a wall with roughness on atomic and larger distances, the collisions are scattered in all directions (e.g. path B), and diffuse resulting in zero net flux along the walls. In the case where many particles are moving together, a viscous velocity profile develops, with an increase in mean velocity from the walls into the pore channels as some net momentum is transferred from particle to particle in the x_3 direction (bottom inset). Where the particle concentration is very low, and path lengths are long compared with the pore dimensions, mutual collisions do not enforce this viscous profile, and so-called 'slip-flow' will occur.

mixture will homogenise over time as a consequence of molecular diffusion. The transport
of each component is in fact coupled by their
mutual interactions in such a way that the net
flux of each component with respect to the
compositional gradient obeys Fick's first law
(Cunningham & Williams 1980). The homogenization rate will be retarded by the presence of
the porous medium relative to bulk conditions,
giving rise to a lower effective molecular diffusivity in the porous medium:

$$\frac{D^M_{\text{bulk}}}{D^M_{\text{por}}} = X^M_{\text{d1}}. \tag{32}$$

In a rarefied gas mixture, the particles travel, on
average, a considerable distance between collisions; in other words, they have a high mean free
path length. If this path length is longer than the
mean pore size (this length ratio is known as the
Knudsen number, N_K) then most collisions will
be with the pore walls. In such a system, when
$N_K > 1$ (the Knudsen region) each gas component will move independently until the compositional gradient is erased. This decoupling leads
to an excess of flux for each component over
that produced by the random buffeting of
molecular diffusion experienced in a system
without walls or where the walls are far apart
(Fick's laws are no longer obeyed).

Viscosity, start up time and flow development.
Now consider the effect of imposing a gradient
of pressure on the system rather than enforcing
isobaric conditions. Under these conditions the
diffusive flow is driven by the gradient in
composition (molecular density) since during
the random motions, more molecules will move
from high to low density (from high to low
pressure) areas than in the reverse direction. If
there are no walls the pressure gradient will
rapidly be removed by molecular diffusion. If
the flow is through a porous medium, then the
effect of collisions on the pore walls will have a
net retarding effect on the flow because on
average, the diffuse reflections will reduce the net
forward momentum of the particles. At the pore
wall itself, the net velocity is zero, and away
from the pore walls, recently reflected particles
will collide with, and slow the molecules that
stream past: this effect is known as viscous
retardation.

After some time (the 'start-up' time) a steady-
state transverse gradient of forwards velocity
emerges, where the flow rate is maximum at the
axis of the pore channels, zero at the walls and
has a 'Poiseuille-type' velocity profile in between.
For a cylindrical channel this velocity profile is

parabolic, but for other sections the profile may
differ significantly according to sectional shape
and longitudinal changes in aperture size (see,
Brown *et al.* 1995 for a good example of this). If
we consider the case of bulk fluid entering a
cylindrical tube, the flow profile will also vary
along the constant tube section, as the viscous
boundary layer takes some distance, as well as
some time, to become fully developed (Skelland
1974). The development profile for the tube is
steady in time and varies smoothly from a flat
velocity profile at the entrance to a parabolic
profile some distance from the entrance (Fig. 5a).
This development distance is a function of fluid
viscosity and sectional area. The fact that
apparent slip occurs at the pore walls does not
imply any extra flow through the tube within the
development distance, rather the decrease in
velocity down to zero at the walls is balanced by

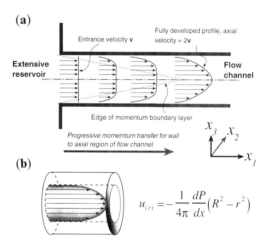

Fig. 5. (a) Start-up flow in a porous medium channel,
here for simplicity assumed to be a slot of infinite
extent in the x_2 direction (normal to the page). To
maintain continuity, flux at each section is constant.
For such plane Poiseuille flow the fully developed
profile has an axial velocity in the x_1 direction of
exactly twice the entrance velocity. Momentum is
transferred from the wall region to the axial part of the
flow within the start-up region, beyond which flow in
the channel is fully developed. These criteria ensure
that Darcy's law for viscous porous medium flow is
valid on the macroscale even through the flow channel
widths vary. (b) Flow of a Newtonian linear-viscous
fluid in an idealized smooth tube illustrating the
development of a parabolic Poiseuille velocity profile
due to the 'no-slip' boundary condition on the
sidewalls. The overall flux is proportional to the
pressure gradient. Hydrodynamic flow per unit
sectional area is much greater in the centres of pore
channels than at the edges, a factor not taken into
account in purely geometrical tortuosity.

an increase in the axial velocity as the fluid progresses along the tube: the development of viscosity along a spatial dimension involves the transfer of momentum from near the walls towards the central part of the flow channels. We can compare this with the time-varying situation during start-up, where the fluid accelerates from rest and momentum is transferred through the fluid in a streamwise direction, and simultaneously in a transverse direction as viscosity becomes manifest.

After the start-up time has elapsed, i.e. at the steady state, the total resistive force integrated across the area of the flow channels must equal the driving force for flow, which is the product of the differential pressure and the sample cross section. Scaling up the size of the flow channels while maintaining their shape will cause an increase in the rate of flow, and in the steepness of the lateral velocity gradients, so that the overall viscous retardation will stay the same (the pressure drop remains balanced). Thus unlike the case for molecular diffusion, where we ignore the pressure drop and consider only concentration terms, the flow rate in the viscous regime is dependent on the size of the flow channels as well as the total cross sectional area of pore channels available for flow (Fig. 5b).

Slip flow and the influence of flow regimes on tortuosity. In the limit of low Knudsen number (liquid or a high pressure gas), flow that is driven by a pressure differential across a medium can be considered to be purely viscous, and providing this viscosity is linear with shear stress

(Newtonian fluid) the flow will obey Darcy's law. At lower pressures, or in constricted throats, the effective viscosity is reduced and transport efficiency is increased by an additional diffusive flux (Igwe 1991). This phenomenon is termed slip flow, or to petroleum engineers, the 'Klinkenberg effect' (Klinkenberg 1941). In the Knudsen regime, particles bounce unhindered from wall to wall, while in a diffusive gas, particles tend to be distributed randomly in the pore space, and in the viscous regime the particle trajectories bunch along the centres of pore channels. Gas transport therefore has the particular property that the measured diffusivity depends upon the pressure of the gas, as the pore space is probed differently by kinetic particles with different mean path lengths (Fig. 6). The effective diffusivity in the Knudsen region is greater than that predicted by extrapolating from molecular diffusion. We have the important result that the measured diffusional tortuosity is lower in the Knudsen region due to an excess of streaming flow in small pores that is not hindered by intermolecular collisions:

$$\frac{D^M_{\text{bulk}}}{D^K_{\text{por}}} = X^K_{\text{dl}} < X^M_{\text{dl}}. \tag{33}$$

Petersen (1958) provides a classic analysis of this problem showing that diffusional tortuosity is lower in the Knudsen regime than in the normal diffusion regime for any given value of the pore body to throat ratio.

Since, in the general case, the streamlines are distributed differently in viscous and molecular flow, we would not expect the tortuosity in these

| Knudsen diffusion | Molecular diffusion | Onset of viscosity |

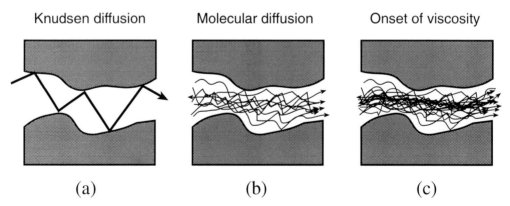

| (a) | (b) | (c) |

Fig. 6. (a) Knudsen diffusion: repeated single random walks of a particle undergoing elastic collisions with the pore walls traces out the tortuosity in this regime, since particles are sparse and the mean path length is long compared with the particle diameter. (b) In the molecular diffusion regime, behaviour can be modelled by simultaneous random walks involving collisions between particles. (c) In the limit of short particle paths, interactions between particles involving short-range attraction leads to the development of viscosity, and to the no-slip boundary condition on pore walls under a gradient of pressure. This leads to the concentration of flow paths towards the centres of pore channels, modifying the average tortuosity.

different flow regimes to be the same either. Since diffusional slip flow increases flux over that with fully developed viscosity, we can argue that hydrodynamic tortuosity is greater than tortuosity retarding molecular flow in the same pore space.

Random walk models, lattice gases and diffusional tortuosity. The kinetic model of gases considers an ensemble of independent particles which move on straight trajectories between collisions, so diffusion in gases can be modelled by using a numerical simulation of a random walk. If we consider this steady-state pattern of random transport with a known path length (unsteady flows will be described below), we can estimate some of the invariant geometrical properties of the medium. The use of stochastic or 'Monte-carlo' simulations of random walkers to model diffusion in porous media was pioneered by Scheidegger (1954) but only became commonplace after significantly greater computing power became available (Evans *et al.* 1980).

Lee & Kozak (1987) describe such a model which they use to derive tortuosity factors for single-phase transport. They define the tortuosity in terms of the average aggregated walk length for particles entering one face of a sample and exiting the opposite face. They used different boundary conditions to establish repulsive, focusing and confining interactions at the pore walls. As predicted above, Lee & Kozak found that quite different effective aggregate path lengths emerge when different mean gas pressures are simulated, i.e., when different step lengths are used in the walk. Chantong & Massoth (1983) and Abbassi *et al.* (1983) give didactic examples and experimental data to support this conclusion. Nakano *et al.* (1987) show that path length also varies locally within the medium according to the presence of bulges and constrictions. Zhang & Seaton (1992, 1994) present similar arguments and show under what conditions it is valid to treat the material as a continuum.

In order that stochastic models should simulate the behaviour of diffusing particles in a realistic manner, we must fulfill the following criteria. First, the geometry and topology of the pore space should be realistic (see section on tortuosity in networks). Secondly, the boundary conditions for wall collisions should be enforced correctly (normally collisions are diffuse rather than specular). Finally, particle interactions must be properly accounted for.

An alternative approach to using structured random walks is the use of cellular automaton or lattice gas simulations (Frisch 1986; Rothman 1988; Qian *et al.* 1992). In these computational models, the statistical properties of the fluid arising from kinetic interactions between particles are simulated by defining a fixed lattice of sites at which the collisions are forced to occur under a simplified set of rules, which ensure that energy and momentum are conserved. When the correct protocol for collisions is enforced, it is found that many different transport processes can be simulated within a single lattice. When the initial particle concentration is changed, by altering the probability that a particular site is occupied, the mean free path between collisions also changes, and this allows different flow regimes to be simulated. The rules describing flow on the macroscopic scale, e.g., Fick's laws and Darcy's law, evolve spontaneously after a certain number of iterations, without having to be prescribed *a priori.*

One advantage of cellular automaton methods is the ability to follow individual 'virtual particles' along their interaction trails through the lattice. By averaging over a number of cycles, it is possible to define the streamlines in the fluid lattice. In this way, the cellular automaton method has been used to investigate the effects of pore geometry on transport processes (Rothman 1988; Kostek *et al.* 1992; Figs 3 & 7). The method is particularly successful because it is easy to visualize changes in the flow regime on the resultant transport patterns. As more work is done in three spatial dimensions (e.g., Chen *et al.* 1991, Cancelliere *et al.* 1993; Knackstedt & Zhang, 1994) it is hoped that a fuller understanding of this aspect of tortuosity will be gained. Indeed Zhang & Knackstedt (1995) have reported a number of simulations designed specifically for this purpose (see p. 327).

Unsteady flow, multi-valued tortuosity, averaged tortuosity, information loss. Consider a collection of particles released instantaneously at time $t = 0$ at the up-flow end of a rock sample (Fig. 8a). The particles will travel through the rock along a random walk constrained by the pore geometry. The time it takes for the particles to traverse the specimen and reach the outflow face is defined by the number of walk steps in the path. If there are sufficient steps to average out random variations, then the distribution of particle arrivals over time will be representative of the distribution of effective lengths of the paths available for flow through the medium. Therefore, in the case of diffusive transport, or random walks, we have to consider a population of tortuosities rather than a single value of tortuosity in the medium (see also Matthews & Spearing 1992). If we move from the discrete to

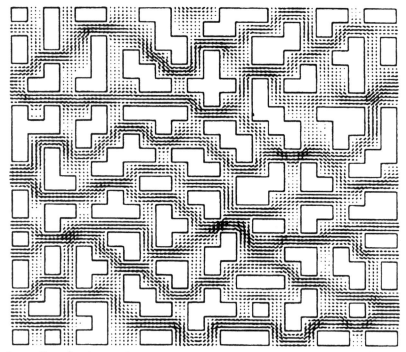

Fig. 7. Simple two-dimensional tortuosity 'maze' model solved for steady, single phase fluid flow solved using a cellular automation method. Reproduced with permission from Rothman (1988). Since the flow pathways can be traced through the medium, the tortuosity can be calculated directly, as has been shown for three-dimensional media by Zhang & Knackstedt (1995). A simpler method may be to calculate the aerosity along a section normal to flow using the intercepted flow vectors, and use this in the equation of Ruth & Suman (1992).

the continuum representation of diffusive flux through the medium, then the rate of particle arrivals can be integrated to give the outflow concentration, and its variation over time can be described by Fick's second law (equation [36]). The temporal structure of the outflow concentration profile carries information about the tortuosity distribution in the medium.

Now consider the case in which we maintain the concentration of particles in the upflow reservoir at a constant level $c1$, and continuously flush the downstream reservoir of particles so that the concentration there is always zero (Fig. 8b). This is classically known as the Wicke-Kallenbach experiment: see Patwardhan & Mann (1991). Having established a constant gradient in concentration, we will after some time arrive at a steady-state flow rate and according to Fick's first law (equation (29)) the number of particle arrivals at the down-flow face will be constant over time. In this instance we cannot learn about the distribution of path lengths as we have lost information about when each particle was released. In this experiment we are only able to deduce the average path length and so the average tortuosity from the

flux-gradient relationship. This distinction between fluxes and concentrations proves crucial to a full understanding of tortuosity (see p. 323).

This argument is developed further in an excellent paper by Tye (1982), who points out that the relationship in equation (18), between electrical-path length tortuosity and formation factor will hold only in the case where all pore channels are of equal equivalent length. In general, the effective conductivity of the porous rock will be equal to a sum of the conductances of each individual (ith) pore channel. In other words, the macroscopic current, I is carried by $i=1$ to n tortuous pathways. From the Cornell & Katz definition of electrical tortuosity, the conductivity and inverse formation factor will be as follows:

$$\sigma_r = \frac{IL}{VA}, \quad \frac{\sigma_r}{\sigma_f} = \frac{1}{F_R} = \sum_{i=1}^{n} \frac{\phi_i}{\tau_{eCK,i}^2}$$

$$\text{or} \quad \sum_{i=1}^{n} = \frac{\phi_i}{X_{e,i}}. \tag{34}$$

Here, I and V are the macroscopic current through, and potential drop across, the sample.

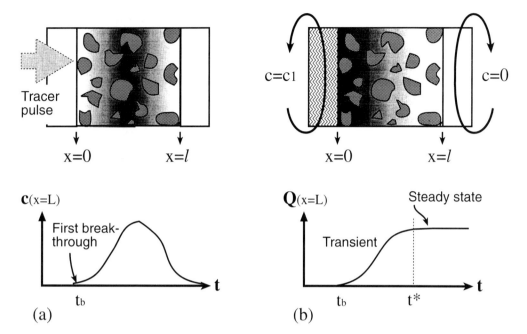

Fig. 8. Thought-experiment considering the effect of multiple flow paths on diffusional tortuosity through a water-saturated porous medium, after Tye (1982). (**a**) Pulses of tracer are introduced to the upflow end and diffuse through the porous sample. Concentration at the outflow end varies over time. Early arrivals follow the least tortuous paths (breakthrough point, t_b) while later arrivals follow the most tortuous paths; the whole spectrum of available paths defining the full outflow curve. (**b**) Concentrations are maintained constant at either end of the sample by circulating tracer from a reservoir at one end and flushing fresh water past the other end. The outflow curve is plotted as downstream flux versus time. The initial, transient part of the curve records a population of increasingly tortuous flow paths through the medium. The plateau marks the steady state flux, from which point (t^*) onwards only a homogenised, average value of tortuosity is recoverable.

From the above discussion it is evident that steady state measurements – such as electrical conductivity – cannot distinguish what type of effective path length (tortuosity) distribution exists in the rock. If we conduct an unsteady diffusion experiment, on the other hand, we can gain some information on this distribution. The presence of pore channels with higher than average tortuosity will be betrayed by the continued egress of tracer particles at long times. Conversely, the presence of unusually straight pore channels will appear as an early breakthrough in the tracer concentration curve.

Dispersion: background. When a liquid carrying a tracer is subject to an overall pressure gradient, material flows due to a combination of diffusion and bulk movement (advection). This type of combined flow is termed dispersive transport, and is a particular case of miscible displacement in which the liquid carries a passive tracer, or the two fluids have equal viscosity (Bear 1969; Scheidegger 1974; Greenkorn 1983).

In order to describe the changes in tracer concentration with time, a combination of the Stokes equation for steady incompressible flow and Fick's second law. The equation is known as the advection-diffusion equation (or convection-diffusion equation), which on the micro-scale is expressed as:

$$\frac{\partial c}{\partial t} + \bar{\mathbf{v}} \cdot \nabla c = D_{\text{por}} \nabla^2 c. \qquad (35)$$

The advective term is the dot product of the mean microscopic velocity (assumed here to be the intrinsic phase average velocity defined by Diedericks & du Plessis 1995) and the local gradient in concentration. Dispersion following equation (35) is often known as Taylor–Aris dispersion (Taylor 1953; Aris 1956). Where the hydrodynamic flow is steady and one-dimensional on the macroscopic scale the right hand side of (35) can be resolved into two components:

$$\frac{\partial c}{\partial t} + \bar{\mathbf{V}} \cdot \nabla c = D_L \frac{\partial^2 c}{\partial x_1^2} + D_T \nabla_T^2 c. \qquad (36)$$

Where $\bar{\mathbf{V}}$ is the macroscopic mean velocity vector, D_L is the effective diffusivity in the longitudinal (x_1) direction and D_T is the effective diffusivity in the direction transverse to the macroscopic flow (x_2 and x_3 directions).

There are two end-members of dispersion. When the diffusivity is zero, transport is purely advective, and tracer particles follow smooth trajectories along the fluid streamlines. When the diffusivity is very great with respect to the advective velocity, the tracer particles essentially undergo pure diffusion. The exact nature of this diffusion in a pore space is dependent on the mean path length of the particles, as discussed below, so the following discussion is limited to the case of passive tracer movement in a viscous fluid where the mean particle path length is very much smaller than the pore size. The parameter describing the relative importance of the two transports in the advection-diffusion equation is the Péclet number (e.g., Bear 1969).

$$Pe = \bar{V}l/D_{bulk} \qquad (37)$$

where l is some characteristic length scale of the medium (pipe diameter or pore size), and $\bar{V}$ is the macroscopic mean fluid velocity vector. When Pe is small the transport is dominated by diffusion, when it is very large, the tracer particles are essentially carried by advection. Note that in equation (37) the coefficient used is the bulk diffusivity, while in the advection–dispersion equation (35, 36) it is the effective diffusivity in the pore space.

Tracer dispersion and tortuosity. Ma & Selim (1994) considered tortuosity during dispersion of a passive tracer (tritiated water) during hydraulic flow through a saturated soil. They used the notion of an effective solute transport length (l_{ed}) and defined the tortuosity as l/l_{ed}, where l is the column length, so that it was equivalent to θ_1 (above). The transport velocity $\bar{V}$ was measured from the water flow rate. The value of dispersive tortuosity was calculated from the mean residence time, and used to see if it affected pure diffusion by the same factor, using two methods of calculation: one based on residence time for a small tracer pulse, and another based on anomalous diffusion for a larger tracer pulse. Diffusional and dispersive tortuosity were found

to be very close in value, X_{d1} ranging from 1.3 to 3 in the silty clay soils tested. Greenkorn (1983) and more recently, Kempers *et al.* (1992) discuss applications of dispersivity measurements for petrophysical appraisal.

Numerical simulations of tracer diffusion and dispersion. In an important paper Koplik *et al.* (1988) extended the theory of random walks in network models to account for the effects of a net advective flow on otherwise diffusive behaviour. Martys (1994) presented some visualisations of flow of a passive tracer in numerical simulations of two-dimensional porous media (Fig. 9). These simulations show that the particle trajectories vary markedly according to the Péclet number. In other words, the average tortuosity and the range of tortuosities in the various pathways is a function of the ratio of diffusive to advective transport, even though the geometrical tortuosity of the pore space must remain constant.

The pore space is explored most fully by the tracer particles when there is a significant amount of diffusion into protected pore space, but this information is best resolved when a considerable component of advective flow stretches out the time base over which observations are recorded. If the flow is purely diffusive then information on pore space with different levels of accessibility is smeared out and homogenized by longitudinal mixing. Therefore, we can conclude that in tracer experiments the maximum amount of information about the pore space is obtained from transport processes at moderate Péclet number when the relative contributions from viscous retardation and diffusive storage are balanced.

Other tortuosity formulations

Dielectrical tortuosity. Bitterlich & Wöbking (1970*a, b*) introduced the concept of capacitative formation factor as a measure of pore structure. The effective capacitance of a rock of porosity ϕ area A and length l is given by:

$$C_r = Y_{pn} \cdot \frac{\phi A}{l^2 \tau_{ew}^2 f^2}. \qquad (38)$$

Here f is the frequency of the electrical signal, and Y_{pn} is an electrolyte constant reflecting the ionic mobility and dielectric properties of the positive and negative ions in the pore fluid. As the capacitance has the dimensions of length, the constant Y has dimensions of acceleration. The factor τ_{ew}^2 is the square of the Wyllie & Rose

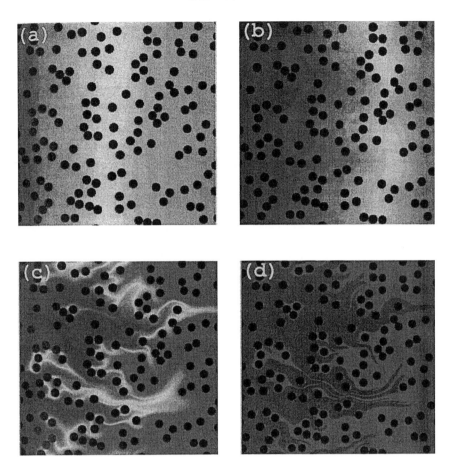

Fig. 9. Numerical computer model of hydrodynamic dispersion in flow through a two-dimensional array of monodisperse discs. The Péclet number increases from zero (pure diffusion (**a**)), through to infinity (pure advection in a viscous hydrodynamic flow (**d**)). Dark indicates high concentration, light indicates medium concentration, mid-grey indicates low concentration. The fronts of constant concentration are increasingly convoluted with higher Péclet number, and mean tortuosity increases accordingly. Reproduced with permission from Martys (1994).

(1950) electrical path-length tortuosity (l_{ee}/l) derived from the 'inclined tube' model of the pore space. The capacitance of two parallel plates is:

$$C = \varepsilon \, \frac{A}{4\pi d} \qquad (39)$$

where ε is the dielectric constant. The effective dielectric constant of the rock then becomes:

$$\varepsilon_r = Y_{pn} \cdot \frac{4\pi \phi A}{l^2 \tau_{ew}^2 f^2} \qquad (40)$$

while the dielectric constant of the pore fluid in bulk is:

$$\varepsilon_f = Y_{pn} \cdot \frac{A}{l^2 f^2}. \qquad (41)$$

By analogy with the formation resistivity factor, Bitterlich and Wöbking define a formation capacitance factor to be (40) divided by (39).

$$F_C = \frac{Y_{pn} \cdot \dfrac{A}{l^2 f^2}}{Y_{pn} \times \dfrac{\phi A}{l^2 \tau_{ew}^2 f^2}},$$

$$\text{so} \quad F_C = \frac{C_f}{C_r} = \frac{\varepsilon_f}{\varepsilon_r} = \frac{\tau_{ew}^2}{\phi}. \qquad (42)$$

The dielectrical tortuosity used here parallels that used by Cornell and Katz (1952) for electrical conduction, as the two formulations are defined relative to the same model of the pore space. According to Parkhomenko (1967), the dielectric constant of a dry sandstone may be about 5 and that of the pure electrolyte about 80 at about 30 kHz and above. The limiting value for saturated rocks is very variable but may be about 40 for a fine-grained rock and apparently much less for a sandstone. The dielectric constants of water and saturated rocks increase with increasing frequency. The dielectric constant in the rock increases less rapidly than that of free water as the d.c. limit is approached. Therefore dielectrical tortuosity can be expected to depend upon frequency.

Work in this field is now generally considered under the topic 'induced polarization' which is a more general term than capacitance and encompasses the effects of various physico-chemical interactions between solute ions and minerals, particularly clays, found in rocks. Bitterlich & Wöbking (1971) attempted to link their capacitative tortuosity with geometrical tortuosity, but their analysis hinges on particulars of the pore structure model that they chose. Given the complexity of the frequency-dependent I.P. response of many rocks, it is unlikely whether a unique and objective measure of geometrical tortuosity can be obtained from this method. More recently, work by the Schlumberger group and others (Johnson et al. 1987) has produced a number of exact results for frequency dependent tortuosities. These results are discussed below, and on pp. 327–329.

Sound propagation in porous media and dynamic permeability. While the mechanisms of seismic wave propagation are of considerable interest to petrophysicists, and in many ways depend, like other transport properties, upon the pore structure of the medium, the discussion here is intended only as a pointer to recent important developments. The principles of sound propagation through porous media were first put in a mathematical context by M. A. Biot (Biot 1962a, b; see also Chang et al. 1988), who identified a fast compressional wave and a shear wave transmitted through the solid framework and carrying the bulk of the energy and a less energetic compressional wave or slow wave travelling through the fluid phase. Biot's approach emphasizes the macro-scale coupling of framework distortion and associated changes in pore fluid pressure as waves pass through the medium, and largely neglects the micromechanics of energy transfer between solid and fluid phases, or the frequency dependence of wave velocity and attenuation.

Johnson and co-workers (Johnson et al. 1982) established a rigorous link between the slow wave propagation in a rigid porous medium and hydrodynamic permeability, as the latter is a special case of fluid movement in the 'direct-current' limit. Accordingly, the extent to which the pore structure permits oscillating flow associated with periodic changes in pressure has become known as the dynamic or 'a.c.' permeability. Tortuosity enters the formulation of dynamic permeability in a similar way to tortuosity for the case of steady flow. The relative degree of retardation and attenuation of pressure waves caused by the structure of the porous medium is found to be dependent on their frequency, because different wavelengths probe different length-scales of the medium. The different probing frequencies are in many ways analogous to the random walkers with different step lengths that were discussed earlier.

Nuclear magnetic resonance methods. In recent years nuclear magnetic resonance (NMR) has become an important tool in petrophysics (Howard & Kenyon 1992; Borgia et al. 1994). NMR imaging is particularly useful for the study of phase distributions within porous rocks, but the resolution that can be achieved currently does not allow the direct visualization of the pore space itself. Fortunately, other NMR techniques enable a number of geometrical and topological characteristics of rocks to be deduced from the study of restricted diffusion stimulated in the medium by pulsed changes in magnetic field intensity (Mitra et al. 1992).

Pulsed-gradient techniques allow a range of length scales to be probed through a variation in the frequency of the magnetic pulses. In the short-time limit, spin diffusion is essentially unrestricted, giving an estimate of the bulk self-diffusion coefficient D_{bulk}. Over longer times, the effect of the pore walls is felt, so that measurement of the mean displacement of particles can be used to estimate D_{pore}. With these data it is then possible to estimate diffusional tortuosity on the micro-scale. Lorenzano-Porras et al. (1994) studied diffusion in fine-grained porous materials using pulse-field gradient spin echo NMR to give values of D_{bulk} and D_{pore}. On the basis of their data, the self-diffusional tortuosity X_{dl} for liquids in packed aggregates such as clean sands ranges between about 1.5 and 3 which is in good agreement with other methods.

The exactness of the analogy between ideal chemical diffusion and diffusion of spin-oriented species is questionable, since the lifetime of the stimulated nuclei is short (usually milliseconds to seconds). This 'limitation' in fact enables NMR to probe effective diffusivity of populations of particles which reduce in number over time, either due to spontaneous spin relaxation, or through interaction with pore walls or reactive particles (known as trapping). Thus effective diffusivity and tortuosity can be measured under ephemeral conditions pertinent to chemical reactions within, say, a porous rock or catalyst structure. Much research has been directed in this area, with the results being compared with random walk simulations in which pore walls can act either as entirely reflective boundaries, or which can be given some lesser probability of trapping the diffusing particles (Mitra et al. 1992).

One interesting result of random-walk simulations is that the mean diffusivity (and so effective tortuosity) calculated over time during the simulations varies strongly according to the topology and surface geometry of the pore space being probed. Indeed, given the definition of the effective diffusivity as the mean-square displacement of the surviving walkers located by position vectors **r**,

$$D(t) \equiv \langle \mathbf{r}^2(t) \rangle_s / 6t \qquad (43)$$

it can be shown that their survival is influenced by the surface curvature of the pore walls as well as the pore size distribution. In certain simple pore geometries trapping can enhance the effective diffusivity, since if particles travelling a short distance are eliminated efficiently, then a greater proportion of the surviving particles will have achieved a greater displacement (Mitra et al. 1992, 1993; Sen 1994; Sen et al. 1994). A pore structure with efficient traps (dead ends) branching off long and straight pore channels will produce an asymptotically increasing diffusivity, as after some time the only surviving particles will be those which travel directly away from the origin. Accordingly, the effective diffusional tortuosity will appear to be rather low. In contrast, without surface relaxation many particles will be confined within re-entrant structures and never diffuse far from the origin (Mitra et al. 1995), skewing the mean displacement, and D(t), to lower values.

It is thus possible to use NMR measurements and random-walk models to estimate pore size and shape distributions, area to volume ratio and diffusional tortuosity in porous media under a range of conditions (Mitra et al. 1995; Helmer et al. 1995; Latour et al. 1995). From these studies we can draw the following conclusions. Firstly, effective diffusivity and tortuosity are functions not only of the topology and geometry of the pore space, but are also dependent upon the path length (see above) and survival characteristics of the diffusing particles. Secondly, diffusivity and tortuosity will be strongly modified if reactions take place between diffusing particles and the solid surfaces or among the fluid components: we cannot simply substitute 'passive' diffusivity values into a pore structure model in order to simulate the progress of a reaction.

Multi-phase flow. Greenkorn (1983) attempted to generalize hydraulic tortuosity for the case of immiscible multiphase flow. Providing that the two phases are continuous, a hydraulic tortuosity can be defined for each following Carman's usage of the term. In the limit that one phase is completely wetting and the other non-wetting, then the tortuosity of the non-wetting phase will define the sinuosity and network topology of the pore-channel skeleton, while the wetting phase tortuosity will be a function of the surface geometry and topology. We have already seen that as the frequency of an applied electric current increases, electrical conduction is progressively localized in the region of pore walls. Therefore electrical measurements may help to predict the distribution and relatively permeability of wetting and non-wetting phases through an analysis of frequency-dependent changes in tortuosity.

Tortuosity in networks

Fundamental topological properties

Interconnected networks have topological properties which make them fundamentally different from one-dimensional models, whatever their complexity, and from models of two or even three dimensions that treat the porous medium as a continuum. The main difference is that interconnectedness is expressed explicitly in network constructs, but is ignored or treated only implicitly in other types of model.

The topological property of a porous body that quantifies the branches and interconnections of the pore space is known as the genus (Brown 1988). While this is clearly defined conceptually for a particular region of porous material, it is evident that except for trivial pore geometries, the genus of a porous material rapidly becomes indeterminate as the volume of investigation increases. Therefore, other measures such as the genus per unit volume (G_v), the genus of a representative unit cell (Adler 1992), or the average connectivity or coordination number are used. The coordination number (denoted by the symbol z; where $z = 2G_v$) is the number of throats emanating from each pore in the skeletal representation of the medium (Dullien 1992). These topological properties can be measured from serial sections (Macdonald *et al.*1986*a, b*; Zhao & Macdonald 1993) or can be computed directly from high-resolution tomographic methods (Thovert 1993).

Network models

Constructing a realistic network. The topological realism of network models has made them very popular as tools for the understanding and prediction of transport processes, particularly at the pore scale where the continuum approach collapses (Owen 1952; Fatt 1956; Van Brakel 1975). Data representing the pore structure and interconnectivity of a porous rock can be mapped into a simplified network model, which in practice usually conforms to a regular two- or three-dimensional grid, but may have a more arbitrary geometrical arrangement. The use of a rigid and periodic network enables discretised versions of the transport equations to be solved using straightforward matrix calculations (Rink & Schopper 1968), whereas network models with a more relaxed arrangement may better capture some of the second-order properties of pore interconnectivity, but require more computational effort. When considering percolation and steady conduction, regular network models seem to recreate essentially the same behaviour as a random network with the same average co-ordination number, and this supposition has been validated by Jerauld *et al.* (1984*a, b*). This correspondence cannot be applied to unsteady or diffusional transport processes, leading a number of workers (e.g., Patwardhan & Mann 1991; Hollewand & Gladden 1992) to use random networks in order to simulate these processes realistically.

Solving the transport equations. In order to derive the scaled transport properties (conductivity, permeability) of a network, a large number of equations have to be solved simultaneously. Therefore it is common to employ approximations that reduce the computational effort. Foremost among these is effective medium theory (Koplik 1982; Koplik *et al.* 1984; David *et al.* 1990). In this method, the pore space is reduced to a primary conducting network that accounts for almost all the flow through the rock. This mapping is transformed into an equivalent network in which the connectivity is the same, but the values of bonds have been homogenized to a single value. One problem with the EMT approach is that it can fail in the limit of a very wide pore size distribution. However, a more immediate concern is

that the homogenization process discards useful information about the range of tortuosities embedded in the network, so that rather than being a useful analogue the network becomes something of a 'black box'. The other popular approach to solving problems on large disordered networks is the renormalization group method (e.g., Madden 1976). This method can cope with wider variations in network properties than EMT, but is also a homogenization procedure.

In order to treat unsteady transport processes, or to assess the range of tortuosities in steady transport paths, it is clear that the network representation has to retain more than just topological realism. Some workers have attempted to solve the fluid flow problem explicitly within a 'stick and ball' type of network model; that is to solve the Navier–Stokes equations within the boundary conditions of an interconnected set of tubes or channels. Koplik (1982) attempted this by specifying a Poiseuille flow law in cylindrical tubes linking circular nodes in a square lattice. Other workers have attempted to create a representational model of a rock that has the same pore size, geometry and connectivity statistics. Bryant *et al.* (1993*a, b*) used a finite-difference method to solve for the fluid flux between pores in a digitized sphere packing that could be generalized to approximate cemented sandstones, and Adler (Adler 1992; Adler *et al.* 1992) has pioneered the direct solution of flow equations in reconstructed three-dimensional porous media. Cellular automaton techniques have been developed for the same purpose (Chen *et al.* 1992; Knackstedt & Zhang 1994).

Tortuosity in simple networks. Once the pore space has been reduced to a skeletal network, the path-length tortuosity can be calculated as a sum of discrete bond lengths. Where the grid is completely regular, it is possible to subsume the effects of tortuosity into a network constant (e.g., Schopper 1966, 1967*a, b*; Rink & Schopper 1968). It is often tempting to view a network tortuosity of this type as a real tortuosity, and this can be a danger when network tortuosity embedded in a model is obscured during the process of up-scaling from the microscopic to the continuum scale. For example Bear (1972) asserts rather dogmatically that the tortuosity (our factor X for any transport process) in any isotropic three-dimensional medium has a value of 3. This value arises because the pore model underlying the up-scaling procedure is a network of mutually-orthogonal tubes, and only one third of the tubes carry flow in any given direction. Bear's argument extends to some more complex geometries (Doyen 1988), but it is easy to find geometries with higher or lower values for X for particular transport processes.

Another approach, originally introduced by Fatt (1956) is to assume that the tubes in the network are circular in section, and to ignore pressure drops at the nodes so that hydraulic and electrical conductance of each bond can be calculated from, respectively Poiseuille's law and Ohm's law. There are a number of ways to calculate tortuosity for such a model. An average electrical tortuosity can be calculated from the relationship between the overall conductance and the porosity, i.e. by calculating the Formation Factor for the model and then dividing by the porosity of the network. Alternatively, the path-length tortuosity of each route contributing to flow through the network can be calculated, and its value weighted according to the amount of flow it carries. In either case, the values for hydraulic tortuosity must be divided by a suitable length-scale to keep tortuosity as a non-dimensional parameter. In the case of simple network models this length scale is simply the tube diameter, so that hydraulic tortuosity is determinate using the second method, since if we ignore the effects at steps and nodes, the representative length-scale for a sequence of tubes in series is the harmonic mean of the diameter distribution (Dullien 1992).

David (1993) was probably the first to employ such a weighting method specifically to investigate tortuosity, although Schopper (1966) had pre-empted at least the methodology. David investigated the effects of reductions in porosity achieved either by decreasing the number of active bonds on the network, or by reducing their width. He found that as the channel width distribution became broader, the flow pattern became more heterogeneous, and importantly more tortuous, with the hydraulic flow being especially focused along a few dominant pathways, a result already known from percolation theory.

In a later study, Bernabé (1995) confirmed the results of David (1993) in a more realistic network where crack-like as well as cylindrical pore channels are considered. Other studies have confirmed that this general pattern holds for different two and three-dimensional geometries. However, there has always been the implicit assumption that the tortuosity somehow represents a ratio of path lengths. The approach described below deals more explicitly with the magnitude and direction of flow in each bond.

Tortuosity and cross-flow. Ruth & Suman (1992) introduced the concept of aerosity,

which can be expressed as the ratio of the area of fluid corrected for the flux divergence to the total sectional area normal to the macroscopic flow direction, in this case the x_1 direction. The aerosity, ξ, differs from the intercepted porosity (Fig. 10) in that the aerosity is calculated as a ratio of the effective area normal to the lines of flux to the total area normal to the cut surface (here subscript 1 refers to planes normal to the x_1 direction, and f refers to the fluid phase, or pore space).

$$\xi = \frac{A_{f1}}{A_1}. \tag{44}$$

The corrected area can be expressed as an integral over the position vectors lying in the pores:

$$A_{f1} = \int_{A_1} \gamma_{(\mathbf{r}_i)} \mathbf{n}_{i(\mathbf{r}_i)} \mathbf{m}_1 \, dA \tag{45}$$

where the function γ is zero everywhere in the solid phase and unity in the fluid phase f, $\mathbf{n}_i$ is a unit vector representing the direction of microscopic flow and $\mathbf{m}_1$ is the unit vector in direction of the macroscopic flow direction. The aerosity can vary between zero and the value of the fractional porosity. Suman & Ruth (1993) established the following relationship between electrical path-length tortuosity, formation factor and aerosity, ξ of a network:

$$\tau_{eSR} = F_R \xi. \tag{46}$$

For a smooth, straight, tube of length l_e inclined at an angle σ to the macroscopic flow along a distance l the aerosity clearly differs from the intercepted porosity by exactly a factor of

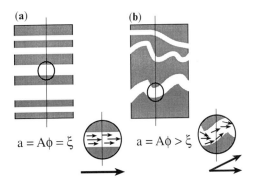

$a = A\phi = \xi$ $a = A\phi > \xi$

Fig. 10. The aerosity and areally-intercepted porosity of (a) a simple parallel tube model, aerosity is equal to areally-intercepted porosity. (b) In a generalized system of tubes, flow lines diverge from the macroscopic direction and aerosity is less than porosity. After Ruth & Suman (1992).

$\cos\sigma$, and so exactly by a factor of l_e/l. Equations (44) to (46) are thus a mathematical derivation of the difference between the Wyllie and Cornell–Katz notions of electrical tortuosity (see p. 305). In an arbitrary pore space the 'effective path length' becomes indeterminate, but at any point, the lines of flux still subtend an angle σ with the macroscopic potential gradient, so that equation (46) still holds on average across any section. Because electrical flux lines carry the same current per unit aerosity regardless of their position in the channel cross sections, we can equate the electrical tortuosity–aerosity relationship with the electrical retardation factor–porosity relationship in the same macroscopic direction.

$$\tau_{eSR} = F_R\xi \quad \text{and} \quad X_e \equiv F\phi \qquad (47)$$

$$\text{so} \quad \frac{X_e}{\phi} = \frac{\tau_{eSR}}{\xi}. \qquad (48)$$

Suman & Ruth (1993) by substituting aerosity for porosity, are calculating the flux density distribution in the pore space. Using Ohm's law and the flux theorem of Gauss, this enables two components of electrical resistance contributing to formation factor to be resolved separately:

$$F_R = \frac{1}{\xi I_1 l_1}\int_{V_f} J_1\,dV$$

$$+ \frac{1}{\xi I_1 l_1 \rho_f}\int_{A_f} V\mathbf{n}_1\,dA. \qquad (49)$$

According to Hulin (1993) the first term is the integral on the fluid volume of the local electrical current J, in the x_1 direction of the macroscopic applied electric field. The second term is the surface integral of the electrical potential on the solid-liquid interface, again acting in the x_1 direction. They can be seen as the voltage drop components associated with local currents parallel to, and perpendicular to the macroscopic flow direction. Ruth & Suman (1992) established a relationship similar to equation (48) for hydrodynamic flow, where the reciprocal of permeability in the x_1 direction can be resolved into 'viscous dissipation' and 'pressure drop' terms.

$$\frac{1}{k_1} = \frac{1}{\xi\mu Q_1 l_1}$$

$$\cdot\left(-\mu\int_{A_f}\frac{\partial\mathbf{w}_1}{\partial\mathbf{r}_j}\,\mathbf{n}_j\,dA + \int_{A_f} P\mathbf{n}_1\,dA\right). \qquad (50)$$

The viscous term has components in local gradient of the flow velocity ($\mathbf{w}_1$), in the x_1

direction, with respect to the coordinate frame of the local flow channel ($\mathbf{r}_j$) described by a unit vector $\mathbf{n}_j$ The overall volumetric flux in the x_1 direction, Q_1, the macroscopic length of the medium, L_1 and the overall aerosity, ξ are constants and common to both terms. Now in a general pore space these two terms are unknown, so we must use some simple cases in order to see their physical meaning.

Ruth & Suman (1992; Suman & Ruth 1993) considered hydrodynamic and electrical flow in a number of idealised sub-networks that possess one of two types of tortuosity, which they termed type-A, with only serial connections and type-B, with parallel interconnections (Fig. 11). Consider fluid flow through the simple model with type-A tortuosity, where l_i is the length of the ith tube, and δ_i is its radius. For this type of network, hydrodynamic tortuosity equals geometrical tortuosity and is simply:

$$\tau_A = \frac{(l_a + l_b + l_c)}{l}. \qquad (51)$$

If the pore space is given network properties, so that cross-flow is possible between parallel flow paths, then a type-B pore space is produced. For the second example in Fig. 11, the unit-cell permeability can be calculated from network theory, and the individual fluxes in each of the five sections (Q_a–Q_e) can be calculated. The tortuosity in this case becomes:

$$\tau_B = \left(\frac{Q_a l_a + Q_e l_e}{\delta_a^2} + \frac{Q_b l_b + Q_d l_d}{\delta_b^2}\right.$$

$$\left.+ \frac{Q_c l_c + Q_e l_e |\delta_b^2 + \delta_a^2|}{\delta_c^2}\right)\frac{(\delta_a^2 + \delta_b^2)}{l}. \qquad (52)$$

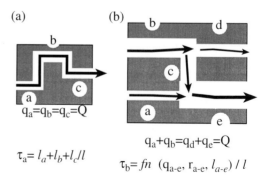

(a)

b

c

a

$q_a=q_b=q_c=Q$

$\tau_a= l_a+l_b+l_c/l$

(b)

b   d

c

a

e

$q_a+q_b=q_d+q_e=Q$

$\tau_b= fn\ (q_{a\text{-}e},\ r_{a\text{-}e},\ l_{a\text{-}e})\,/\,l$

Fig. 11. (a) A simple network model consisting of three inter-linked tubes defining type-A tortuosity. (b) A five component network model defining type-B tortuosity. Adapted from Ruth & Suman (1992).

Thus it is seen that the notion of hydrodynamic tortuosity as a simple ratio of lengths is oversimplified, even for a five-element unit cell. Crucially, the hydrodynamic tortuosity is a function of flux and not simply path length and the magnitude of τ_B depends upon the amount of cross-flow in the medium. This, in turn depends on the ratio of the sizes $(\delta_a - \delta_e)$ of the five tubes. If the pressures at both ends of the cross-bond (tube c), are equal then there will be no cross flow and the type-B tortuosity will decrease (possibly to zero) even though the pore space is more complicated than the simple three tube model. The size of the cross-tube then becomes irrelevant for the overall permeability, as the channel is stagnant pore space. If the parallel channels differ in size then there will be lateral gradients in pressure along tube c and so a certain amount of cross-flow will take place.

The preceding argument can be extended to a more realistic petrophysical model by considering larger networks, and relaxing the conditions that the tubes linking nodes be straight and cylindrical in section. The use of sinuous rather than straight tubes introduces additional type-A tortuosity to the model at the level of individual bonds, but if the tube conductances remain matched, need not have any effect on the type-B tortuosity. Changing the tube cross section has no effect on type-A tortuosity provided the section remains constant along each length segment.

Ruth & Suman (1992) found that type-A tortuosity depends only on the pressure term in the flow equation, whereas type-B tortuosity has components arising from the pressure term, due to the overall sinuosity of the channel segments, and components in the viscous term arising from cross flow. The magnitude of the pressure term determining cross-flow and τ_B is very small if the pore size distribution is restricted, while it can become large if the pore size distribution is wide, provided that there are at least some larger channels linking laterally to narrower channels. Exactly the same arguments apply to electrical flow, with the amount of cross flow depending on the lateral gradients in electrical potential between linked nodes: type-A tortuosity only includes potential terms of equation (49), while type-B networks also have tortuosity contributions from the current term (Suman & Ruth 1993).

The aerosity concept applies to general pore spaces as much as to discrete networks, and by definition is a measure of the cross-flow (or ratio of flux terms to potential terms) at each point. Where we have a bulge or constriction in a flow channel, the streamlines will diverge or converge, so that we generate cross-flow and thus tortuosity varies continuously rather than in discrete units associated with particular channels. This cross-flow can be treated in exactly the same way as type-B tortuosity on our simple network, since the equations (49) and (50) are blind to its particular origin.

Particle tracking on a network. Torelli & Scheidegger (1971) and Torelli (1972) used a simple numerical algorithm to track tracer particles in a network of tubes which are assumed to have a plane Poiseuille velocity distribution. This produces the well known result that within the tubes themselves, the longitudinal dispersivity, D_L is of the order of the mean flow velocity (and so scales with the power unity), while for the whole network D_{por} scales approximately as $\bar{v}^{1.2}$. The reduction in the overall longitudinal dispersivity (i.e. reduction in the width of the tracer distribution over time) arises because at the tube junctions the particles can swap from a high to a low velocity streamline, or vice-versa, so that more and more of the particles will tend to travel at a velocity close to the mean. The distribution will not asymptotically approach D scaling with $\bar{v}^{1.0}$ because the fundamental variation in microscopic velocity is spatially correlated on the scale of the tubes and is not random.

Freidman & Seaton (1995) and Zhang & Seaton (1994) have shown that random walks in networks have to be handled very carefully if they are to provide realistic estimates of diffusivity and diffusional tortuosity. The shape, size, wall roughness, length and sinuosity of the individual bonds are all important. This contrasts with the case of estimating effective tortuosity for steady conduction phenomena, when the bonds can be treated as 'black box' conducting elements.

Comparing different measures of tortuosity

Diffusional versus electrical tortuosity

The Laplacian analogy. Carman (1956) recognized that diffusion (of gas or ions) and electrical conduction are closely related phenomena: they are both solved by a Laplace equation describing a scalar field of, respectively, chemical concentration and electrical charge density,

$$\nabla^2 \psi(\mathbf{r}) = 0 \qquad (53)$$

with the boundary condition that the field strength resolved in the direction of the solid walls vanishes at fluid solid boundaries; i.e., where $\mathbf{U(r)} \equiv -\nabla\psi$ we have $\mathbf{U} \cdot \hat{\mathbf{n}} = 0$. For electrical flow, ψ is the electrical potential field and $\mathbf{U}$ is the electrical field intensity, while for chemical diffusion ψ is the concentration and $\mathbf{U}$ the concentration gradient. Martys (1994) used this mathematical analogy in a numerical model to plot idealized electrical and diffusional pathways in a two-dimensional model pore space (Fig. 12).

From equation (31c), the retardation of diffusion due to the porous medium can be expressed as:

$$\frac{D_{\text{bulk}}}{D_{\text{por}}} = \frac{X_{\text{d2}}}{\phi} \qquad (54)$$

where D_{por} is effective diffusion coefficient measured in the porous sample and D_{bulk} is the diffusion coefficient in the unconfined fluid, in the same way that the formation factor,

$$F_{\text{R}} = \frac{\sigma_{\text{f}}}{\sigma_{\text{r}}} = \frac{X_{\text{e}}}{\phi} \qquad (55)$$

expresses the ratio of electrical conductivity of the pore fluid and saturated rock.

Klinkenberg (1961) suggested that the retardation factors, or tortuosities, X_{d} and X_{e}, were the same. He presented some data to suggest that electrical flow and diffusional flow of gas were inhibited, over that expected for open flow, to the same extent in a range of porous media. However, at this time no rigorous relationship was derived, and only a single direct comparison was given, which showed a discrepancy of about ten percent

between measured electrical and diffusional tortuosities. Van Brakel (1975) also pondered the similarity between Fick's laws and the equations of electrical flow, but concluded that the available experimental evidence suggested that the tortuosities in the two cases were different. Berner (1980 and references therein), Epstein (1989), Mitchell (1993) and Boudreau (1996) discuss some of the reasons for this difference in terms of the chemical activity of sediment particles, but below we consider the case of passive transport in an inert porous medium.

The flux-gradient relationship. It is important to point out that the diffusional processes under consideration by Klinkenberg refer to steady-state diffusion under an established concentration gradient, whereas the great majority of diffusion processes of relevance to petrophysicists occur in the unsteady-state, where the concentration gradient and local concentrations change over time: often we can only apply Fick's second law rather than the basic flux-gradient relationship expressed in Fick's first law. Now if we re-examine this relationship in the parallel cases of electrical and chemical flow we derive a surprising result.

We wish to draw an analogy between steady electrical and chemical diffusive flows. In the first case we have Ohm's law $V = IR$. In a slab or rock of area A, length l and conductivity σ the resistance is

$$R = \frac{l}{A\sigma} \qquad (56)$$

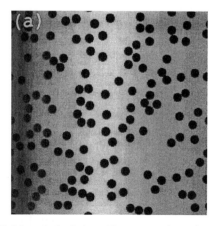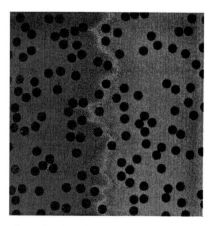

Fig. 12. Numerical solution of flow pattern in random two-dimensional model media consisting of monodisperse discs with a porosity of 0.75. (**a**) Diffusive flux; dark indicates high concentration, light indicates low concentration. (**b**) Electrical flux (dark, high potential; light, low potential). The patterns are essentially the same because the flows are derived from a solution of the Laplace equation with the same boundary conditions. Reproduced with permission from Martys (1994).

and so the current (flux of current in amperes = coulombs per second) can be expressed in terms of the macroscopic potential gradient $\Delta V/l$:

$$I_{\text{por}} = -\sigma_{\text{por}} \frac{\Delta V}{l} A. \qquad (57)$$

We now compare equation (57) with Fick's first law in the same one-dimensional steady case of flow through the slab. Lets say that Δc represents a change in concentration in moles per litre over a distance l metres, and D_{por} is the effective diffusion coefficient in metres squared per second. If we are interested in the diffusive flux, $Q_{\text{D,por}}$, in moles per second, then we must multiply not by the total area of the slab but by the area of the slab through which the liquid actually flows, i.e., the product of the total area and the porosity.

$$Q_{\text{D,por}} = -D_{\text{por}} \frac{\Delta c}{L} A\phi. \qquad (58)$$

This illustrates the difference between current I in Ohms law, which is a flux and the quantity $\mathbf{q}_{\text{D}}$ in our original statement of Fick's first law (equation (29)) which is a flux density. Now suppose we replace the rock by a filled volume of fluid of the same dimensions, and keep the same concentration gradient. In order to work out the flux in moles per second we replace the effective diffusivity with the bulk diffusivity of the liquid, and multiply by the entire area of the slab:

$$Q_{\text{D,bulk}} = -D_{\text{bulk}} \frac{\Delta c}{L} A. \qquad (59)$$

Performing the same substitution for electrical flow we have:

$$I_{\text{bulk}} = -\sigma_{\text{bulk}} \frac{\Delta V}{L} A. \qquad (60)$$

While the electrical formation factor is normally expressed in terms of the conductivity ratio it can equally be expressed as the ratio of currents or fluxes:

$$F_{\text{R}} = \frac{\sigma_{\text{bulk}}}{\sigma_{\text{por}}} = \frac{I_{\text{bulk}}}{I_{\text{por}}}. \qquad (61)$$

The inverse of the formation factor is then a measure of the efficiency of electrical transport in the system, an argument developed in a key paper by Herrick & Kennedy (1994; see p. 307). If we try to express the ratio of bulk and porous medium diffusivities in the same way, then we find,

$$\frac{Q_{\text{D,bulk}}}{Q_{\text{D,por}}} = \frac{1}{\phi} \frac{D_{\text{bulk}}}{D_{\text{por}}}. \qquad (62)$$

Now consider the micro-mechanism of electrical conduction in an electrolyte-saturated medium. The electrical potential gradient causes ions and electrons to flow though the porous medium, and after a brief transient period they establish a steady diffusive flux, (the macroscopic current) which is proportional to the number of charge carriers per second flowing through the volume occupied by fluid. If the charge carriers follow purely diffusive behaviour, implicit in Laplace's equation, then we can apply equations (59) and (60) simultaneously to yield:

$$F_{\text{R}} = \frac{1}{\phi} \frac{D_{\text{bulk}}}{D_{\text{por}}}. \qquad (63)$$

Given the identity $X \equiv F_{\text{R}}\phi$ we also establish

$$X_{\text{e}} = \frac{D_{\text{bulk}}}{D_{\text{por}}} \qquad (64)$$

and so sensibly X_{d} should be defined in terms of fluxes and not corrected for porosity, so that:

$$X_{\text{e}} = \frac{D_{\text{bulk}}}{D_{\text{por}}} \equiv X_{\text{d1}}. \qquad (65)$$

In this derivation we have relied upon assumptions about the transport of ions through the pore space and have used a volume and time-averaged flux in simple one dimensional case. A more rigorous proof can be found in Koplik et al. (1988). Extension of the proof to unsteady diffusion and dispersive transports has not so far been attempted. However, by inspection it is easy to rationalise that, (1) dispersive tortuosity should be considered in terms of flux and not concentration, and (2) in the limit of high Péclet number, i.e., transport dominated by advection, the flux-based tortuosity for the dispersion of tracer must equal a flux-based tortuosity for the hydrodynamic flow, which we seek in the following section.

While the derivation of equation (63) is rigorous, one must be careful about how it is applied (see Epstein 1989). Matthews & Spearing (1992) derived diffusional tortuosity in a sandstone using the time taken for steady-state flux to be reached in experiments using a variety of hydrocarbon gases. From Fick's second law it can be deduced that the effective diffusivity after time t is related to the mean square displacement of the diffused particles (Schwartz et al. 1991, 1994a).

$$D_{\text{eff}}(t) = \frac{\langle |\mathbf{r}|(0) - \mathbf{r}(t)|^2 \rangle}{6t} \qquad (66)$$

and so for one-dimensional diffusion through a sample of length l, when the flux has reached the steady state the effective diffusivity in the pore space is:

$$D_{por} = \frac{l^2}{6t^*}. \qquad (67)$$

Here t^* is the time to reach steady flux conditions. The molecular diffusivities (D_{bulk}) were derived by Matthews & Spearing from thermodynamic data, so that the effective tortuosity, according to one's preferred definition, can be calculated according to equations (31a–c). Matthews & Spearing also measured the formation factor of the sandstones (porosity about 11%), and calculated an electrical retardation factor, X_e, of 5.9. Using the data of Matthews & Spearing uncritically, we find that the values of X_{d1} for methane diffusion into atmospheric nitrogen, are very high, ranging from 32 to 121, while their values for the porosity-biased diffusional retardation factor (X_{d2}) range from about 2 to over 13. The X_{d2} values are much closer in value to the electrical retardation factor than X_{d1} and also seem more reasonable, given the other properties of the sandstone samples. The explanation of the apparent discrepancy is very instructive. Matthews & Spearing had measured the gas outflow rate (flux) from the sandstone cores, rather than point values for concentration in the pore space. Accordingly, they used the porosity-corrected formulation, equation (31c), to obtain the correct retardation factor (here X_{d2}) for these boundary conditions.

Electrical versus hydraulic (hydrodynamic) tortuosity

Hydrodynamic tortuosity. In the light of the previous section in which we defined electrical and diffusional tortuosities as a ratio of fluxes we see that a distinction arises between the static hydraulic tortuosity we have considered hitherto and a hydrodynamic tortuosity factor, which we can denote X_h, that accounts for retardation of fluid flow due to the complexities of the transport path.

In the Kozeny–Carman equation, the properties of the pore space that control fluid flow are encapsulated into four factors: total porosity, hydraulic radius, a shape factor and a tortuosity factor. This subdivision is artificial because it arises from the idealized notion of 'equivalent channels' in a capillary model of the pore space (Dullien 1992). Of these four factors, only porosity can be measured easily. Specific surface

area can be measured with some difficulty, but again its value is dependent upon the method employed and the model used to reduce the test data (e.g., Igwe 1991; Lowell & Shields 1991). This leaves the shape factor and tortuosity as essentially indeterminate quantities in all but trivial pore spaces: in general one cannot be calculated without knowing the value of the other. By contrast, the very definition of the original Wyllie electrical tortuosity means that all the pore structure factors, aside from porosity, are encapsulated in the single flux-retardation factor X_e. Therefore X_e is a definite and measurable property however complex the pore space may be.

Even in a purely one-dimensional case, where we eliminate tortuosity from consideration, the viscous flow of fluid depends on the pore channel size and cross sectional shape while electrical flow, being purely diffusive, is only dependent on the total cross-sectional area available to conduct the current. The hydraulic radius takes account of these pore size effects, and so X_e is more closely analogous to the Kozeny constant (C_k) than it is to the Carman hydraulic tortuosity factor (T_{hC}). Accordingly, we can expect X_h to combine the characteristics of a path length tortuosity and a shape factor.

Early comparisons of electrical and hydraulic 'path-length' tortuosities. Wyllie & Rose (1950) in their derivation of electrical tortuosity acknowledged that their term was analogous rather than identical to the Carman hydraulic tortuosity because while electrical conduction is proportional to the square of flow channel radius, hydraulic conduction is proportional to this radius raised to the fourth power (Fig. 13). Carman (1956) discusses whether the 'electrical path lengths' envisaged by Wyllie & Rose (1950) were related to the 'effective hydraulic path lengths' he used to define his hydraulic tortuosity. He surmised that the electrical and hydraulic tortuosities were probably similar, but acknowledged that they were not proven to be identical, i.e,

$$l_{ee} \approx l_{eh} \quad \text{and so} \quad \tau_{eW} \approx \tau_{hK}$$
$$\text{and} \quad T_{eW} \approx T_{hC}. \qquad (68a\text{–}c)$$

Wyllie himself was circumspect about the physical meaning of electrical tortuosity. He regarded the effective electrical path length as being indeterminate, and certainly not equal to the shortest geometrical path confined within the fluid. Wyllie & Spangler (1952) stated that electrical tortuosity 'has no precise physical

(a)

Electrical Flow

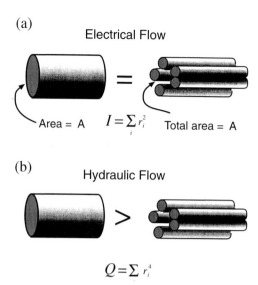

Area = A $I = \sum_i r_i^2$ Total area = A

(b)

Hydraulic Flow

$$Q = \sum r_i^4$$

Fig. 13. (a) The electrical conductance of a single channel is the same as a bundle of smaller channels providing the total section available to carry current is the same. (b) In the case of hydrodynamic flow, the single larger channel carries very much more fluid than the bundle of smaller tubes.

significance except that implicit in the derivation of equations such as (19) above'. They identified hydraulic tortuosity with the electrical tortuosity on the assumption that carriage of electrical current makes use of the same pore space as the hydraulic transport, while they conceded that no rigorous relationship had been established.

Demonstration of the non-identity of electrical and hydraulic pathways. The equations for electrical flow and hydrodynamic flow differ fundamentally in form. Viscous advective flow reduces to a solution consisting of a vector field of particle velocities, which due to viscous retardation are forced to zero at the pore walls and increase gradually towards the centres of the flow channels. Electric flow is solved using a Laplace equation to produce a scalar field of electrical charge density, where the electrical field vanishes suddenly at the pore walls. Therefore it should come as no surprise that electrical and hydrodynamic flow-lines have different patterns in the same pore space. Brown (1989) was one of the first to demonstrate this using simple graphical representations of flow through two-dimensional apertures of varying width. The fluid flow lines tend to become more strongly bunched along preferred paths

than do the lines of electrical flux (Fig. 14a, b). Martys & Garboczi (1993), Schwartz *et al.* (1994*b*) and Martys (1994) used computer rendering to illustrate the same point for two-dimensional particle arrays (Fig. 14c, d).

Dullien (1992) reported calculations of the formation factor and permeability of simple networks with a bivariate channel radius distribution. The electrical tortuosity is calculated from equation (15) and the hydraulic tortuosity by assuming Poiseuille flow so that $r_h = r_{(bond)}$ and the shape factor $= 2$ in the Kozeny–Carman equation. In each case the electrical tortuosity was found to be less than the hydraulic tortuosity by a small factor. The same general result had been predicted by Schopper as early as 1966.

David (1993) used simple two-dimensional network representations to contrast electrical and hydrodynamic pathways on a lattice of channels of variable cross-section. In a given distribution of pore channel radii, the hydraulic flow is more heterogeneously distributed than the electrical flow because of the stronger dependence of the former on the channel radius. David demonstrated that so long as the bond distribution is irregular, the average path length is greater for hydraulic flow than for electrical flow in the same network, so that, in terms of mean path length, the electrical tortuosity was always less than the hydraulic tortuosity on the network. This relationship is not a simple one: if the porosity is reduced by the complete removal of bonds, then in the limit of a single remaining pathway (i.e. at the percolation threshold) the electrical and hydraulic path lengths must be the same (see also Bernabé 1995). This is a paradoxical result, because up to this point, the trend is for electrical and hydrodynamic tortuosity to diverge.

Direct calculation of electrical and hydrodynamic tortuosities. In theory, a purely hydrodynamic tortuosity could be calculated by vector analysis of streamlines determined in numerical simulations of two-or three dimensional viscous flows through model porous media. We have already hinted that this may be possible for cellular automaton models in simple two-dimensional models like those of Rothman (1988), since the calculations keep track of the progress of fluid elements and so map out the transport vectors directly. Zhang & Knackstedt (1995) have used the lattice gas automaton method to track both electrical and fluid-flow pathways in three dimensions, and in doing so have essentially solved our problem of 'the effective path' by

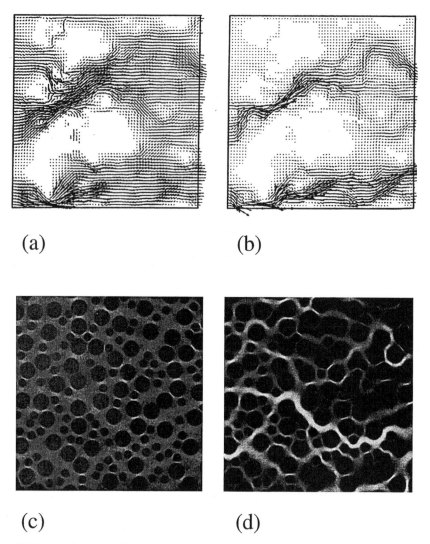

(a)   (b)

(c)   (d)

Fig. 14. (a, b) Numerical solution of flow equations in the plane of a single fracture, from Brown (1989; copyright AGU). (a) Electrical flow, (b) fluid flow. White areas mark asperities. Note fluid flow is more concentrated and diverges more (is more tortuous) than the electrical flow in the same region. (c, d). Two-dimensional model medium consisting of bidisperse discs with a porosity of 0.50, reproduced with permission from Martys & Garboczi (1992). Whiter areas have greater flux density. The electrical paths (c) are diffuse but show some concentration at the throats, while the hydrodynamic fluid flow (d) is strongly focused along a few preferred channels.

direct computation. Each particle track is weighted for its contribution to the overall flux by dividing each path length by its associated travel time. For each of n particles, the individual tortuosity is, according to the definition (11) the Kozeny hydraulic tortuosity for the track through the model:

For $i = 1$ to n; $\tau_{hK,i} = \dfrac{l_{h,i}}{l}$ (69)

the average tortuosity is calculated, and squared to give a hydrodynamic tortuosity factor.

$$T_{hZK} = \left[\frac{\sum\limits_{i=1}^{n} \dfrac{\tau_{hK,i}}{t_i}}{\sum\limits_{i=1}^{n} \dfrac{1}{t_i}} \right]^2 .$$ (70)

Note that all the particles (or at least a very large, and randomly sampled number of particles) must be tracked to make this simulation yield a valid estimate of average tortuosity, since each starting point represents an unique streamline. The swift particles at the centre of the flow channels contribute more to the flux than the slow fluid elements adjacent to the pore walls. The same calculations can be performed on the electrical paths, which give the equivalent electrical path lengths according to the definition of Cornell & Katz (1953).

$$\tau_{eCK,i} = \frac{l_{eCK,i}}{l} \qquad (71)$$

$$T_{eZK} = \frac{\left[\sum\limits_{i=1}^{n} \dfrac{\tau_{eCK,i}}{t_i}\right]^2}{\sum\limits_{i=1}^{n} \dfrac{1}{t_i}}. \qquad (72)$$

In their preliminary simulations, the electrical and hydraulic paths are clearly different, for example showing path length tortuosities, τ_e and τ_h of 1.27 and 1.75 respectively in a medium of 35% porosity. The corresponding flux-weighted average path lengths, which we can call T_{eZK} and T_{hZK} differed by an amount that varied from a few percent to a factor of ten as the porosity of the model medium was decreased.

This appears to be the first time that weighted particle tracking has been used to measure tortuosity in generalized rather than network-mapped media. Matthews & Spearing (1992) had previously derived electrical tortuosities similar in value to those calculated by Zhang & Knackstedt (1995) using a slightly different weighting method on random walks through a cubic network model representing sandstone. Note that Tye (1982) considered the flux of each path in accordance with its transit time, thereby enforcing the condition $\tau_{e,i} = X_{e,i}$.

Tortuosity as a measure of flow efficiency

Electrical efficiency. The formation factor is the most usual way to present data on electrical conductivity of rocks. Recently, Herrick & Kennedy (1994) suggested that electrical properties could be expressed in a modified form, which they termed electrical efficiency. Although this is only a small adjustment to existing theory, it allows a very clear treatment of the problem. In their analysis, the actual conductivity of the rock is compared with the conductivity of a single conducting channel having the same porosity as the rock. As the most efficient means of conducting electricity through a body is maximize the surface cross section for a given volume, Herrick & Kennedy deduce that the ideal channel is a right cylinder of radius r_0 with a volume equal to the pore volume of the rock. We therefore have, for a tube of length l in a prism of rock of total sectional area A:

$$\pi r_0^2 l = \phi A l \qquad (73)$$

$$\pi r_0^2 = \phi A \qquad (74)$$

$$r_0 = \sqrt{\frac{\phi A}{\pi}}. \qquad (75)$$

The conductivity of this tube is then:

$$\sigma_{max} = \frac{a\sigma_f}{l} \qquad (76)$$

where a is the cross-sectional area of the tube, which from stereological considerations has an area of ϕA. The electrical efficiency of this tube is defined to be 1. We can relate the ratio of resistances which describe the formation factor of a rock to the ratio of conductances which describe its electrical efficiency.

$$E = \frac{\sigma}{\sigma_{max}} = \frac{\sigma}{\sigma_f \phi} = \frac{1}{F_R \phi}. \qquad (77)$$

From the definition of generic electrical tortuosity we find that electrical efficiency is simply the reciprocal of the electrical retardation factor:

$$X_e = F_R \phi \quad \text{and so} \quad E \equiv \frac{1}{X_e}. \qquad (78)$$

Electrical conduction is dependent only on the total cross-sectional area available for flow (strictly aerosity), and not on the size or the cross-sectional shape of the flow channel providing it is a right-prism oriented directly across the pore space. Thus, electrical efficiency of a bundle of straight-walled tubes of any size and sectional shape arranged in parallel is the same, and equal to 1 if their aggregate porosity is constant. The same argument applies to the electrical conductivity and efficiency of a sphere pack, which remains the same for any size of spheres providing the porosity and packing geometry are the same. We can generalize this reasoning to the conductivity, resistivity and electrical efficiency of any pore space which we shrink or expand uniformly. Therefore, electrical efficiency, and by extension, X_e, is independent of porosity (by definition) and is also independent of the length scale of the pores.

The power developed (energy dissipated per unit time) in an electrical circuit is well known to be the product of current and potential difference. In a continuum, such as a porous medium saturated with a conducting fluid, the equivalent relationship rigorously using non-equilibrium thermodynamics. Such an expression for power, Θ is given by (Herrick & Kennedy 1994):

$$\Theta = \int_V \mathbf{J} \cdot \mathbf{E} \, dV \qquad (79)$$

where $\mathbf{J}$ and $\mathbf{E}$ are vector fields of the current density and the electrical potential, respectively (i.e., $\mathbf{J}$ is in amperes per square metre and $\mathbf{E}$ is equal to $-\mathbf{grad}\,\Phi$ where Φ is the scalar electrical potential under isothermal conditions, Yeung & Mitchell 1993). The efficiency can then be expressed as power generated in the rock relative to the power dissipated in the ideal tube:

$$E = \frac{\Theta_{rock}}{\Theta_{tube}}. \qquad (80)$$

We have seen in a previous section that the most efficient path is the one in which electrical current in maximized, and is related directly by Herrick & Kennedy to their electrical efficiency. We have from the conservation of energy and Ohm's Law:

$$\Theta = \frac{V^2}{R}. \qquad (81)$$

Where V is a constant potential across the sample, we must minimize R to maximize the power, and this means minimising the formation factor. Also, on the macroscopic scale,

$$\Theta = VI \qquad (82)$$

so the current must be maximized to generate the maximum power for a given potential difference.

As power is a scalar quantity, the total power dissipated is equal to the sum of the power dissipation in each element considered in any order. Power is developed on a continuous basis throughout the conducting pore fluid and Herrick & Kennedy derive an expression for local efficiency, e, in terms of the local product of $\mathbf{J}$ and $\mathbf{E}$ in the locality of the ith position vector.

$$e_{(\mathbf{r}_i)} = \frac{(\mathbf{J} \cdot \mathbf{E})_{(\mathbf{r}_i),rock}}{(\mathbf{J} \cdot \mathbf{E})_{(\mathbf{r}_i),tube}}. \qquad (83)$$

The global efficiency can then be considered to be the average of the local efficiencies, and Herrick & Kennedy suggest that we could use a discrete form of this relationship to calculate the individual efficiencies of bonds within a network representation of a medium, i.e.,

$$E_{net} = \frac{1}{n}\left(\sum_{i=1}^{n} e_{(\mathbf{r}_i)}\right). \qquad (84)$$

Since the most efficient tube against which we normalize has a completely uniform distribution of $\mathbf{J}$ and $\mathbf{E}$ in which the unit vectors are directed in the direction of macroscopic potential gradient, the local efficiency represents the difference between the magnitude and direction of the actual electrical current and potential vectors, and the corresponding ideal 'unit' vectors, at each point. Two interesting points now emerge. Firstly, we can envisage a situation where the local efficiency can be greater than unity because the flow concentrates the current and potential drop in a particular place; such a situation is remarked upon for the network models of Bernabé (1995), and Bernabé & Revil (1995). Secondly, the idea that current alone at each point defines the efficiency is not strictly true. Instead the current can be considered to exactly represent efficiency in macroscopic transits though the entire medium from face to face only because in each case the potential drop along every path is the same, and equal to the overall potential drop. The corollary to this is that along every complete tortuous path the current defines the efficiency, as we showed above from a completely different standpoint.

Efficiency of hydrodynamic flow. Herrick & Kennedy (1994) show that we can define electrical efficiency in terms of conductivity or in terms of energy dissipation (power) and get identical results. We can use the same approach for hydrodynamic flow by examining the energy dissipation, which in each fluid element is the product of the local displacement rate and the local shear force (e.g. see Happel & Brenner 1973 for a first-principles treatment).

The theorem of minimum energy dissipation (Happel & Brenner 1973) means that fluid or electricity will always find the most efficient pathway though a medium. For a given flow rate, the lines of flux are arranged so that the potential drop is the smallest possible, and in any pore space there is only one solution to the flow equations, which is always the most efficient. Parney & Smith (1995) demonstrate this in simulations of fluid flow in discrete fracture networks. As the interconnections become more restricted, flow occurs along a path which is more geometrically tortuous on the average, but which makes the best use of the wider channels so that the velocity in the direction of the overall flow gradient is still the maximum possible. If the flow took shorter but slower routes, the overall geometrical tortuosity would be reduced, (ultimately to unity in square networks), but the flow would be less efficient.

In the case of steady incompressible hydrodynamic flow, the only supply of energy to the system is the pressure drop, so the energy loss in a particular element is generally expressed as loss of pressure $(-\Delta P)$ associated with flow. Pressure is an expression of potential energy, which can be expressed in units of joules per cubic metre of fluid, so the work done is the pressure drop multiplied by the flowing volume. The local power dissipated under isothermal conditions $(\eta_{h,(r)})$ is the product of the pressure gradient vector and the local volumetric flux vector, just as the local electrical power is the product of electrical potential drop and current. So we have:

$$\eta_{h,(r_i)} = -(\Psi_{(r_i)} \cdot \mathbf{u}_{(r_i)}) \qquad (85)$$

where $\Psi_{(r)}$ is the pressure potential field and $\mathbf{u}_{(r)}$ is the local seepage velocity (identical to the volumetric flux per unit area) at i points defined by position vectors $\mathbf{r}$. Energy loss is considered on trajectories along which the gradient of the pressure potential is maximum, i.e., the macroscopic flow direction is the reference direction for the vectors. In porous media flow, we also deal with the local velocity, which is exactly analogous to the volume-flux per unit area (see equation 20 of Ruth & Suman 1992). So we could replace $\mathbf{u}_{(r)}$ in equation (83) with $\bar{w}_i a_i$, where a_i is the local intercepted unit area in the direction normal to local flow and $\bar{w}_i$ is the local average streamwise velocity vector (see Diedericks & du Plessis 1995). We do not use the dot product locally as the power dissipation is not dependent on the global flow direction, only its magnitude. If we force the component of velocity to be in the direction of the macroscopic flow by using a fixed rectilinear reference frame for our volume elements, then the intercepted area must be normalized to maintain the magnitude of the scalar power product. By inspection the correct normalization factor is equal to the aerosity employed by Ruth & Suman (1992).

Microscopically: $\bar{w}_i = \dfrac{Q_i}{a_i}$ and macroscopically,

$$\bar{V}_1 = \frac{Q_1}{\xi A_1} \equiv \mathbf{u}. \qquad (86a,b)$$

A similar treatment can be found in the monograph of Coussy (1995), who shows that the correct energy balance for viscous hydrodynamic flow can be derived from first principles using this approach. He defines a macroscopic retardation factor χ with respect to flux density:

$$\chi = \frac{\langle (\mathbf{v}_{(r)})^2 \rangle}{(\mathbf{V}_{(r)})^2} \qquad (87)$$

where $\mathbf{v}$ is the microscopic relative velocity and $\mathbf{V}$ is the macroscopic relative velocity (vector fields defined by position vectors $\mathbf{r}$ in a fixed reference frame). Coussy's retardation factor is a function of the geometry of the medium, and of the viscosity of the fluid (the inclusion of a length-scale correction would obviate this requirement; see p. 331), its value being unity for a completely filled fluid volume (porosity $= 1.0$) and tending towards infinity as the connected porosity diminishes to zero.

The mathematics employed by Coussy is rather complicated, since it is not explicitly formulated for the limits to which we have adhered hitherto, namely slow steady flow in a rigid medium. However, we can highlight the main point raised. In the Zhang & Knackstedt method, paths are corrected with respect to the overall flux in the tortuous medium, rather than the flux in the ideal medium, as is the case in the general efficiency formulation. Therefore, if all the paths are of the same length but some are slower than others, as a consequence of lateral velocity gradients enforced by viscosity, then we still have tortuosity equal to unity, even though efficiency is not necessarily equal to 1. This may be one of the attributes we seek in a tortuosity measure, but it does not allow us to derive a rigorous formulation of hydrodynamic efficiency. In order to pursue the Herrick & Kennedy electrical model more strictly, we should in the hydrodynamic case perform a piece-wise comparison retaining the local fluxes, by analogy with equation (83), rather than using local velocities, when we normalize with respect to the 'efficient' medium.

Characteristic length scales

Defining a suitable length scale. Given that the most efficient configuration for hydrodynamic transport, at a given porosity is a single straight cylinder of radius equal to:

$$r_0 = \sqrt{\frac{\phi A}{\pi}}. \qquad (88)$$

The inverse formation factor of this tube is simply ϕ the porosity, while the permeability (e.g., Bernabé 1991) is equal to:

$$k_{\text{tube}} = \frac{\phi r_0^2}{8}. \qquad (89)$$

Thus in order to define permeability, we must also consider a length-scale characteristic of the medium. If we choose this to be the tube radius, then by reducing pore size we simply reduce the scaling of the Poiseuille velocity distribution for our 'most efficient' case. Alternatively, we could use a weighted length scale of some kind. We could compare our real flux distribution with a Poiseuille distribution that has been normalized using a global length scale. This is equivalent to replacing our single tube with a number of identical tubes that, in order to maintain constant porosity, have a smaller radius.

If instead we shrink the tubes non-uniformly, there are two end-member outcomes. In the case where the tubes undergo serial changes in dimension, the efficiency decreases markedly, clearly to zero in the limit of vanishingly narrow throats. In a distribution of cylindrical tubes of arbitrary radii, arranged in parallel, the efficiency increases over a uniform distribution, towards the limit where we have one large tube and numerous vanishingly small ones (efficiency is unity). While the first scenario is as valid for electrical efficiency as for fluid flow, the second scenario only applies to hydrodynamic flow. Therefore our weighting function must be a very cleverly devised 'length scale' that takes into account pore sizes according to their position on a connected flow path rather than simply reflecting their absolute values.

In the Kozeny–Carman equation, the length-scaling is manifest as the hydraulic radius, r_h. The hydraulic radius is a static, scalar measurement that does not embody any information about pore connectivity or pore size distribution. It is not surprising, therefore, that the Kozeny–Carman equation fails to predict permeability in media with a wide range of pore sizes. This has prompted a number of workers to try and derive a length-scale which is more meaningful for hydrodynamic transport.

Katz & Thompson (1986, 1987) introduced a fractal percolation model to predict the permeability of disordered porous media. In invasion percolation, a non-wetting fluid will first connect from one face of the sample to the other only when the driving pressure is sufficient to penetrate the smallest pore throat, of radius r_c on the most efficient conducting pathway. With various assumptions it can be deduced that the

permeability should vary with the length of this critical pore radius according to the relation:

$$k_{\text{KT}} = \frac{2}{113F} r_c^2. \qquad (90)$$

For reasons of space this equation cannot be discussed further here. The interested reader is referred to Le Doussal (1989), Thompson (1991) Kamath (1992) Sahimi (1993, 1994), Nelson (1994) and Bernabé (1995) for critiques.

For conduction of electrical currents, Johnson, Schwartz and co-workers (Johnson *et al.* 1986, Schwartz *et al.* 1989) introduced a dynamic length scale that is a function of the electrical field intensity in the medium.

$$\Lambda_e = 2 \frac{\int |\mathbf{E}|^2 \, d\phi}{\int |\mathbf{E}|^2 \, dA}. \qquad (91)$$

The numerator is the square of the electrical field magnitude integrated over the pore fluid volume, and this is divided by the square of the magnitude of the same field directed out of the fluid. Since the electric field vanishes where it is directed out at the pore walls, enclosed pore space does not contribute to Λ_e, which therefore becomes weighted heavily towards the pore throats. In the case where the porosity decreases by uniform growth of the solid phase, the length scale is related to the formation factor, and so Archie's exponent m by the relationship:

$$\frac{2}{\Lambda_e} = \frac{d \ln F_R}{d \ln \phi} \frac{S_0}{\phi} \equiv m(\phi) \frac{S_0}{\phi}. \qquad (92)$$

Where S_0 is the specific surface area. This means, physically, that Λ_e is a modified hydraulic radius (compare with equation 12). Schwartz *et al.* (1989) derived the following equation for permeability:

$$k_{\text{SSJ}} = \frac{\Lambda_e^2}{8F}. \qquad (93)$$

Schwartz *et al.* (1993) tested the above equation in a number of numerical simulations. We can see that equations (90) and (93) both have the same form: neither expresses permeability as a function of porosity or has any adjustable parameter and both express permeability in terms of inverse formation factor. In the case of the 'SSJ' equation (93), such a correction is necessary, because although the length scale may be a good measure of pore throat size, it remains a scalar property and requires information on connectivity from another transport property. This equation is particularly relevant to our current discussion as it could equally well be expressed in terms of X_e and ϕ to give

$$k_{SSJ} = \frac{\Lambda_e^2 \phi}{8 X_e}. \tag{94}$$

If we compare (94) with (89) we find that when $\Lambda_e = r_0$, $k_{SSJ} = k_0$ when $X_e = 1$ (i.e. for a tube); and also $\Lambda_e = r_h$ because the flux vanishes in all directions except along the axis of the single tube after probing a distance of its radius. This is a valid, if trivial, result. If the SSJ weighting function is universally applicable, then Λ_e must be able to correct for the known discrepancy between electrical and hydrodynamic tortuosity. Bernabé (1995) examined this point, and found that the Λ_e was in general quite successful as a predictor of the formation factor and permeability of network models having a variety of pore size distribution functions. However, he found that the results diverged from predicted values if the model medium had a variety of pore sectional shapes. Bernabé corrected his results by the using a multiplication factor. This factor, suggested by A. Revil, is the ratio of the local electrical field intensity to the macroscopic field of electrical potential in each bond of the network. This correction factor is analogous to that employed by Herrick and Kennedy to account for local efficiency when deriving an equation for overall efficiency.

Hilfer (1991) defined a number of pore structure parameters on a local scale using the porosity autocorrelation function (derived from computer image analysis) as a starting point, and weighting the result according to the higher order statistics of the medium. Results were obtained for electrical conduction in the low and high frequency limits, in order that the length scale may be estimated from the dielectrical response of the sample rather than exhaustive petrographic measurement. Hilfer's 'local porosity' approach is very promising as it is testable, links measurable properties, and suggests a way that the lambda parameter may be estimated from image analysis (Boger et al. 1992).

Length scales and the viscosity problem.
Bernabé & Revil (1995) developed a length scale analogous to that of Johnson et al. (1986), but employing the local pressure field rather than the local electrical field, and again weighting with respect to the macroscopic pressure field. The derivation of this length scale uses the local viscous energy dissipation, in a way similar to that already examined, but falls short of our requirement to derive global hydrodynamic

efficiency from a continuous integration of local efficiency. This is because the local pressure gradient field can only be specified on a scale of observation over which lateral velocity gradients arising from viscous drag are averaged. This is not a problem in a network model as each discrete bond can be given a lumped flux, rather than a bundle of flux lines, so long as the overall bond conductance is honoured. However the 'viscosity problem' highlighted in the discussion of equations (86) and (89) remains in the general case of local averaging.

Various other length scales have been proposed along similar lines. Martys & Garboczi (1992) suggest a weighting based on the range of the microscopic pore fluid velocity correlation function. This formulation appears to be essentially similar in form to that suggested by Coussy (1995) since the length scale corrects for viscosity. Avellaneda & Torquato (1991) derived a length scale from viscosity and self-diffusion times, that can be used to derive permeability from NMR measurements (see Banavar & Schwartz 1987; Sahimi 1993).

Rigorous relationships between electrical and hydrodynamic flow

In our earlier discussion of diffusion we noted that viscosity was manifest only after flow has developed to honour the 'no slip' boundary condition. With any instantaneous change of pressure at a flow boundary, the propagation of the pressure perturbation through the porous medium also obeys the Laplace equation, and so is termed pressure 'diffusion' (the Biot slow wave is a diffusive pressure front). Thus, in the limit of high frequencies, it can be shown that the fluid behaves as if it has no viscosity, since the viscous profile is never developed. Inviscid flow and high frequency oscillatory flow both experience a limiting 'slip-flow' tortuosity which is usually known as α_∞ (Johnson et al. 1987).

Under these circumstances, the tortuosity felt by the fluid particles is the same as the steady diffusional tortuosity. We already have the result that the pure diffusional tortuosity is the same as the electrical tortuosity in the d.c. limit and so the conjecture of Brown (1980) that the hydrodynamic tortuosity in the high frequency limit is equal to the electrical tortuosity in the low frequency limit is strongly supported by the work of Johnson et al. (1987), i.e.,

$$X_e = \alpha_\infty. \tag{95}$$

Soon after the lambda parameter was introduced, a rigorous link was established between

Λ_e and the limiting high-frequency tortuosity (Johnson et al. 1987; Rubinstein & Torquato 1989). Subsequently, the upper and lower bounds on permeability in the low frequency limit have been successively narrowed (Torquato & Lu 1990; Avellaneda & Torquato 1991).

An analysis of electrical and hydrodynamic coupling by Pride (1994) has resulted in an expression for Λ_e directly in terms of the Stokes flow geometry field and so, ultimately, the microscopic fluid-pressure gradient field and microscopic electrical gradient field can be related. Pride (1994) defined the steady permeability, k in terms of the fundamental pore structure parameters α_∞, Λ_e, and ϕ:

$$k = \frac{\phi \Lambda_e^2}{\varpi \alpha_\infty} \qquad (96)$$

where ϖ is a constant. Thus, using the results of Johnson et al. (1987) the widely-held conjecture,

$$k = \frac{\phi \Lambda_e^2}{\varpi X_e} \equiv \frac{\Lambda_e^2}{\varpi F_R} \qquad (97)$$

would appear to hold. The residual factor ϖ is found experimentally or by simulation to vary between 4 and 8 (Smeulders et al. 1992; Saeger et al. 1995), whereas from comparison with (92) we should always expect a value of exactly 8 for ϖ if the asymptotic tortuosity reflects the true hydrodynamic tortuosity that has been correctly weighted for the difference between Λ_e and Λ_h. Apparently, we still need a little fine tuning to remove our 'fudge factor', ϖ, from the transport equation.

Li et al. (1995) also used electrokinetic coupling to relate electrical and hydrodynamic transport (see Yeung & Mitchell 1993 and Pride 1994 for more details of this aspect of non-equilibrium thermodynamics). The flow of fluid in a medium results in a streaming potential while the flow of an electric current in the same medium produces an electro-osmotic pressure difference. Expressed as flux-gradient-density relations the coupling equations are,

$$\mathbf{J} = -\sigma_0 \nabla \Phi - L_{12} \nabla \Phi \qquad (98)$$

$$(\mathbf{J_h} =) \quad \mathbf{u} = -L_{21} \nabla \Phi - \frac{k_0}{\mu} \nabla P. \qquad (99)$$

Here, Φ is the electrical potential field, σ_0 the pure electrical conductivity of the rock in the absence of coupling, L_{ij} are coupling coefficients (indices 1 and 2 denote hydraulic and electrical properties, respectively), and $\mathbf{u}$ is the seepage velocity discussed on p. 301.

Onsager's reciprocal relation states that the coupling coefficients in the two cases are equal

and we can give this cross-coupling coefficient the symbol Π, i.e.,

$$L_{12} = L_{21} = \Pi. \qquad (100)$$

The entropy dissipation per unit volume, or equivalently the heat dissipation under isothermal conditions, must also be the same in either case, so that:

$$-\mathbf{J} \cdot \nabla \Phi = -\mathbf{u} \cdot \nabla P. \qquad (101)$$

Note that this is the same as equating the 'power terms' as defined by Herrick & Kennedy (1994) and given in equations (79) and (83). Where there is only pure electro-osmosis or pure streaming potential we can define the coupling coefficients in our rock sample as:

$$\zeta_{str} = \frac{\Pi}{\sigma_c}; \qquad \zeta_{eos} = \frac{\Pi \mu}{k_c} \qquad (102)$$

so that the measured or 'coupled' permeability k_c and the 'coupled' conductivity σ_c in the rock can be related:

$$k_c = \mu \sigma_c \frac{\zeta_{str}}{\zeta_{eos}}. \qquad (103)$$

Note that some workers use an electrokinetic coupling coefficient, $Z = dV/dP$, equivalent to the induced voltage per applied pressure drop (volts per pascal) or the current generated per unit of seepage velocity (ampere seconds per metre); see Jouniaux & Pozzi (1995). Now the rock conductivity and permeability measured in the absence of coupling are:

$$\sigma = \sigma_c (1 - \zeta_{eos} \zeta_{str}) \qquad (104)$$

and

$$k = k_c (1 - \zeta_{eos} \zeta_{str}). \qquad (105)$$

So we can relate the electrical conductivity and hydrodynamic permeability (in the d.c. limit) in the same rock if we can determine the pure coupling coefficients. In the absence of anomalous surface conduction or other effects that may upset our application of the Onsager relation, we have:

$$k = \mu \sigma \frac{\zeta_{str}}{\zeta_{eos}}. \qquad (106)$$

The 'unified' length scale derived from this approach we can term Λ_{eh} and is given by:

$$\Lambda_{eh}^2 = 8 \mu \sigma_f \frac{\zeta_{str}}{\zeta_{eos}}. \qquad (107)$$

While equation (106) is apparently exact, an excellent approximation can be made by ignoring the difference between the pure (uncoupled) and measured parameters, i.e. between k_c and k, and σ_c and σ. Very good agreement between measured permeability and that predicted from electrokinetic measurement (based on frequency response) was found by Li *et al.* (1995) for samples of both sandstone and limestone in the range 10^{-14} to 10^{-11} m^2 (10 mD to 10 D). Most of the error arises from the difficulty in making low-frequency electrical conduction measurements in real as opposed to ideal electrolytes. These problems are amplified in materials with a strong surface charge and large surface area (e.g., clays and shales). Some theoretical and practical points relevant to the treatment of electro-osmosis in soils, sediments and rocks can be found in the works by Yin *et al.* (1996); Finno *et al.* (1996) and Jouniaux & Pozzi (1995).

If equation (106) holds in the general case it is an extremely powerful result, and could lead to new practical methods of measuring permeability with streaming potentials. Permeability is only related to tortuosity indirectly through the formulation for the frequency-impedance characteristics of the rock, used to derive the coupling coefficients, which should produce a final relationship of the form given by equation (97). It can be argued that the coupling coefficients embody more information that the formation factor, and so our small discrepancy in using $X_e(\alpha_\infty)$ instead of 'α_0', the limiting d.c. fluid flow tortuosity equivalent to X_h, can thereby be eliminated.

Discussion

The tortuosity problem

Tortuosity is not an easy concept to understand. Tye (1982) summarized the 'tortuosity problem' as follows:

'Although the concept of tortuosity is simple, it is not well understood, and the literature is often misleading. There is an unfortunate tendency to allocate to tortuosity all the differences in behaviour between the real system and some model chosen for simplicity and mathematical tractability. Under these circumstances tortuosity is often only an adjustable parameter used to improve the fit between the predictions of the model and the real data.... Much greater precision of thought is necessary if the concept of tortuosity is to become an aid rather than a hindrance...'.

More fundamentally, it is arguable whether tortuosity really exists as a fundamental attribute of the pore space, in the same way that porosity is a definite property, or whether it is merely invoked as a 'fudge factor' in phenomenological descriptions of transport processes. This disquiet was articulated memorably by I. Gates in a discussion of the pioneering work of Fatt (1956):

The use of the [network] model eliminates the 'bugger factor', sometimes termed tortuosity, which is used to make theoretical calculations on the bundle of tubes model fit experimental data.

Gates was reasoning that the configuration of a network model can account for tortuosity explicitly without the need for adjustable parameters, an argument espoused forcefully by Dullien, who stated in his 1992 monograph 'The concept of a tortuosity factor is limited to one-dimensional models of the transport process'.

Dullien showed that if the network model successfully incorporates both the parallel and the serial type of pore non-uniformities, then a discrete tortuosity factor no longer enters the transport equations. Commonly, the required pore structure information embodying tortuosity is embedded in the petrophysical measurements (formation factor, image analysis, mercury injection) that condition a network model of the 'stick and ball' type (e.g., Matthews *et al.* 1993; Bernabé 1995).

While network models may exclude tortuosity as a discrete factor, Suman and Ruth have shown that both type-A and type-B tortuosity remain as intrinsic properties of the network. Schopper (1966) and Rink & Schopper (1968) used a 'network constant' to correct their electrical network models to account for the directed cross-flows forced by their square lattice. Numerous other examples could be cited where tortuosity is used in the construction of a model, whether it belongs to the capillary network or continuum families, but does not enter the final transport equations as an explicit factor (e.g., see Van Brakel 1975; Dullien 1992).

In a similar way, equations (90) of Katz & Thompson (1986), and (93) of Schwartz *et al.* (1989), which are derived from basic physical principles, also have a measure of tortuosity embedded within them through the incorporation of the formation factor. Formation factor and electrical tortuosity are linked inextricably: not only by definition, as we saw on p. 305, but also, as explained in the fourth section, through the principle of flow efficiency.

Finally, we can consider models where the detail of representation is sufficient to reconstruct a virtual pore space that serves as a template for the direct solution of the transport equations (e.g., Adler *et al.* 1992; Bryant *et al.* 1993*a*; Knackstedt & Zhang 1994). Such models require no adjustable parameters such as a characteristic length scale or tortuosity factor. However, this cannot be used as an argument to disprove the existence of rigorously definable tortuosity measures: it would be as reasonable to disprove the existence of porosity, as this parameter does not need to be specified either.

Tortuosity and path length revisited: how long is a piece of fractal string?

It is confusing to define tortuosity in terms of path lengths, because what are, geometrically the shortest paths, are not necessarily the least 'tortuous' in terms of the efficiency of flow. This is evident from flow simulations in networks (e.g. Ruth & Suman 1992; Suman & Ruth 1993; David 1993; Bernabé 1995) and in representational flow models (e.g., Parney & Smith 1995). Nevertheless, flow efficiency can still be expressed in this way providing that the 'paths' are correctly weighted for the flux that they carry. The discrete-element approach of Zhang & Knackstedt realizes the effect of diversion of the flow trajectory increments from a straight line by tracking a virtual particle from the inflow to the outflow face across which the overall potential drop is known. The local potential gradients and current densities are therefore averaged along a particle track (even if this is not exactly the same as a real line of flux). However, only if we assume, like Cornell & Katz (1953), that as we increase the length of flow paths, we decrease their number (or cross section) in exact proportion are we able to integrate along paths in order to get the global efficiency. Herrick & Kennedy (1994) instead perform a volume integration of scalar quantities, and the rigour of this method is more easily justified.

Tye (1982) tried to sidestep the problem of actual particle trajectories by defining his 'individual flow channels' on an *a priori* basis such that each contributed $1/X_{e,i}$ or e_i of the total flow. Following Cornell & Katz (1953) he then forced the identity $\tau_{ew,i} = \sqrt{X_{e,i}}$. Following this argument a step further, we are forced to make some assumption about the space taken up by the paths in order to integrate over cross-sectional area and thereby retrieve the total flux. The 'space' taken by the paths (or their flux density) is accounted for by Suman & Ruth in their concept of aerosity, so that it has been argued that some of the 'ethereal attributes of tortuosity' are simply transferred to the aerosity (Hulin 1993).

The subdivision of the total conducting pore volume into flow channels is achieved in mathematical solution of the flow equations by defining a series of streamlines, or electrical lines of flux, and orthogonal to these, equipotential surfaces. It is found that in general these are topologically complex structures. However they are subdivided, the flow tubes or 'strings' carrying equal increments of flux must change in sectional area, shape, and orientation, and must also branch and rejoin, so that it is meaningless to attribute to them a particular 'tortuous' length. Delnick & Guidotti (1990) extended the analysis of Tye and found that except in special (and usually trivial) cases, a population of space-filling conduction paths that are expressible as Euclidean objects cannot produce a power law relationship between porosity and conductivity that has a non-integer exponent. Since most rocks have a cementation exponent in Archie's law that lies somewhere between 1.2 and 2.8, the 'flow strings' must be non-Euclidean and presumably fractal objects. Accordingly, they do not have a definite length, or even an integral dimension. Such a finding is not unexpected, since the pore structure of sandstones has long been considered to be fractal in nature (Thompson 1991 and references therein), and fractal structures have been found to underlie diffusion and quasi-static percolation in disordered media (Hulin 1993; Sahimi 1993, 1994).

Summary and conclusions

Tortuosity means different things to different people. We can consider four classes of tortuosity. Firstly, in any pore structure we can define the geometrical tortuosity, (page 300), as an objective characteristic of the pore structure. In theory the geometrical tortuosity could be determined from petrographical analysis, but we cannot use measurements of a transport property to find the geometrical tortuosity, except for certain very specific, and usually trivial pore structures. The second class of tortuosity measures are 'retardation factors' extracted from the transport properties of the porous medium. Electrical tortuosity (page 305), and the diffusional tortuosity (page 307) are examples. Thirdly, we have tortuosity parameters that enter into some simplified construct of a real pore space, such as a network model. Finally, we have tortuosity measures which are nothing more than adjustable correction factors in an empirical model. None of the four classes of tortuosity is exclusively correct or incorrect, rather each is distinct and they cannot and must not be used interchangeably.

Within the second class of definitions, which has been the main concern of this paper, it is possible to link some of the retardation factors in a rigorous way. The linkage depends on being able to identify the equations that underlie the physics of the transport process being considered. Different tortuosities can be compared if we convert the transport to an overall flux, and compare the efficiency of the transport in our sample with an idealized case that has maximum efficiency. This is relatively straightforward for macroscopically one-dimensional steady electrical and diffusional fluxes, and we can use this methodology to prove that the electrical retardation factor X_e and diffusional tortuosity X_{dl} are identical in the steady state. This approach can be extended to steady hydrodynamic flows if we consider the relative rates of viscous energy dissipation, but there are no rigorous results relating hydrodynamic tortuosity to other retardation factors except in the asymptotic limit of high frequency.

Tortuosity can be calculated in a model pore space by using weighted random walks or by the analysis of stream vectors that take into account the distribution of flux as well as the length of flow pathways. Using such methods it is possible to show that electrical and hydrodynamic tortuosity are different within the same pore space, a result which has been known for some time from calculations on networks. We can generalize to say that if the physics underlying the flow is different then the tortuosity is different.

Tortuosity is not necessarily a 'bugger factor', but the fudge element is often increased by the introduction of unnecessary corrections, or by artificial sub-divisions into 'geometrical'

and 'kinematic' (constrictedness) components. Recent advances in porous media characterization have lessened the need for tortuosity factors to appear explicitly in transport equations. Nevertheless, it is apparent that the underlying concept of tortuosity often remains embedded in the remaining parameters, such as formation resistivity factor or 'effective porosity'. Indeed, Pride (1994) has shown that electrical tortuosity (X_e) is one of four fundamental properties of a porous medium that are measurable, and rigorously interrelated, the others being porosity (ϕ), steady hydrodynamic permeability (k) and the electrical length-scale lambda (Λ_e). This rigorous definition does not involve the concept of an effective path length.

Steady-state tortuosity represents an average of transport through all available flow pathways. The details of pore structure are only resolved if we consider unsteady transport processes. Particles of fluid or tracer transported through a porous medium act as probes of the pore space, and this information can be recovered through temporal analysis of inflow and outflow pressures, concentrations, etc. Particles traversing paths of increasingly greater tortuosity exit the sample at increasingly longer times. Important information is lost if all petrophysical measurements use steady flow processes: unsteady and oscillating flows sample a wider range of pore structure parameters, and so should be better for predicting transport properties.

This work was conducted at the University of Leeds while employed by Rock Deformation Research and subsequently under a NERC-funded research project (GST/02/835). Patient reviews and editorial comments have significantly improved the quality of the paper. While I have attempted to be as rigorous as possible in synthesizing previous work, I accept full responsibility for any errors.

Appendix: *Annotated list of symbols*

Symbol	Meaning	Dimensions	Eqn.	Notes and reference
Latin symbols				
a	Area	L^2		
a	Constant in modified Archie's Law	Dimensionless	(25)	Winsauer *et al.* (1952)
b	Structure constant	Dimensionless	(28)	Katsube *et al.* (1991)
c	Concentration	ML^{-3}	(29)	
d	As subscript, diffusional	None		
e	Local electrical efficiency	Dimensionless	(83)	
e	As subscript, electrical	None		
f	As subscript, fluid	Dimensionless	(1)	Adler (1992)

Appendix: *Continued*

Symbol	Meaning	Dimensions	Eqn.	Notes and reference
f	Frequency	T^{-1}	(38)	Bitterlich & Wöbking (1970a)
g	Gravitational accleration	LT^{-2}	(2)	
g	As subscript, geoemetrical	None		
h	Elevation	L	(2)	
h	As subscript, hydraulic	None		
i	index sum variable	None		
k	Hydraulic permeability	L^2	(2)	
k_i	Principle permeability in x_i direction	L^2		
k_0	Permeability in absence of electro-osmotic coupling	L^2	(99)	
k_c	Permeability in the presence of electro-osmotic coupling	L^2	(16)	
k_{tube}	Permeability of an ideal tube	L^2	(89)	
k_{KT}	Permeability from percolation model	L^2	(90)	Katz & Thompson (1987)
k_{SSJ}	Permeability from electrical length-scale	L^2	(93)	Schwartz et al. (1989)
l	(characteristic) length	L	(1, 37)	
l_{min}	Straight-line length between points	L	(1)	
l_h	Hydraulic path length	L		
l_{ed}	Effective diffusional path length	L		
l_{ee}	Effective electrical path length	L		
l_{eh}	Effective hydraulic path length	L	(7)	Kozeny (1927)
m	Cementation exponent	Dimensionless	(22)	Archie (1942)
$\mathbf{m}_1$	Unit vector in direction of macroscopic flow (x_1 direction)	None	(45)	
n	Index sum constant	None		
$\mathbf{n}_i$	Unit vector in direction of microscopic flow	None	(45)	
$\mathbf{n}_j$	Unit vector	None		
$\mathbf{q}_D$	Diffusional flux density	$ML^{-2}T^{-2}$, vector	(29)	
r	Channel radius	L		
r	As subscript, rock	None		
r_0	Radius of cylinder	L	(2, 73)	
r_{bond}	Effective radius of a bond in a network model	L		
r_{eff}	Channel equivalent effective radius	L	(3)	
r_h	Hydraulic radius	L	(5)	
$\mathbf{r}$	Position vector	L, vector	(1)	
s	As subscript, surviving random walkers	None	(43)	
t	Time	T	(29)	
t_i	Transit time for ith flow pathway	T		
t^*	Time to reach steady state	T	(67)	
$\mathbf{u}$	Seepage velocity	LT^{-1}, vector	(2)	Also known as specific flux, Darcy velocity
$\mathbf{v}$	Macroscopic fluid velocity vector	LT^{-1}, vector	(87)	Coussy (1995)

Appendix: *Continued*

Symbol	Meaning	Dimensions	Eqn.	Notes and reference
$\bar{\mathbf{v}}$	Instrinsic phase average velocity	LT^{-1}, vector	(9, 35)	Phillips (1991)
$\mathbf{w}_i$	Local streamwise velocity	LT^{-1}, vector	(86a)	
x	Spatial co-ordinate; x_1, x_2, x_3 mutually orthogonally directions	None		x_1 is macroscopic flow direction
z	Co-ordination number	Dimensionless		Brown (1988)
A	Cross-sectional area	L^2	(2)	
A_1	Total intercepted area normal to x_1 direction	L^2	(45)	
A_{fl}	Intercepted fluid-filled area normal to x_1 direction (45)	L^2	(45)	
C	Capacitance	$I^2T^4M^{-1}L^{-2}$	(39)	Farads (I = electrical current)
C_f	Fluid capacitance	$I^2T^4M^{-1}L^{-2}$	(41)	Bitterlich & Wöbking (1970a)
C_r	Rock capacitance	$I^2T^4M^{-1}L^{-2}$	(38)	Bitterlich & Wöbking (1970a)
C_K	Kozeny constant	Dimensionless	(5)	Kozeny (1927)
D	Diffusion or dispersion coefficient	LT^{-2}	(29)	Diffusivity, dispersivity
D_{eff}	Effective diffusivity	LT^{-2}	(66)	
D_{por}	Effective molecular diffusivity of porous medium	LT^{-2}	(31)	
D_{bulk}	Molecular self-diffusion coefficient	LT^{-2}	(31)	
D_L	Effective diffusivity	LT^{-2}	(66)	
D_L	Longitudinal dispersivity	LT^{-2}	(36)	
D_T	Transverse dispersivity	LT^{-2}	(36)	
E	Electrical efficiency	Dimensionless	(77)	Herrick & Kenedy (1994)
$\mathbf{E}$	Electrical field strength	$MLI^{-1}T^{-3}$	(79)	Volts per unit length
F_R	Formation resistivity factor	Dimensionless	(13)	Archie (1942)
G	Electrical geometry factor	Dimensionless	(27)	Perez-Rosales (1982)
G_v	Genus per unit volume	L^{-3}		Zhao & Macdonald (1993)
I	Macroscopic electrical current	I	(34)	Amperes
$\mathbf{J}$	Electrical current density	IL^{-3}, vector	(79)	Current per unit volume
$\mathbf{J}_D$	Mass flux density	$ML^{-2}T^{-1}$, vector		
K	As superscript, Knudsen	None	(33)	
L	As subscript, longitudinal	None	(36)	
L_{ij}	Coupling co-efficients: subscripts 1 = electrical, 2 = hydrodynamic	Various	(98)	Li et al. (1995), see (100) and definition of Π
M	As superscript, molecular	None	(32)	
N_K	Knudsen number	Dimensionless		Ratio of mean path length and pore diameter
P	Pressure	MLT^{-2}	(2)	Joules per unit volume
Pe	Péclet number	Dimensionless	(37)	Ratio of advective to diffusional flux
Q	Volumetric flow rate	L^3T^{-1}	(2)	
Q_i	Volumetric flow rate in ith tube of a network model	L^3T^{-1}	(52)	
Q_D	Mass diffusive flux	MT^{-1}		

Appendix: *Continued*

Symbol	Meaning	Dimensions	Eqn.	Notes and reference
R	Macroscopic resistance	$ML^2I^{-2}T^{-3}$	(81)	Ohms
S'	Electrical constrictedness factor	Dimensionless	(21)	Dullien (1992)
S_0	Specific surface area	L^{-1}	(13)	
T	As subscript, transverse	None	(36)	
T	Tortuosity factor	Dimensionless		Nominally L^2/L^2
T_{hC}	Carman hydraulic tortuosity factor	Dimensionless	(17)	Carman (1937)
T_{eg}	Geometrical component of electrical tortuosity factor	Dimensionless	(21)	Dullien (1992)
T_{eZK}	Electrical tortuosity factor	Dimensionless	(72)	Zhang & Knackstedt (1995)
T_{hZK}	Hydraulic tortuosity factor	Dimensionless	(70)	Zhang & Knackstedt (1995)
V	Macroscopic electrical potential difference	$ML^2I^{-1}T^{-3}$		Volts
$\mathbf{V}$	Macroscopic fluid velocity vector	LT^{-1}, vector	(87)	At the large scale
$\bar{\mathbf{V}}$	Macroscopic mean fluid velocity vector	LT^{-1}, vector	(37)	In a representative elementary volume
X	Retardation factor	Dimensionless		Dullien (1992)
X_{d1}	Diffusional retardation factor	Dimensionless	(31b)	
X_{d2}	Diffusional retardation factor	Dimensionless	(31c)	
X_{d3}	Diffusional retardation factor	Dimensionless	(31d)	
X_e	Electrical retardation factor	Dimensionless	(17)	
X_h	Hydrodynamic retardation factor	Dimensionless		
Y_{pn}	Ionic mobility constant	$I^2T^2M^{-1}L^{-2}$	(38)	Bitterlich & Wöbking (1970a)

Greek symbols

Symbol	Meaning	Dimensions	Eqn.	Notes and reference
α_∞	Acoustic or alternating flow tortuosity in high frequency limit or for inviscid flow	Dimensionless	(95)	Johnson et al. (1987)
β	Shape factor	Dimensionless	(8)	Kozeny (1927)
γ	Binary phase function; 0 = solid, 1 = fluid	None	(46)	Ruth & Suman (1992)
δ	Diameter of bond in network model	L	(52)	
ε	Dielectric constant	$I^2T^4M^{-1}L^{-3}$	(39)	Or permittivity capacitance per unit length
ε_f	Fluid dielectric constant	$I^2T^4M^{-1}L^{-3}$	(41)	Bitterlich & Wöbking (1970a)
ε_r	Rock dielectric constant	$I^2T^4M^{-1}L^{-3}$	(40)	Bitterlich & Wöbking (1970a)
ζ_{eos}	Electro-osmotic coupling coefficient	ITL^{-3}	(102)	Charge per unit volume, Li et al. (1995)
ζ_{str}	Streaming potential coupling coefficient	$L^3I^{-1}T^{-1}$	(102)	volume per unit charge, Li et al. (1995)
η	Hydrodynamic power dissipation per unit volume	$ML^{-1}T^{-1}$	(85)	
θ	Reciprocal diffusional tortuosity	Dimensionless		Used by Berner (1980) and others for X_{d1}
θ_1	Defined by flux	Dimensionless	(31a)	
θ_2	Corrected for porosity	Dimensionless	(31c)	

Appendix: *Continued*

Symbol	Meaning	Dimensions	Eqn.	Notes and reference
θ_3	Corrected for porosity and blind pore space	Dimensionless	(31d)	
μ	Dynamic viscosity	$ML^{-1}T^{-1}$	(2)	
ρ	Density	ML^{-3}	(2)	
ρ_f	Fluid resistivity	$MLI^{-2}T^{-3}$	(16)	Resistance per unit length
ρ_r	Rock resistivity	$MLI^{-2}T^{-3}$	(16)	
ξ	Aerosity	Dimensionless	(44)	Ruth & Suman (1992)
σ	An arbitrary angle	Degrees		
π	The numerical constant 3.1415...		(39)	
σ_c	Fluid conductivity in the presence of electro-osmotic coupling	$I^2T^3M^{-1}L^{-3}$	(16)	Conductance per unit length
σ_0	Fluid conductivity in the absence of electro-osmotic coupling	$I^2T^3M^{-1}L^{-3}$	(16)	
σ_f	Fluid conductivity	$I^2T^3M^{-1}L^{-3}$	(16)	
σ_r	Rock conductivity	$I^2T^3M^{-1}L^{-3}$	(16)	
σ_{max}	Maximum possible conductivity	$I^2T^3M^{-1}L^{-3}$		
τ	Path length tortuosity	Dimensionless		All are nominally L/L
τ_{eCK}	Electrical path length tortuosity	Dimensionless	(19)	Cornell & Katz (1953)
τ_{etK}	'True' electrical tortuosity	Dimensionless	(28)	Katsube *et al.* (1991)
τ_{eW}	Electrical path length tortuosity	Dimensionless	(18)	Wyllie & Rose (1950)
τ_{eSR}	Electrical flux corrected tortuosity	Dimensionless	(46)	Suman & Ruth (1993)
τ_g	Geometrical tortuosity	Dimensionless	(1)	Adler (1992)
τ_{hK}	Hydraulic path length tortuosity	Dimensionless	(17)	Kozeny (1927)
τ_{hZK}	Hydraulic path length tortuosity	Dimensionless	(70)	Zhang & Knackstedt (1995)
τ_A	A-type tortuosity	Dimensionless	(51)	Ruth & Suman (1992)
τ_B	B-type tortuosity of Ruth and Suman	Dimensionless	(52)	Ruth & Suman (1992)
ϕ	Porosity	Dimensionless	(5)	
χ	Reciprocal viscous hydrodynamic tortuosity factor	Dimensionless, scalar or tensor	(87)	Coussy (1995)
ψ	Generic scalar potential field	Various	(53)	
ϖ	Arbitrary constant	Dimensionless	(96)	
Z	Electrokinetic coupling coefficient	ITL^{-1}		dV/dP, Journiaux & Pozzi (1995)
Θ	Electrical power dissipated per unit volume	$ML^{-1}T^{-1}$	(79)	
Λ_d	Diffusional length scale (trapping length)	L		
Λ_e	Electrical length scale	L	(91)	Johnson *et al.* (1986)
Λ_h	Hydrodynamic length scale	L	(93)	
Λ_{eh}	Putative unified electrical-hydrodynamic length scale	L	(107)	
Π	Electro-osmotic/streaming potential cross-coupling coefficient	IT^2M^{-1}	(100)	Li *et al.* (1995). Note: electrokinetic coupling coefficient is dV/dP (seepage velocity per unit current; potential per unit pressure)
Φ	Electrical potential field	$ML^2I^{-1}T^{-3}$	(79)	Volts
Ψ	Scalar field of pressure potential	MLT^{-2}	(85)	Pascals

References

ABBASI, M. H., EVANS, J. W. & ABRAMSON, I. S. 1983. Diffusion of gases in porous solids: Monte Carlo simulations in the Knudsen and ordinary diffusion regimes. *American Institute of Chemical Engineers Journal*, **29**, 617–624.

ADLER, P. M. 1992. *Porous Media: Geometry and Transports*. Butterworth-Heinemann.

——, JACQUIN, C. G. & THOVERT, J.-F. 1992. The formation factor of reconstructed porous media. *Water Resources Research*, **28**, 1571–1576.

ARCH, J. & MALTMAN, A. J. 1990. Anisotropic permeability and tortuosity in deformed wet sediments. *Journal of Geophysical Research*, **95**, 9035–9047.

ARCHIE, G. E. 1942. The electrical resistivity log as an aid in determining some reservoir characteristics. *Transactions of the American Institute of Mechanical Engineers*, **146**, 54–67.

ARIS, R. 1956. On the dispersion of a solute flowing through a tube. *Proceedings of the Royal Society of London*, **A235**, 67–77.

AVELLANEDA, M. & TORQUATO, S. 1991. Rigorous link between fluid permeability, electrical conductivity and relaxation times for transport in porous media. *Physics of Fluids*, **A3**, 2529–2540.

AZZAM, M. I. S. and DULLIEN 1977. Flow in tubes with periodic step changes in diameter. A numerical solution. *Chemical Engineering Science*, **32**, 1445.

BANAVAR, J. & SCHWARTZ, L. M. 1987. Magnetic resonance as a probe of permeability in porous media. *Physical Review Letters*, **58**, 1411–1414.

BEAR, J. 1969. Hydrodynamic dispersion. *In*: DE WEIST, D. M. (ed.) *Flow Through Porous Media*. Wiley, 109–199.

——1972. *Dynamics of Fluids in Porous Media*. Elsevier, New York.

BENZI, R., SUCCI, S. & VERGASSOLA, M. 1992. The lattice Boltzmann equation: theory and applications. *Physics Reports*, **222**, 145–197.

BERNABÉ, Y. 1991. Pore geometry and pressure dependence of the transport properties in sandstones. *Geophysics*, **56**, 436–446.

——1995. The transport properties of networks of cracks and pores. *Journal of Geophysical Research*, **100**, 4231–4241.

—— & REVIL, A. 1995. Pore-scale heterogeneity, energy dissipation and the transport properties of rocks. *Geophysical Research Letters*, **22**, 1529–1532.

BERNER, R. A. 1980. *Early Diagenesis: A Theoretical Approach*. Princeton University Press.

BIOT, M. A. 1962a. Generalized theory of acoustic propagation in porous dissipative media. *Journal of the Acoustical Society of America*, **34**, 1254–1264.

——1962b. Mechanics of deformation and acoustic propagation in porous media. *Journal of Applied Physics*, **33**, 1482–1498.

BITTERLICH, V. W. & WÖBKING, H. 1970a. Eine Method zur direkten Bestimmung der sogenannten "Geometrischen Tortuosität" (A method for the direct determination of geometrical tortuosity). *Zeitschift Für Geophysik*, **36**, 607–620.

—— & ——1970b. Eine neue gefundliche Größe und ihre Bedeutung für die Bestimmung der Porosität und der elecktrischen Tortuosität von Gesteinen. (A new petrofabric quantity and its significance for determining porosity and electrical tortuosity of rocks). *Gerlands Betreit Geophysik*, **79**, 185–195.

—— & ——1971. Betrachten zur electriktrischen Tortuosität von Gesteinen (Studies of the electrical tortuosity of rocks). *Geologie*, **20**, 887–895.

BOGER, F. 1992. Microstructural sensitivity of local porosity distributions. *Physica A*, **187**, 55–70.

BORGIA, G. C., FANTAZZINI, P., GORE, J. C., SMITH, M. E. & STRANGE, J. H. (eds). 1994. Proceedings of the second international meeting on recent advances in MR applications to porous media. *Magnetic Resonance Imaging*, **12**, 161–377.

BOURDREAU, B. P. 1996. The diffusive tortuosity of fine-grained unlithified sediments. *Geochimica et Cosmochimica Acta*, **60**, 3139–3142.

BROWN, J. S. 1980. Connection between formation factor for electrical resistivity and fluid-solid coupling factor in Biot's equations for acoustic waves in fluid-filled porous media. *Geophysics*, **45**, 1269–1275.

BROWN, R. 1988. *Topology: A Geometric Account of General Topology, Homotopy Types and the Fundamental Groupoid*. Ellis Horwood, NY.

BROWN, S. R. 1989. Transport of fluid and electric current through a single fracture. *Journal of Geophysical Research*, **89**, 9429–9438.

——, STOCKMAN, H. W. & REEVES, S. J. 1995. Applicability of the Reynolds equation for modeling fluid flow between rough surfaces. *Geophysical Research Letters*, **22**, 2537–2540.

BRYANT, S., CADE, C. & MELLOR, D. 1993a. Permeability predictions from geologic models. *American Association of Petroleum Geologists Bulletin*, **77**, 1338–1350.

——, S. L., KING, P. R. & MELLOR, D. W. 1993b. Network-model evaluation of permeability and spatial correlation in a real random sphere packing. *Transport in Porous Media*, **11**, 53–70.

CANCELLIERE, A., CHANG, C., FOTI, E., ROTHMAN, D. H. & SUCCI, S. 1990. The permeability of a random medium: comparison of simulation with theory. *Physics of Fluids*, **A2**, 2085–2087.

CARMAN, P. C. 1937. Fluid flow though granular beds. *Transactions of the Institute of Chemical Engineers*, **50**, 150–166.

——1956. *Flow of Gases Through Porous Media*. Butterworth, London.

CHANG, S. K., LIU, H. L. & JOHNSON, D. L. 1988. Low frequency tube waves in permeable rocks. *Geophysics*, **55**, 519–527.

CHANTONG, A. & MASSOTH, F. E. 1983. Restrictive diffusion in aluminas. *American Institute of Chemical Engineers Journal*, **29**, 725–731.

CHEN, S., DIEMER, K., DOOLEN, G. D., EGGERT, K., FU, C., GUTMAN, S. & TRAVIS, B. J. 1991. Lattice gas automata for flow through porous media. *Physica*, **D47**, 72–84.

COATES, G. & DUMANOIR, J. 1973. A new approach to log-derived permeability. *Proceedings, Society of Professional Well-Log Analysts*, Paper R, 1–28.

CORNELL, D. & KATZ, D. L. 1953. Flow of gases through consolidated porous media. *Industrial and Engineering Chemistry*, **45**, 2145–2152.

COUSSY, O. 1995. *Mechanics of Porous Continua*. John Wiley.

CRANK, J. 1975. *The Mathematics of Diffusion*, 2nd edition. Oxford.

CUNNINGHAM, R. E. & WILLIAMS, R. J. J. 1980. *Diffusion in Gases and Porous Media*. Plenum.

DAVID, C. 1993. Geometry of flow paths for fluid transport in rocks. *Journal of Geophysical Research*, **98**, 12267–12278.

——, GUEGEN, Y. & PAMPOUKIS, G. 1990. Effective medium theory and network theory applied to the transport properties of rock. *Journal of Geophysical Research*, **95**, 6993–7005.

DIEDERICKS, G. P. J. & DU PLESSIS, J. P. 1995. On tortuosity and aerosity tensors for porous media. *Transport in Porous Media*, **20**, 265–279.

DELNICK, F. M. & GUIDOTTI, R. A. 1990. Ionic-conduction in porous-media. *Journal of the Electrochemical Society*, **137**, 11–16.

DOYEN, P. M. 1988. Permeability, conductivity and pore geometry of sandstone. *Journal of Geophysical Research*, **93**, 7729–7740.

DULLIEN, F. A. L. 1992. *Porous Media: Fluid Transport and Pore Structure*. Second edition. Academic Press, San Diego.

EPSTEIN, N. 1989. On tortuosity and the tortuosity factor in flow and diffusion in porous media. *Chemical Engineering Science*, **44**, 777–779.

ETRIS, E. L. 1991. *Investigations into the control of pore/throat structure on capillary pressure imbibition and its relevance to resistivity index and permeability of reservoir rocks*. PhD Thesis, University of South Carolina.

EVANS, J. W., ABBASI, M. H. & SARIN, A. 1980. A Monte Carlo simulation of the diffusion of gases in porous solids. *Journal of Chemical Physics*, **72**, 2967–2973.

FATT, I. 1956. The network model of porous media. Parts I, II, & III. *Transactions, American Institute of Mining Engineers*, **207**, 144–181. Discussion by GATES, I., JOSENDAHL, V. A., ROSE, W. D., WITHERSPOON, P. A. & CARPENTER, D. W. with reply by author.

FINNO, R. J., CHUNG, K., YIN, J. & FELDKAMP, J. R. 1996. Coefficient of permeability from AC electro-osmosis experiments: 2: Results. *Journal of Geotechnical Engineering*, **122**, 346–354.

FREIDMAN, S. P. & SEATON, N. A. 1995. A corrected tortuosity factor for the network calculation of diffusion coefficients. *Chemical Engineering Science*, **50**, 897–900.

FRISCH, U., HASSLACHER, B. & POMEAU, Y. 1986. Lattice-gas automata for the Navier-Stokes equation. *Physical Review Letters*, **56**, 1505–1508.

GOLIN, Y. L., KARYARIN, V. E., POSPELOV, B. S. & SEVEDKIN, V. I. 1992. Pore tortuosity estimates in porous media. *Soviet Electrochemistry*, **28**, 87–91.

GREENKORN, R. A. 1983. *Flow Phenomena in Porous Media*. Marcel Dekker, New York.

HAPPEL, J. & BRENNER, H. 1973. *Low Reynolds Number Hydrodynamics: With Special Applications to Particulate Media*, 2nd Edition. Prentice-Hall, NY.

HAVLIN, S. & BEN-AVRAHAM, D. 1987. Diffusion in disordered media. *Advances in Physics*, **36**, 695–798.

HELMER, K. G., HURLIMANN, M. D., DE SWIET, T. M., SEN. P. N. & SOTAK, C. H. 1995. Determination of ratio of surface area to pore volume from restricted diffusion in a constant field gradient. *Journal of Magnetic Resonance*, **A115**, 257–259.

HERRICK, D. C. & KENNEDY, W. D. 1994. Electrical efficiency – A pore geometric theory for interpreting the electrical properties of reservoir rocks. *Geophysics*, **59**, 918–927.

HILFER, R. 1991. Geometric and dielectric characterization of porous media. *Physical Review*, **B44**, 60–75.

——1992. Local-porosity theory for flow in porous media. *Physical Review*, **B45**, 7115–7121.

HOLLEWAND, M. P. & GLADDEN, L. F. 1992. Modeling of diffusion and reaction in porous catalysts using a random 3-dimensional network model. *Chemical Engineering Science*, **47**, 1761–1770.

HOWARD, J. J. & KENYON, W. E. 1992. Determination of pore size distribution in sedimentary rocks by proton nuclear magnetic resonance. *Marine and Petroleum Geology*, **9**, 139–145.

HULIN, J. P. 1993. Comments on "Formation factor and tortuosity of homogeneous porous media" by R. Suman and D. Ruth. *Transport in Porous Media*, **12**, 291–292.

IGWE, G. J. I. 1991. *Powder Technology and Multiphase Systems: Gas Permeametry and Surface Area Measurement*. Ellis Horwood, N.Y.

JERAULD, G. R., HATFIELD, J. C., SCRIVEN, L. E. & DAVIS, H. T. 1984a. Percolation and conduction on Voronoi and triangular networks: a case study in topological disorder. *Journal of Physics C: Solid State Physics*, **17**, 1519–1529.

——, SCRIVEN, L. E. & DAVIS, H. T. 1984b. Percolation and conduction on Voronoi and rectangular networks: a second case study in topological disorder. *Journal of Physics C: Solid State Physics*, **17**, 3429–3439.

JOHNSON, D. L. KOPLIK, J. & SCHWARTZ, L. M. 1986. New pore size parameter characterizing transport in porous media. *Physical Review Letters*, **57**, 2564–2567.

——, ——, DASHEN, R. 1987. Theory of dynamic permeability and tortuosity in fluid-saturated porous-media. *Journal of Fluid Mechanics*, **176**, 379–402.

——, Plona, T. J., SCALA, C., PASIERB, F. & KOJIMA, H. 1982. Tortuosity and acoustic slow waves. *Physical Review Letters*, **49**, 1840–1844.

JOUNIAUX, L. & POZZI, J.-P. 1995. Streaming potential and permeability of saturated sandstones under triaxial stress: consequences for electrotelluric anomalies prior to earthquakes. *Journal of Geophysical Research*, **100**, 10197–10209.

KAMATH, J. 1992. Evaluation of accuracy of estimating air permeability from mercury injection data. *Society of Petroleum Engineers Formation Evaluation*, **7**, 304–310.

KATSUBE, T. J., MUDFORD, B. S. & BEST, M. E. 1991. Petrophysical characteristics of shales from the Scotian Shelf. *Geophysics*, **56**, 1681–1689.

KATZ, A. J. & THOMPSON, A. H. 1986. Quantitative prediction of permeability in porous rock. *Physical Review*, **B34**, 8179–8181.

—— & ——1987. Prediction of rock electrical conductivity from mercury injection measurements. *Journal of Geophysical Research*, **92**, 599–607.

KEMPERS, L. J. T. M., HAAS, H. & GROENEWEG, H. K. 1992. Experimental observations of dispersive mixing zones in the presence of a viscosity contrast and density contrast. *In*: QUINTARD, M. & TODOROVIC, M. (eds) *Heat and Mass Transfer in Porous Media*. Elsevier, 1525–1536.

KLINKENBERG, L. J. 1941. The Permeability of Porous Media to Liquids and Gases. *API Drilling & Production Practice*, 200–236.

——1961. Analogy between diffusion and electrical conductivity in porous rocks. *Bulletin of the Geological Society of America*, **62**, 559–564.

KNACKSTEDT, M. A. & ZHANG, X. 1994. Direct evaluation of length scales and structural parameters associated with flow in porous-media. *Physical Review*, **E50**, 2134–2138

KOPLIK, J. 1982. Creeping flow in two-dimensional networks. *Journal of Fluid Mechanics*, **119**, 219–247.

——, LIN, C., & VERMETTE, M. 1984. Conductivity and permeability from microgeometry. *Journal of Applied Physics*, **56**, 3127–3131.

——, REDNER, S. & WILKINSON, D. 1988. Transport and dispersion in random networks with percolation disorder. *Physical Review*, **A37**, 2619–2636.

KOSTEK, S., SCHWARTZ, L. M. & JOHNSON, D. L. 1992. Fluid permeability in porous media: comparisons of electrical estimates with hydrodynamical calculations. *Physical Review*, **B45**, 186–195.

KOZENY, J. 1927. Uber kapillare Leitung des Wassers im Boden. *Sitzungberichte der Akadämie der Wissenschaftung in Wein Abteilung IIa*, **136**, 271–301.

LATOUR, L. L., KLEINBERG, R. L., MITRA, P. P. & SOTAK, C. H. 1995. Pore-size distributions and tortuosity in heterogeneous porous media. *Journal of Magnetic Resonance*, **A112**, 83–91.

LE DOUSSAL, P. 1989. Permeability versus conductivity for a porous medium with a wide variation of pore sizes. *Physical Review*, **B39**, 4816–4819.

LEE, P. H. & KOZAK, J. J. 1987. Calculation of the tortuosity factor in single-phase transport through a structured medium. *Journal of Chemical Physics*, **86**, 4617–4627

LI, S. X., PENGRA, D. B. & WONG, P.-Z. 1995. Onsager reciprocal relation and the hydraulic conductivity of porous media. *Physical Review*, **E51**, 5748–5751.

LORENZANO-PORRAS, C. F., CARR, P. W. & MCCORMICK, A. V. 1994. Relationship between pore structure and diffusion tortuosity of ZrO2 colloidal aggregates. *Journal of Colloid and Interface Science*, **164**, 1–8.

LOWELL, S. & SHIELDS, J. E. 1991. *Powder Surface Area and Porosity*. Chapman & Hall, London.

MA, L. & SELIM, H. M. 1994. Tortuosity, mean residence time and deformation of tritium breakthroughs from soil columns. *Soil Science Society of America Journal*, **58**, 1076–1085.

MACDONALD, I. F., KAUFMANN, P. & DULLIEN, F. A. L. 1986a. Quantitative image analysis of finite porous media. 1. Development of genus and pore map software. *Journal of Microscopy*, **144**, 277–296.

——, —— & ——1986b. Quantitative image analysis of finite porous media. 2. Specific genus of cubic lattice models and Berea sandstone. *Journal of Microscopy*, **144**, 297–316.

MACDONALD, M. J., CHU, C.-F., GUILLOT, P. P. & NG, K. M. 1991. A generalized Blake-Kozeny equation for multisized spherical particles. *American Institute of Chemical Engineers Journal*, **37**, 1583–1588.

MADDEN, T. R. 1976. Random networks and mixing laws. *Geophysics*, **41**, 1104–1125.

MARTYS, N. S. 1994. Fractal growth in hydrodynamic dispersion through random porous media. *Physical Review*, **E50**, 335–342

—— & GARBOCZI, E. J. 1992. Length scales relating to the fluid permeability and electrical conductivity in random two-dimensional model porous media. *Physical Review*, **B46**, 6080–6090.

MATTHEWS, G. P. & SPEARING, M. C. 1992. Measurement and modelling of diffusion, porosity and other pore level characteristics of sandstones. *Marine and Petroleum Geology*, **9**, 146–154.

——, MOSS, A. K., SPEARING, M. C. & VOLAND, F. 1993. Network calculation of mercury intrusion and absolute permeability in sandstone and other porous media. *Powder Technology*, **76**, 95–107.

MITCHELL, J. K. 1993. *Fundamentals of Soil Behaviour*. Second edition. Wiley, New York.

MITRA, P. P., LATOUR, L. L., KLEINBERG, R. L. & SOTAK, C. H. 1995. Pulsed-field-gradient NMR measurements of restricted diffusion and the return-to-the-origin probability. *Journal of Magnetic Resonance*, **A114**, 47–58.

——, SEN, P. P. & SCHWARTZ, L. M. 1993. Short-time behaviour of the diffusion coefficient as a geometrical probe of porous media. *Physical Review*, **B47**, 8565–8574.

——, ——, —— & LE DOUSSAL, P. 1992. Diffusion propagator as a probe of the structure of porous media. *Physical Review Letters*, **68**, 3555–3558.

NAKANO, Y., IWAMOTO, S., YOSHINAGA, I. & EVANS, J. W. 1987. The effect of pore necking on Kundsen diffusivity and collision frequency of gas molecules with pore walls. *Chemical Engineering Science*, **42**, 1577–1583.

NELSON, P. H. 1994. Permeability-porosity relationships in sedimentary rocks. *The Log Analyst*, May–June 94, 38–62.

NORTHRUP, N. A., KULP, T. J. & ANGEL, S. M. 1992. Imaging of interstitial velocity fields and tracer distributions in a refractive index-matched porous media. *In*: QUINTARD, M. & TODOROVIC, M. (eds) *Heat and Mass Transfer in Porous Media*. Elsevier, 157–177.

OLSEN, H. W. 1962. Hydraulic flow through saturated clays. *In*: *Proceedings, 9th National Conference on Clays and Clay Minerals. Clays and Clay Minerals*, **9**, 131–162.

OWEN, J. E. 1952. The resitivity of a fluid-filled porous body. *Petroleum Transactions, American Institute of Mining Engineers*, **195**, 169–174.

PARKHOMENKO, E. I. 1967. *Electrical Properties of Rocks*. Plenum, N.Y.

PARNEY, R. & SMITH, L. 1995. Fluid velocity and path length in fractured media. *Geophysical Research Letters*, **22**, 1437–1440.

PATWARDHAN, A. V. & MANN, R. 1991. Effective diffusivity and tortuosity in Wicke-Kallenbach experiments: Direct interpretation using stochastic pore networks. *Chemical Engineering Reserach & Design*, **69**, 205–207.

PEREZ-ROSALES, C. 1982. On the relationship between formation resistivity factor and porosity. *Society of Petroleum Engineers Journal*, **45**, 531–536.

PETERSEN, E. E. 1958. Diffusion in a pore of varying cross-section. *American Institute of Chemical Engineers Journal*, **4**, 343–345.

PEURRUND, L. M., RASHIDID, M. & KULP, T. J. 1995. Measurement of porous medium velocity fields and their volumetric averaging characteristics using particle tracking velocimetry. *Chemical Engineering Science*, **50**, 2243–2253.

PHILLIPS, O. M. 1991. *Flow and Reactions in Permeable Rocks*. Cambridge.

PRIDE, S. 1994. Governing equations for the coupled electromagnetics and acoustics of porous media. *Physical Review*, **B50**, 15678–15696.

QIAN, Y. H., d'HUMIERES, D. & LALLEMAND, P. 1992. Lattice BGK models for Navier-Stokes equation. *Europhysics Letters*, **17**, 479–484.

RINK, M. & SCHOPPER, J. R. 1968. Computation of network models of porous media. *Geophysical Prospecting*, **16**, 277–294.

—— & ——1976. Pore structure and physical properties of porous sedimentary rocks. *Pure and Applied Geophysics*, **114**, 273–284.

ROTHMAN, D. H. 1988. Cellular-automation fluids: a model for flow in porous media. *Geophysics*, **53**, 509–518.

RUBINSTEIN, J. & TORQUATO, S. 1989. Flow in random porous media: mathematical formulation, variational principles and rigorous bounds. *Journal of Fluid Mechanics*, **206**, 25–46.

RUTH, D. & SUMAN, R. 1992. The role of microscopic cross flow in idealized porous media. *Transport in Porous Media*, **7**, 103–125.

SAEGER, R. B., SCRIVEN, L. E. & DAVIS, H. T. 1995. Transport processes in periodic porous media. *Journal of Fluid Mechanics*, **299**, 1–15.

SAHIMI, M. 1993. Flow phenomena in rocks: from continuum models to fractals, percolation, cellular automata and simulated annealing. *Reviews of Modern Physics*, **65**, 1393–1501.

——1994. *Applications of Percolation Theory*. Taylor & Francis, London.

SATTERFIELD, C. N. & SHERWOOD, T. K. 1963. *The Role of Diffusion in Catalysis*. Addison-Wesley, Reading Mass.

SCHEIDEGGER, A. E. 1954. Statistical hydrodynamics in porous media. *Journal of Applied Physics*, **25**, 994–1001.

——1974. *The Physics of Flow Through Porous Media*. University of Toronto Press.

SCHOPPER, J. R. 1966. A theoretical investigation of the formation factor/permeability/porosity relationship using a network model. *Geophysical Prospecting*, **14**, 301–314.

——1967a. Experimentelle Methoden und eiene Apparatur zur untersuchung der Beziehungen zwischen hydraulishen und electrischen Eigenschaften Loser und kunstlich verfestige poroser Medien. *Geophysical Prospecting*, **15**, 651–701.

——1967b. A theoretical study on the reduction of statistical pore system parameters to measurable quantities. *Geophysical Prospecting*, **15**, 262–287.

SCHWARTZ, L. M., SEN, P. N. & JOHNSON, D. L. 1989. Influence of rough surfaces on electrolytic conduction in porous media. *Physical Review*, **B40**, 2450–2458.

——, MARTYS, N., BENTZ, D. P., GARBOCZI, E. J. & TORQUATO, S. 1993. Cross-property relations and permeability estimation in model porous media. *Physical Review*, **E48**, 4584–4591.

——, SEN, P. N. & MITRA, P. P. 1994a. Nuclear magnetism and transport in porous media. *Magnetic Resonance Imaging*, **9**, 657–662.

——, WILKINSON, D. J., KOSTEK, S., JOHNSON, D. L. & BANAVAR, J. R. 1991. Simulations of pulsed field gradient spin-echo measurements in porous media. *Magnetic Resonance Imaging*, **12**, 241–244.

——, AUZERAIS, F., DUNSMUIR, J., MARTYS, N., BENTZ, D. P. & TORQUATO, S. 1994b. Transport and diffusion in 3-dimensional composite media. *Physica*, **A207**, 28–36.

SEN, P. N. 1994. Diffusion in a periodic porous medium with surface relaxation. *Physica*, **A211**, 387–392.

SEN, P. P., SCHWARTZ, L. M., MITRA, P. P. & HALPERIN, B. I. 1994. Surface relaxation and the long-time diffusion coefficient in porous media: periodic geometries. *Physical Review*, **B48**, 215–225.

SHACKELFORD, C. D. & DANIEL, D. E. 1991. Diffusion in saturated soil. I: Background. *Journal of Geotechnical Engineering*, **117**, 467–484.

SHEPARD, J. S. 1993. Using a fractal model to compute the hydraulic conductivity function. *Soil Science Society of America Journal*, **57**, 300–306.

SKELLAND, A. H. P. 1974. Diffusional mass transfer. John Wiley, N.Y.

SMEULDERS, D. M. J., EGGELS, R. L. G. M. & VAN DONGEN, M. E. H. 1992. Dynamic permeability: reformulation of theory and new experimental and numerical data. *Journal of Fluid Mechanics*, **245**, 211–227.

SUMAN, R. & RUTH, D. 1993. Formation factor and tortuosity of homogeneous porous media. *Transport in Porous Media*, **12**, 185–206.

TAYLOR, G. I. 1953. The dispersion of soluble matter in solvent flowing slowly through a tube. *Proceedings of the Royal Society of London*, **A219**, 186–203.

THOMPSON, A. H. 1991. Fractals in Rock Physics. *Annual Reviews of Earth and Planetary Science*, **19**, 137–262.

THOVERT, J. F. 1993. Computerised characterisation of the geometry of real porous media: their discretization, analysis and interpretation. *Journal of Microscopy*, **170**, 65–79.

TORELLI, L. 1972. Computer simulations of the dispersion phenomena occurring during flow through porous media using a random maze model. *Pure and Applied Geophysics*, **96**, 75–88.

—— & SCHEIDEGGER, A. E. 1971. Random maze models of flow through porous media. *Pure and Applied Geophysics*, **89**, 32–44.

TORQUATO, A. S. & LU, B. 1990. Rigorous bounds on the fluid permeability: effect of polydispersivity in grain size. *Physics of Fluids*, **A2**, 487–490.

TYE, F. L. 1982. Tortuosity. *Chemistry and Industry*, **182**, 322–326.

VAN BRAKEL, J. 1975. Pore space models for transport phenomena in porous media. Review and evaluation with special emphasis on capillary liquid transport. *Powder Technology*, **11**, 205–236.

WALSH, J. B. & BRACE, W. F. 1984. The effect of pressure on porosity and the transport properties of rocks. *Journal of Geophysical Research*, **89**, 9425–9431.

WINSAUER, W. O., SHEARIN, H. M., MASSON, P. H. & WILLIAMS, M. 1952. Resitivity of brine-saturated sands in relation to pore geometry. *American Association of Petroleum Geologists Bulletin*, **36**, 253–277.

WITT, K.-J. & BRAUNS, J. 1983. Permeability anisotropy due to particle shape. *Journal of Geotechnical Engineering*, **109**, 1181–1187.

WORTHINGTON, P. F. 1993. The uses and abuses of the Archie equations, 1: The formation factor-porosity relationship. *Journal of Applied Geophysics*, **30**, 215–228.

WYLLIE, M. R. J. & GREGORY, A. R. 1955. Fluid flow through unconsolidated porous aggregates. Effect of porosity and particle shape on Kozeny–Carman constants. *Industrial and Engineering Chemistry*, **47**, 1379–1388.

—— & ROSE, W. D. 1950. Some theoretical considerations related to quantitative evaluation of the physical characteristics of reservoir rock from electric log data. *Transactions, American Institute of Mechanical Engineers*, **189**, (*Journal of Petroleum Technology*, **2**), 105–118.

—— & SPANGLER, M. B. 1952. Application of electrical resitivity measurements to the problem of fluid flow in porous media. *American Association of Petroleum Geologists Bulletin*, **36**, 359–403.

YEUNG, A. T. & MITCHELL, J. K. Coupled fluid, electrical and chemical flows in soil. *Géotechnique*, **43**, 121–134.

YIN, J., FINNO, R. J., FELDKAMP, J. R. & CHUNG, K. 1996. Coefficient of permeability from AC electro-osmosis experiments: 1: Theory. *Journal of Geotechnical Engineering*, **122**, 346–354.

ZHANG, L. & SEATON, N. A. 1994. The application of continuum equations to diffusion and reaction in pore networks. *Chemical Engineering Science*, **49**, 41–50.

ZHANG, X. & KNACKSTEDT, M. A. 1995. Direct simulation of electrical and hydrodynamic tortuosity in porous solids. *Geophysical Research Letters*, **22**, 2333–2336.

ZHAO, H. Q. & MACDONALD, I. F. 1993. An unbiased and efficient procedure for 3-d connectivity measurement as applied to porous media. *Journal of Microscopy*, **172**, 157–162.

Feature recognition and the interpretation of images acquired from horizontal wellbores

J. C. LOFTS[1], J. BEDFORD[2], H. BOULTON[2]
J. A. VAN DOORN[3] & P. JEFFREYS[4]

[1] Schlumberger-GeoQuest, Loirston House, Wellington Road, Altens,
Aberdeen AB12 3BH, UK
[2] Schlumberger-GeoQuest, Schlumberger House, Gatwick, West Sussex, RH6 ONZ, UK
[3] Schlumberger-GeoQuest, The Hague
[4] Schlumberger-GeoQuest, Qatar

Abstract: The acquisition of borehole resistivity and acoustic images has become increasingly common in horizontal wells drilled for hydrocarbons. The appearance, and later interpretation of these images is somewhat different to that of conventional vertical wellbore images. Shallow-dipping bedding planes appear as long drawn-out sinusoids in unrolled workstation images whilst sub-vertical fractures appear as low-amplitude sinusoids, the opposite to what geologists are used to in vertical wells. Due to the sub-parallel orientation of the wellbore with respect to the bedding, such intersections will produce unusual features that can be misleading and difficult to interpret. The mechanics of drilling, and of acquiring image data in a horizontal plane also leads to the creation of certain artefacts or noise in images which must be identified correctly. The most common sedimentary and tectonic features imaged in horizontal wellbores are described and discussed here as a reference for geoscientists.

Horizontal and highly deviated wells have become increasingly common as drilling technology has advanced. The prohibitive cost of coring large sequences of horizontal wellbore means that imaging with logging tools, has become the most cost-effective, viable method of their geological–petrophysical evaluation.

The main objectives in the evaluation of horizontal wellbores, and thus of borehole imaging, are two-fold. Firstly, it has become increasingly common practice to use borehole imaging for target zone evaluation. This is the test of whether the wellbore has been drilled into the correct horizon (Hurley *et al.* 1994) and, once identified, it is used to identify the horizon laterally for optimum recovery. Secondly, borehole imaging is used for the identification of zones of open and closed (often vertical) fracturing where such zones are important for perforation and production. The nature of the well trajectory is such that fracture zones are frequently intersected at a high angle of incidence to the strike of the fracture. Such fracture zones are ideal targets for fracture density analysis using high resolution images. In addition to these primary objectives, we must not forget the conventional uses of imaging devices that are well established in vertical wellbores, such as the evaluation of structural dip, the identification of sedimentary structures and related facies, and the interpretation of palaeo-transport information.

With the increasing use of horizontal imaging we need to be able to correctly interpret the acquired images. Conventionally, images are displayed on a workstation where the geologist is able to view the borehole image as an unwrapped cylinder (Fig. 1a). Here, the orientation reference of the unrolled image relies on a bearing relative to North and an arbitrary point is taken as the reference for plotting (such as the first pad of a resistivity tool). North is then plotted to the far left hand side of the image. At any point down the wellbore we can orientate the features with respect to north using measurements from the sophisticated inclinometry device accompanying the imaging tool. With highly deviated and horizontal wellbores alike, it is no longer meaningful to orientate the workstation image towards the north. Referencing of the 'unwrapped' workstation image is made to the borehole axis, where convention is to take the top the hole as our reference point. Top-of-hole is now to the far left and far right (Fig. 1b).

It is the former, vertical image, referenced to north that interpreters are used to viewing yet with horizontal images we have a complete change, or reversal, in spatial geometry when we view an image on a workstation. This can and often does confuse interpretation. Instead of drilling at a high apparent angle to bedding planes, drilling is now sub-parallel in many

From Lovell, M. A. & Harvey, P. K. (eds), 1997, *Developments in Petrophysics*, Geological Society Special Publication No. 122, pp. 345–365.

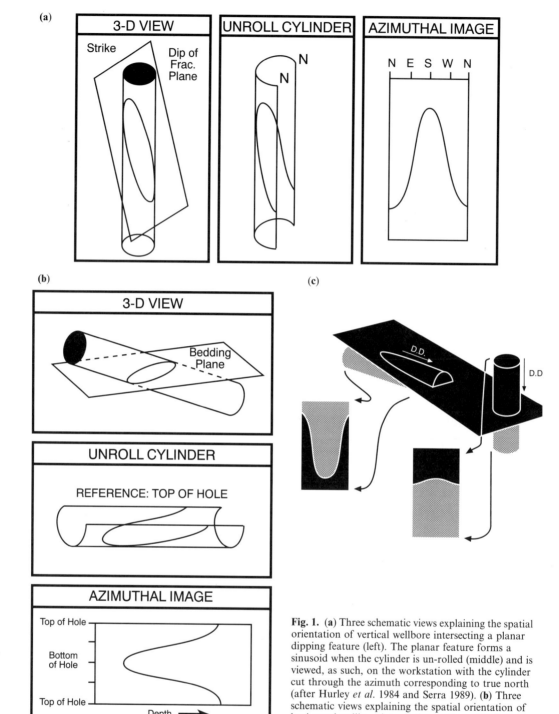

Fig. 1. (a) Three schematic views explaining the spatial orientation of vertical wellbore intersecting a planar dipping feature (left). The planar feature forms a sinusoid when the cylinder is un-rolled (middle) and is viewed, as such, on the workstation with the cylinder cut through the azimuth corresponding to true north (after Hurley *et al.* 1984 and Serra 1989). (b) Three schematic views explaining the spatial orientation of horizontal wellbore intersecting a planar dipping feature (left). The planar feature forms a sinusoid when the cylinder is unrolled (middle) and is viewed, as such, on the workstation with the cylinder cut through the azimuth corresponding to the top of the hole. (after Hurley *et al.* 1984 and Serra 1989). (c) Schematic comparison of the different intersections that a vertical and horizontal wellbore make with a shallow dipping plane (top right). Illustrated, are the workstation images that are formed by the intersection of such (bottom left). Intersection by the horizontal wellbore produces a long drawn out, high amplitude sinusoid while the vertical wellbore intersection forms a tight, low amplitude sinusoid. DD indicates the drilling direction.

(a)

(b)

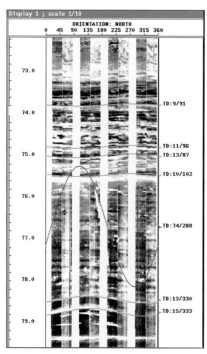

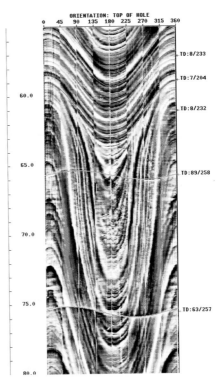

Fig. 2. (**a**) An example of a number of different bedding surfaces and a fracture surface intersected by a vertical wellbore. The steepest amplitude sinusoid is the vertical fracture (74° magnitude, blue sinusoid) while the shallower-angle bedding (10–15°, green sinusoid) are the lowest amplitude sinusoids. Drilling direction is towards base of figure. Scale is 1:10, depth is in feet. Image is referenced to north on the far left. South is therefore dead centre. As with all the resistivity images shown throughout this paper, colour representation ranges from white, representing a high resistivity to black which represents conductive features. The two fractures identified are thus resistive. (**b**) An example of a number of different shallow angle bedding planes intersected by a truly horizontal wellbore. The lowest angle features (less than 7°) are the long drawn out sinusoids covering most of the image. Slightly steeper bedding of 7–8° is present over the top quarter of the image. Two very steep fractures (89° and 63°) are also present (light blue sinusoids). Drilling direction is towards base of figure. Scale is 1:10, depth is in feet. Bitsize 6 inches. Image is referenced to top-of-hole on the far left. Bottom-of-hole is therefore dead centre.

instances and this greatly changes the images we view on the workstation.

The emphasis of this paper is to highlight the change in the workstation representation of features commonly interpreted in vertical wellbores, as a result of this change in spatial geometry in a horizontal wellbore. The aim, therefore, is to describe commonly imaged features in way that the interpreter can begin to better appreciate and conceptualise features which, because of their sub-parallel orientation with respect to the wellbore, often produce strange unwrapped workstation images. We include examples of simple features such as bedding, lithology boundaries, fracturing and we describe more complex features such as cross

bedding-bedding relationships, undulating bedding, complex bedding-lithology relationships and faulting. We also describe some of the more common non-geological artefact features caused by the image acquisition procedure which may be confusing or of conjecture to the interpreter. Only with the correct interpretation of all features in horizontal images will the full benefits and objectives be realised.

Imaging within horizontal wellbores is predominated by the use of resistivity imaging devices. These are similar to conventional dipmeter tools except that they have numerous additional electrode 'buttons' to detect electrical current and provide raw data that can be processed to yield high-resolution images of

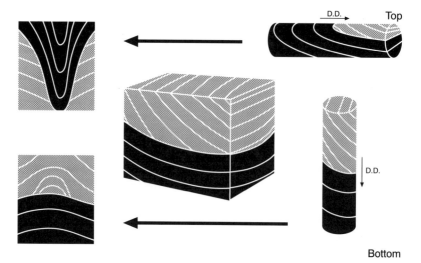

Fig. 3. Schematic comparison of the intersection that a vertical and a horizontal wellbore make with a simple geological feature (illustrated centre), here a group of foresets on-lapping a flat lying bounding surface. Illustrated right, are cylindrical schematics representing the vertical (bottom right) and horizontal (top right) borehole wall intersection with the feature. Illustrated left, are schematic representations of workstation (unwrapped) images. Intersection of a flat lying feature by a horizontal wellbore (top left) produces a long drawn out, high amplitude sinusoid, while the vertical wellbore (bottom left) forms a low amplitude sinusoid. DD indicates the drilling direction.

(b)

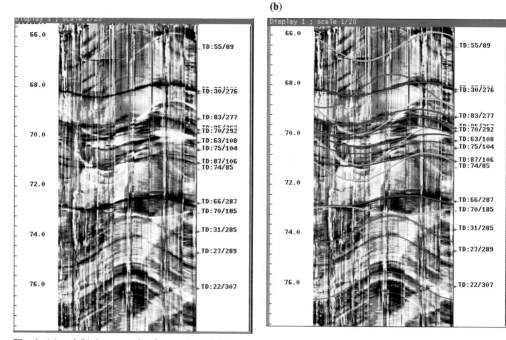

Fig. 4. (a) and (b) An example of a number of different shallow angle bedding and fracture planes intersected by a horizontal wellbore. The steepest amplitude sinusoids are shallow angle bedding (in red, 20–30° magnitude in b) while the steepest features, here vertical conductive 'open' fractures (dark blue) are the smallest amplitude sinusoids. A population of slightly shallower angle (50–80°) resistive 'closed' fractures (light blue) show sinusoids with amplitudes between the two extremes. The vertical striations observed are produced by mud cake smearing which obscures the electrode buttons. Drilling direction is towards base of figure. Scale is 1:20, depth is in feet. Bit size 6 inches. Image is referenced to top-of-hole on the far left. Bottom-of-hole is therefore dead centre.

the borehole wall on a scale of millimetres. In addition to these resistivity-imaging devices, the introduction of new imaging tools such as the azimuthal laterolog tools and ultrasonic tools, means that there are now a number of alternative tools for interpretation. The majority of this paper cites electrical images from the FMI tool (Fullbore Formation MicroImager). This is the latest generation of electrical imaging device capable, in a 6.25 inch wellbore, of producing up to 94% coverage with a vertical resolution of approximately 5 mm (0.2 inches) and sampling rate of 2.5 mm (0.1 inches). This is currently the highest resolution, most commonly run device for horizontal imaging and consequently, the best for the illustration of common features. The interpretation and discussion of features here should also be applicable to the other imaging devices acquired horizontally and a few examples are given. For further reference on resistivity devices, the reader is directed to Ekstrom *et al.* (1987), Serra (1989) and Safinya *et al.* (1991).

Spatial orientation

Vertical versus horizontal imaging

Figure 1c illustrates the intersection of a dipping planar feature with a circular wellbore. This produces an ellipsoid section. The trace of this feature on a planar surface such as the way we view an image on a workstation (by unfolding the cylinder and unflattening the image) produces a sine wave as observed in Figs 1a and 1b. The low apex of the sinusoid will represent the azimuth direction. (Calculation of dip is simply: $\tan^{-1}$ (X/borehole diameter) where X = amplitude of the sinusoid). In vertical wellbores, the height of the sine wave is dependant on the magnitude of dip of the feature. The steeper the plane of feature, the higher the amplitude of the sinusoid will be. Hence, shallow angle bedding will produce low angle sinusoids and horizontal bedding will produce almost a straight line. Similarly, as the dip increases, the sinusoid will elongate and the sine wave will increase in amplitude. Truly vertical dipping features will produce two vertical lines 180° apart.

Within horizontal wellbores the converse is true, as is illustrated in Fig. 1c. Shallow dipping features appear as long drawn out sinusoids and vertical features as tight low amplitude sinusoids. Whilst this may be readily apparent from Fig. 1 it is a source of confusion to the interpreter especially when the orientation reference has changed from being relative to north to being

relative to top-of-hole and then the cylinder is unwrapped. (NB: although images are referenced to top-of-hole it is still possible to calculate a true dip orientation from the apparent dip.)

Deviated wells between vertical and horizontal will produce sinusoids that will vary in amplitude as a direct result of the deviation of the wellbore between the two extremes. As a comparison, Fig. 2 illustrates the relationship of simple low-angle bedding and steep fracturing within a vertical borehole (Fig. 2a) and within a horizontal borehole (Fig. 2b). Note the change in reference used to orientate these images.

For the interpreter, it is more meaningful for the features that are commonly observed within

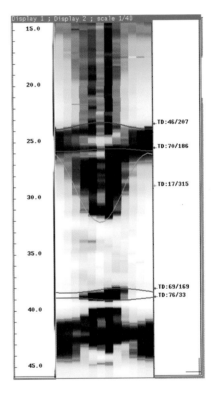

Fig. 5. Feature determination in a horizontal wellbore from an azimuthal resistivity imager device (ARI*). The vertical resolution is lower than the FMI* tool (approx. 1 inch for the ARI* tool) but an interpretation of planar features is still possible. Shallow angle bedding (drawn-out green sinusoid) and fracture planes ('tight' dark blue sinusoids) are illustrated. Drilling direction is towards base of figure. Scale is 1 : 40, depth is in feet. Bit size 5.6 inches. Image is referenced to top-of-hole on far left. Bottom-of-hole is therefore dead centre.
* Denotes a Mark of Schlumberger throughout.

(a)

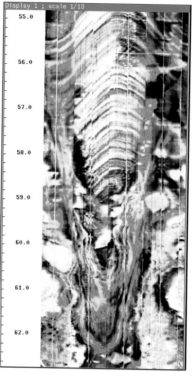

(b)

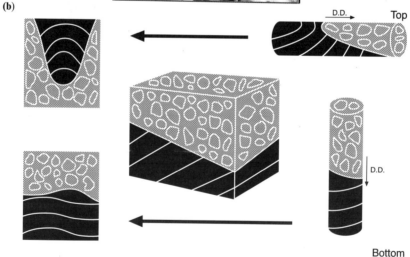

Fig. 6. (**a**) Sandstone–breccia contact surface. This is an example of the intersection of two contrasting lithologies. Although the appearance is complex, the image reveals a thinly laminated sandy unit underlying a matrix supported brecciated clastic unit. The contact surface between the two is very shallow angle but drawn out here as a high-amplitude sinusoid and is probably erosive. Drilling direction is towards base of figure. Scale is 1:10, depth is in feet. Bit size 6 inches. Image is referenced to top-of-hole on far left. Bottom-of-hole is therefore dead centre. (**b**) Schematic illustrating the breccia–sandstone contact surface (centre) within a horizontal and vertical wellbore. Note the quite different appearance on a workstation image as illustrated top and bottom left. DD indicates the drilling direction.

images to be described and discussed in terms of geological features. These are now described below.

Recognition of features

Simple geological features

Bedding. Figure 2b illustrates the variety of surfaces that can be present in truly horizontal wellbore. The lowest angle surfaces are the longer drawn out sinusoids covering most of the image. They indicate that it has taken approximately 12 ft to penetrate a single low-angle bedding plane of the order of 2° dip, whilst it takes approximately 4 ft to drill through a 8° dipping bedded unit at the top of the image. In contrast, the 2 vertical 'resistive' fractures (light blue sinusoids) are of very low amplitude. This is also illustrated in Fig. 3, which shows a schematic comparison of the intersection that a vertical and a horizontal wellbore will make with a simple geological feature (here a group of sedimentary foresets on-lapping a flat lying bedding surface). Intersection of a flat lying feature and a horizontal wellbore produces a long drawn out, high amplitude sinusoid, while the vertical wellbore intersection forms a tight, low amplitude sinusoid. This has implications when we consider bed thickness. The beds imaged here will represent apparent bed thickness and not necessarily true bed thickness.

Fractures. Often the primary target of a horizontal well is to locate and produce from highly fractured intervals within a target zone horizon. Identification of these fractures (as shown in Fig. 4a) is paramount. Figure 4a shows an image over a zone of fracturing which contains two fracture populations, a set of resistive 'healed' fractures (light blue sinusoids in Fig. 4b) and a set of conductive 'open' fractures (dark blue sinusoids) as interpreted in Fig. 4b. Contrary to vertical wellbore interpretations where fractures are of high amplitude, the fractures here are of low amplitude. The steeper conductive set (c. 83–85° dip) showing the lowest amplitudes and the resistive set (c. 70–74° dip) a slightly higher amplitude. Sedimentary dip, at 20–30°, is also visible towards the base of the image and shows high amplitude sinusoids.

An image from an azimuthal laterolog imaging device, the ARI* (Azimuthal Resistivity Imager), was logged through a horizontal wellbore in Fig. 5. Although the resolution of this device is less than that of the FMI* tool, both

low angle bedding plane (green) and steeper fractures (dark blue) are visible as low and high amplitude sinusoids respectively.

Lithological boundaries. Within vertical wells, images displaying contact surfaces between marked lithologies are easy to interpret because the low angle nature of the surface is represented as a 'flat' low amplitude sinusoid with a change in image facies either side of the boundary. Within a highly deviated or horizontal wellbore, the sub-parallel nature of the intersection draws such contact surfaces into high amplitude events often spanning many feet and the actual contact surface often becomes confused. Figure 6a

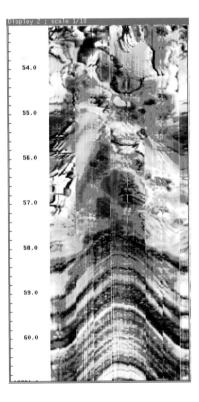

Fig. 7. Sandstone–clay contact surface. This is an example of the intersection of two contrasting lithologies. Although the appearance is complex, the image reveals a thinly laminated sandy unit underlying a claystone. The contact surface between the two is a very shallow angle but drawn out here into a sinusoid of moderate amplitude. The patches imaged within the claystone are caused by the complex current paths within the claystone lithology (current gather and dispersion). Drilling direction is towards base of figure. Scale is 1 : 10, depth is in feet. Bit size 6 inches. Image is referenced to top-of-hole on the far left. Bottom-of-hole is therefore dead centre.

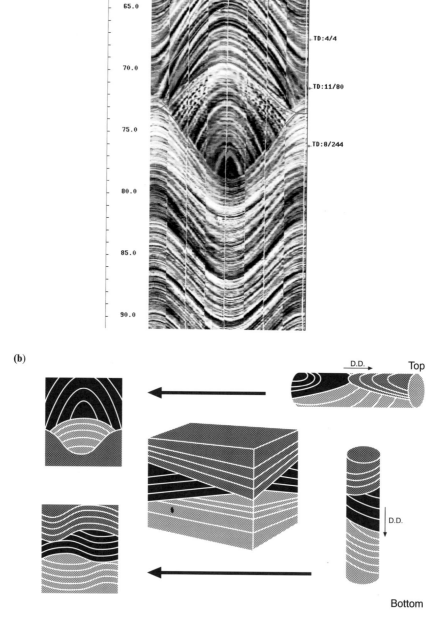

Fig. 8. (a) This example displays three changes in bedding orientation within a 10 ft interval. From the top of the image the bedding changes from a northerly azimuth through to a easterly azimuth at xx72 ft and to a westerly azimuth immediately below. Here, the near horizontal wellbore is seen intersecting three differently orientated, low angle, finely bedded sand units. Drilling direction is towards base of figure. Scale is 1:40, depth is in feet. Bit size 6 inches. Image is referenced to top-of-hole on the far left. Bottom-of-hole is therefore dead centre.
(b) Schematic illustrating the azimuth changes seen in (a) (centre) within a horizontal and vertical wellbore. Note the quite different appearance on a workstation image as illustrated top and bottom left. DD indicates the drilling direction.

shows an image over such a boundary (a boundary between a cross-bedded sandstone overlain and possibly truncated by a matrix supported breccia horizon) where penetration of the boundary takes approximately 4–5 ft. This feature is surprisingly similar to a steep angle fault surface viewed in a vertical wellbore. Figure 6b illustrates the different workstation images formed by such a contact surface in a vertical and horizontal wellbore. Figure 7 shows an image over a more classical shale-sandstone boundary. Here, the well is deviated about 70° resulting in a decrease in the magnitude of the clay-sandstone contact which is approximately 15° here.

Complex geological features

Some features when unwrapped as workstation images look complicated. More often than not, they result from reasonably simple features that become complicated because of the plane of intersection with the wellbore. These more complex features are now discussed.

Cross bedding. A change in the orientation of bedding surfaces can often lead to confusing images. Figure 8a displays an image with three changes in bedding orientation within a 10 ft interval. From the top of the image the bedding changes from a northerly azimuth through to a easterly azimuth at xx72 ft and to a westerly azimuth immediately below. Notice that it takes up to 8 ft to penetrate one shallow-angle bedding surface. Here, the near horizontal wellbore is seen intersecting three differently orientated low-angle, finely bedded sand units. These represent localised changes in deposition and it is possible to identify first, second and third order erosive (bounding) surfaces (trough cross-bedding associated with crescentic dune structures). This is illustrated schematically in Fig. 8b which also shows the different workstation images formed in a vertical and horizontal wellbore.

Cross bedding: Foresets. The intersection of dune foresets and an erosive surface are imaged in Fig. 9a at xx00.0 and xx02.5 ft. There are two sets of foresets in the image, one set towards the top and one set towards the base (red sinusoids) with dip magnitudes of 19–24°. These are separated by shallower angle surfaces (green sinusoids) that probably represent the dune apron–interdune surfaces. The cross cutting relationship between first and

second bounding surfaces is visible at xx99.5 ft and xx03.0 ft and is illustrated schematically in Fig. 9b.

The thickness of the lower-angle surfaces (between xx00.0 and xx03.0 ft) is approximately 3 ft. This should be regarded as the apparent thickness of the unit. The true thickness of this unit is likely to be much less but here it appears exaggerated because of the plane of the borehole intersection. This has large implications for interpretation. What may be identified from an image as a possible permeability barrier in a horizontal well (like a shale horizon) maybe interpreted as being thicker than it actually is. Consequently, it may be given a much higher priority as a barrier to fluid flow than is warranted. It is therefore important to consider apparent and true thickness when interpreting horizontal images.

It is not uncommon for a horizontal well to penetrate laterally, two separate sets of beds for a considerable distance. Figure 9c illustrates where the borehole has intersected two sets of bedding surfaces from xx80 ft to an erosive surface at xx88 ft. Steeper dipping beds, interpreted as dune slip face deposits, appear below the erosion surface (centre and lower part of the image) and are separated from an upper, less steep set of dune apron surfaces ('lobes' to the left and right upper part of the image. This is illustrated schematically in Fig. 9d.

Undulating bedding (bulls eyes). In it is not uncommon that a horizontal well will be drilled to trace one particular horizon and then used to identify that horizon laterally. Ideally, once the horizon is located, its upper and lower surface should never be intersected as the borehole is drilled along the target horizon. Invariably, because of the naturally undulating nature of even the most flat bedding surfaces, or more noticeably because of faulting, the borehole will pass out of the bedding surface at a particular undulation and then re-enter the bedding surface almost immediately. This gives rise to a variety of 'bulls-eye' or 'wood grain' effects that look much like the grain effect seen in sawn wood. This is imaged in Fig. 10a and illustrated schematically in Fig. 10b. Figure 11 illustrates a true bulls-eye effect where the bedding surface is highly undulating and is possibly formed by wet sediment deformation.

Figure 12 shows the intersection of two different lithologies, an overlying poorly sorted breccia and a underlying finely laminated sandstone. These contact surfaces are typical

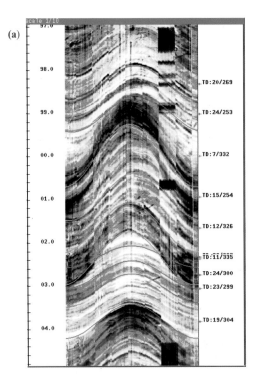

(a)

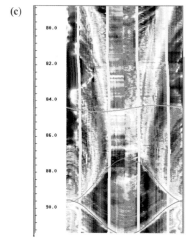

(c)

Fig. 9. (a) Cross-bedding. The intersection of dune foresets and first order erosive surfaces. Two sets of foresets (dune slip face) are present, one at the top of the image and one at the base (red sinusoids) with dip magnitudes of 19–24°. These are separated by a series of shallower dune apron surfaces (4 green sinusoids 12–14°). The cross cutting relationship of the first order erosive surface is visible at xx02.5 ft. The loss in image data to the right of the image (vertical stripe) is due to the malfunction of one of the imager pads. Drilling direction is towards base of figure. Scale is 1:10, depth is in feet. Bit size 6 inches. Image is referenced to top-of-hole on the far left. Bottom-of-hole is therefore dead centre. **(b)** Schematic illustrating the cross bedding (centre) within a horizontal and vertical wellbore. Note the quite different appearance on a workstation image as illustrated top and bottom left. DD indicates the drilling direction. **(c)** The borehole here intersects two sets of bedding surfaces that are bound by an erosive surface at xx88 ft (green sinusoid). A set of shallow (<10 degrees) dipping dune apron surfaces (the 'lobes') to the left and right of the upper part of the image are interpreted from xx80 ft to the erosive surface at xx88 ft. They are interpreted to be the result of small scale migrating crescentic dunes features. Below the erosion surface, steeper dipping beds (>15°, red sinusoid), interpreted as dune slip face deposits are present (center and lower part of the image). The cyan sinusoid is interpreted as a resistive fracture. The right hand side of the 3D cylinder plot represents the top-of-hole. This is displayed here to show the erosion surface (green line). Drilling direction is towards base of figure. Scale is 1:20, depth is in feet. Bit size 6.5 inches. Image is referenced to top-of-hole on the far left. Bottom-of-hole is therefore dead centre. **(d)** Schematic illustrating image seen in (c). Note the quite different appearance on a workstation image as illustrated top and bottom left. DD indicates the drilling direction.

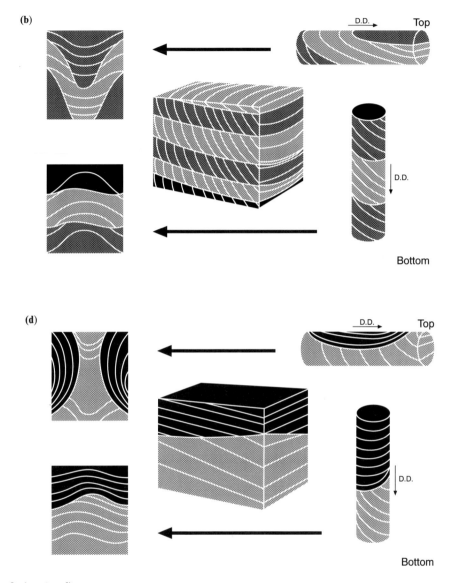

Fig. 9. (*continued*)

surfaces that need to be interpreted in target zone evaluation. Figure 13 indicates a bulls-eye interpreted from a horizontal azimuthal resistivity (ARI*) image. Although the vertical and horizontal resolution is less than the FMI* image (8 inches (vertical) as opposed to 0.2 inches), the bulls-eye feature is still visible.

Deformational structures: drape. Figure 14a illustrates the effect that can be caused by the post or syndepostional deformation of sediments (drape) when intersected by a horizontal wellbore. The image is interpreted to be the result of the drape of sediments around a nodular feature which could be a primary clastic depositional feature, such as a boulder, or a secondary diagenetic feature. A half 'bulls-eye' feature is produced and is notable under the nodular feature. The 3D cylinder shows the ramping of the underlying sediment around the feature. This is illustrated in Fig. 14b.

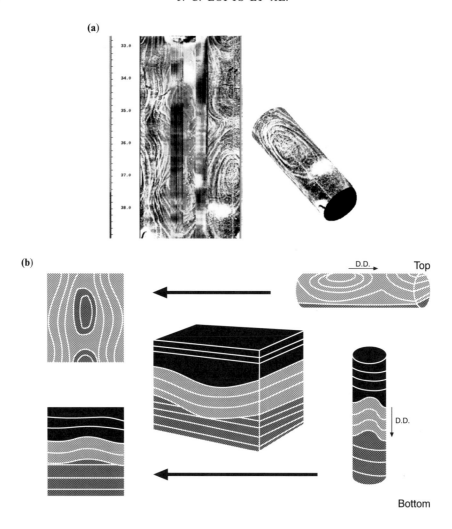

Fig. 10. (a) These image effects are produced by the intersection of the wellbore with thinly bedded undulating sand layers. The result is the production of a bulls-eye effect. The 3D cylinder representation illustrates the planar features that are intersected by the borehole wall. The dark areas towards the centre are due to malfunctioning pad flaps. Drilling direction is towards base of figure. Scale is 1:10 , depth is in feet. Bit size 6 inches. Image is referenced to top-of-hole on the far left. Bottom-of-hole is therefore dead centre. (b) Schematic illustrating undulating bedding surfaces (centre) within a horizontal and vertical wellbore. Note the quite different appearance on a workstation image as illustrated top and bottom left. DD indicates the drilling direction.

Structural displacement: faulting. Much like fracturing, it is highly desirable if not imperative, to identify structural displacements such as faults. Arguably, it becomes harder in horizontal wellbores to locate and accurately quantify sub-vertical fault azimuths by virtue of the decreased amplitude of the sinusoid. Figure 15a shows a small displacement along a fracture. This feature has a dip magnitude in the order of 80°. If this feature were any steeper, it would be less likely

that an accurate dip azimuth and thus true azimuth of the plane of displacement could be computed. This is because the sinusoid picked on the workstation image representing the planar feature would appear as more-or-less a straight line.

Figure 16a shows another example of a fault in a highly deviated well. The fault here is produced by a series of small dislocations, suggesting the presence of a fault zone rather

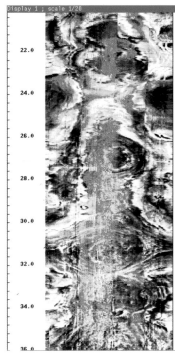

Display 1 ; scale 1/20

22.0

24.0

26.0

28.0

30.0

32.0

34.0

36.0

Fig. 11. Bulls-eye effect: (deformed bedding). A further example of the effects produced by the passage of a wellbore through undulating bedding surfaces. The near horizontal wellbore is here seen intersecting a low angle highly undulating finely bedded sand unit, possibly soft sediment deformed. A true bulls-eye effect is displayed. Drilling direction is towards base of figure. Scale is 1 : 20, depth is in feet. Bit size 6 inches. Image is referenced to top-of-hole on far left. Bottom-of-hole is therefore dead centre.

than a single planar surface. It is not hard, however, to envisage a sinusoid that would fit and describe the mean orientation of the fault zone. The fault is seen to bring into abrupt contact two widely differing sedimentary facies, a relatively homogeneous sand, showing features due to diagenetic effects, overlying well-defined high-angle dune foreset units. These fault features strongly contrast the long drawn out sinusoids indicative of faulting commonly observed in vertical wellbores (Fig. 16b).

Breakout features. Borehole breakout is common in horizontal wells but it does not form for entirely the same reasons that breakout forms in vertical wellbores. This is because breakout in vertical wells is attributed to being caused as a result of the dominant *in situ* horizontal stress direction (Cox 1970; Gough &

Bell 1981). Within a horizontal wellbore however, the dominant stress is vertical (i.e., gravity squeezing down on the borehole). This gives rise to a change in the orientation of the breakout which is observed at the sides of the borehole wall and not at the top and bottom of the borehole. This is imaged in Fig. 17 where two dark stripes can be seen running along the image. These are formed by conductive mud filling the space produced by the breakout and they are located at the sides of the borehole.

Acquisition features

There are a number of features commonly observed in horizontal images that are the direct result of the drilling and log acquisition. These may lead to confusion and debate during interpretation and are hence worthy of note. The features described here are those most commonly observed within highly deviated wellbores. The reader is directed to Bourke (1989) for a description of other non-geological artefact features common in vertical wellbores and of which most also apply to highly deviated wells. These features are summarized in Table 1.

Table 1. *Summary of the recognized artefacts commonly seen on resistivity images*

Acquisition artefacts
  Tool stick*
  Button death
  Mud smear
  Signal loss due to lubricants*
  Loss of pad contact*

Borehole wall artefacts
  Bad hole condition and breakout*
  Stand off
  Mud cake
  Tool marks*
  Pipe wallow and keyseating

Processing artefacts
  Window length
  Extremes of condition
  Rescaling
  Depth match (old tools)

Derived image artefacts
  Fracture plane aureoles
  Cementation mottling

* Features discussed in this paper.
Classification based on Bourke (1989).

Fig. 12. The bulls-eye effect seen here is produced by the intersection of a low angle bedding surface between two widely differing lithologies, an overlying poorly sorted breccia and an underlying finely laminated sandstone. These contact surfaces are typical surfaces that need to be interpreted in target zone evaluation. Drilling direction is towards base of figure. Scale is 1:10, depth is in feet. Bit size 6 inches. Image is referenced to top-of-hole on the far left. Bottom-of-hole is therefore dead centre.

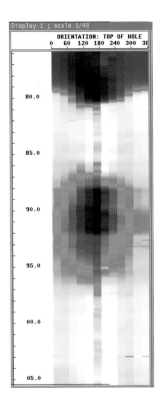

Fig. 13. This figure displays a bulls-eye feature interpreted from a horizontal azimuthal resistivity (ARI*) image. Drilling direction is towards base of figure. Scale is 1:40, depth is in feet. Bit size 6 inches. Image is referenced to top-of-hole on the far left. Bottom-of-Hole is therefore dead centre.

Tool Stick. The most common non-geological feature observed in horizontal wellbores is tool stick. This is often corrected for during image processing but sometimes when the tool stick is very severe the result will be a blurred interval on the image. This is easily recognized on images as a zone of blurring across the image with sharp boundaries, and is not often a source of confusion during interpretation. It can be easily diagnosed by a check of the cable speed, cable tension, head tension and accelerometer curves. Such curves are useful for log quality control especially over the areas where the imaging sonde stands static during the pipe conveyed logging process employed for highly deviated wells.

Scarring and scratching. Although not exclusive to horizontal wellbores, diagonal scarring marks along the borehole are very common. These are often caused by the reaming process used to clean and maintain the wellbore surface. They can also be caused by the 'catching' of logging tools, centralising devices or the drill pipe. Figure 18 illustrates the three different diagonal scars observed in a horizontal well.

Keyseating. Keyseating is not a feature exclusive to highly deviated wells, but it is extremely common. It is characterized by a mark on the base of the borehole wall produced by the weight and rotation of the drill string. Within horizontal wells, gravity plays a great part on

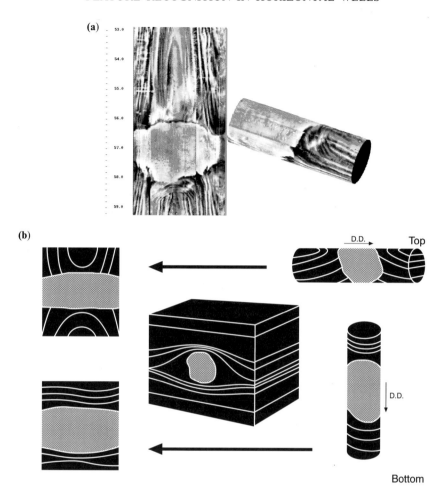

Fig. 14. (**a**) An example of the deformation of sediment around a nodular feature that could be a primary clastic deposition such as a boulder or a secondary, diagenetic feature. Here, the surrounding bedding has deformed around the feature and is intersected by the well bore. This produces a partial bulls-eye feature notable under the feature. The 3D cylinder shows the ramping of the underlying sediment around the feature. Drilling direction is towards base of figure. Scale is 1:10, depth is in feet. Bit size 6 inches. Image is referenced to top-of-hole on the far left. Bottom-of-hole is therefore dead centre. (**b**) Schematic illustrating compactional drape around a nodular feature (centre) within a horizontal and vertical wellbore. Note the quite different appearance on a workstation image as illustrated top and bottom left. DD indicates the drilling direction.

forcing the tool string and bit to lie on the base of the borehole which ultimately leads to various degrees of keyseating. This feature is easy to identify as a single streak that runs down the centre of the image (the horizontal workstation image is referenced as top-of-hole to the edges of the image therefore the base is at the centre). Hole azimuth can be used to verify a keyseat.

Loss of pad contact. A very common artefact seen in highly deviated wellbores is loss of pad

contact in pad tools such as the FMI*. This is especially common at the top of the hole where gravity is playing a large part in positioning the tool towards the low side of the hole. Tools such as the FMI* are equipped with powered arms but sometimes this artefact is unavoidable. Figure 19 illustrates the blurring and darkening of an image due to loss of pad contact at the top of the hole (seen at the far left and right of the image). Equally, for non-pad tools centralized within the borehole (such as acoustic image tools) the eccentralization of the tool, for the

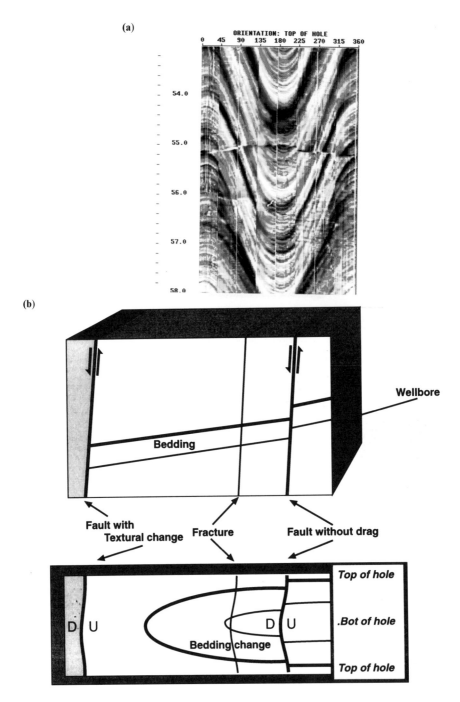

(a)

ORIENTATION: TOP OF HOLE

0   45   90   135   180   225   270   315   360

54.0

55.0

56.0

57.0

58.0

(b)

Wellbore

Bedding

Fault with
Textural change

Fracture

Fault without drag

D U

Bedding change

D U

Top of hole

.Bot of hole

Top of hole

Borehole Image

Fig. 15. (a) A small-scale fault intersected by a horizontal wellbore. Although the feature was 'opened' at the time of fracturing it is now sealed and is resistive. This is evidenced by the resistive, light coloured appearance of the sinusoid. The sense of displacement is interpreted to be vertical, with a displacement of the upper section of the image downwards with respect to the lower section. Displacement is likely to be in the order of inches. Drilling direction is towards base of figure. Scale is 1:10, depth is in feet. Bit size 6 inches. Image is referenced to top-of-hole on the far left. Bottom-of-hole is therefore dead centre. (b) A Schematic illustrating the sense of displacement of a normal vertical fault surface in a horizontal well and its representation in an unrolled workstation image. The down thrown side is to the left of the fault plane. DD indicates the drilling direction.

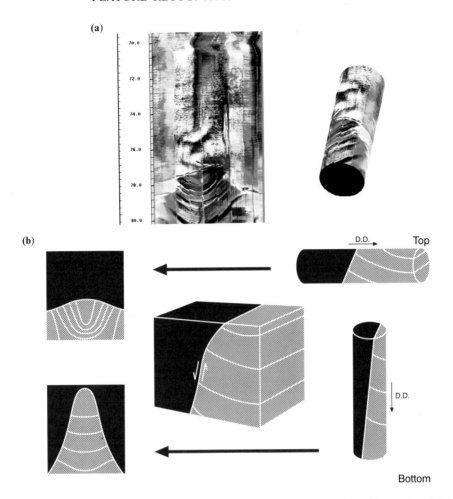

Fig. 16. (a) An example of a fault in a highly deviated well. The fault here is produced by a series of dislocations, suggesting the presence of a fault zone rather than a single 'plane' although the sinusoid representing the mean orientation of the fault could easily be envisaged over the boundary at xx77–78 ft. The fault is seen to bring into abrupt contact two widely differing sedimentary lithologies, a relatively homogeneous sand unit (possibly blurred because of the faulting), overlying a well defined high angle dune foreset unit. The 3D cylinder illustrates the vertical orientation of the fault zone. Drilling direction is towards base of figure. Scale is 1 : 20, depth is in feet. Bit size 6 inches. Image is referenced to top-of-hole on the far left. Bottom-of-hole is therefore dead centre. (b) Schematic illustrating faulting (centre) within a horizontal and vertical wellbore. Note the quite different appearance on a workstation image as illustrated top and bottom left. DD indicates the drilling direction.

same reasons of gravity, may produce similar artefacts. These can be largely compensated for during image processing.

Image degradation. It is increasingly common with highly deviated wells to use a lubricant in the water-based drilling fluid to aid the drilling process. Unfortunately, the use of this lubricant, if not carefully monitored, can lead to a decrease in signal to noise ratio and thus image degradation. An attempt is often made to flush the borehole in preparation for logging. Fortunately,

this only appears in patches (depending on mud cake and porosities) but it may hinder interpretation. The spotty nature of the image in Fig. 20 is attributed to the use of oil-based lubricant.

Discussion

Dip estimation: precision and accuracy

Within vertical wellbores, one question asked concerns the accuracy and precision of the dip

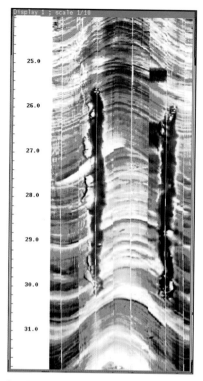

Fig. 17. An example of borehole breakout in a horizontal wellbore. The two dark vertical stripes represent voids filled with conductive muds which are produced by borehole breakout and very typically, they are located at the sides of the wellbore. Drilling direction is towards base of figure. Scale is 1 : 10, depth is in feet. Bit size 6 inches. Image is referenced to top-of-hole on far left. Bottom-of-hole is therefore dead centre.

azimuth and magnitude of planes that were of a high angle and thus sub-parallel to the wellbore. These are commonly fractures and fault features whose sinusoids extended many feet along the wellbore. It has been made clear that we see the opposite occurring within horizontal wells, where shallow-angle bedding now has this sub-parallel intersection with the borehole wall and is now the long drawn-out feature. Arguably, the question of accuracy of dip calculation now

becomes more important. A few degrees of error in orientating a vertical fracture at 70° is relatively insignificant compared to a bedding plane which maybe less than 5° true dip when structural dip is the objective. If vertical fractures are the primary objective then the importance will most likely be secondary. There is clearly a need to quantify further this problem and its importance will ultimately depend on the primary objective of the imaging. Current research needs to address the problems of both accuracy and precision of feature picking within horizontal (and vertical) wells. Statistics that meaningfully quantify accuracy and precision in this respect need to be firmly established so that a confidence level can be applied to an interpretation.

Bed boundary effects

Horizontal wells which penetrate sub-horizontal layers characteristically produce anisotropy within the borehole image, i.e., when part of the well is within one formation (e.g., shale) and part of the borehole is in a different lithology (e.g., sandstone). These azimuthal differences in the measurement that occur over the zone adjacent to such a lithology contact do, in theory, give an indication of the wellbores proximity to the boundary. Such information is valuable to the interpretation of the other formation evaluation logs such as density and neutron. These have a relatively deep investigation distance and as a result, become difficult to interpret in horizontal wells leading up to and over a lithological boundary, due to the mix of lithologies from which the measurement is taken. Clearly, the response of the borehole image will depend upon the type of imaging tool (electrical or acoustic) and its depth of investigation, but in general it is minimal. There may even be a case then for re-processing the image data over such lithological boundaries. This could be by either single pad equalization (in the case of resistivity pad devices), which would preserve directional information, or by using different equalization window lengths in order to emphasise the effect. The image response could then be related back

Fig. 18. Drilling induced spiral scars on the borehole wall appear as conductive diagonal stripes on the workstation image. These are likely to have been produced by reaming of the borehole (**a**) or the 'catching' of part of the drill string (**b**). In (**a**) and (**b**) there is approximately one rotation per foot. (**c**) drilling induced scars in a horizontal wellbore as seen by ultrasonic acoustic imaging device, The UBI*. (This is an amplitude display where the darker areas correspond to lower amplitudes, which in turn, correspond to areas of enlarged borehole diameter.) Drilling direction is towards base of figure. Scale is 1 : 20, 1 : 10 and 1 : 10 respectively. Depth is in feet. Bit size 6 inches. Image is referenced to top-of-hole on the far left. bottom-of-hole is therefore dead centre.

(a)

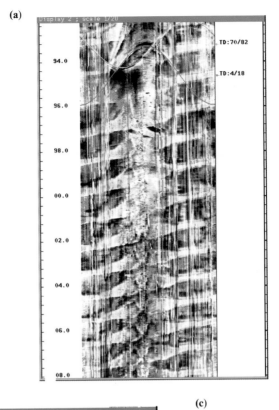

(b)

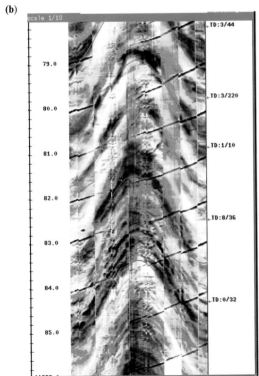

(c)

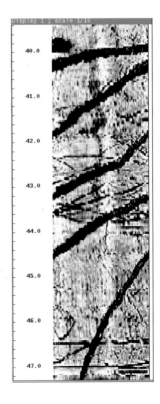

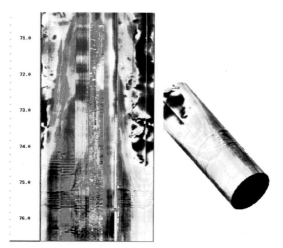

Fig. 19. The blurring and darkening at the sides of this image are due to the loss of pad contact at the top of the hole, and is more easily visualized in the 3D cylinder view. On the workstation image this is seen at the far left and right of the image. Drilling direction is towards base of figure. Scale is 1:10, depth is in feet. Bit size 6 inches. Image is referenced to top-of-hole on the far left. Bottom-of-hole is therefore dead centre.

to the deeper investigation tools to fully understand in turn their log response over lithological boundaries.

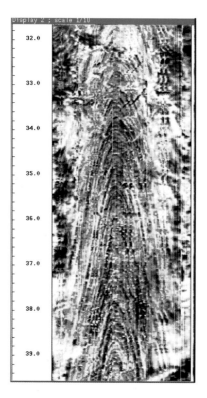

Summary and conclusions

A clear interpretation of borehole images in horizontal wells is important for target-zone evaluation and fracture identification. However, an interpreters first encounters with images from these wells can be disconcerting. The reason is that the interpreter is presented with an entirely new spatial geometry and an almost 'opposite' perspective of the features and fractures conventionally displayed in vertical wells. To fully understand, interpret and thus exploit the wealth of knowledge within high resolution images from deviated wells, this contribution identifies the need for a careful re-think of how planar features intersect a horizontal wellbore. Other salient points include the following.

(1) Horizontal wells are becoming increasingly common and borehole imaging provides the only cost effective means of acquiring accurate structural and sedimentary information. Therefore images have never been more important although the error tolerance, for example when targeting productive layers with

Fig. 20. Example of the speckled effect on images caused by the decrease in signal/noise ratio, attributed to the use of oil based lubricants in the drilling process. This is also imaged in the lower part of the image in Fig. 7. Drilling direction is towards base of figure. Scale is 1:10, depth is in feet. Bit size 6 inches. Image is referenced to top-of-hole on the far left. Bottom-of-hole is therefore dead centre.

horizontal wells, is far smaller than that of vertical wells.

(2) Care must be taken with the estimation of true bed thickness as must the estimation of azimuth and dip magnitude of both near vertical and extremely flat lying features.

(3) Horizontal wells provide a unique opportunity to measure lateral changes in layer boundary orientation in place and can validate the interpretation of other logs by the preservation of directional micro-resistivity data and subsequent application to the log response of deeper logs.

(4) Some ambiguity does remain as to the origin and thus interpretation of all imaged features. These should where possible, be calibrated to local knowledge and core data. Ultimately, as is the case without calibration, it is only by measuring, combining and then mapping the true orientations of all features that a picture emerges. A workstation analysis of borehole images is therefore essential in the case of highly deviated and horizontal wells.

The authors wish to thank all of the companies who have allowed kind use of their image data. A. Fenwick is also thanked for final preparation and checking of the manuscript.

References

BOURKE, L. T. 1989. Recognising artefact images of the Formation MicroScanner. *SPWLA 13th Annual Logging Symposium, June, Paper WW.*

COX, J. W. 1970. The high resolution dipmeter reveals dip related borehole and formation characteristics. *The Log Analyst*, Sept–Oct, 220–227.

EKSTROM, M. P., DAHAN, C. A., CHEN, M. Y., LLOYD, P. M. & ROSSI, D. J. 1987. Formation imaging with micro electrical scanning arrays, *The Log Analyst*, May–June, **28**, 294–306.

GOUGH, D. I. & BELL, J. S. 1981. Stress orientations from oil well fractures in Alberta and Texas. *Canadian Journal of Earth Science*, **18**, 638–645.

HURLEY, N. F., THORN, D. R., CARLSON, J. L. & EICHELBERGER, S. L. W. 1994. Using Borehole Images for target zone evaluation in horizontal wells. *American Association of Petroleum Geologists, Bulletin*, **78**, 238–246.

SAFINYA, K. A., LE LAN, P., VILLEGES, M. & CHEUNG, P. S. 1991. *Improved formation imaging with extended micro electrical arrays.* Society of Petroleum Engineers, Special Paper 22726.

SERRA, O. 1989. *Formation MicroScanner image interpretation.* Schlumberger Educational Services, Houston Texas.

Scattering attenuation as a function of depth in the upper oceanic crust

D. GOLDBERG & Y. F. SUN

Borehole Research Group, Lamont-Doherty Earth Observatory of Columbia University,
Palisades, NY 10964, USA

Abstract: The *in situ* attenuation is computed through 2.1 km of the upper oceanic crust in the vicinity of ODP Hole 504B on the southflank of the Costa Rica Rift in the eastern equatorial Pacific. The results strongly tie crustal properties to seismic measurables and observed geological structures: we find that the attenuation can be used to define seismic layer boundaries and is closely related to vertical heterogeneity. The *in situ* attenuation Q^{-1} consists of both intrinsic and scattering contributions, but is dominated by the scattering attenuation unless porosities are near zero, when it approaches typical estimates from seismic refraction studies. The attenuation is analytically modeled by multiple backscattering from heterogeneities observed in a sonic V_p log and is found to decrease step-wise from $Q = 25$ to $Q > 300$ between the top of seismic layer 2A and a sharp discontinuity at 1.3 km depth. These changes correspond with heterogeneities at 1.0–1.3 m and at 10.0 m wavelengths that are associated with fracturing and the structure of pillow basalts and lava flows in seismic layers 2A and 2B. Although seismic velocity studies suggest that the layer 2–3 boundary also occurs at about 1.3 km, the large variation in Q (140 to 460) below this depth indicates that a seismically homogeneous layer 3 has not been reached in Hole 504B. From the observed results, we derive an empirical relationship between attenuation and porosity $Q^{-1} = Q_0^{-1}\,e^{\beta\phi}$, where $Q_0^{-1} = 0.004$, $\beta = 25$, that may be applicable at other oceanic crust locations and useful for constraining seismic inversion models.

Historically, most studies of the seismic properties of the seafloor have used measurement techniques related to wave velocities and travel time. Numerous authors have inverted seismic data for velocity profiles as a function of depth (e.g., Ewing & Purdy 1982; Purdy 1987; Vera 1989; Harding *et al.* 1989) and modelled Earth materials to predict the velocity below the seafloor (Toksoz *et al.* 1976; Walsh & Grosenberg 1979; Fryer & Wilkens 1988; Moos 1990). Prior to ocean drilling, investigations of the elastic properties of the uppermost oceanic crust relied almost exclusively on the results of seismic refraction experiments (e.g. Houtz & Ewing 1976). Raitt (1963) and then Houtz & Ewing (1976) described the structure of the oceanic crust by discriminating *in situ* refraction velocities and velocity gradients among seismic layers 2A, 2B, 2C and 3. This past emphasis on refraction and ocean bottom seismic methods has revealed considerable information about the P-wave velocity structure within the limited resolution of the seismic wavelength (Spudich & Orcutt, 1980; Ewing & Purdy 1982; Purdy 1987). Detrick *et al.* (1994) interpret velocity analyses to show that the seismic layer 2–3 boundary is a gradational transition in elastic properties within the sheeted dykes of layer 2, not a lithological transition from sheeted dykes to gabbro.

Ophiolite complexes have provided another means to describe the structure and petrology of the ocean crust and compressional and shear wave velocity measurements on ophiolite samples have been correlated to seismic layers (Fox *et al.* 1973; Christensen & Salisbury 1982). This work indicates that the seismic layer model varies with both depth and age in the ocean crust and often within a few tens of kilometers spatially. From comparisons of laboratory data with seismic measurements, the observed increase in velocity in the crust is generally considered to be a function of diagenetic alteration products replacing basalt and void spaces, off-axis volcanism, or a decrease in the number of cracks (Fox *et al.* 1973). Because of the difference in scale between laboratory experiments on rock samples and seismic data which measure aggregate properties, estimates of elastic properties often differ. Understanding the geological processes involved, the *in situ* properties, and their evolution with time and depth in the crust is critical to interpreting ophiolite samples or seismic data with the seismic layer model.

Although expensive, drilling directly into the ocean crust provides a 'ground truth' for testing any seismic interpretation or geological model. The lithologies and structures encountered by drilling are complex and its *in situ* properties, such as mineralogy and composition, porosity, permeability, fracture density and fracture aspect ratio, are common objectives in most locales. Several past studies have used elastic properties measured in Ocean Drilling Program (ODP) drill holes to investigate the structure

From Lovell, M. A. & Harvey, P. K. (eds), 1997, *Developments in Petrophysics*, Geological Society Special Publication No. 122, pp. 367–375.

of the oceanic crust as a function of depth. The vertical distribution of fractures, alteration zones, and volcanic structures in young ocean crust were studied by Moos *et al.* (1986) in ODP Hole 504B, which lies at $1°13.611'$ N, $83°43.818'$ W on the south flank of the Costa Rica Rift in the eastern equatorial Pacific. They found that acoustic velocities are quite different for pillow, massive units, and their fractured equivalents, as in each the spatial distribution of aspect ratios of cracks and pores is different. Also at this site, investigators have inferred the orientation of structural features from conventional acoustic logs (Anderson *et al.* 1985) and attempted to measure the horizontal velocity anisotropy from oblique seismic surveying (Stephen 1983; Little & Stephen 1985). Because of the similarity between the sonic log wavelength and the size of the geologic features around the borehole, variations in sonic log energy have been shown to be consistent with changes in alteration, lithology, and fracturing (Goldberg *et al.* 1985; Cheng *et al.* 1986; Goldberg & Zinszner 1988).

The seismic model of the oceanic crust can be further improved by constraining another *in situ* measurable as a function of depth: the seismic attenuation. Seismic attenuation is usually quantified by using quality factor Q, the inverse of attenuation, where low Q corresponds to strong attenuation of acoustic or seismic waves. Wepfer & Christensen (1991) among others (Goldberg *et al.* 1992; Swift & Stephen 1992) modelled the structure of the oceanic crust using laboratory measurements of velocity and Q. The models indicate that velocity and Q increase with depth through layer 2, but Q decreases in layer 3. Most models predict that layer 3 consists of gabbro. Estimates of Q in layer 3 from amplitude modeling of seismic refraction data, however, are high ($Q \approx 300$) and conflict with the low Q (≈ 8–20) for gabbro measured in the laboratory (Spudich & Orcutt 1980; Vera *et al.* 1990; Lindwell 1991). This has been explained by variation in crustal composition and heterogeneity (Swift & Stephen 1992; Goldberg *et al.* 1992). Because the seismic attenuation is closely related to the intensity of rock heterogeneity, also an indicator of the deformation/alteration history, a Q profile in a drill hole provides a proxy measurement for physical changes as a function of depth. The attenuation profile is therefore diagnostic of differentiation during accretion and the intensity of later deformation and alteration.

In this study, we investigate the structure of the ocean crust by estimating seismic attenuation as a function of depth. Since average core recovery is about 35%, continuous downhole logs with fine vertical resolution provide an essential means to obtain a continuous attenuation-depth profile. *In situ* logs provide 0.6 m resolution of velocity in a continuous profile without large errors resulting from core disturbance or dilatancy and without depth uncertainty. Using the acoustic velocity log recorded in Hole 504B, we predict the *in situ* attenuation through the upper 2.1 km of the oceanic crust.

Attenuation estimation

Attenuation is an inherent property of wave propagation. For any acoustic, elastic, or electromagnetic wave, the attenuation causes energy loss of the propagating wave into the medium. Attenuation is usually quantified by using the quality factor Q which is inversely proportional to it. Attenuation can also be meaningfully divided into the sum of intrinsic and extrinsic components. The former is related to the interaction between the propagating wave and the porous medium, including its saturating fluids; while the latter is due to the scattering loss from heterogeneity in the medium. The distinction of intrinsic and extrinsic attenuation has been described previously in applications mainly relating to earthquake seismology and is important when the scale length of heterogeneity is comparable to or larger than the propagating wavelength (Richards and Menke 1983; Wu 1985).

Many methods and theories have been developed to quantify attenuation and its physical mechanisms. Prior studies of attenuation span the fields of granular material science (e.g., Nowick & Berry 1972), hydrocarbon exploration (e.g. Bourbie *et al.* 1987), and seismology (e.g., Aki & Richards 1980), but no single model addresses all the physical mechanisms that affect a propagating wave or the measurement of its properties. Seismic attenuation in oceanic crust has been investigated in the past by amplitude modeling of seismic refraction and reflection data (Spudich & Orcutt 1980; Mithal & Mutter 1989; Vera *et al.* 1990; Lindwell 1991) and by laboratory experiments (Wepfer & Christensen 1991; Goldberg *et al.* 1991, 1992). Solomon (1973) has used frequency-dependent Q to quantify the extent of partial melting in the upper mantle. Rutledge & Winkler (1989) and Swift & Stephen (1992) have tried to estimate the *in situ* attenuation through crustal sequences using the spectral ratio method with VSP (vertical seismic profile) data, but typically large discrepancies in attenuation are observed between seismic, laboratory, log, and VSP estimates (Goldberg & Yin 1994). In this work, we estimate the scattering attenuation as a continuous depth profile using the method outlined by Wu (1982, 1985).

Multiple backscattering theory

The scattering attenuation, the energy lost by seismic propagation through a heterogeneous medium, is often significantly greater than intrinsic attenuation, the energy lost by conversion to heat (Toksoz and Johnston 1981; Ganley 1981; Richards & Menke 1983). Based on the models of Chernov (1960) and Aki & Richards (1980), Wu (1982, 1985) developed a model for multiple backscattering that accounts for intrinsic and extrinsic attenuation in weakly heterogeneous media. Using Wu's model and the observed distribution of heterogeneities from a sonic velocity log, Goldberg & Yin (1994) estimated large seismic attenuation ($Q \approx 10$) in oceanic gabbros drilled in ODP Hole 735B in 1987. This model corroborated the large seismic attenuation estimated in the same hole from laboratory experiments ($Q \approx 20$) (Goldberg et al. 1992) and from vertical seismic profiling ($Q \approx 8$) (Swift & Stephen 1992).

Attenuation is a measure of the total energy loss as a seismic wave propagates through a solid medium. The intrinsic component of attenuation results in a loss of energy that decays exponentially in both frequency and propagation distance; the scattering component is the result of heterogeneity in the formation and strongly depends on the propagating wavelength. Wu's model for seismic attenuation describes these additive components and consists of a linear combination of scattering and intrinsic losses,

$$\xi = \xi_s + \xi_i, \qquad (1)$$

where $\xi =$ total extinction coefficient, $\xi_s =$ scattering attenuation coefficient, $\xi_i =$ intrinsic absorption coefficient.

The model accounts for the attenuation due to multiple scattering in a weakly heterogeneous medium through a distance x with a velocity distribution that varies around a mean value, v_0, with a small range of variation, δv. Through the fluctuation $g(x) = -\delta v(x)/v_0$, the statistical property of the medium can be described by a correlation function

$$N(|x_1 - x_2|) = \langle g(x_1)g(x_2) \rangle / \langle g^2 \rangle, \qquad (2)$$

in which $\langle \cdot \rangle$ represents the ensemble average. In this model, an isotropic random medium is

divided into many slices of equal thickness in the direction of the incident plane wave. The thickness of each slice, a, represents a typical correlation length of the velocity variation. For a plane incident wave of wavenumber k_0, with energy flux I_0, the transmitted energy flux through $x(x \gg a)$ inside the medium can be expressed approximately in the form of $I = I_0 \exp(-\xi x)$. The scattering attenuation Q_s^{-1} and coefficient ξ_s can then be written as

$$Q_s^{-1}(k_0) = \xi_s/k_0$$
$$= 2\langle g^2 \rangle k_0 (F_c(\sqrt{2}k_0) - F_c(2k_0)), \quad (3)$$

where $F_c(k)$ is the Fourier-cosine transform of the correlation function,

$$F_c(k) = \int_0^\infty N(x) \cos(kx) \, dx. \qquad (4)$$

Equation (3) is the fundamental expression of the scattering model. It shows that once the correlation function is given for a random medium, the spectrum of attenuation can be analytically calculated. For a simple correlation function, such as the exponential $N(x) = \exp(-x/a)$, closed-form solutions can be obtained and show that Q_s^{-1} reaches its maximum value of about $0.3\langle g^2 \rangle$ at $k_0 \approx 1$. A Gaussian correlation function of the general form $N(x) = \exp[-(x/a)^2]$ generates a corresponding maximum Q_s^{-1} value of about $0.6\langle g^2 \rangle$ at $k_0 \approx 1.5$. Therefore, the relationship between the correlation length a and the incident wavelength λ_0 can be approximated by

$$a \cong \lambda_0/5, \qquad (5)$$

for distributions between an exponential and a Gaussian. In general, the number, location, and magnitude of peaks in the Q_s^{-1} spectrum depend on the form and complexity of the correlation function $N(x)$, which is representative of the heterogeneity in the medium. Using data from ODP Hole 504B, we compute $N(x)$ from field observation of formation heterogeneity assuming that the fluctuation profile is well represented by the sonic velocity log.

Model limitations

In the multiple backscattering model, a random medium is assumed to have smoothly varying velocities. Scattering effects of velocity gradients or P to S mode conversions are also assumed to be negligible. Because this study focuses on only compressional wave modes with velocities varying by 15 to 20%, these assumptions are reasonable. Also, the backscattered energy from each element is assumed to be entirely lost. Although both forward- and backward-scattered energy can be scattered in reverse directions, the magnitude of this effect is proportional to the square of the magnitude of backscattered energy (Wu 1982). This second-order contribution is significant only when $k_0 \approx 1$ and $|g| \approx 1$ and does not greatly reduce the attenuation predicted using Wu's model with these data.

In Aki & Richards' (1980) derivation of seismic scattering due to inhomogeneity, it is assumed that single scattering effects are valid only in the 'far field' region, although Wu's model assumes single scattering between neighboring elements. This may cause Wu's model to slightly overestimate the attenuation, but not significantly more than using a full-wavefield 3D scattering model (Tang & Burns 1993; Jannaud et al. 1993). Although the oceanic crust is certainly variable in 3D, the shortcomings of a 1D multiple scattering model do not significantly affect the interpretation of it in the vertical dimension (depth). The 1D estimates can be used reliably to relate the statistics of the medium to the associated attenuation spectrum and to compare the changes in attenuation versus depth in the oceanic crust.

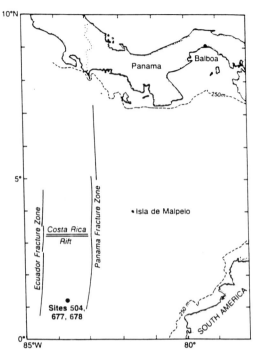

Fig. 1. Location of Hole 504B in the eastern equatorial pacific (after Becker *et al.* 1988).

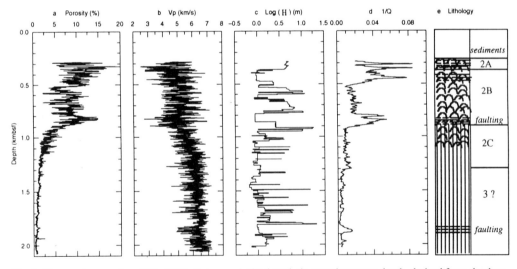

Fig. 2. Basement properties at Site 504B. Column a is the downhole porosity versus depth, derived from the deep-resistivity log obtained on ODP Leg 148. Column b is the sonic velocity log (V_p) obtained on ODP Leg 148. Column c is the dominant heterogeneity length (H), on a logarithmic scale. Two major heterogeneity lengths occurs at 1.0–1.3 m and 10.0 m, indicating the existence of internal structures and/or structural alternations at these lengths. Column d is the attenuation ($1/Q$) versus depth, which reveals the Q structure of the upper oceanic crust. Major discontinuities in the $1/Q$ profile occur at 465 m, 800 m, 900 m, 1300 m, and 1850 m bsf. Column e shows the lithostratigraphy. Note the different characteristics of *in situ* Q and H corresponding to different lithology and geological events (see text for details).

Results

Depth profiles in Hole 504B

Hole 504B was drilled in the eastern equatorial Pacific through a thick sequence of intensely fractured and altered pillow basalts, lava flows, and sheeted dikes during seven drilling legs over the last 16 years (Becker *et al.* 1988; Dick *et al.* 1992; Alt *et al.* 1993). Figure 1 shows the location of the site (after Becker *et al.* 1988). Because the average core recovery is only 35%, numerous downhole measurements have provided a primary means to measure continuous physical and chemical profiles (Becker *et al.* 1985, 1988; Pezard 1990; Dick, *et al.* 1992; Alt *et al.* 1993). Figure 2 shows the characteristic stratigraphy (column e) based on the core and downhole data (Honnorez *et al.* 1983; Becker 1985; Alt *et al.* 1985; Pezard 1990) and the downhole porosity (column a), derived from the deep-resistivity log following the procedure given by Becker (1985) and Pezard (1990), assuming that all the pores and porous cracks are connected and filled with sea water. Porosity reaches >20% in the fractured and altered rocks above 800 m below sea floor (bsf) and decreases to 0.3–0.7% below 2100 m bsf in the sheeted dykes.

In Fig. 2, the results are shown using Wu's methodology on the most recently acquired sonic velocity log from 0.3 to 2.1 km bsf in Hole 504B. The velocity log (column b) was recorded with the Schlumberger long-spaced tool with 0.1524 m (0.5 ft) vertical sampling interval (same as the porosity log) and increases from 2500 m s^{-1} to >6800 m s^{-1} at the bottom of the hole. Using the method outlined above, the continuous spectrum of attenuation Q^{-1} is computed over a 0.6–6095 m range of incident wavelengths (or correlation lengths) for overlapping 76.2 m (500 point) depth intervals of the sonic log. Figure 3 (a–c) illustrates the computation procedure for a 75 m interval from 329 to 405.2 m bsf: (a) the sonic velocity log over the interval; (b) the smoothed (with a five-point least-square filter) correlation function $N(x)$ from the velocity log over the interval; and (c) the computed attenuation spectrum Q^{-1} using the analytical transformation in Eq. 3. Peaks observed in (c) represent the maximum Q^{-1} at particular wavelengths of heterogeneity in this depth interval.

In Fig. 4, a continuous three-dimensional attenuation spectrum is derived by overlapping intervals such as shown in Fig. 3 over the entire logged interval in Hole 504B. The resulting spectrum illustrates the internal structure of the

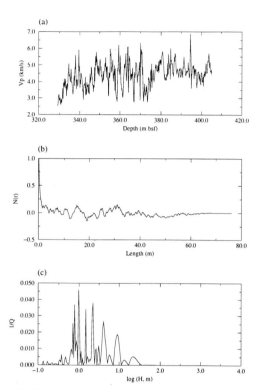

Fig. 3. Example of continuous *in situ* attenuation computation method: (**a**) Sonic V_p log. (**b**) Plot of $N(r)$ which is the correlation function from the sonic V_p log in (a). (**c**) Continuous *in situ* attenuation spectrum, $1/Q$.

vertical heterogeneity in Hole 504B. The complex nature of layer 2A and 2B is evident, consisting of both large massive flow units and intensely fractured intervals with differing spectral signatures. Layers 2A, 2B and fractured intervals above 900 m bsf can be easily identified by the occurrence of large spectral peaks. At any specific depth, peaks in this spectrum indicate the maximum attenuation due to scattering at the dominant wavelength of heterogeneity. Picking these peaks versus depth gives both the *in situ* heterogeneity scale H and attenuation Q^{-1} profiles. These curves are plotted as a function of depth in Fig. 2 (columns c and d) and the derived *in situ* logs can be used to describe the internal structure of the oceanic crust. H is low and Q^{-1} is high when scattering attenuation is greatest in intensely fractured, thin and porous alternating layers; H is high and Q^{-1} is low in more massive homogeneous intervals, which scatter little seismic energy.

As shown in Fig. 2 (column d), crustal heterogeneity in Hole 504B occurs predominantly at wavelengths of 1.0–1.3 m and 10.0 m.

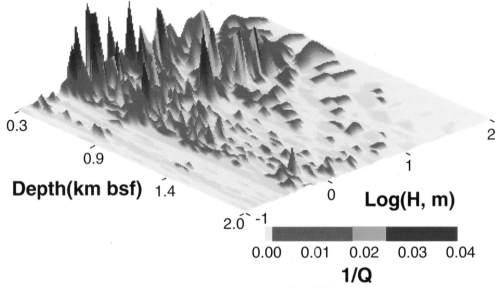

Fig. 4. 3D high-resolution image of *in situ* Q and H v. depth for Hole 504B.

Scattering attenuation from these hetero-geneities decreases exponentially with depth by a factor >10, and at sharp discontinuities, from the top of layer 2A to the bottom of the hole. Q values are as low as 10 (and average about 25) in layer 2A above 465 m bsf and increase in successive layers with depth to >300 below 2.1 km bsf. Major boundaries in the Q^{-1} profile occur at 465 m, 800 m, 900 m, 1300 m, and 1850 m bsf which can be related to the contacts between petrological and seismic lithologies inferred from core and downhole studies (Fig. 2, column e). Table 1 shows the resulting estimates of the average Q structure of the oceanic crust in Hole 504B.

The high attenuation (low Q) observed in layer 2A and in the uppermost part of seismic layer 2B correlates with the occurrence of alternating morphologies of pillow basalts, massive lava flows, and fractured sheeted dykes (e.g. Moos *et al.* 1986). Q^{-1} remains rela-tively constant throughout layer 2B, except for high attenuation in a faulted zone from

Table 1. *Average Q structure of the upper oceanic crust*

Layer	Av. Q	Std Dev.
2A	25	±8
2B	60	±13
2C	110	±23
3?	>300	±160

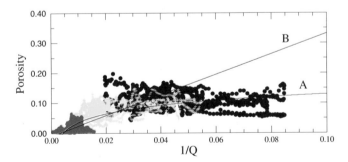

Fig. 5. Porosity, ϕ, versus attenuation, $1/Q$. The porosity data is from Fig. 2 column a and $1/Q$ from Fig. 2 column d. Black dots represent data from layer 2A and the top portion of layer 2B; yellow dots, data from the lower portion of layer 2B; and red dots, data from layer 2C and below. Line A is the result of exponential regression. Line B results from linear regression. The difference between the linear and exponential estimation shows that scattering from large heterogeneities such as lava flows in layer 2A and the fault zone in layer 2B significantly increases the *in situ* attenuation (see text for details).

800–900 m bsf (Pezard 1990). Q^{-1} decreases and then remains relatively constant with depth in layer 2C having lower porosity, higher velocity and velocity gradient, and thick homogeneous intervals. The decrease in Q^{-1} below 1300 m bsf roughly coincides with the layer 2–3 boundary proposed from seismic reflection and refraction velocity gradients (Vera et al. 1990; Detrick et al. 1994). In this region below 1300 m bsf, average $Q > 300$, however, Q varies between 140 and 460. A thin, low Q interval near 1850 m bsf may be attributed to induced hydrothermal fracturing during drilling (Alt et al. 1993). About 100 m above the bottom of the hole, maximum V_p reaches c. 6.7 km s^{-1} and maximum Q reaches c. 450. The standard deviation of Q below 1300 m bsf is >2 times larger than observed in the overlying layers and indicates that layer 3 is not as seismically homogeneous as expected from seismic velocity studies and laboratory attenuation results (e.g., Raitt 1963; Houtz & Ewing 1976; Wepfer & Christensen 1991; Detrick et al. 1994). Our explanation is that the expected seismically uniform layer 3 has not been reached at this depth in Hole 504B.

In summary, variations in structural morphologies on both pore and macroscopic scales most importantly control the attenuation and can be deduced from the Q^{-1} profile estimated by scattering from velocity heterogeneities. Consistent and distinct boundaries in the Q^{-1} profile correspond with greater or less heterogeneity within the oceanic crust and support the conclusion that seismic layers in young ocean crust are not petrologic boundaries, but structural boundaries.

Attenuation versus porosity

Using the results shown in Fig. 2, the relationship between porosity and the derived attenuation reveals the different, additive attenuation mechanisms that are dominant in different layers of the oceanic crust. In Fig. 5, an exponential regression between in situ attenuation and porosity ϕ (decimal) gives

$$Q^{-1} = Q_0^{-1} e^{\beta\phi}, \qquad (6)$$

where $Q_0^{-1} = 0.004$, $\beta = 25$. At low porosities deep in the crust this relationship yields $Q^{-1} = Q_0^{-1}$, close to the intrinsic attenuation of crustal rocks estimated both in laboratory experiments (e.g. Wepfer & Christensen 1991) and from seismic refraction surveys (Spudich & Orcutt 1980; Vera et al. 1990; Lindwell 1991). At high porosities in the uppermost portions of

layer 2, scattering attenuation predominates and decreases with depth as lateral heterogeneity is reduced due to the closure of fractures and the absence of morphologic changes (pillow fragments and lavas flows). This porosity-attenuation relationship may be useful to evaluate attenuation from a porosity profile or to predict porosity from measured seismic attenuation at other oceanic crust sites. Such a prediction would help significantly to constrain seismic inversions for porosity estimation.

A linear relationship, representative of oceanic crust with low porosities $\phi < 8\%$, is also derived,

$$Q^{-1} = Q_i^{-1} + Q_p^{-1}\phi, \qquad (7)$$

where $Q_i^{-1} = 0.0025$ and $Q_p^{-1} = 0.29$. The difference between the linear and exponential prediction is most significant at $\phi > 8\%$ in the shallow crust, where scattering from large heterogeneities such as lava flows in layer 2A and the fault zone within layer 2B increases the in situ attenuation up to two times more than in surrounding intervals. Further comparisons between log-derived predictions and seismic estimates of attenuation would determine whether the linear slope Q_p^{-1} accurately describes the relationship between porosity and in situ seismic attenuation in higher porosity and other layered, heterogeneous environments. This linear porosity-attenuation relationship obtained from Hole 504B may then be useful in other heterogeneous environments.

Conclusions

The in situ attenuation in the ocean crust, consisting of both intrinsic and scattering contributions, is high because it is dominated by scattering effects from heterogeneities. Using Wu's analytical model for multiple backscattering from heterogeneities and data from the observed sonic V_p log in Hole 504B, both in situ heterogeneity and attenuation profiles are estimated and used to describe the internal structure of the oceanic crust: H is low and Q^{-1} is high when scattering attenuation is greatest in intensely fractured, thin and porous alternating layers; H is high and Q^{-1} is low in more massive homogeneous intervals, which scatter little seismic energy. As a consequence of scattering, attenuation decreases exponentially with porosity and their relationship in this hole is $Q^{-1} = Q_0^{-1} e^{\beta\phi}$, where $Q_0^{-1} = 0.004$ and $\beta = 25$. A similar relationship may be applicable at other crustal sites. As porosity decreases and depth increases, attenuation approaches $Q_0^{-1} = 0.004$, typical of estimates from seismic refraction studies. Q^{-1} varies step-wise with depth in relatively isotropic and distinct layers that reflect changes in crustal morphology, decreasing on average from $Q = 25$ to $Q > 300$ between the top of seismic layer 2A and a sharp discontinuity at 1.3 km depth. Scattering occurs primarily from heterogeneities at 1.0–1.3 m and at 10.0 m wavelengths that are associated with fracturing and the structure of pillow basalts and lava flows. The large variation in Q to this depth and below indicates that the ocean crust is not seismically isotropic and that seismic layer 3 has probably not been reached in Hole 504B.

We thank two anonymous reviewers for their critical and helpful comments. D.G. and Y.F.S. were supported for this report under NSF contract JOI 66-84. Lamont-Doherty contribution number 5499.

References

AKI, K. & RICHARDS, P. G. 1980. *Quantitative Seismology*. Vols I & II, W. H. Freeman and Company.

ALT, J. C., LAVERNE, C. & MUEHLENBACKS, K. 1985. Alteration of the upper oceanic crust: mineralogy and processes in Deep Sea Drilling Project Hole 504B, Leg 83. *In*: ANDERSON, R. N., HONNOREZ, J., BECKER, K. *ET AL*. (eds) *Initial Reports of the Deep Sea Drilling Project*. US Government Printing Office, Washington, **83**, 217–241.

——, KINOSHITA, H., STOKKING, L. B. & SHIPBOARD SCIENTIFIC PARTY 1993. Site 504. *In*: ALT, J. C., KINOSHITA, H., STOKKING, L. B. *ET AL*. (eds) *Proceedings of the Ocean Drilling Program, Initial Reports*. Ocean Drilling Program, College Station, Texas, **148**, 27–121.

ANDERSON, R. N., O'MALLEY, H. & NEWMARK, R. L. 1985. Use of geophysical logs for quantitative determination of fracturing, alteration, and lithostratigraphy in the upper oceanic crust. *In*: ANDERSON, R. N., HONNOREZ, J., BECKER, K. *ET AL*. (eds) Initial Reports of the Deep Sea Drilling Project. US Government Printing Office, Washington, **83**, 443–478.

BECKER K. 1985. Large-scale electrical resistivity and bulk porosity of the oceanic crust, Deep Sea Drilling Project Hole 504B, Costa Rica Rift. *In*: ANDERSON, R. N., HONNOREZ, J., BECKER, K. *ET AL*. (eds) *Initial Reports of the Deep Sea Drilling Project*. US Government Printing Office, Washington, **83**, 419–427.

——, SAKAI, H. & SHIPBOARD SCIENTIFIC PARTY 1988. Site 504: Costa Rica Rift. *In*: BECKER, K., SAKAI, H. *ET AL*. (eds) *Proceedings of the Ocean Drilling Program, Initial Reports, Part A*. Ocean Drilling Program, College Station, Texas, **111**, 35–251.

BOURBIE, T., COUSSY, O & ZINSZNER, B. 1987. *Acoustics of Porous Media*. Gulf Publishing Company.

CHENG, C. H, WILKENS, R. H. & MEREDITH, J. A. 1986. Modeling of full waveform acoustic logs in soft marine sediments. *Transactions of the Society of Professional Well Log Analysts, 27th Annual Logging Symposium*, Paper LL.

CHERNOV, L. A. 1960. *Wave Propagation in a Random Medium*. McGraw-Hill, New York.

CHRISTENSEN, N. I. & SALISBURY, M. H. 1982. Lateral heterogeneity in the seismic structure of the oceanic crust inferred from velocity studies in the Bay of Islands ophiolite, Newfoundland. *Geophysical Journal of Royal Astronomical Society*, **68**, 675–688.

DETRICK, R., COLLINS, J., STEPHEN, R. & SWIFT, S. 1994. *In situ* evidence for the nature of the seismic layer 2/3 boundary in oceanic crust. *Nature*, **370**, 288–290.

——, HONNOREZ, J., BRYAN, W. B., JUTEAR, T. & SHIPBOARD SCIENTIFIC PARTY 1988. *In*: DETRICK, R., HONNOREZ, J., BRYAN, W. B., JUTEAU, T. *ET AL*. (eds) *Proceedings of the Ocean Drilling Program, Initial Reports, Part A*. Ocean Drilling Program, College Station, Texas, **106/109**.

DICK, H. J. B., ERZINGER, J., STOKKING, L. B. & SHIPBOARD SCIENTIFIC PARTY 1992. Site 504. *In*: DICK, H. J. B., ERZINGER, J., STOKKING, L. B. *ET AL*. (eds) *Proceedings of the Ocean Drilling Program, Initial Reports*. Ocean Drilling Program, College Station, Texas, **140**, 37–200.

EWING, J. I. & PURDY, G. M. 1982. Upper crustal velocity structure in the Rose area of the East Pacific Rise. *Journal of Geophysical Research*, **87**, 8397–8402.

FOX, P. J., SCHREIBER, E. & PETERSON, J. J. 1973. The geology of the oceanic crust: compressional wave velocities of oceanic rocks. *Journal of Geophysical Research*, **78**, 5155–5172.

FRYER, G. J. & WILKENS, R. H. 1988. Porosity, aspect ratio distributions, and the increase of seismic velocity with age in young oceanic crust. *EOS, Transactions of the American Geophysical Union*, **69**, 1323.

GANLEY, D. C. 1981. A method for calculating synthetic seismograms which include the effects of absorption and dispersion. *Geophysics*, **46**, 1100–1107.

GOLDBERG, D. & YIN, C.-H. 1994. Attenuation of P-waves in oceanic crust: Multiple scattering from observed heterogeneities. *Geophysical Research Letters*, **21**, 2311–2314.

—— & ZINSZNER, B. 1988. Acoustic properties of altered peridotite at Site 637 from laboratory and sonic waveform data. *In*: BOILLOT, G., WINTERER, E. L. *ET AL*. (eds) *Proceedings of the Ocean Drilling Program, Scientific Results*. Ocean Drilling Program, College Station, Texas, **103**, 269–276.

——, MOOS, D. & ANDERSON, R. N. 1985. Attenuation changes due to diagenesis in marine sediments. *Transactions of the Society of Professional Well Log Analysts, 26th Annual Logging Symposium*, Paper KK.

——, BADRI, M. & WEPFER, W. 1991. Ultrasonic attenuation measurements in gabbros from Hole 735B. In: VON HERZEN, R. P., ROBINSON, P. T. ET AL. (eds) Proceedings of the Ocean Drilling Program, Scientific Results. Ocean Drilling Program, College Station, Texas, 118, 253–259.

——, —— & ——1992. Acoustic attenuation in oceanic gabbro. Geophysical Journal International, 111, 193–202.

HARDING, A. J., ORCUTT, J. A. KAPPUS, M. E., VERA, E. E., MUTTER, J. C., BUHL, P., DETRICK, R. S. & BROCHER, T. M. 1989. The structure of young oceanic crust at 13°N on the East Pacific Rise from expanding spread profiles. Journal of Geophysical Research, 94, 163–196.

HONNOREZ, J., LAVERNE, C., HUBBERTEN, H.-W., EM-MERMAN, R. & MUEHLENBACHS, K. 1983. Alteration processes in layer 2 basalts from Deep Sea Drilling Project Hole 504B, Costa Rica Rift. In: CANN, J. R., LANGSETH, M. G., HONNOREZ, J., VON HERZEN, R. P., WHITE, S. M. ET AL. (eds) Initial Reports of the Deep Sea Drilling Project. US Government Printing Office, Washington, 69, 509–546.

HOUTZ, R., & EWING, J. 1976. Upper crustal structure as a function of plate age. Journal of Geophysical Research, 81, 2490–2498.

JANNAUD, L., ADLER, P. & JACQUIN, C. 1993. Spectral analysis and inversion of experimental codas. Geophysics, 58, 408–418.

LINDWELL, D. A. 1991. Old Pacific crust near Hawaii: a seismic view. Journal of Geophysical Research, 96, 8191–8203.

LITTLE, S. A. & STEPHEN, R. A. 1985. Costa Rica Rift seismic experiment, Deep Sea Drilling Project Hole 504B, Leg 92, In: ANDERSON, R. N., HONNOREZ, J., BECKER, K. ET AL. (eds) Initial Reports of the Deep Sea Drilling Project. US Government Printing Office, Washington, 83, 517–528

MITHAL, R. & MUTTER, J. C. 1989. A low-velocity zone with the layer 3 region of 118 Myr old oceanic crust in the western North Atlantic. Geophysical Journal, 97, 275–294.

MOOS, D. 1990. Petrophysical results from logging in DSDP Hole 395A, ODP Leg 109. In: DETRICK, R., HONNOREZ, J., BRYAN, W. B., JUTEAU, T. ET AL. (eds) Proceedings of the Ocean Drilling Program, Scientific Results. Ocean Drilling Program, College Station, Texas, 106/109, 237–253.

——, GOLDBERG, HOBART, M. A. & ANDERSON, R. N. 1986. Elastic wave velocities in layer 2A from full waveform sonic logs at Hole 504B, In: LEINEN, M., REA, D. K. ET AL. (eds) Initial Reports of the Deep Sea Drilling Project, US Government Printing Office, Washington, 92, 563–570.

NOWICK, A. S. & BERRY, B. S. 1972. Anelastic Relaxation in Crystalline Solids. Academic Press, New York.

PEZARD, P. A. 1990. Electrical properties of mid-ocean ridge basalt and implications for the structure of the upper oceanic crust in Hole 504B. Journal of Geophysical Research, 95, 9237–9264.

PURDY, G. M. 1987. New observations of the shallow seismic structure of young oceanic crust. Journal of Geophysical Research, 92, 9351–9362.

RAITT, R. W. 1963. The crustal rock. In: HILL, M. N. (ed.) The Sea, 3. Interscience, New York, 85–102.

RICHARDS, P. G. & MENKE, W. 1983. The apparent attenuation of a scattering medium. Bulletin of the Seismological Society of America, 73, 1005–1021.

RUTLEDGE, J. T. & WINKLER, H. 1989. Attenuation measurements from vertical seismic profile data: Leg 104, Site 642. In: ELDHOMN, O., THIEDE, J., TAYLOR, E. ET AL. (eds) Proceedings of the Ocean Drilling Program, Scientific Results. Ocean Drilling Program, 104, 965–972.

SOLOMON, S. C. 1973. Shear wave attenuation and melting beneath the Mid-Atlantic Ridge. Journal of Geophysical Research, 78, 6044–6059.

SPUDICH, P. & ORCUTT, J. A. 1980. A new look at the seismic velocity structure of the oceanic crust. Reviews of Geophysics and Space Physics, 18, 627–645.

STEPHEN, R. A. 1983. The oblique seismic experiment on Deep Sea Drilling Project Leg 70. In: CANN, J. R., LANGSETH, M. G., HONNOREZ, J., VON HERZEN, R. P., WHITE, S. M. ET AL. (eds) Initial Reports of the Deep Sea Drilling Project. US Government Printing Office, Washington, 69, 301–308.

SWIFT, S. A. & STEPHEN, R. A. 1992. How much gabbro is in oceanic seismic layer? Geophysical Research Letters, 19, 1871–1874.

TANG, X. & BURNS, D. 1993. Seismic scattering and velocity dispersion due to heterogeneous lithology. Transactions of the Society of Exploration Geophysicists, 63, 824–827.

TOKSOZ, M. N. & JOHNSTON, D. H. 1981. Seismic Attenuation. Society of Exploration Geophysicists. Reprint Series, 2.

——, CHENG, C. H. & TIMUR, A. 1976. Velocities of seismic waves in porous rocks. Geophysics, 41, 621–645.

VERA, E. E. 1989. Seismic structure of 0 to 4.5 M.Y.-old oceanic crust between 9° and 13° N on the East Pacific Rise. PhD thesis, Columbia University, New York.

——, MUTTER, J. C., BUHL, P., ORCUTT, J. A., HARDING, A. J., KAPPUS, M. E., DETRICK, R. S. & BROCHER, T. M. 1990. The structure of 0 to 0.2 m.y old oceanic crust at 9° N on the East Pacific Rise from Expanded Spread Profiles. Journal of Geophysical Research, 95, 15 529–15 556.

WALSH, J. B., & GROSENBAUGH, M. A. 1979. A new model for analyzing the effect of fractures on compressibility. Journal of Geophysical Research, 84, 3532–3536.

WEPFER, W. W. & CHRISTENSEN, N. I. 1991. Q structure of the oceanic crust. Marine Geophysical Researches, 13, 227–237.

WU, R.-S. 1982. Attenuation of short period seismic waves due to scattering. Geophysical Research Letters, 9, 9–12.

——1985. Multiple scattering and energy transfer of seismic waves – separation of scattering effect from intrinsic attenuation. Theoretical modeling. Geophysical Journal of the Royal Astronomical Society, 82, 57–80.

An application of the moiré method to a study of local strains during rock failure in tension

N. PASSAS, C. BUTENUTH & M. H. DE FREITAS

Engineering Geology Group, Department of Geology, Imperial College of Science, Technology and Medicine, Prince Consort Road, London SW7 2BP, UK

Abstract: The hoop tension test has proved to be a reliable and sensitive test to measure rock strength in tension and provides a convenient method for studying petrophysical aspects of rock failure. However, to make best use of this test it is necessary to be able to compare the strains associated with measured strengths. This paper describes initial experiments to measure by moiré fringes the distribution of strain over the surface of specimens of sandstone, failed under tension in the hoop tension test. The method utilizes the optical interference between two grids of similar size, one on the surface of the specimen which is deformed with the specimen during the test, and another remote from the specimen which remains undeformed. The moiré fringes produced and the strains so measured, were obtained so as to be compared with those obtained from previous investigations on similar samples using holographic and strain gauge measurements. The fringes acquired can be transformed into iso-deformation lines based on the geometrical relationships of the interference grids, to reveal the distribution of strain over the specimen surface observed.

The hoop tension test tests rock under extension (Xu *et al.* 1988; John *et al.* 1990) and can measure the tensile strength of rock (Al-Samahiji 1992; Butenuth *et al.* 1993, 1994). The test differs from that of the conventional hoop test applied to rock (Hardy & Jayaraman 1970) by using two rigid platens which are separated within the rock annulus to cause failure. The stress pattern developed during this test has been investigated extensively (John *et al.* 1990; Gentier *et al.* 1991; Al-Samahiji 1992) but the distribution of strain with the samples has not been investigated to the same extent. Strain gauges have been placed at different locations on hoops of aluminium (Gentier *et al.* 1991) and revealed that in a hoop of elastic material loaded in this way both tensile and compressive regions can exist (Fig. 1); however, the distribution of strain over the sample remained unknown. Knowledge of this distribution is needed if an accurate comparison between samples tested under different environments (e.g. wet and dry) and with different boundary conditions (e.g. size of sample) is to be made. One attempt to achieve this distribution of strain used single beam reflective holography (Al-Samahiji 1992) to reveal the strain over the surface of a hoop. These holographic investigations were designed to compare qualitatively the pattern of strain within hoops of sandstone with the pattern of stress revealed from studies using isotropic and elastic materials via photoelasticity and finite element analysis (John *et al.* 1990; Al-Samahiji 1992): no quantification of the strains observed

holographically was attempted. We now know that such a task can be achieved in rock with the aid of the moiré method. Moiré fringes can be generated whenever two similar, but not necessarily identical, arrays of either equally spaced lines or dots are arranged so that one array can be viewed through the other. Because considerable insight into the strain on a surface, and by implication the body bounded by the surface, can be gained by studying the pattern and geometry of moiré fringes, the theory supporting the technique has been well established and the practical application of the technique well refined so as to measure both displacements and strains. Two commonly used methods for analysing moiré fringes were developed in the late 1940s. One utilizes the geometry of the interfering grids relative to the fringes they produce and was first presented in print by Tollenaar (1945); it is known as the 'geometrical approach' and a complete analyses of strain using this approach has been presented by Morse *et al.* (1960). Soon after Tollenaar, Weller & Shepard (1948) presented an alternative form of analyses designed for using the moiré fringes as a means for measuring displacement: this approach, known as the 'displacement field approach', can be used for both small displacements, as in the elastic range of brittle solids, and larger displacements (Sciammarella & Durelli 1961). In these analyses the distance between the specimen grid and the reference grid can change (indeed it is this change in distance which is the essence of the displacement field approach); however the analyses

From Lovell, M. A. & Harvey, P. K. (eds), 1997, *Developments in Petrophysics*, Geological Society Special Publication No. 122, pp. 377–388.

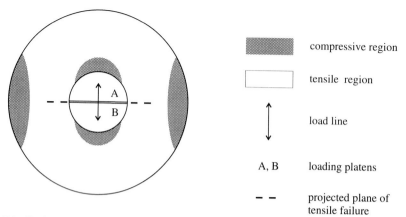

Fig. 1. Distribution of tensile and compressive regions in a hoop which fails under the Hoop Tension Test as found by strain gauges placed at different locations on the surface of aluminium hoops; (after Gentier *et al.* 1991).

assume the grids remain at all locations planar. Theocaris (1964*a, b*) extended the use of moiré fringes into the analyses of out-of-plane displacements so that strains on a surface which is curving (as in bending) could be more accurately calculated. A useful review of the analytical methods available is provided by Dally & Riley (1991).

The method was applied to a material (Penrith sandstone), the bulk mechanical properties of which have been carefully defined by the authors and their colleagues (Al-Samahiji 1992; Butenuth *et al.* 1993). The samples of Penrith sandstone tested were almost isotropic and hoops of the material were studied using the moiré method in an attempt to attribute to each point over a hoop surface a strain value. This paper describes the deformations that were observed using the moiré fringe method, the strains calculated from each of the fringes and from these the iso-deformation lines which can be drawn over the whole surface of the rock sample tested.

Moiré fringes: background

The arrays used to produce moiré fringes, frequently referred to as grids, can be a series of straight parallel lines, a series of radial lines emanating from a point, a series of concentric circles, or a pattern of dots. In stress-analysis work, ideal arrays consist of straight parallel lines; opaque bars with transparent interspaces of equal width are the most commonly used. Image analysis techniques are normally employed to measure changes in spacing between the intersecting points of a grid network before and after loading, and from these data, displacements and strains can be determined.

When two grids are overlaid, by either mechanical or optical means, moiré fringes are produced. The two grids, the specimen grid which is applied by printing a grid on the specimen and the reference grid, are quite frequently identical, and matched when the specimen is in the undeformed state. In many cases this grid matching is difficult to achieve and therefore fringes of apparent deformation may appear at the beginning of testing. The deformation induced from loading produces new fringes and changes the strain value of any pre-existing fringes.

Definitions and conventions

The number of lines per unit length normal to them is referred to as the density of a grid. The centre-to-centre distance between the grid lines is referred to as the pitch of the grid (p for the reference and p' for the specimen grid) and is the reciprocal of the density. The direction perpendicular to the lines of the reference grid is referred to as the primary direction while the direction parallel to the lines of the reference grid is referred to as secondary direction.

Moiré fringes are produced either by a rigid body like rotation of the specimen grid with respect to the reference grid or by changes in pitch of the specimen grid as a result of load-induced deformations. At any point over the surface of a stressed specimen, these two effects can occur simultaneously. Figure 2a illustrates a fringe pattern produced by rotation of the specimen grid relative to the reference grid. The information available from such a pattern is the angle of inclination ϕ of the fringes with

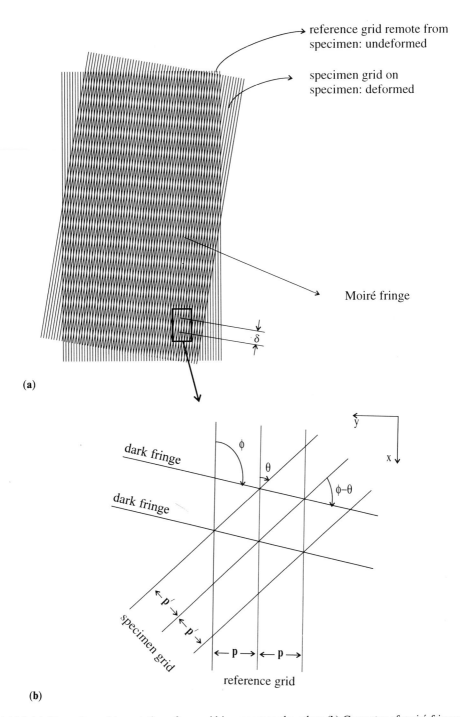

Fig. 2. (a) Moiré fringes formed by rotation of one grid in respect to the other; (b) Geometry of moiré fringes.

respect to the lines of the undeformed reference grid, and the distance between fringes δ.

The choice of analyses one uses depends in part upon the quality of the grid that can be employed. The displacement field approach would have been very useful for our work but requires a dense grid for resolving the small strains encountered in rock failed under tension. The geometrical approach was applicable to the scale of grid that we could use and was employed. The quantity to be measured at the points of interest over the surface is the angle of rotation θ of the specimen grid with respect to the lines of the reference grid. This method of analysis, which is very convenient when information is desired at only selected points, gives rotations and strains that are average values between two fringes: such analyses should therefore be limited to uniform strain fields or to very small regions of nonuniform fields (Dally & Riley 1991).

Relationships between p, p', and θ with respect to the lines of the reference grid, can be obtained from a geometric analysis of the intersections of the lines of the two grids, as shown in Fig. 2b. Here the opaque bars of the fringes in Fig. 2a are defined by their centrelines, and it is these that are referred to as 'fringes' in Fig. 2b. The geometry of these lines shows that

$$\frac{p}{\sin(\pi - \theta)} = \frac{p'}{\sin(\phi - \theta)}. \qquad (1)$$

Once the deformed specimen pitch p' has been determined, the component of strain in a direction perpendicular to the lines of the reference grid can be written as

$$\epsilon = \frac{p' - p}{p} \qquad (2)$$

where positive values of ϵ correspond to extensional strain fields whilst negative values correspond to compressive strain fields.

Fig. 3. The angles (a) (**a**) and (b) (**c**) were measured in each hoop to check the parallelism between the top and bottom (**a**), and inner and outer (**b**), surfaces of the specimens. The height, inner and outer diameter of each hoop were measured at three locations in each sample separated by 120° degrees rotation. Angle (a) is defined as:

$$\tan(a) = \frac{\text{higher height value} - \text{lower height value}}{\text{average hoop outer diameter}}$$

and angle (b) as:

$$\tan(b) = \frac{1' - 1}{\text{average hoop height}}.$$

Experimental procedure

Six identical hoops of porous sandstone were failed with the hoop tension test. In this test two semi cylindrical steel platens are placed in the internal hollow of the hoop sample and

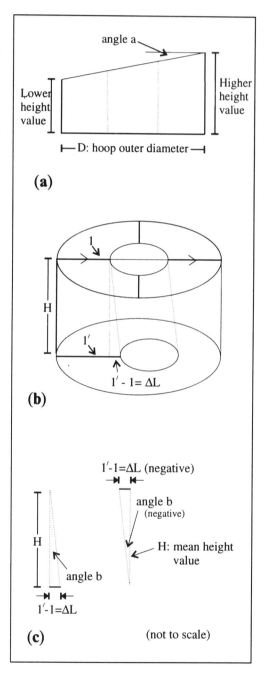

Table 1. *Main characteristics of the samples tested*

Sample no.	Outer diameter (cm)	Inner diameter (cm)	Height (cm)	Angle a (degrees)	Angle b (degrees)	Time to failure (s)	Force to failure (N)
STA1	10.076	4.454	8.508	0.03	−0.15	225.8	4731
STA2	10.124	4.451	8.520	0.17	0.03	267.8	5618
STA3	10.150	4.460	8.593	2.29	0.33	245.5	5147
STA6	10.100	4.447	8.507	0.24	−0.46	208.4	4365
STA8	10.105	4.454	8.517	0.12	0.04	234.0	4905
STA10	10.087	4.451	8.500	0.07	−0.34	245.5	5144

displaced using an hydraulic jack, to cause ultimate failure in the plane of platen opening (Fig. 1). The sandstone samples studied come from the Lower Permian of NW England near Penrith (Bowes quarry). It is a coarse-grained well-sorted orthoquarzite (as defined by Waugh 1970) with a high degree of mineralogical and textural maturity. A distinctive feature of this rock is the predominance of rounded quartz grains, usually surrounded by a patina of iron, with the space between them being partially filled by quartz overgrowths to the grains; subordinate feldspar and rock fragments are also present. The specimens exhibited minor variation in grain size in the direction normal to bedding with coarser bands varying from 2 mm to 3 cm in maximum thickness. Because of this, all samples tested were oriented with their bedding orthogonal to the induced plane of tensile failure. The six hoops of sandstone tested came from the same block. Each sample had an outer diameter of approximately 10 cm, an inner diameter of approximately 4.4 cm and a height of approximately 8.5 cm. Particular care was taken to ensure that the samples tested had their top and bottom surfaces and their inner and outer surfaces parallel to each other. This was checked by measuring the angles (a) and (b) for each sample (Fig. 3), to check the parallelism between the top and the bottom surface of each hoop and the parallelism between the inner and the outer surface of each hoop. Angle (a) for the hoops tested was in the range between 0.03 and 0.24°, except sample STA3 which had (a) angle at 2.29°, and angle (b) was in the range −0.46–0.33° (Table 1).

A grid of black parallel lines with a density of 25 lines per linear 2.54 cm, and each of 0.2 mm thickness, was printed on the top surface of each rock hoop. The lines were applied using silk-screen printing and the same process used to print the reference grid on a non reflective glass plate to be positioned above the rock hoop. The experimental set up is illustrated in Fig. 4.

Due to the porous nature of the rock tested it was difficult to form the lines on the specimen so that they were perfectly parallel, because the ink penetrated the pores of the rock. As a result of this, at the beginning of each test the ideal overlapping of the reference grid over the specimen grid was imperfect and fringes could be seen before any deformation had been induced in the specimens. This meant that at the beginning of each test a pattern of fringes could be recognised and a value for apparent strain could be attributed to them. This was taken into account when analysing the deformation patterns acquired at the end of each test.

The dry hoops were tested at a loading rate of $230\,000\,\text{kPa}\,\text{h}^{-1}$ which gave failure in 3 to 4 minutes; a reasonable duration for a tensile

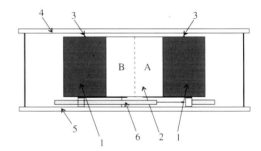

Fig. 4. Vertical section through the experimental set up used to detect the moiré effect as seen on the upper surface of sandstone hoops failed under the Hoop Tension Test: (1) Rock hoop seen in vertical section oriented along the loading line as shown in Fig. 1; (2) Semi-cylindrical platens in vertical section as seen along the loading line as shown in Fig. 1; (3) Specimen grid printed on top surface of hoop; (4) Reference grid printed on glass plate supported above top surface of the hoop and co-planar with it; (5) Basal low friction plate on which the hoop is supported; (6) Differential transformers oriented to measure displacement in the direction of the loading line. To see the fringes the specimen grid (3) is observed from above, through the reference grid (4) to reveal the interference of the two grids.

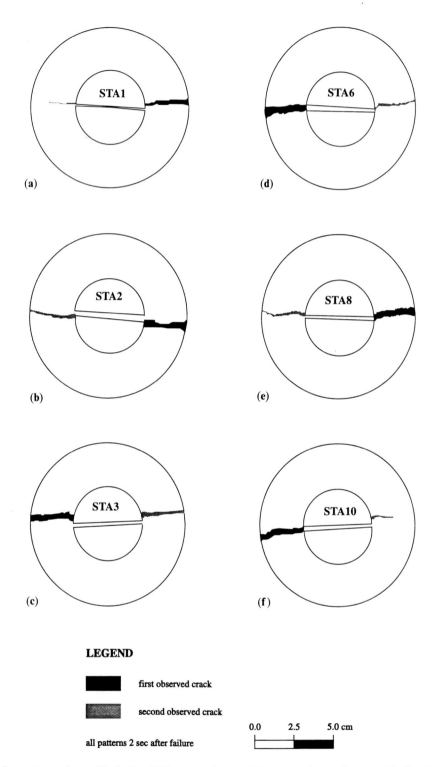

Fig. 5. Failure patterns observed in the Penrith Sandstone hoops which were used to study the moiré effect during failure in the hoop tension test.

laboratory test (Brown 1981). The failure patterns produced due to each test are presented in Fig. 5. One of the samples tested, namely STA1, was initially loaded to a pressure which did not cause ultimate failure. Immediately after this initial loading the same sample was again loaded, this time to ultimate failure. Panchromatic photographs were taken using a Pentax Super A camera with motor drive and macro lense of 50 mm focal length with time intervals of 3–5 s during each test in order to detect the development and evolution of the deformation pattern produced. A typical example of the sets of pictures obtained is given in Fig. 6 where the specimen grid and reference grid are parallel to the loading line (Fig. 1) and normal to the plane of platen opening. Because the moiré effect can detect strain only in the primary grid direction (i.e., orthogonal to the direction of the reference grid), three of the rock specimens were tested to measure strains normal to the loading line and three parallel to it. In this way it was possible to see the deformation normal and parallel to the platen opening.

Analytical procedure

Analysis of the experimental results essentially consists of comparing the pictures obtained at the beginning and at the end of each test. To do so a picture at time zero must be used. Due to the time interval between pictures it was difficult to capture the exact time of the crack development. Therefore, two pictures were considered at the end of each test: the last one taken just before failure and the first one taken after failure (Fig. 6). The load induced deformation resulted either in the production of new fringes or in the change of strain value of the existing ones, or both. The apparent deformation at the beginning of each test had to be quantified and subtracted from the deformation observed at the end of each test to reveal the true deformation. For this to be achieved each one of the pictures chosen for analyses was magnified to 21 cm × 29.7 cm using a scanner (Scangal 1988) prior to being inserted into an image analyser (Jandel 1993) for the angles θ and ϕ (Fig. 2b) to be determined. In the image analyser it was possible to mark the area covered by each fringe, and draw a line along the centre of this band together with the value found for the corresponding strain (ϵ). Each one of the lines so produced represented an iso-deformation line. From the strain value of each iso-deformation line at the end of the test the corresponding value of the same line which pre-existed at the

beginning of the test was subtracted to reveal the deformation due to the tension induced. Fringes produced where there was no pre-existing fringe at the beginning of a test were allocated their strain value, unchanged.

Because the grid used was of relatively low density due to the porous nature of the rock surface observed there were some hoops for whose surface the patterns of strain could not be clearly identified. The practical problem with our work was that the fringes prior to failure were too far apart, leaving large areas of the specimen surface within an interfringe area: this was not the case after failure (Fig. 6). For the strain pattern to be extended over the whole surface of the hoop, contour drawing software, namely SURFER, (Keckler 1994) was used based on the measured strain values from the observation points on the hoop. This enables comparisons between samples to be made. The iso-deformation lines so obtained for each specimen and the strain patterns acquired by this procedure are presented in Figs 7 and 8.

Discussion of experimental results

The strain patterns of Figs 7 and 8 just before and after each test look different and reflect the fact that once each sample is broken then the pieces move in respect to themselves and the reference grid so producing dramatic changes in the strain patterns detected. This movement may reveal rotation, extension and, or compression of the specimen grid in respect to the reference one. The patterns produced provide evidence for the post-failure behaviour of the rock, however, this was not analysed. The strain patterns acquired just before failure reflect the measurement of rock response to the stresses applied and were analysed using the methods described earlier. These pre-failure patterns must be considered with some caution as we know from the theory of elasticity and from strain gauge measurements that the field of strain is not uniform, yet in extending the strain pattern in the way described the assumption is made that uniformity exists. We do not know the magnitude of this potential error at this scale, nevertheless it is instructive to compare the strains so calculated with the behaviour of the sample at failure.

ϵ_y strains

The patterns of failure (Fig. 5a, b, c) for samples STA1, STA2 and STA3, where strains are

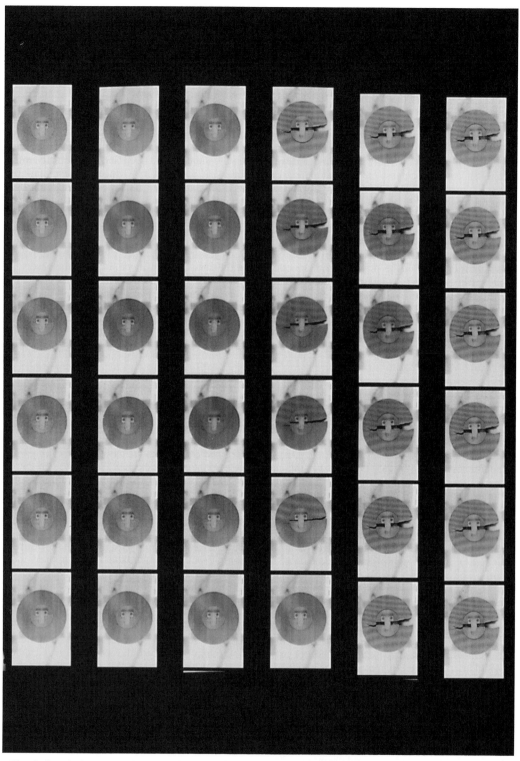

Fig. 6. A typical example of the sets of pictures obtained (sample STA10) where the specimen and reference grid are normal to the loading line and parallel to the platen opening. The outer diameter of the hoop is 10 cm. The analysis involved three pictures: one at the beginning of the test (top row; first picture from the left) and two other pictures, one before and one after the failure (fourth row; first two pictures from the left).

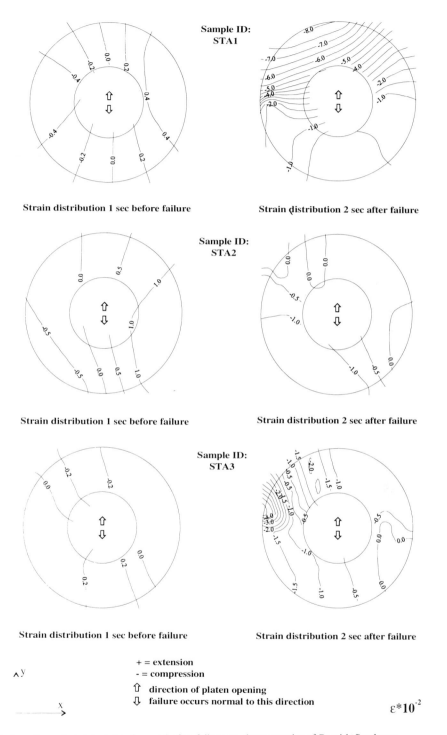

Fig. 7. Strain patterns (ϵ_y) detected before and after failure on three samples of Penrith Sandstone.

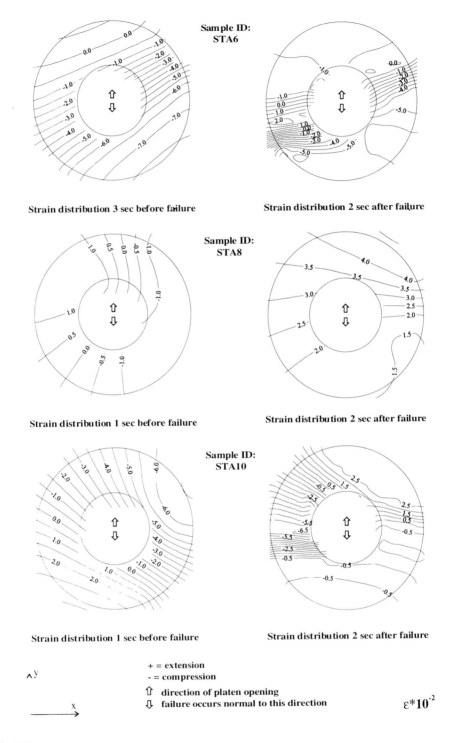

Fig. 8. Strain patterns (ϵ_x) detected before and after failure on three samples of Penrith Sandstone.

detected parallel to the loading line and normal to platen opening, consist of two fractures one of which occurred after the other was partially or fully developed. In samples STA1 and STA2 (Fig. 7) the first fracture (Fig. 5a & b) did not occur directly opposite the platen opening while the second ones initiated at the point where the platen opening meets the rock and developed parallel to the platen opening. In sample STA3 neither fracture initiated at the point where the platen opening meets the rock (Fig. 5) but still developed parallel to the platen opening. Although all the fractures observed occur parallel to the platen opening, their position with respect to the platen opening reflect small scale variations between the samples, a fact which may account for the different strain patterns acquired.

In both samples STA1 and STA2 (Fig. 7) the first fracture (Fig. 5a & b) occurs in a region of tensile strain. Further, despite the uncertainties described with the projection of strains into large interfringe areas, the maximum tensile strain (ϵ_y) so calculated for samples STA1 and STA2, intersected the internal boundary of the hoop at exactly the point where the first crack was observed to have initiated. Note that the ϵ_y strains in this region seem to be lower for STA1 and it was this sample that failed after it had been loaded once before to a stress that did not produce ultimate failure. The fringes seem to be sensitive to sample history. In sample STA3 the first fracture (Fig. 5c) occurred in a region where compressive ϵ_y strain appeared, contrary to what is expected (Fig. 7). The inner surface of this sample was not in full contact with the cylindrical platens; a few millimetres of rock projected above the platens. As a result of the above and the fact that this sample had not its top and bottom surfaces perfectly parallel as in the other samples involved in this investigation, the hoop of rock failed in tension except that part which was above the platens and on which was printed the specimen grid. The fringes suggest that this protruding part was bent in compression by the separation of the broken hoop below it.

ϵ_x *strains*

The patterns of failure (Fig. 5d, e & f) for samples STA6, STA8 and STA10 are from samples where the grids were oriented to observe strains normal to the loading line; the hoops tend to thin in this direction as elongation progresses along the direction of the load line until failure. Compressive strains might thus be

expected in the region where the first fracture is being initiated, and this is seen in STA6 and STA8 and just about seen in STA10. The same is true for the second fracture with the exception of STA8. Thus again it is seen that despite the uncertainties expressed about the projection of strains, they provide the correct sign. Although two fractures occur, one occurred after the other was partially or fully developed, and all of them propagated parallel to the platen opening: the first fractures occur opposite the platen opening while the second ones initiated at a point almost opposite. Additionally, in sample STA10 the second fracture developed more slowly than it developed in samples STA6 and STA8. Hence the strain patterns acquired for each reflect its progress to failure.

Conclusions

Petrographical studies that require a knowledge of strain can be completed using moiré fringes: the experiments described here demonstrate that despite the practical difficulties associated with using fine grids on porous rock surfaces, such fringes can be remarkably sensitive methods of recording the evolution of failure within a sample and the relationship of failure to applied loads. It must be remembered that the strains measured are those sustained at the surface of the body tested, as are those from strain gauges and holography, but unlike these methods, moiré fringes require no contact with the sample, although the sample must be visible during the test. They provide an elegant and simple method for studying the change in strain which accompanies a change in sample composition, fabric, size and environment, and enhance the value of experiments in petrophysical studies. Our particular application was for observing rock strain at different scales to support the development of an analytical solution to rock strength based on force and displacement, and for this the moiré method has been instructive.

References

AL-SAMAHIJI, D. K. 1992. *Experimental Investigation of Rock Failure in Extension.* PhD Thesis, Imperial College, London.
BROWN, E. T. (ed.) 1981. *Rock Characterization Testing and Monitoring. ISRM Suggested Methods.* Pergamon Press. Great Britain.
BUTENUTH, C., DE FREITAS, M. H., AL-SAMAHIJI, D. K., PARK, H.-D., COSGROVE, J. W. & SCHETELIG, K. 1993. Technical note: observations on the measurement of tensile strength using the

hoop test. *International Journal for Rock Mechanics Mining Sciences and Geomechanical Abstracts*, **30**, 157–162.

——, ——, PARK, H.-D., SCHETELIG, K., VAN LENT, P. & GRILL, P. 1994. Observations on the use of hoop test for measuring the tensile strength of anisotropic rock. *International Journal for Rock Mechanics Mining Sciences and Geomechanical Abstracts*, **31**, 733–741.

DALLY, J. W. & RILEY, W. F. 1991. Moiré methods. *In: Experimental Stress Analysis*. 3rd Edition. McGraw-Hill Book Co., 389–423.

GENTIER, S., POINCLOU, C. & BERTRAND, L. 1991. *Essai de Traction sur Anneaux (Hoop Tension Test), Application aux Schists Ardoisiers*. Commisariat a l' Energie Atomique, France.

HARDY, H. R. JR. & JAYARAMAN, N. I. 1970. An investigation of methods for the determination of the tensile strength of rock. *In: Proceedings of the 2nd Congress of the International Society for Rock Mechanics*, **III**, 85–92.

JOHN, S. J., AL-SAMAHIJI, D. K., DE FREITAS, M. H., COSGROVE, J. W., CLARKE, B. A., LOE, N. & TANG, H. 1990. Stress analysis of a unidirectionally loaded hoop specimen. *In: Proceedings of the 7th International Congress of the International Society for Rock Mechanics*, 513–518.

KECKLER, D. 1994. *SURFER for Windows, Contouring and 3D Surface Mapping*. Golden Software, Inc., 809 14th Street, Golden, Colorado 80401–1866, USA.

MORSE, S., DURELLI, A. J. & SCIAMMARELA, C. A. 1960. Geometry of moiré fringes in strain analysis. *Journal of the Engineering Mechanics Division.*

Proceedings of the American Society of Civil Engineers, **86 (EM4)**, 105–126.

JANDEL SCIENTIFIC (ed.) 1993. *SigmaScan/Image, Measurement Software for Windows: User's Manual*. USA.

SCANGAL 1988. *Scanning Gallery Plus, Version A.03.00*, Hewlett-Packard Co. and Microsoft Corporation.

SCIAMMARELLA, C. A. & DURELLI, A. J. 1961. Moiré fringes as a means of analysing strains. *Journal of the Engineering Mechanics Division. Proceedings of the American Society of Civil Engineers*, **87 (EM1)**, 55–74.

THEOCARIS, P. S. 1964a. Isopachic patterns by the moiré method. *Experimental Mechanics*, **4**, 153–159.

——1964b. Moiré patterns of isopachics. *Journal of Scientific Instrumentation*, **41**, 133–138.

TOLLENAAR, D. 1945. *Moiré: Interferentieverschijnselen bij Rasterdruk*. Amsterdam Instituut voor Grafische Techniek, Amsterdam.

WAUGH, B. 1970. Petrology, provenance and silica diagenesis of the Penrith Sandstone (Lower Permian) of NW England. *Journal of Sedimentary Petrology*, **40**, 1226–1240.

WELLER, R. & SHEPARD, B. M. 1948. Displacement measurement by mechanical interferometry. *Proceedings of the Society for Experimental Stress Analysis*, **VI**, 35–38.

XU, S., DE FREITAS, M. H. & CLARKE, B. 1988. The Measurement of Tensile Strength of Rock. *Proceedings International Symposium of International Society for Rock Mechanics, Madrid: Rock Mechanics and Power Plants*, vol. **1**, 125–132.

Index